Immunopharmacology of Epithelial Barriers

THE HANDBOOK OF IMMUNOPHARMACOLOGY

Series Editor: Clive Page
King's College London, UK

Titles in this series

Cells and Mediators

Immunopharmacology of
Eosinophils
(edited by H. Smith and R. Cook)

The Immunopharmacology of Mast
Cells and Basophils
(edited by J.C. Foreman)

Lipid Mediators
(edited by F. Cunningham)

Immunopharmacology of
Neutrophils
(edited by P.G. Hellewell and
T.J. Williams)

Immunopharmacology of
Macrophages and Other
Antigen-Presenting Cells
(edited by C.A.F.M. Bruijnzeel-
Koomen and E.C.M. Hoefsmit)

Adhesion Molecules
(edited by C.D. Wegner,
forthcoming)

Immunopharmacology of
Lymphocytes
(edited by M. Rola-Pleszczynski,
forthcoming)

Immunopharmacology of Platelets
(edited by M. Joseph, forthcoming)

Systems

Immunopharmacology of the
Gastrointestinal System
(edited by J.L. Wallace)

Immunopharmacology of Joints
and Connective Tissue
(edited by M.E. Davies and
J. Dingle)

Immunopharmacology of the Heart
(edited by M.J. Curtis)

Immunopharmacology of Epithelial
Barriers
(edited by R. Goldie)

Immunopharmacology of the Renal
System
(edited by C. Tetta)

Immunopharmacology of the
Microcirculation
(edited by S. Brain)

Drugs

Immunotherapy for Immune-
related Diseases
(edited by W.J. Metzger,
forthcoming)

Immunopharmacology of AIDS
(forthcoming)

Immunosuppressive Drugs
(forthcoming)

Glucocorticosteroids
(forthcoming)

Angiogenesis
(forthcoming)

Immunopharmacology of Free
Radical Species
(forthcoming)

Immunopharmacology
of
Epithelial Barriers

edited by

Roy Goldie
University of Western Australia
Perth, Australia

ACADEMIC PRESS
Harcourt Brace and Company, Publishers
London   San Diego   New York
Boston   Sydney   Tokyo   Toronto

ACADEMIC PRESS LIMITED
24/28 Oval Road
London NW1 7DX

United States Edition published by
ACADEMIC PRESS INC.
San Diego, CA 92101

This book is printed on acid-free paper

A catalogue record for this book
is available from the British Library

ISBN 0–12–288030–7

Typeset by J&L Composition Ltd, Filey, North Yorkshire
Printed and bound in Great Britain by The Bath Press, Avon

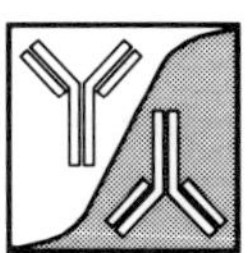

Contents

3. Adhesion Molecules and the Modulation of Mucosal Inflammation 71

Robert H. Gundel and L. Gordon Letts

4. Innervation of the Airway Mucosa: Structure, Function and Changes in Airway Disease 85

Peter K. Jeffery

5. Inflammatory Mediators and Modulation of Epithelial/Smooth Muscle Interactions 119

Douglas W.P. Hay, Stephen G. Farmer and Roy G. Goldie

6. Airway Epithelial Inflammation and its Functional Consequences 147

Paul M. O'Byrne and Ellinor Ädelroth

7. Immunopharmacology of Epithelial Cell–Virus Interactions 159

David B. Jacoby

8. Gastroduodenal Defence: Role of Epithelial Factors 197

Thomas A. Miller, Gregory S. Smith and Jose C. Barreto

9.   Role of the Epithelium in Defence of the Small and Large Intestine   213

A.G. Cummins and I.C. Roberts-Thomson

10.   Immunoreactivity and Inflammation in Renal Epithelium   225

D. Gwyn Williams

11. *Immunological Barriers in the Eye*   241

Jerry Y. Niederkorn

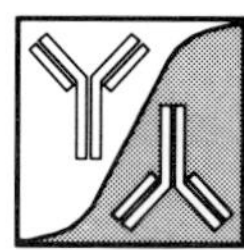

Contributors

E. Ädelroth
Department of Pulmonary Medicine
University of Umeå,
Umeå
Sweden

J.C. Barreto
Department of Surgery
The University of Texas Medical School
Houston
Texas TX 77030
USA

P.M. O'Byrne
Department of Medicine
Health Sciences Centre
McMaster University
Hamilton L8N 3Z5
Ontario
Canada

A.G. Cummins
Department of Gastroenterology
The Queen Elizabeth Hospital
28 Woodville Road
Woodville 5011
South Australia
Australia

S.G. Farmer
Senior Research Pharmacologist
Zeneca
Wilmington
Delaware 19897
USA

R.G. Goldie
Department of Pharmacology
University of Western Australia
Perth
Nedlands 6009
Western Australia
Australia

R.H. Gundel
Department of Pharmacology
Boehringer Ingelheim Pharmaceuticals Inc
Ridgefield
Connecticut 06877
USA

D.W.P. Hay
Department of Pharmacology
SmithKline Beecham Pharmaceuticals
King of Prussia
Pennsylvania PA 19406–0939
USA

D.B. Jacoby
Division of Pulmonary and Critical Care Medicine
Johns Hopkins Asthma and Allergy Centre
301 Bayview Boulevard
Baltimore, Maryland 21205
USA

P.K. Jeffery
Department of Lung Pathology
National Heart and Lung Institute
Brompton Hospital
London SW3 6NP
UK

L.G. Letts
Department of Pharmacology
Boehringer Ingelheim Pharmaceuticals Inc
Ridgefield
Connecticut 06877
USA

G. Mayrhofer
Department of Immunology
University of Adelaide
North Terrace
Adelaide, South Australia 5000

T.A. Miller
Department of Surgery
The University of Texas Medical School
Houston
Texas TX 77030
USA

J.Y. Niederkorn
University of Texas
South Western Medical Centre
Dallas
Texas TX 75235–9057
USA

I.C. Roberts-Thomson
Department of Gastroenterology
The Queen Elizabeth Hospital
28 Woodville Road
Woodville 5011
South Australia
Australia

G.S. Smith
Department of Surgery
The University of Texas Medical School
Houston
Texas TX 77030
USA

J.H. Widdicombe
Department of Physiology
Cardiovascular Research Institute
University of California Medical School
San Francisco
CA 94143–0130
USA

D.G. Williams
Department of Nephrology
United Medical and Dental Schools
Hunt's House
Guys Hospital
London SE1 9RT
UK

Series Preface

The consequences of diseases involving the immune system such as AIDS, and chronic inflammatory diseases such as bronchial asthma, rheumatoid arthritis and atherosclerosis, now account for a considerable economic burden to governments worldwide. In response to this, there has been a massive research effort investigating the basic mechanisms underlying such diseases, and a tremendous drive to identify novel therapeutic applications for the prevention and treatment of such diseases. Despite this effort, however, much of it within the pharmaceutical industries, this area of medical research has not gained the prominence of cardiovascular pharmacology or neuropharmacology. Over the last decade there has been a plethora of research papers and publications on immunology, but comparatively little written about the implications of such research for drug development. There is also no focal information source for pharmacologists with an interest in diseases affecting the immune system or the inflammatory response to consult, whether as a teaching aid or as a research reference. The main impetus behind the creation of this series was to provide such a source by commissioning a comprehensive collection of volumes on all aspects of immunopharmacology. It has been a deliberate policy to seek editors for each volume who are not only active in their respective areas of expertise, but who also have a distinctly *pharmacological* bias to their research. My hope is that *The Handbook of Immunopharmacology* will become indispensable to researchers and teachers for many years to come, with volumes being regularly updated.

The series follows three main themes, each theme represented by volumes on individual component topics.

The first covers each of the major cell types and classes of inflammatory mediators. The second covers each of the major organ systems and the diseases involving the immune and inflammatory responses that can affect them. The series will thus include clinical aspects along with basic science. The third covers different classes of drugs that are currently being used to treat inflammatory disease or diseases involving the immune system, as well as novel classes of drugs under development for the treatment of such diseases.

To enhance the usefulness of the series as a reference and teaching aid, a standardized artwork policy has been adopted. A particular cell type, for instance, is represented identically throughout the series. An appendix of these standard drawings is published in each volume. Likewise, a standardized system of abbreviations of terms has been implemented and will be developed by the editors involved in individual volumes as the series grows. A glossary of abbreviated terms is also published in each volume. This should facilitate cross-referencing between volumes. In time, it is hoped that the glossary will be regarded as a source of standard terms.

While the series has been developed to be an integrated whole, each volume is complete in itself and may be used as an authoritative review of its designated topic.

I am extremely grateful to the officers of Academic Press, and in particular to Dr Carey Chapman, for their vision in agreeing to collaborate on such a venture, and greatly hope that the series does indeed prove to be invaluable to the medical and scientific community.

C.P. Page

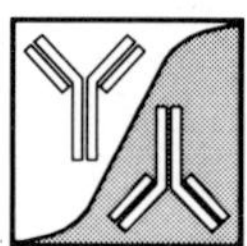

Preface

Immunopharmacology was in its infancy 10 years ago, as was our appreciation of the functions of the various epithelial structures in the human body. Today, immuno-pharmacology is one of the most active and rapidly expanding of the biological sciences, integrating as it does the dizzying complexities of immunology with the excitement of pharmacology and the prospect of safe, therapeutic modulation of immune function.

The realisation that epithelial tissues are not simply passive barriers to the absorption of materials into internal environments, has also triggered an exponential growth in investigations of mucosal functions, including their active and passive protective roles. How then does one construct a useful reference work concerning these burgeoning fields, without burying the reader with information? The additional challenge was to avoid compiling a series of chapters which simply provided essentially similar information about epithelia of various types. That approach would inevitably result in the production of yet another library-based paperweight. The alternative was to challenge the reader to begin at the beginning and to advance his or her understanding of epithelial function in small but critical steps, with each of the early chapters progressively adding important foundation information. Subsequent chapters concerning particular epithelia including respiratory, gastrointestinal, renal and ocular epithelia might then be better appreciated, with the whole volume providing a relatively comprehensive but integrated treatise of epithelial function and its immunopharmacology.

Several important epithelial tissues have been briefly discussed in so-called foundation chapters, but not subsequently dealt with in more depth in separate chapters. This was a deliberate attempt to hold the book to a manageable and hopefully more readable size. Previous and future volumes in this series have compre-hensively described areas not included in the present volume.

Epithelia are highly organized but complex structures, subserving numerous functions including immunologic defence. The use of pharmacologic tools in these systems is increasing and the result is improved understanding of epithelial immunobiology. It is hoped that this volume will serve as an appropriate starting point at which the clinical pulmonologist and the research scientist can obtain an appreciation of some aspects of epithelial immunopharmacology as they are currently understood.

I would like to thank each of the authors for their contributions to this volume. I am also most grateful to Mrs Alison Wallace for her tireless efforts on my behalf in retyping and proofreading each manuscript. I trust that this volume will serve as a tool useful for more than weighting paper. The combined efforts of all concerned in its production suggest that this should be so.

Roy G. Goldie

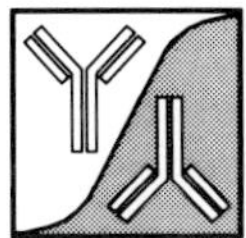

1. Structure, Growth and Repair of Epithelia

Jonathan H. Widdicombe

1. Introduction

Epithelia are the sheets of cells which comprise the skin and line the tubes and ducts which open to the exterior. As early as the 16–32-cell blastula tight junctions are present between all the cells (Biggers *et al.*, 1988), thus conferring polarity on the ectoderm and allowing it to perform the two major functions of an epithelium: acting as a barrier, and performing directional transport of solutes and water. In epithelia of endodermal (airways, gut and associated glands) or mesodermal origin (e.g. kidney and genital tracts), formation of mature tight junctions coincides with the development of lumens (Montesano *et al.*, 1975; Humbert *et al.*, 1976; Schneeberger *et al.*, 1978; Luciano *et al.*, 1979). Thus, very early in embryonic life, the ectoderm and the linings of the various tubular organs consist of functioning epithelia, whose integrity is uninterrupted during the further development into the adult.

The importance of epithelia is reflected in the dramatic physiological and pathological changes which result on alteration of their function. Menstruation and some features of asthma, for instance, are associated with a breakdown of epithelial barrier function. Alterations in epithelial salt and water transport underlie a host of diseases, including cholera and cystic fibrosis.

2. Structure

2.1 BASIC DESIGN

What are the structural features which allow epithelia to perform their two major functions of providing a barrier, and transporting solutes and water? Tight junctions are perhaps the most obvious and most important. These specializations surround the apical portions of epithelial cells, and serve to hold the cells together and to restrict transepithelial movement of solutes. They have been likened to the plastic strips that hold together a six pack of beer (Diamond, 1977), and mark the transition between two membranes of completely different structure and function. On the luminal side of the tight junction lies the apical membrane. On the blood side is the basolateral membrane. Between the cells, lined by the basolateral membranes, are the lateral intercellular spaces (LIS). Solutes cross epithelia either by passing through both the

Immunopharmacology of Epithelial Barriers
ISBN 0–12–288030–7

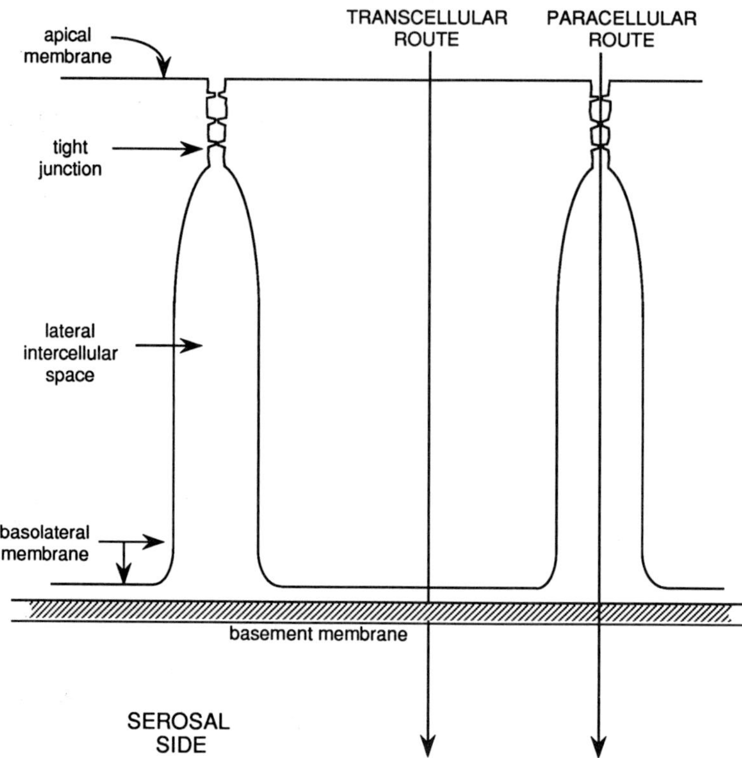

Figure 1.1 Basic structure of epithelia, and routes of transepithelial permeation.

apical and basolateral cell membranes (the transcellular pathway), or by diffusing across the tight junctions and along the LIS (the paracellular pathway). These basic features of epithelial structure are illustrated in Fig. 1.1.

Within this basic design, differences in the shape of cells produce considerable variation in form between epithelia (Fawcett, 1986). The cells may be highly flattened to produce a squamous epithelium (Fig. 1.2A). When the height and width of the cells is approximately the same, the epithelium is called cuboidal (Fig. 1.2B). Columnar epithelia are those in which the cells are considerably higher than wide (Fig. 1.2C). In these three types of epithelia, all cells stretch across the entire epithelium; apical membranes make contact with the lumen, and basal membranes of the same cells abut on the basement membrane. In stratified epithelia, several layers of cells may be connected together by desmosomes (Fig. 1.2E). In such epithelia, the cells of the outermost layer have no contact with the basement membrane. The basal cells in stratified epithelia are generally cuboidal in shape, but the outermost layers may be either squamous or columnar in form, providing stratified squamous and stratified columnar epithelia, respectively. The former is typical of epidermis. The latter is relatively uncommon, being found in the pharynx, urethra and two or three other restricted locations (Fawcett, 1986). In pseudo-

stratified columnar epithelia, all cells make contact with the basement membrane, but not all reach the lumen. Accordingly, nuclei are found at many different depths within the epithelium, giving an appearance similar to a true stratified epithelium (Fig. 1.2D).

Another source of variation between epithelia lies in the types of cells present. Figure 1.3 shows the different cell types of airway epithelium. However, the same basic classification applies to other epithelia. Epithelial cells can initially be divided into those which make contact with the lumen and those which do not. Many epithelia lack the latter. In others, "basal" cells lie on the basement membrane between the foot processes of epithelium-spanning cells. In stratified epithelia, the basal cells form a continuous layer. Cells with apical membranes can be divided into two or three types. First, there is a very large variety of macromolecule-secreting cells. These cells are characterized by relatively undifferentiated apical membranes and by the presence of membrane-bound secretory granules in their apical portions. Among other compounds, the secretory granules may contain mucins, surfactant or neuropeptides. Second, a large proportion of cells in most epithelia are involved in transepithelial transport of solutes and water. Epithelial cells specialized for solute transport frequently show marked increases in the surface area of their apical and/or basolateral membranes. In

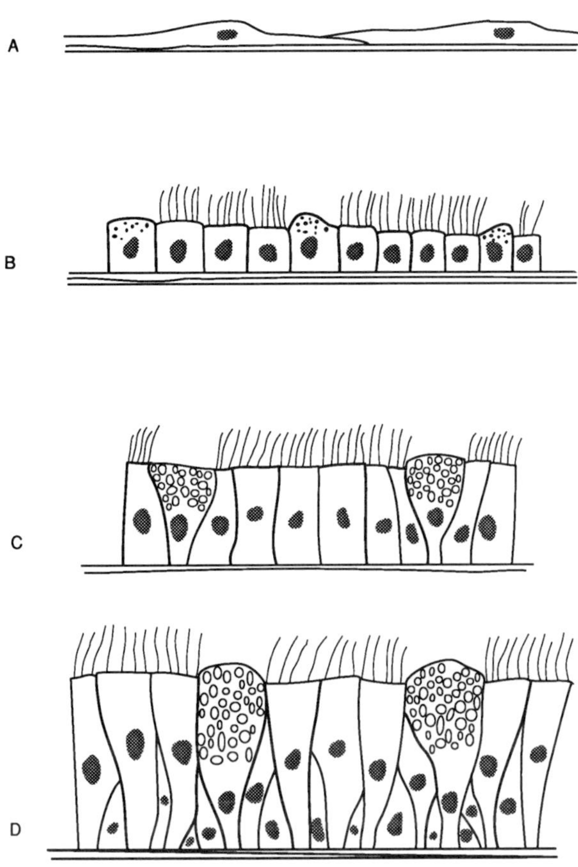

Figure 1.2 Types of epithelia. (A) Squamous. (B) Cuboidal. (C) Columnar. (D) Pseudostratified. (E) Stratified squamous.

apical membranes this increase is achieved by increasing the numbers and length of the microvilli (Fig. 1.4A). Basolateral membrane area is amplified by increasing the degree of interdigitation and infolding of membranes from adjacent cells (Fig. 1.4B). Specialized solute-transporting cells generally show increased numbers of mitochondria. Ciliated cells form a special subcategory of solute-transporting cells. These cells have apical microvilli and transport solutes. In addition, their cilia propel mucus and particulate matter, such as coal dust or ova.

Cells of non-epithelial origin (e.g. nerve fibres and a variety of white blood cells) are often found interspersed among the epithelial cells (Jeffery and Reid, 1981). One such group of cells, of particular importance to the immune response, is the dendritic cell/Langerhans cell lineage (dendritic cells are immature Langerhans cells; the latter can be distinguished by the presence of typical cytoplasmic inclusions, the Birbeck granules). Langerhans cells in the skin are found in the suprabasal layers, have cell bodies about 10 μm in diameter, and comprise 2–4% of the total epidermal cell population (Breathnach, 1965; Wolff and Stingl, 1983). Dendritic cells were first described in human lung by Holt *et al.* (1985). They are 25–40 μm in length and 2–8 μm in width with two to three slender cytoplasmic processes (Sertl *et al.*, 1986), which may extend to the airway lumen (Holt and Schon, 1987). Langerhans cells have also recently been demonstrated in human airways, and the numbers of these and of dendritic cells are markedly increased in the parenchyma of heavy smokers, though their numbers within the epithelium are unaltered (Soler *et al.*, 1989). Dendritic and Langerhans cells are derived from a mobile pool of bone marrow-derived precursor cells (Katz *et al.*, 1979), and are the only cells in skin and airway epithelium which have type II antigens of the MHC (Stingl *et al.*, 1978; Sertl *et al.*, 1986). Their major role is as antigen-presenting cells, i.e. they take up antigens and present them together with type II MHC antigens to T lymphocytes thereby initiating the T cell immune response. In suspensions of isolated epidermal and airway cells, this response is prevented by pretreatment with anti-type II MHC serum (Stingl *et al.*, 1981; Sertl *et al.*, 1986).

2.2   MEMBRANE POLARITY

The apical and basolateral membranes of epithelia are entirely different in their structure and composition. The apical membrane frequently possesses cilia and microvilli, specialized organelles lacking from the basolateral membrane. In addition, as illustrated in Fig. 1.4A, the apical membrane is often coated with a pronounced glycocalyx (Ito, 1965). By contrast, the basolateral membrane is relatively unspecialized except for areas of cell-to-cell or cell-to-substratum contact (tight and gap junctions, desmosomes and hemidesmosomes). Not only are the membranes structurally different, but there are differences in their lipids (Dragsten *et al.*, 1981; van Meer and Simons, 1986) and integral membrane proteins (e.g. receptors and transport proteins). It is an asymmetrical distribution of transport proteins that underlies directional solute movement across epithelia. Epithelia transport a wide range of organic solutes, as well as inorganic ions ranging in size from H^+ to PO_4^{4-}. Active Na transport and active secretion of Cl are described here briefly to illustrate how protein localization to one or other membrane can effect net solute transfer. Except in epithelia lining the CNS (choroid plexus and retinal pigment epithelium), epithelial Na/K ATPase is confined

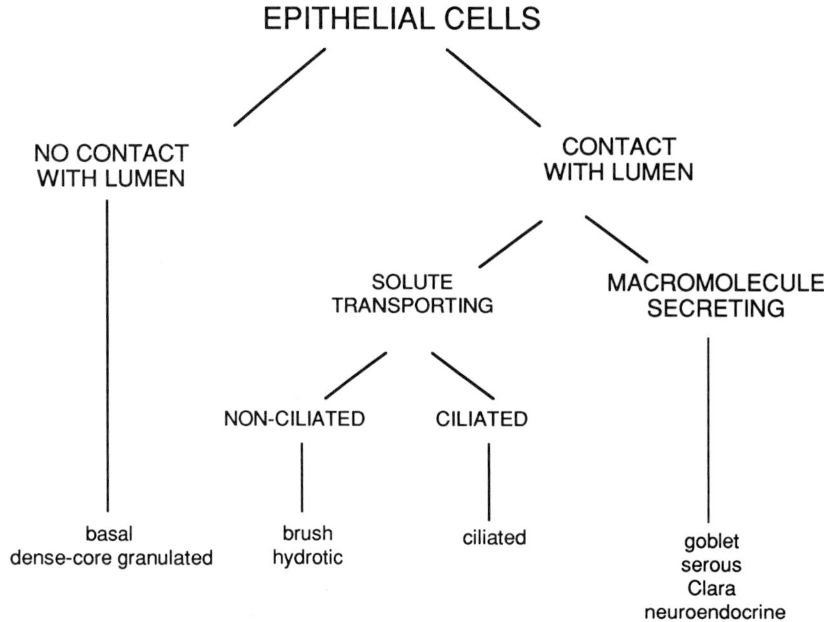

Figure 1.3 Types of epithelial cell found in the airways.

to the basolateral membrane (Ernst and Mills, 1980). This membrane, in general, is also K-selective (i.e. the predominant selectivity of the ion channels present is to K). In Na-absorbing epithelia (Koefoed-Johnsen and Ussing, 1958), the major transport protein of the apical membrane is an amiloride-sensitive Na channel. As illustrated in Fig. 1.5A, net movement of Na into the cell occurs by passive diffusion through the Na channels down both electrical and chemical concentration gradients. The Na entering across the apical membrane is actively extruded from the cell by the basolateral Na/K ATPase. The K pumped in by this enzyme recycles across the basolateral membrane via the K channels. The turnover rates of these various transport proteins are closely linked so as to prevent marked changes in intracellular ion content and cell volume (Schultz, 1981). In Cl-secreting epithelia (Frizzell *et al.*, 1979), in addition to K channels and the Na/K ATPase, the basolateral membrane contains a Na/K/2Cl co-transporter, whereas the apical membrane contains Cl channels (Fig. 1.5B). In such epithelia, entry of Cl across the basolateral membrane is by co-transport with Na. The energy in the transmembrane concentration gradient for Na allows Cl to be accumulated within the cells to a level greater than that predicted for passive distribution according to the apical membrane potential (ψ_a). Accordingly, there is a net exit of Cl across the apical membrane, which is Cl-selective. The Na that enters by co-transport with Cl is removed from the cells by the basolaterally located Na/K ATPase. As in Na-absorbing epithelia, the K pumped in by the Na/K ATPase recycles through the basolateral K channels.

2.3   CELL-TO-CELL CONTACTS

There are four types of organelle where membranes of adjacent epithelial cells are closely apposed (Fig. 1.6). (Farquhar and Palade, 1963; Revel and Karnovsky, 1967; Friend and Gilula, 1972). At the cell apices, completely surrounding each cell, are ZO, or tight junctions. The function of these organelles is to hold cells together, and also to regulate transepithelial movement of solutes. Immediately below the ZO are the ZA. These organelles are believed to play a role in cell adherence during formation of the ZO. Together with a third component, the macula adherens or desmosome, the ZO and ZA make up the "tight junctional complex" (Farquhar and Palade, 1963). Finally, gap junctions are regions of very close membrane apposition where small solutes (e.g. ions, second messengers, ATP) pass directly between cells.

Tight junctions (or ZO) mark the point of transition from basolateral to apical membrane. At one time it was thought that tight junctions physically kept the apical and basolateral cell membranes apart. However, they are now thought to be a reflection of epithelial polarity rather than the cause. Thus, single epithelial cells in culture show some degree of apical and basolateral membrane specialization even in the absence of neighbours (Rodriguez-Boulan *et al.*, 1983; Vega-Salas *et al.*, 1987). Also, T_{84} cells (a cell line taken from human colon) early in culture can show discontinuity in the tight junctions' circumferential seal (Madara *et al.*, 1987), though structural polarity of the membrane surface exists at these sites. In thin sections, tight junctions consist of zones where the lateral membranes of adjacent cells are closely apposed for distances of 0.1–0.6 µm. Within this zone, there may be

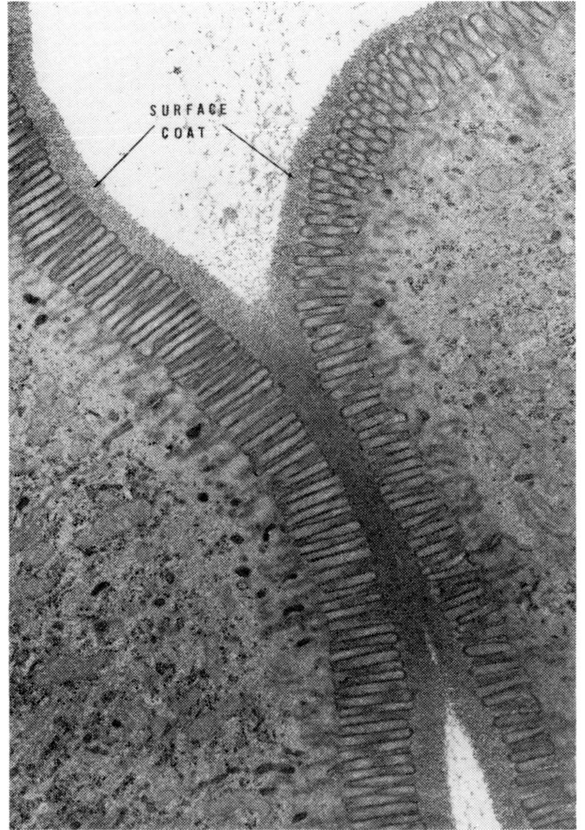

A

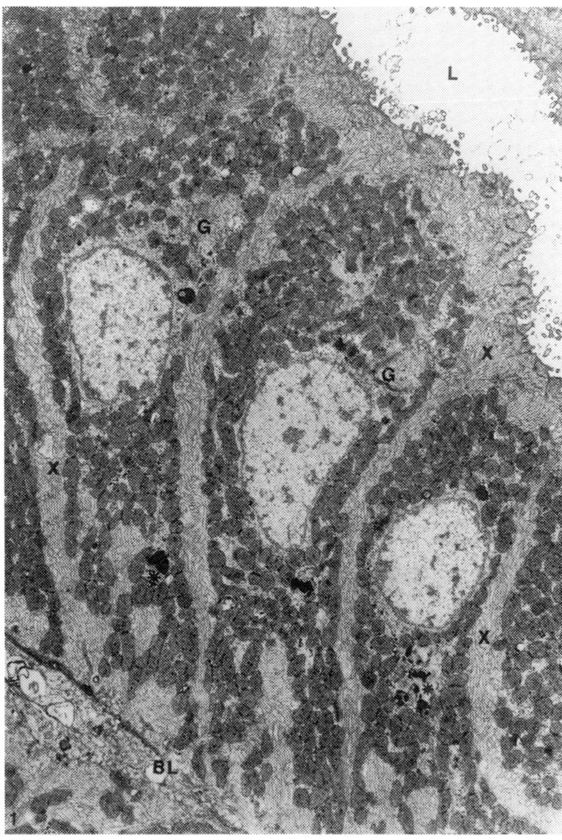

B

Figure 1.4 (A) Microvilli on the apical membrane of cat intestinal epithelium. Note the presence of a pronounced glycocalyx. × 10 400. (Reproduced, with permission, from Ito, 1965.) (B) Extreme amplification of basolateral membrane in epithelium from elasmobranch rectal gland, an organ specialized for secretion of Cl. L, lumen; BL, basal lamina; G, Golgi apparatuses; X, areas of basolateral membrane infolding. × 3100. (Reproduced, with permission, from Ernst *et al.* (1981). J. Membr. Biol. 58, 101–114.)

up to eight sites at which the outer membrane leaflets of the neighbouring membranes appear to fuse (see Fig. 1.6A). In transmission electron micrographs, a penta-laminar double-membrane structure is seen at these sites. In freeze–fracture electron micrographs, these points of fusion are seen as an anastomizing network of ridges in the E (external) face and furrows on the P (protoplasmic) face (see Fig. 1.6B). At one time the tight junctional strands were believed to be mainly lipidic in nature (Da Silva and Kachar, 1982). However, protein components of the ZO are being identified, and current opinion holds that the strands consist predominantly of integral membrane proteins (Stevenson *et al.*, 1988). Claude (1978) has established a logarithmic relationship between the numbers of tight-junctional strands and the electrical resistance of an epithelium. A linear relationship might have been expected, as resistances in series add up arithmetically. Cereijido *et al.* (1989) have suggested that the strands are electrically non-conductive, but contain aqueous pores which flicker open and closed. However, because of the

anastomoses between strands, a set of open pores in one strand may be electrically isolated from the pores in a neighbouring strand. This model yields the observed logarithmic relationship between strand number and electrical resistance. The concept of pores which open and close is supported by studies in which Ca-free medium causes large increases in paracellular conductance with little change in the ultrastructure of tight junctional strands (Martinez-Palomo *et al.*, 1980). Tight junctional resistance is regulated by cAMP, Ca^{2+}, and protein kinase C (Madara, 1990b). When dying cells are extruded from an epithelium, tight junctional complexes seem to move down the lateral walls of the dying cell, thereby maintaining epithelial integrity (Madara, 1990a). Migration of PMNs across epithelia during inflammation occurs via the tight junctions, and results in substantial increases in tight junctional permeability, which can be localized with morphologically detectable macromolecular tracers to the points where PMNs are actively crossing the tight junction (Nash *et al.*, 1987). Following large-scale migration,

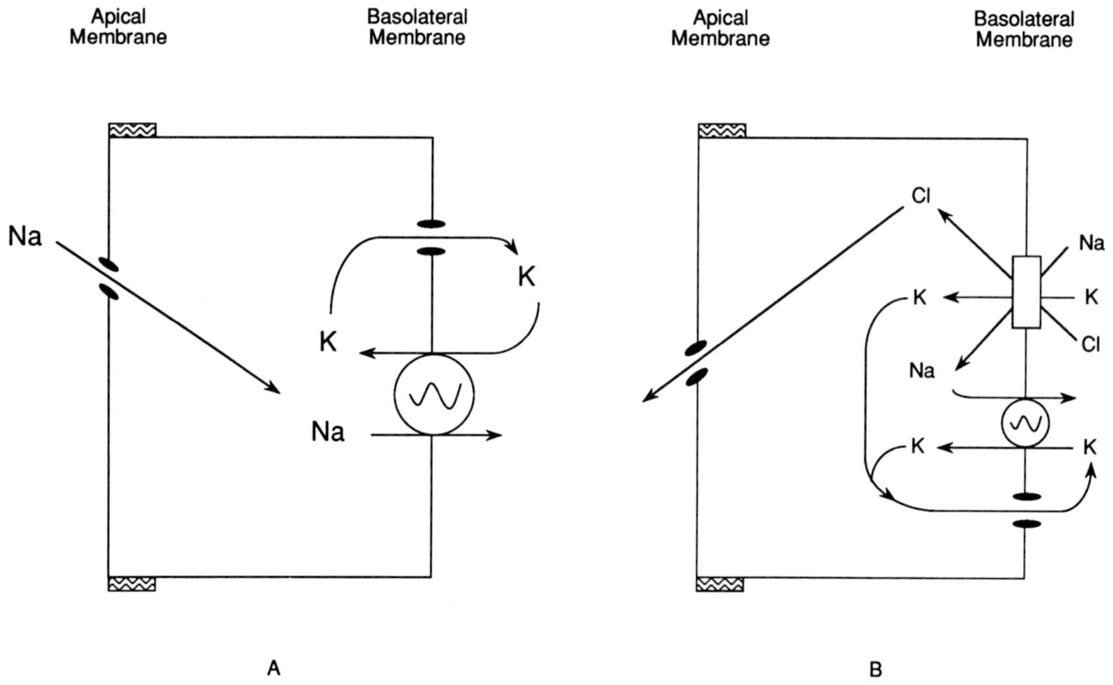

Figure 1.5 Models for (A) Na absorption and (B) Cl secretion across airway epithelia. See text for details.

the tight junctions remain damaged for some time, with transepithelial resistance taking between 4 and 18 h to recover (Nash *et al.*, 1987).

The ZA or intermediate junctions (so called because in the junctional complex they often lie between the ZO and the maculae adherentes) are sites from 0.3 to 0.5 μm in depth encircling the cell just below the ZO (Farquhar and Palade, 1963). Here, bands of actin filaments and other cytoskeletal elements circle the cell and attach to the inside of the cell membrane. This ring may be connected by cytoskeletal elements to the points of membrane fusion in the ZO. Alterations in tight junctional permeability have been linked to changes in the adjacent cytoskeleton, and agents which affect the cytoskeleton (e.g. cytocholasin B) alter tight junctional permeability (Madara, 1989). It is believed that CAMs are associated with the ZA, and are important in holding adjacent lateral membranes together during tight junctional formation (and possibly maintenance) (Gumbiner, 1987). The intercellular space at the ZA is approximately 25 nm across and is filled with homogeneous amorphous material of low density (Farquhar and Palade, 1963).

Spot desmosomes, or maculae adherentes (Schwartz *et al.*, 1990) are common just below the ZA, but are also found scattered all over the basolateral membrane. They are believed to play a role in holding epithelial cells together. At these organelles, adjacent cell membranes approach to within approximately 30 nm of one another. The material within the gap between cells is known as desmoglea, and consists of a central dense stratum (the

central disc) connected to the membranes by lateral cross-bridges. On the cytoplasmic side of the cell membranes at the desmosome are dense plaques (~0.3 μm in diameter) to which intermediate filaments attach. In epithelia these filaments are composed of keratin (Franke *et al.*, 1978). The proteins responsible for linking the keratin filaments to the membrane are beginning to be characterized (Jones and Green, 1991). The complex network of desmosomes and associated keratin filaments may be the primary determinant of cell shape and epithelial form.

Gap junctions are sites where adjacent cell membranes come to within 2–3 nm of one another. Appropriate staining reveals a septilaminar structure, which has a total thickness of approximately 18 nm (two unit membranes of 7.5 nm thickness, with a 3 nm gap) (McNutt and Weinstein, 1970). Though very closely apposed, the membranes are not entirely fused, as extracellular markers such as lanthanum can penetrate the gap. Freeze–fracture of gap junctions shows a typical hexagonal array, with a centre-to-centre spacing between neighbouring subunits of about 9 nm. At the centre of each hexagon are a pair of channels, one in each membrane, whose central aqueous pores are lined up so as to allow diffusion of solute between cells (Loewenstein, 1981). Molecules and ions of up to approximately 2 nm in diameter are able to pass through the channels between cells. The number of hexagonal subunits (connexons) in a gap junction vary from two or three to several hundred. Evidence that these structures truly represent cell-to-cell coupling comes from

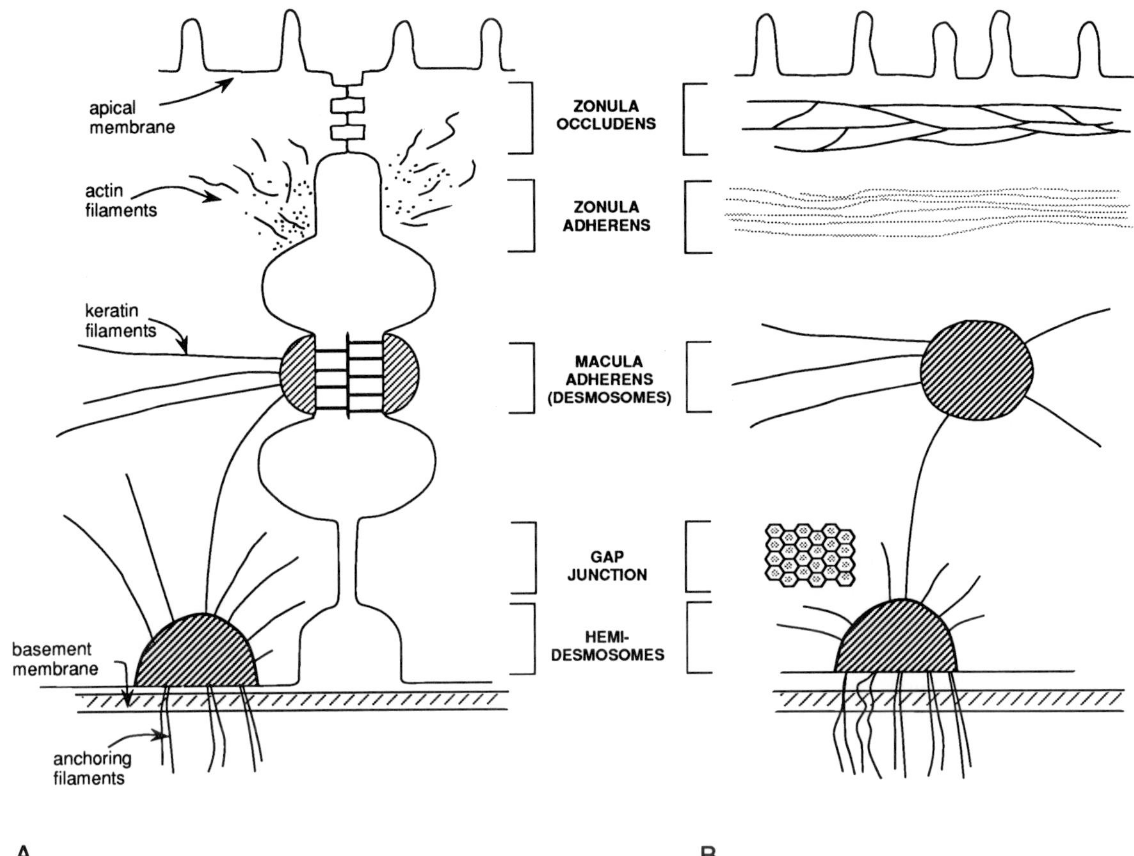

Figure 1.6 Cell-to-cell and cell-to-substratum contacts in epithelia (A) in cross-section and (B) viewed face-on. Drawings are not to scale, but are instead designed to illustrate major structural features of the different junctions. See text for further details.

studies in which the flow of electrical current between cells increased in a quantal fashion as the numbers of connexions between fusing cells increased (Loewenstein, 1981).

2.4 HEMIDESMOSOMES AND ANCHORING FILAMENTS

Hemidesmosomes are relatively small plaque-bearing (0.4 μm diameter) domains on basal cell membranes adjacent to the basement membrane. Though structurally resembling half desmosomes, they do not generally react with antibodies raised against desmosomal components (Schwartz *et al.*, 1990). From the hemidesmosomes, bundles of anchoring filaments pass through the basement membrane and terminate among the collagen fibres of the submucosa (Kawanami *et al.*, 1979; Ellison and Garrod, 1984).

2.5 CYTOSKELETON

Bundles of intermediate filaments (6–11 nm thick) composed mainly of keratin pass between desmosomes and from desmosomes to the nuclear envelope (Franke *et al.*, 1978; Fey *et al.*, 1984). As discussed above, epithelial cells also contain an actin cytoskeleton, which is a prominent component of the tight junctional complex. Actin filaments are also found running longitudinally along the cores of microvilli. In addition, epithelial cells contain a network of microtubules, which do not appear to be directly associated with the cell membranes (Bre *et al.*, 1987). The microtubular network is probably involved in intracellular trafficking of endosomes. Linkage of integral membrane proteins to the cytoskeleton either directly or via other proteins (such as ankyrin and spectrin) is believed to play an important role in the maintenance of epithelial polarity (Nelson, 1989).

2.6   LATERAL INTERCELLULAR SPACE

The LIS are the spaces between cells. They are bounded by the lateral membranes, start just below the tight junctions, and end at the basement membrane. They can be dilated maximally by very small hydrostatic pressures (~4 mmHg) (Spring and Hope, 1979). During active fluid absorption by epithelia, pumping of solutes into the LIS is believed to raise their osmolarity, draw water into them, increase their hydrostatic pressure and thereby cause them to dilate (Diamond, 1979). The state of dilation of the LIS, in response to either fluid transport or submucosal oedema, may profoundly influence the permeability of the paracellular pathway. Smulders, Wright and Tormey (Smulders *et al.*, 1972; Wright *et al.*, 1972), for instance, found that collapse of gallbladder LIS in response to osmotic fluid flows led to a 125% increase in transepithelial resistance and marked declines in the permeability to water and sucrose. Extreme dilation of the LIS in response to submucosal hydrostatic pressures over 5 cmH$_2$O can lead to disruption of tight junctions and increases in permeability to molecules as large as proteins (van Os *et al.*, 1979; Kondo *et al.*, 1992). It has been speculated (Kondo *et al.*, 1992) that the widespread submucosal oedema seen in asthma could produce sub-epithelial pressures of this magnitude, and contribute to the exfoliation of cells seen in this disease.

3. *Development, Maintenance and Repair of Epithelia*

Epithelial cells divide and differentiate during development, to replace dying cells in an otherwise healthy epithelium, and during regeneration from wound healing. It is important to distinguish these three processes because they show differences in the rate of division and in the pathways of cellular differentiation. The information presented below will concentrate on airway epithelium, with brief descriptions of events in the skin and intestine. For further details on cell kinetics in these and other epithelia the reader is referred to the comprehensive treatise by Wright and Alison (1984).

3.1   THE CELL CYCLE

Studies on cell proliferation are based on the concept of a cell cycle, which can be divided into several phases (Fig. 1.7). Actively dividing cells show a period (the S phase) of DNA synthesis prior to mitosis (the M phase). A prolonged gap (G$_1$) is found between mitosis and the next period of DNA synthesis, and may occupy up to 80% of the cell cycle. A briefer gap (G$_2$) exists between the S and M phases of the cycle. Only a fraction (the growth fraction) of a tissue's cells are proliferating (cycling) at any given time. Cells leave the cycle in one of three ways. First, they can die. Secondly, they may become terminally differentiated, in which case they become incapable of any further division. Ciliated cells are generally agreed to be in this category. Lastly, they can enter a resting state (G$_0$), from which they remain capable of rejoining the cycle in response to growth factors. Inhibitory factors (chalones) may be responsible for preventing division and causing cells to enter G$_0$. There are several indices of cell proliferation (Ayers and Jeffery, 1988). The mitotic index is the ratio of cells in mitosis to the total number of nucleated cells. As the M phase accounts for only about 5% of the cell cycle, this index is very small, and hard to measure reliably in epithelia, such as those of airways, which have a small growth fraction and a long cell cycle duration. A more commonly used measure of proliferation is the LI. Here, cells are exposed briefly to tritiated thymidine, which is

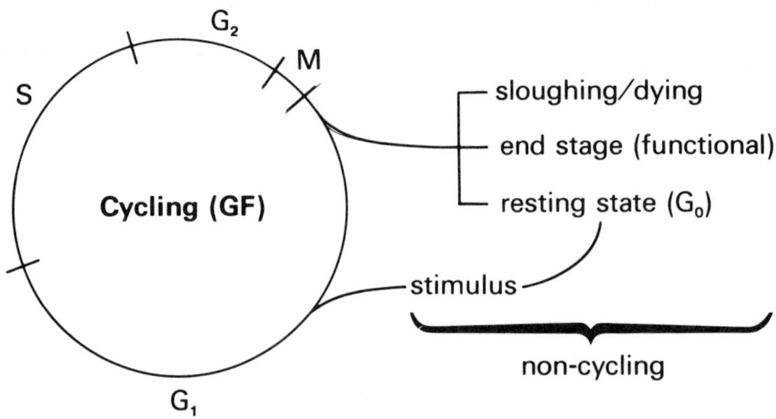

Figure 1.7 The cell cycle. Cells in the growth fraction (GF), go successively through DNA synthesis (S), a gap phase (G$_2$), followed by mitosis (M), and another gap phase (G$_1$) prior to another round of DNA synthesis. Cells may leave the cycle in several ways. Those in the resting state (G$_0$) can be induced to rejoin the cycle by growth factors and other stimuli. (Reproduced, with permission, from Ayers and Jeffery (1988).)

incorporated into DNA by cells in the S phase. Labelled cells are then counted by autoradiography. The LI is generally greater than the mitotic index, as the S phase accounts for a greater part of the cell cycle than the M phase. With each cell division, the radiation in the nuclei is halved, resulting in a 50% decrease in grain density on autoradiographs. The time to the first halving of average grain density gives a value for the duration of the S, G and M phases combined. The time taken for successive halvings gives the duration of the entire cell cycle (Wright and Alison, 1984; Ayers and Jeffery, 1988). Combining pulse and continuous labelling with [^{3}H]- and [^{14}C]thymidine provides information from which can be derived the duration of all the individual components of the cell cycle, as well as the growth fraction and rates of cell birth and loss (Wright and Alison, 1984).

3.2 DEVELOPMENT

In early fetal life, the only type of epithelial cell present in the airways is a non-differentiated columnar cell, which shows little endoplasmic reticulum, a small Golgi apparatus, no secretory granules and an apical membrane that lacks microvilli or cilia (Leeson, 1961; McDowell *et al.*, 1985a; Plopper *et al.*, 1986). In large airways, the first differentiated cell to appear is the neuroendocrine cell, followed by ciliated cells, then secretory cells, with basal cells appearing last (Plopper *et al.*, 1986). It is believed that all the differential cell types are initially derived from the undifferentiated columnar cells (McDowell *et al.*, 1985b). In developing hamster trachea, mitoses (as reflected by the incorporation of [^{3}H]thymidine) are somewhat more frequent in secretory cells than basal cells, but absent from neuroendocrine and ciliated cells (McDowell *et al.*, 1985b). Cells with both secretory granules and cilia have been seen under the electron microscope, suggesting that secretory cells can differentiate into ciliated. However, division of basal cells is thought to produce only further basal cells (McDowell *et al.*, 1985b).

During fetal development, the rate of cell division in large airway epithelium is much greater than after birth. Thus, McDowell and co-workers (McDowell *et al.*, 1985b; Otani *et al.*, 1986) found in the hamster trachea that on day 10 of gestation (the time at which the tracheal rudiment becomes distinct from the oesophagus) approximately 5% of cells were undergoing mitosis. By the day of birth (day 16), this figure had fallen to approximately 0.3%, and remained at this level for the next 7 days postnatally.

In general, in animals with longer gestational periods, the differentiated cell types of the trachea appear relatively earlier in fetal life (McDowell *et al.*, 1985a; Plopper *et al.*, 1986). Though all the major differentiated cell types are present at birth, the degree of postnatal development is highly variable. In the ferret (Leigh *et al.*, 1986), mouse (Kawamata and Fujita, 1983) and rat (Smolich *et al.*,

1976), for instance, there are dramatic changes in the relative numbers of the different cell types after birth with a progressive replacement of secretory by ciliated cells. On the other hand, in the hamster (Otani *et al.*, 1986) and rhesus monkey (Plopper *et al.*, 1986), the cells do not change their relative numbers after birth, but undergo rapid cytodifferentiation and functional maturation.

In smaller airways, the only differentiated cell types are Clara and ciliated. The exact time of appearance of these cell types during development is not known. Thus, it is uncertain whether ciliated cells are the direct progeny of the initial undifferentiated columnar cells, or whether they develop from Clara cells. In cell culture or during wound healing, it is well established that Clara cells can differentiate into ciliated cells (Evans *et al.*, 1978; Brody *et al.*, 1987).

The human skin epidermis starts as a single layer of cells, the stratum germinativum. Mitoses in this layer are initially 'horizontal', serving merely to increase the area of skin as the fetus enlarges. At about 35 days of fetal life, 'vertical' mitoses start to produce a second layer, the periderm. This is a transitory layer, disappearing by 160 days in the human fetus. At about 60 days, intermediate layers (stratum intermedium), which eventually become the spiny layer of the adult, start to form between the stratum germinativum and the periderm. Early in fetal life, all these layers show mitoses. The stratum germinativum shows predominantly vertical mitoses which contribute cells to the overlying layers and increase the degree of stratification. The cells in the stratum intermedium show mainly horizontal mitoses which contribute to lateral growth of these layers. Gradually during fetal life, cell production becomes restricted to the basal layer (stratum germinativum), and this situation persists into adulthood.

The various portions of the gastrointestinal tract all develop from a simple tube of columnar epithelium. This stratifies early in fetal life, and glands then develop from clefts within the stratified cells layers. Once a gland is formed, further glands are produced by a process of budding. These buds come off the neck of gastric glands and from the base of the crypt in the intestine.

3.3 MAINTENANCE

As reviewed by Ayers and Jeffery (1988), the cell turnover of healthy airway epithelium is much slower than that of most other epithelia. The rate of entry into mitosis is 0.14–0.3 mitoses per 1000 cells per hour. LIs range from about 0.1 to 2%. By contrast, the LI for stratified squamous epithelia averages about 10%, and is as high as 50% in parts of the intestinal epithelium (Wright and Alison, 1984). Cell cycle time for mucous and goblet cells in the hamster (as determined by the time for nuclear grain counts to halve) is about 100 h, with the great majority of the cycle being occupied by G_1. The estimated turnover times (the hypothetical time for the number of epithelial cells to double in the absence of cell death) are correspondingly slow, ranging from 20 to 200 days. In oral epithelium, by contrast, the turnover time is approximately 10 days (Wright and Alison, 1984).

When considering the pathways of cell differentiation in the airways, there is only one generally accepted rule that applies to all conditions and airway regions: ciliated cells are terminally differentiated and do not divide (Ayers and Jeffery, 1988). In the smaller airways, Clara cells are the progenitor cells, developing into either ciliated or further Clara cells, but not other cell types (Brody *et al.*, 1987). In the large airways, it was long held that basal cells were the stem cells under all circumstances. However, the low level of proliferation in airways creates difficulties in testing this experimentally. Accordingly, most studies of airway cell kinetics studies were performed on injured or developing epithelium, where proliferative rates may be elevated five- to 50-fold (Ayers and Jeffery, 1988). These studies cast considerable doubt on a role for basal cells, and instead implicated immature secretory cells (small mucous granule cells) as the main progenitor or stem cells of the airway epithelium. Other arguments against a role for basal cells as stem cells are that they appear later during development than ciliated and secretory cells (Plopper *et al.*, 1986), they are absent from the smaller airways where Clara cells function as the stem cells (Evans *et al.*, 1978; Brody *et al.*, 1990), and Evans and Plopper (1988) having convincingly provided an alternative role, that of epithelial anchors, for basal cells. However, the elegant work of Breuer *et al.* (1990) has recently established a major role for basal cells in renewal of healthy airway epithelium under steady-state conditions. Following injection of [³H]thymidine into hamsters, they looked for labelled nuclei in the same small zone of epithelium in animals killed over the next 14 days. About 500000 cells were examined, and in agreement with earlier workers they found an average LI of approximately 0.15%. Though basal cells made up only 6% of the total cells, their LI (~1.5%) was the highest of any cell type. The proliferative capacity (the product of the LI and the fraction of the total

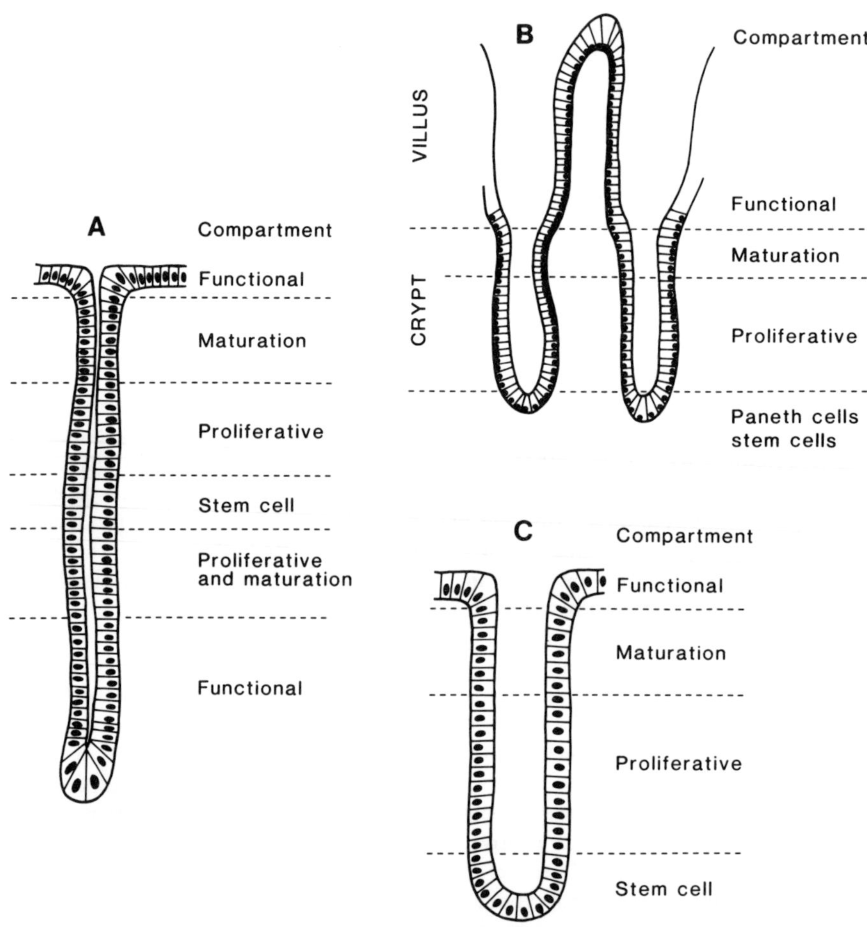

Figure 1.8 Diagrammatic representations of the kinetic compartments of (A) gastric glands, (B) small intestinal crypts and (C) colonic crypts. (Reproduced, with permission, from Wright and Alison (1984).)

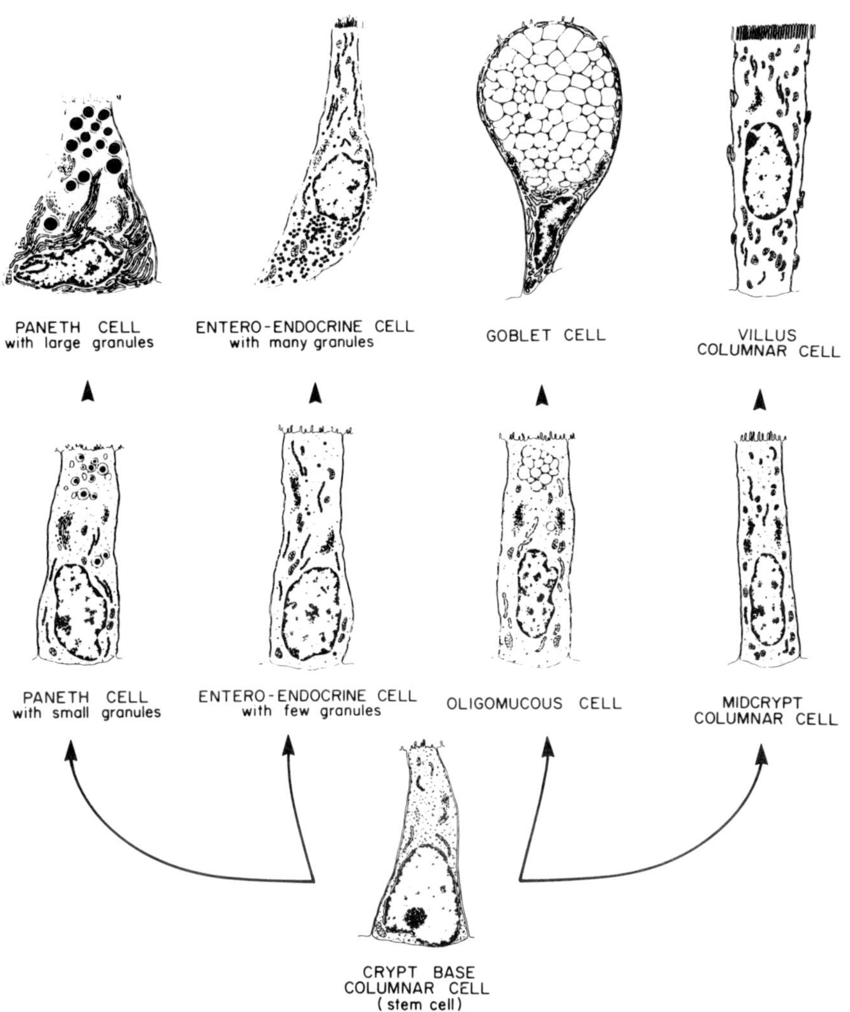

Figure 1.9 Cell lineages in the small intestine. All differentiated cell types are derived from columnar stem cells at the bottom of the crypts. (Reproduced, with permission, from Wright and Alison (1984); original drawing by Professor C.P. LeBlond.)

cells) was approximately equal for basal and the commoner secretory cells. With time, after addition of thymidine, the number of labelled basal cells declined, while the numbers of labelled secretory and ciliated cells increased, suggesting conversion of basal cells into these other types. Also suggesting a stem cell role is the finding that pure preparations of basal cells inoculated into tracheae denuded of their own epithelium produce all the cell types of differentiated adult epithelium (Inayama *et al.*, 1988).

In the skin, the basal cells (stratum germinativium) form the proliferative compartment. Newly produced progeny of the basal cells push older cell layers towards the surface, where they eventually keratinize and desqua-

mate. The granular layer (stratum granulosum) marks the transition between the viable cells below and dead, keratinized cells above. The granules in the cells of the stratum granulosum may represent organelles undergoing destruction in lysosomes and/or an early event in the synthesis of keratin. The outermost, horny corneocytes lack nuclei, which are lost in the granular layer.

The gut epithelium is much more complicated than either the skin or the airway epithelium. There are important regional differences in structure, and a wide variety of different cell types in each region. The gastric mucosa shows a multitude of glands containing chief or zymogen cells (which secrete pepsin), parietal or oxynctic cells (acid-secreting), neck mucous cells, and argentaffin cells (which secrete serotonin and other hormones). The epithelium of the small intestine is thrown up into numerous villi, with crypts in between. Columnar cells in the crypts form the bulk of the proliferating cell population. In addition, there are Paneth cells (of uncertain function), goblet cells and endocrine cells. The large intestine lacks the villi of the small intestine, but does contain glands (crypts), which differ from those of the small intestine in their greater numbers of goblet cells. At the bottom of the large intestinal glands are proliferating, relatively undifferentiated columnar cells and occasional endocrine (argentaffin) cells. Stem cells are found at the bases of glands in the small and large intestine, and near the top of the glands in the gastric mucosa (Fig. 1.8). It is believed that all specialized cell types are derived directly from the undifferentiated stem cells in the progenitor regions of the glands (Fig. 1.9). The LIs in these progenitor regions are high (10–50%) and cells are continuously migrating up the gland to the surface, where they form the absorptive cells of the villus in the small intestine, and mucus-secreting cells in the colon and gastric mucosa. In addition, in the gastric glands, there is a migration of cells down the crypt to form specialized peptic and parietal cells.

3.4   REPAIR AND REGENERATION

Damage to airway epithelium of experimental animals by noxious gases, carcinogens, mechanical trauma, elastase, and viral and bacterial infections leads to dramatic (three- to 60-fold) increases in both the mitotic index and LI (Ayers and Jeffery, 1988). These changes probably represent a decrease in the duration of the cell cycle. In nearly all cases there is increasing evidence that the predominant progenitor cells which respond to epithelial damage in the upper airways are mucous cells. In rat bronchi exposed to cigarette smoke for 7 days, for instance, 40% of cells labelled with [³H]thymidine were mucous cells; no mucous cells were labelled in unexposed animals (Ayers and Jeffery, 1982). The authors concluded that pre-existing serous cells differentiated into mucous cells, which then divided to form further mucous cells. Again in rat bronchi, the primary cells involved in epithelial regeneration after NO_2 exposure are also the secretory cells (Evans *et al.*, 1986). With mild mechanical damage, a layer of basal cells is left on the basement membrane, which divide to produce predominantly mucous cells (Lane and Gordon, 1974). Where denudation is complete, secretory cells at the edge of the denuded region flatten out and migrate into the injured area (Keenan *et al.*, 1982a,b,c). Eventually an area of stratified squamous epithelium is produced which consists predominantly of small mucous granule cells. Over the course of the next few days, normal mucociliary ultrastructure is regained.

The regenerative response of airway epithelium is metaplastic in that it generates an epithelium of abnormal morphology. Two types of metaplasia can be distinguished (Basbaum *et al.*, 1990). Mucous cell metaplasia is found in human bronchitis and is characterized by submucosal gland hypertrophy and increased numbers of mature goblet cells in the surface epithelium. This form of metaplasia can be induced in rats by chronic exposure to SO_2 or cigarette smoke (Jeffery and Reid, 1981; Lamb and Reid, 1986; Lamb and Reid, 1986). Elastase induces a similar morphology in hamster airways (McDowell *et al.*, 1983), but here the changes occur without alteration in mitotic rate, and are believed to be due to development of goblet cells, and accumulation of mature mucous granules. Squamoid (epidermoid) metaplasia is the usual response to carcinogens, mechanical injury or vitamin A deficiency (Basbaum *et al.*, 1990).

In bronchioles, Clara cells are the sole stem cell differentiating in response to injury into further Clara cells and ciliated cells (Evans *et al.*, 1978). In alveoli, in response to injury, type II cells divide and differentiate into type I cells (Evans *et al.*, 1975). These pathways of differentiation are almost certainly the same as those involved in maintenance of the normal healthy epithelia.

Compared to experimental animals, little is known about the response of human airway epithelium to damage. However, it is believed that epithelial damage plays a major role in the hyper-reactivity seen in asthma and other inflammatory airway diseases. A number of mechanisms may be involved, including increased access of allergens to submucosal mast cells, exposure of sensory nerve terminals, increased release of chemotactic agents (Rennard *et al.*, 1991), reduced release of epithelium-derived inhibitory factor (Vanhoutte, 1988) and reduced levels of enkephalinase (Jacoby *et al.*, 1988). In support of a role for epithelial damage in airway hyper-reactivity is the well-established finding that asthma is exacerbated by viral infections (Minor *et al.*, 1974), which may cause complete exfoliation of airway epithelium in experimental animals (Jacoby *et al.*, 1988).

A number of studies have documented epithelial damage in asthma. Naylor (1962) found sheets of desquamated epithelial cells (Creola bodies) in 3% of sputum samples from non-asthmatic subjects. However, in asthmatics, these cell sheets were found in 42% of sputum samples. Furthermore, the number of sheets per sample was increased 10-fold during attacks. More recently, Beasley *et al.* (1989) have shown a significant correlation between hyper-responsiveness to spasmogens and the numbers of desquamated epithelial cells in sputum from asthmatics. Autopsy material from patients dying in status asthmaticus also suggests widespread areas of desquamation and metaplastic regeneration (Dunnill, 1960). Biopsy material from asthmatic airways, as visualized by both scanning and transmission electron microscopy, essentially confirms the autopsy results (Laitinen *et al.*, 1985). To measure the permeability of human airway epithelium *in vivo*, the extracellular **marker**

^{99m}Tc-labelled DTPA has been delivered as an aerosol and its clearance from the lungs measured with a gamma camera, the assumption being that ^{99m}Tc-labelled DTPA is lost from the lungs by diffusion across the epithelium. Early studies, however, showed no change in the clearance of DTPA in asthmatics, but increased clearance in heavy smokers (Elwood *et al.*, 1983; Kennedy *et al.*, 1984). Later, Ilowite *et al.* (1989) pointed out that DTPA was lost not only by transepithelial diffusion, but also by entrainment in the mucus streaming towards the mouth.

When correction was made for this latter route (by measuring clearance of ^{99m}Tc-labelled albumin), the rate of transepithelial movement of DTPA in asthmatics was found to be twice normal. These functional studies thus confirm the presence of the epithelial damage revealed by microscopy, and are also consistent with the increased levels of blood proteins in sputum of asthmatics (Brogan *et al.*, 1975). Unfortunately, histological analysis of autopsy material is complicated by postmortem degenerative changes, and biopsies create mechanical damage of the

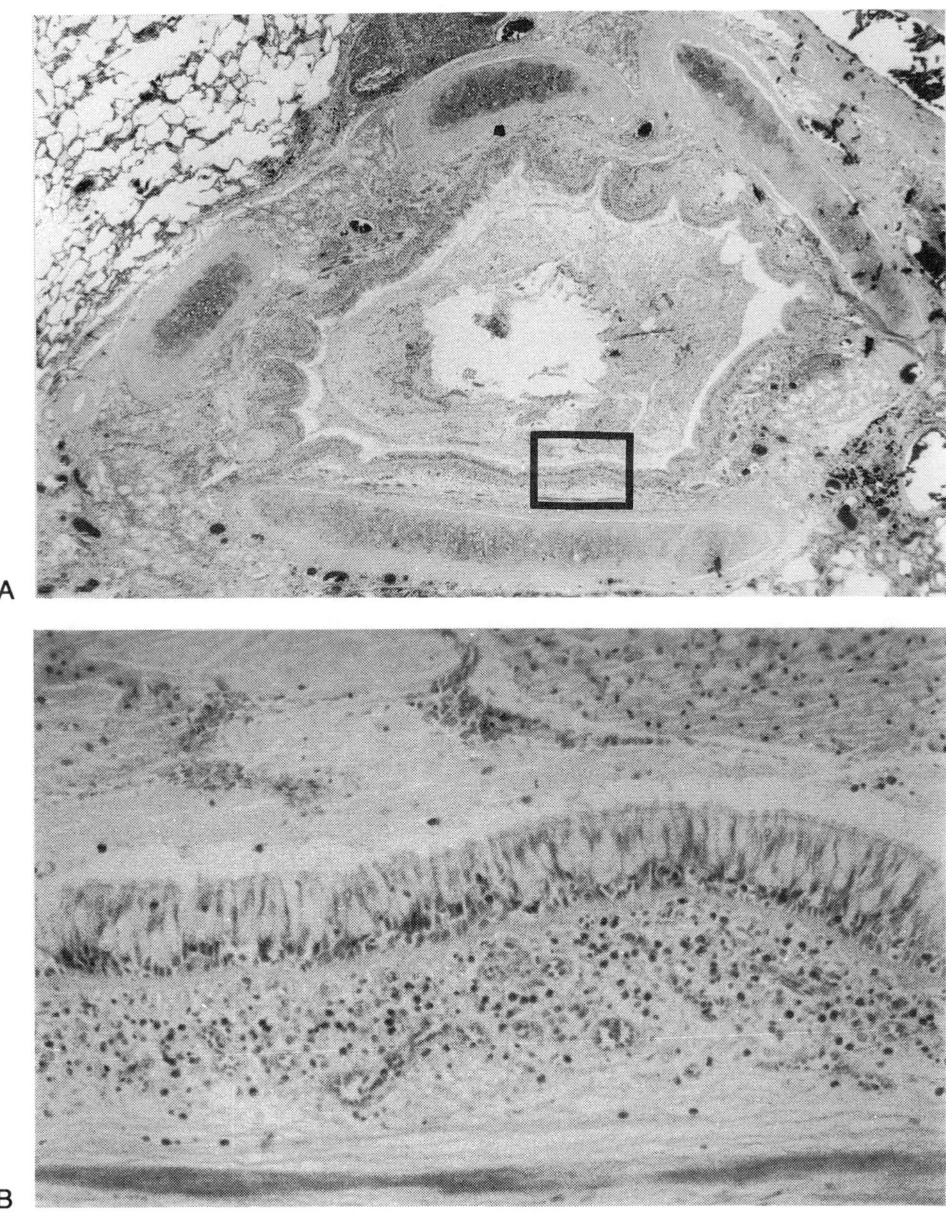

Figure 1.10 Apparently normal bronchial epithelium from the region of a mucus plug in a patient dying of status asthmaticus. (A) cross-section of bronchus with mucus plug. The area in the rectangular box is shown at a highter magnification in part (B). Note the apparently normal epithelium, and infiltration of the lamina propria and epithelium with inflammatory cells. (Courtesy of Dr W.E. Finkbeiner.)

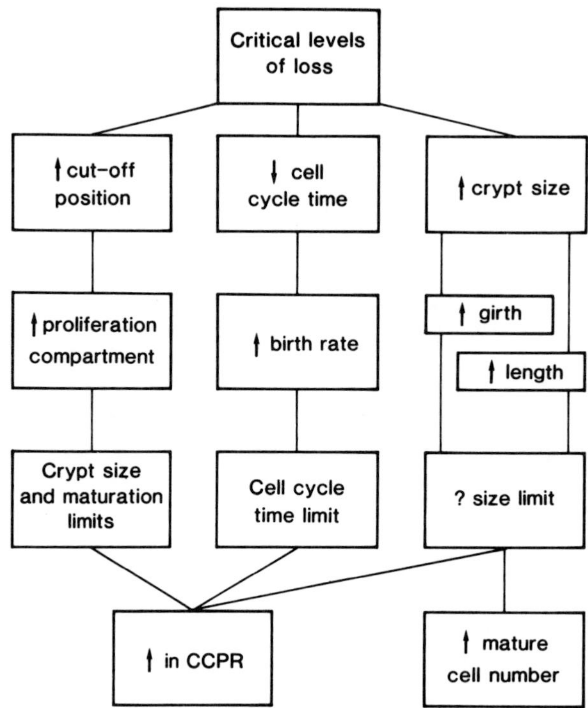

Figure 1.11 Mechanisms of compensatory crypt reactions in response to increased cell loss. See text for futher details. CCPR, crypt cell proliferation rate. (Reproduced, with permission, from Wright and Alison (1984).)

epithelium (Jeffery *et al.*, 1989). Furthermore, none of the studies on the structure of airway epithelium in asthma has applied morphometric techniques, and, though areas of epithelial damage can be found in asthmatics, other areas may appear absolutely normal (Fig. 1.10). Thus, the critical question as to the exact percentage of epithelium in asthma that is denuded or damaged is unanswered.

In the skin, the first response to injury is migration of basal cells into the injured region (Potten and Allen, 1977). Then there is a lag phase of 30–49 h (in human epidermis), before both basal and non-basal cells (Argyris, 1976) in the region of the injury start to show an increased LI. The increase in labelling peaks at about 96 h (Bishop and Cox, 1974). This increase in the LI may represent a decreased cell cycling time of proliferating cells, rather than recruitment of cells into the cycle from G_0 (Wright and Alison, 1984). After reviewing a considerable body of literature, Wright and Alison (1984) made the following conclusions about the skin's proliferative response to wounding. The lag phase is due to depletion of the basal cell layer by migration. This depletion stimulates proliferation of the remaining cells by an unknown mechanism, which could involve either local reduction in a chalone concentration, or the emergence of chalone-unresponsive stem cells. As cell numbers increase during proliferation, then chalone concentration increases (and/or the cells regain res-

ponsiveness to chalones) until proliferation returns to its normal rate.

Wright and Alison (1984) have identified three factors in the proliferative response to injury of gut epithelium (Fig. 1.11). The first is a change in the 'cut-off' position. This is the average cell position along a crypt at which a cell enters its final cell cycle. It is a measure of the size of the proliferative compartment. Because the growth faction (the proportion of proliferating cells) is approximately 60% in the crypts of the total cells, alteration of this fraction can at most increase crypt cell production by 1.4–fold. The second proliferative response is a decrease in the cell cycle time. However, crypt cells are already cycling very rapidly (cell cycle time 40 h), and this mechanism can at most double the rate of crypt cell production. The third response is to increase the population of crypt cells by increasing the size of the crypts. This last seems to be potentially the most potent way of increasing the numbers of proliferating cells, with the crypt population increasing by four-fold in certain diseases. With increasing cell damage, the order in which these proliferative responses are activated is decreased cell cycle time, increase in the growth fraction and finally expansion of the crypt population. Severe damage (e.g. after irradiation) necessitating a sudden and large increase in the production of new cells activates all three proliferative responses. The factors controlling these responses are perfectly understood. One possibility is that cell destruction removes negative feedback from chalones. Evidence also exists for stimulation of intestinal cell production by food in the lumen, and by systemic and local humoral agents.

4. *References*

Argyris, T.S. (1976). Kinetics of epidermal production during epidermal regeneration following abrasion in mice. Am. J. Pathol. 83, 329–340.

Ayers, M.M. and Jeffery, P.K. (1982). In "Cellular Biology of the Lung" (eds G. Bonsignore and G. Cumming), pp 33–59. Plenum Press, New York.

Ayers, M.M. and Jeffery, P.K. (1988). Proliferation and differentiation in mammalian airway epithelium. Eur. Respir. J. 1, 58–80.

Basbaum, C.B., Madison, J.M., Sommerhoff, C.P., Brown, J.K. and Finkbeiner, W.E. (1990). Receptors on airway gland cells. Am. Rev. Respir. Dis. 142, S141–S144.

Beasley, R., Roche, W.R., Roberts, J.A. and Holgate, S.T. (1989). Cellular events in the bronchi in mild asthma and after bronchial provocation. Am. Rev. Respir. Dis. 139, 806–817.

Biggers, J.D., Bell, J.E. and Benos, D.J. (1988). Mammalian blastocyst: transport functions in a developing epithelium. Am. J. Physiol. 139, C419–C432.

Bishop, S.C. and Cox, J.F. (1974). Autoradiographic study of the DNA synthetic activity in human skin cultured *in vitro*: effects of stripping and the inhibition by glucosamine. J. Invest. Derm. 62, 74–79.

Bre, M.H., Kreis, T.E. and Karsenti, E. (1987). Control of microtubule nucleation and stability in Madin–Darby canine kidney cells: the occurrence of noncentrosomal, stable detyrosinated microtubules. J. Cell Biol. 105, 1283–1296.

Breathnach, A.S. (1965). The cell of Langerhans. Int. Rev. Cytol. 18, 1–28.

Breuer, R., Zajicek, G., Christensen, T.G., Lucey, E.C. and Snider, G.L. (1990). Cell kinetics of normal adult hamster bronchial epithelium in the steady state. Am. J. Respir. Cell. Mol. Biol. 2, 51–58.

Brody, A.R., Hook, G.E., Christensen, T.G., Lucey, E.C. and Snider, G.L. (1987). The differentiation capacity of Clara cells isolated from the lungs of rabbits. Lab. Invest. 57, 219–229.

Brogan, T.D., Riley, H.C., Neal, L. and Yassa, J. (1975). Soluble proteins of bronchopulmonary secretions from patients with cystic fibrosis. Thorax 30, 72–79.

Cereijido, M., Gonzalez-Mariscal, L. and Contreras, G. (1989). Tight junction: barrier between higher organisms and environment. NIPS 4, 72–76.

Claude, P. (1978). Morphological factors influencing transepithelial permeability: a model for the resistance of the zonula occludens. J. Membr. Biol. 39, 219–232.

Da Silva, P. and Kachar, B. (1982). On tight-junction structure. Cell 28, 441–450.

Diamond, J. (1977). The epithelial junction: bridge, gate and fence. Physiologist 20, 10–18.

Diamond, J.M. (1979). Osmotic water flow in leaky epithelia. J. Membr. Biol. 51, 195–216.

Dragsten, P.R., Blumenthal, R. and Handler, J.S. (1981). Membrane asymmetry in epithelia: is the tight junction a barrier in the plasma membrane? Nature 294, 718–722.

Dunnill, M.S. (1960). The pathology of asthma, with special reference to changes in the bronchial mucosa. J. Clin. Pathol. 13, 27–33.

Ellison, J. and Garrod, D.R. (1984). Anchoring filaments of the amphibian epidermal–dermal junction traverse the basal lamina entirely from the plasma membrane of hemidesmosomes to the dermis. J. Cell Sci. 72, 163–172.

Elwood, R.K., Dennedy, S., Belzberg, A., Hogg, J.C. and Pare, P.D. (1983). Respiratory mucosal permeability in asthma. Am. Rev. Respir. Dis. 128, 523–527.

Ernst, S.A. and Mills, J.W. (1980). Autoradiographic localization of tritiated ouabain-sensitive sodium pumpsites in ion transporting epithelia. J. Histochem. Cytochem. 28, 72–77.

Evans, M.J. and Plopper, C.G. (1988). The role of basal cells in adhesion of columnar epithelium to airway basement membrane. Am. Rev. Respir. Dis. 138, 481–482.

Evans, M.J., Cabral, L.J., Stephens, R.J. and Freeman, G. (1975). Transformation of alveolar type II cells to type I cells following exposure to NO_2. Exp. Mol. Pathol. 22, 142–150.

Evans, M.J. Cabral-Anderson. L.J. and Freeman, G. (1978). Role of the Clara cell in renewal of the bronchiolar epithelium. Lab. Invest. 38, 648–655.

Evans, M.J., Shami, S.G., Cabral-Anderson, L.J. and Dekker, N.P. (1986). Role of nonciliated cells in renewal of the bronchial epithelium of rats exposed to NO_2. Am. J. Pathol. 123, 126–133.

Farquhar, M.G. and Palade, G.E. (1963). Junctional complexes in various epithelia. J. Cell. Biol. 17, 375–385.

Fawcett, D.W. (1986). "A Textbook of Histology", W.B. Saunders, Philadelphia.

Fey, E.G., Wan, K.M. and Penman, S. (1984). Epithelial cytoskeletal framework and nuclear matrix-intermediate filament scaffold: three-dimensional organization and protein composition. J. Cell. Biol. 98, 1972–1984.

Franke, W.W., Weber, K., Osborn, M., Schmid, E. and Freudenstein, C. (1978). Antibody to prekeratin. Decoration of tonofilament-like arrays in various cells of epithelial character. Exp. Cell Res. 116, 429–445.

Friend, D.S. and Gilula, N.B. (1972). Variations in tight and gap junctions in mammalian tissues. J. Cell Biol. 53, 758–777.

Frizzell, R.A., Field, M. and Schultz, S.G. (1979). Sodium-coupled chloride transport by epithelial tissues. Am. J. Physiol. 236, F1–F8.

Gumbiner, B. (1987). Structure, biochemistry, and assembly of epithelial tight junctions. Am. J. Physiol. 253, C749–758.

Holt, P.G. and Schon, H.M. (1987). Localization of T cells, macrophages and dendritic cells in rat respiratory tract tissue: implications for immune function studies. Immunology 62, 349–56.

Holt, P.G., Degebrodt, A., Venaille, T., O'Leary, C., Krska, K., Flexman, J., Farrell, H., Shellam, G., Young, P., Penhale, J., Robertson, T. and Papadimitriou, J.M. (1985). Preparation of interstitial lung cells by enzymatic digestion of tissue slices: preliminary characterization by morphology and performance in functional assays. Immunology 54, 139–146.

Humbert, F., Montesano, R., Perrelet, A. and Orci, L. (1976). Junctions in developing human and rat kidney: a freeze-fracture study. J. Ultrastruct. Res. 56, 202–214.

Ilowite, J.S., Bennett, W.D., Sheetz, M.S., Groth, M.L. and Nierman, D.M. (1989). Permeability of the bronchial mucosa to ^{99m}Tc-DTPA in asthma. Am. Rev. Respir. Dis. 139, 1139–1143.

Inayama, Y., Hook, G.E., Brody, A.R., Cameron, G.S., Jetten, A.M., Gilmore, L.B., Gray, T. and Nettesheim, P. (1988). The differentiation potential of tracheal basal cells. Lab. Invest. 58, 706–717.

Ito, S. (1965). The enteric surface coat of cat intestinal microvilli. J. Cell Biol. 27, 475–491.

Jacoby, D.B., Tamaoki, J., Borson, D.B. and Nadel, J.A. (1988). Influenza infection causes airway hyperresponsiveness by decreasing enkephalinase. J. Appl. Physiol. 64, 2653–2658.

Jeffery, P.K. and Reid, L.M. (1977). In "Lung Biology in Health and Disease" (eds J.D. Brain, D.F. Proctor and L.M. Reid), pp 193–245. Marcel Dekker, New York.

Jeffery, P.K. and Reid, L.M. (1981). The effect of tobacco smoke, with or without phenylmethyloxadiazole (PMO), on rat bronchial epithelium: a light and electron microscopic study. J. Pathol. 133, 341–359.

Jeffery, P.K., Wardlaw, A.J., Nelson, F.C., Collins, J.V. and Kay, A.B. (1989). Bronchial biopsies in asthma. An ultrastructural, quantitative study and correlation with hyperreactivity. Am. Rev. Respir. Dis. 140, 1745–1753.

Jones, J.C. and Green, K.J. (1991). Intermediate filament–plasma membrane interactions. Curr. Opin. Cell. Biol. 3, 127–132.

Katz, S., Tamaki, K. and Sachs, D.H. (1979). Epidermal Langerhans cells are derived from cells originating in the bone marrow. Nature 282, 324–326.

Kawamata, S. and Fujita, H. (1983). Fine structural aspects of the development and aging of the tracheal epithelium of mice. Arch. Histol. Japon. 46, 355–372.

Kawanami, O., Ferrans, V.J. and Crystal, R.G. (1979). Anchoring fibrils in the normal canine respiratory system. Am. Rev. Respir. Dis. 120, 595–611.

Keenan, K.P., Combs, J.W. and McDowell, E.M. (1982a). Regeneration of hamster tracheal epithelium after mechanical injury. I. Focal lesions: quantitative morphologic study of cell proliferation. Virchows Arch. 41, 193–214.

Keenan, K.P., Combs, J.W. and McDowell, E.M. (1982b). Regeneration of hamster tracheal epithelium after mechanical injury. II. Multifocal lesions: staphmokinetic and autoradiographic studies of cell proliferation. Virchows Arch. 41, 215–229.

Keenan, K.P., Combs, J.W. and McDowell, E.M. (1982c). Regeneration of hamster tracheal epithelium after mechanical injury. III. Large and small lesions: comparative stathmokinetic and single pulse and continuous thymidine labeling autoradiographic studies. Virchows Arch. 41, 231–252.

Kennedy, S.M., Elwood, R.K., Wiggs, B.J.R. and Pare, P.D. (1984). Increased airway mucosal permeability of smokers. Am. Rev. Respir. Dis. 129, 143–148.

Koefoed-Johnsen, V. and Ussing, H.H. (1958). The nature of the frog skin potential. Acta Physiol. Scand. 42, 298–308.

Kondo, M., Finkbeiner, W.E. and Widdicombe, J.H. (1992). Changes in permeability of dog tracheal epithelium in response to hydrostatic pressure. Am. J. Physiol. 262, 176–182.

Laitinen, L.A., Heino, M., Laitinen, A., Kava, T. and Haahtela, T. (1985). Damage of the airway epithelium and bronchial reactivity in patients with asthma. Am. Rev. Respir. Dis. 131, 599–606.

Lamb, D. and Reid, L. (1968). Mitotic rates, goblet cell increase and histochemical changes in mucus in rat bronchial epithelium during exposure to sulphur dioxide. J. Pact. Bact. 96, 97–111.

Lane, B.P. and Gordon, R.E. (1974). Regeneration of rat epithelium after mechanical injury. I. The relationship between mitotic activity and cellular differentiation. Proc. Soc. Exp. Biol. Med. 145, 1139–1144.

Leeson, T.S. (1961). The development of the trachea in the rabbit with particular reference to its fine structure. Anat. Anz. 110, 214–223.

Leigh, MW., Gambling, T.M., Carson, J.L., Collier, A.M., Wood, R.E. and Boat, T.F. (1986). Postnatal development of tracheal surface epithelium and submucosal glands in the ferret. Exp. Lung Res. 10, 153–169.

Loewenstein, W.R. (1981). Junctional intercellular communication: the cell-to-cell membrane channel. Physiol. Rev. 61, 829–899.

Luciano, L., Thiele, J. and Reale, E. (1979). Development of follicles and of occluding junctions between the follicular cells of the thyroid gland. J. Ultrastruct. Res. 66, 164–181.

McDowell, E.M., Combs, J.W. and Newkirk, C. (1983). Changes in secretory cells of hamster tracheal epithelium in response to acute sublethal injury: a quantitive study. Exp. Lung Res. 4, 227–243.

McDowell, E.M., Newkirk, C. and Coleman, B. (1985a). Development of hamster tracheal epithelium: I. A quantitative morphologic study in the fetus. Anat. Rec. 213, 429–447.

McDowell, E.M., Newkirk, C. and Coleman, B. (1985b). Development of hamster tracheal epithelium: II. Cell proliferation in the fetus. Anat. Rec. 213, 448–456.

McNutt, N.S. and Weinstein, R.S. (1970). The ultrastructure of the nexus. A correlated thin-section and freeze–cleave study. J. Cell Biol. 47, 666–671.

Madara, J.L. (1989). Loosening tight junctions. Lessons from the intestine. J. Clin. Invest. 83, 1089–1094.

Madara, J.L. (1990a). Maintenance of the macromolecular barrier at cell extrusion sites in intestinal epithelium: physiological rearrangement of tight junctions. J. Membr. Biol. 116, 177–184

Madara, J.L. (1990b). Warner–Lambert/Parke–Davis Award Lecture. Pathobiology of the intestinal epithelial barrier. Am. J. Pathol. 137, 1237–1281.

Madara, J.L., Stafford, J., Dharmsathaphorn, K. and Carlson, S. (1987). Structural analysis of a human intestinal epithelial cell line. Gastroenterology 92, 1133–1145.

Martinez-Palomo, A., Meza, I., Beatty, G. and Cereijido, M. (1980). Experimental modulation of occluding junctions in a cultured transporting epithelium. J. Cell Biol. 87, 736–745.

Minor, T.E., Dick, E.C., Demeo, A.N., Ouelette, J.J., Cohen, M. and Reed, C.E. (1974). Viruses as precipitants of asthmatic attacks in children. J. Am. Med. Assoc. 227, 292–298.

Montesano, R., Friend, D.S., Perrelet, A. and Orci, L. (1975). *In vivo* assembly of tight junctions in fetal rat liver. J. Cell Biol. 67, 310–319.

Nash, S., Stafford, J. and Madara, J.L. (1987). Effects of polymorphonuclear leukocyte transmigration on the barrier function of cultured intestinal epithelial monolayers. J. Clin. Invest. 80, 1104–1113.

Naylor, B. (1962). The shedding of the mucosa of the bronchial tree in asthma. Thorax 17, 69–72.

Nelson, W.J. (1989). In "Functional Epithelial Cells in Culture", (eds K.S. Matlin and J.C. Valentich) pp 3–42. Alan R. Liss, New York.

Otani, E.M., Newkirk, C. and McDowell, E.M. (1986). Development of hamster tracheal epithelium: IV cell proliferation and cytodifferentiation in the neonate. Anat. Rec. 214, 183–192.

Plopper, C.G., Alley, J.L. and Weir, A.J. (1986). Differentiation of tracheal epithelium during fetal lung maturation in the rhesus monkey, Macaca mulata. Am. J. Anat. 175, 59–71.

Potten, C.S. and Allen, T.D. (1977). Ultrastructure of cell loss in intestinal mucosa. J. Ultrastruct. Res. 60, 272–277.

Rennard, S.I., Beckman, J., Daughton, D., Ertl, R.F., Koyama, S., Rickard, K., Romberger, D., Rubenstein, I., Shoji, S., Sisson, J., Spurzen, J., Thompson, A.B., Von Essen, S. and Robbins, R.A. (1991). In "The Airway Epithelium. Physiology, Pathophysiology, and Pharmacology" (eds S.G. Farmer and D.W.P. Hay), pp 117–133. Marcel Dekker, New York.

Revel, J.P. and Karnovsky, M.J. (1967). Hexagonal airways of subunits in intercellular junctions of the mouse heart and liver. J. Cell. Biol. 33, C7–C10.

Rodriguez-Boulan, E., Padkiet, K.T. and Sabatini, D.D. (1983). Assembly of enveloped viruses in MDCK cells: polarized budding from single attached cells and from clusters of cells in suspension. J. Cell Biol. 96, 866–874.

Schneeberger, E.E., Walters, D.V. and Olever, R.E. (1978). Development of intercellular junctions in the pulmonary epithelium of the foetal lamb. J. Cell Sci. 32, 307–324.

Schultz, S.G. (1981). Homocellular regulatory mechanisms in sodium-transporting epithelia: avoidance of extinction by "flush through". Am. J. Physiol. 241, F579–F590.

Schwartz, M.A., Owaribe, K., Kartenbeck, J. and Franke, W.W. (1990). Desmosomes and hemidesmosomes: constitutive molecular components. Ann. Rev. Cell Biol. 6, 461–491.

Sertl, K., Takemura, T., Tschachler, E., Ferrans, V.J., Kaliner, M.A. and Shevach, E.M. (1986). Dendritic cells with antigen-presenting capability reside in airway epithelium, lung parenchyma, and visceral pleura. J. Exp. Med. 163, 436–451.

Smolich, J.J., Stratford, B.F., Maloney, J.E. and Ritchie, B.C. (1976). Postnatal development of the epithelium of larynx and trachea in the rat: Scanning electron microscopy. J. Anat. 124, 657–673.

Smulders, A.P., Tormey, J.M. and Wright, E.M. (1972). The effect of osmotically induced water flows on the permeability and ultrastructure of the rabbit gallbladder. J. Membr. Biol. 7, 164–197.

Soler, P., Moreau, A., Basset, F. and Hance, A.J. (1989). Cigarette smoking-induced changes in the number and differentiated state of pulmonary dendritic cells/Langerhans cells. Am. Rev. Respir. Dis. 139, 1112–7.

Spring, K.R. and Hope, A. (1979). Dimensions of cells and lateral intercellular spaces in living Necturus gallbladder. Fed. Proc. 38, 128–133.

Stevenson, B.R., Anderson, J.M. and Bullivant, S. (1988). The epithelial tight junction: structure, function and preliminary biochemical characterization. Mol. Cell. Biochem. 83, 129–45.

Stingl, G., Katz, S.I., Shevach, E.M., Wolff-Schreiner, E. and Green, I. (1978). Detection of Ia antigens on Langerhans cells in guinea pig skin. J. Immunol. 127, 1707–1713.

Stingl, G., Gazze-Stingl, I.A., Aberer, W. and Wolff, K. (1981). Antigen presentation by murine epidermal Langerhans cells and its alteration by ultraviolet B light. J. Immunol. 127, 1707–1713.

van Meer, G. and Simons, K. (1986). The function of tight junctions in maintaining differences in lipid composition between the apical and the basolateral cell surface domains of MDCK cells. EMBO J. 5, 1455–1464.

van Os, C.H., Wiedner, G. and Wright, E.M. (1979). Volume flows across gallbladder epithelium induced by small hydrostatic and osmotic gradients. J. Membr. Biol. 49, 1–20.

Vanhoutte, P.M. (1988). Epithelium-derived relaxing factor(s) and bronchial reactivity. Am. Rev. Respir. Dis. 138, S24–S30.

Vega-Salas, D.E., Salas, P.J., Gundersen, D. and Rodriguez, B.E. (1987). Formation of the apical pole of epithelial (Madin–Darby canine kidney) cells: polarity of an apical protein is independent of tight junctions while segregation of a basolateral marker requires cell–cell interactions. J. Cell Biol. 104, 905–916.

Wolff, K. and Stingl, G. (1983). The Langerhans cell. J. Invest. Derm. 80, 17s-21s.

Wright, E.M., Smulders, A.P. and Tormey, J.M. (1972). The role of the lateral intercellular spaces and solute polarization effects in the passive flow of water across the rabbit gallbladder. J. Membr. Biol. 7, 198–219.

Wright, N. and Alison, M. (1984). "The Biology of Epithelial Cell Populations". Clarendon Press, Oxford.

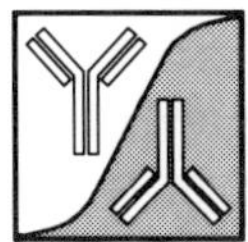

2. *Epithelial Disposition of Antigen*

Graham Mayrhofer

1. *General Perspective*

The defence of the epithelial surfaces of the body against microbial pathogens is essential for survival. To a considerable extent, this is achieved by innate mechanisms involving epithelial integrity, fluid flow and the presence of antimicrobial factors in the secretions at the various surfaces, backed up by non-specific cellular defences in the sub-epithelial tissues (Mayrhofer, 1984). However, transient acute infections of mucosal surfaces are common and involve colonization by pathogens, with or without penetration of the epithelium. Recovery from established infections and development of immunity against reinfection requires a specific adaptive immune response. The study of congenital and acquired immunodeficiency syndromes in recent years has provided unparalleled opportunities to assess the importance of adaptive immunity in resistance to systemic and mucosal infections.

It has emerged that an intact humoral immune response can complement the innate cellular defences quite

Immunopharmacology of Epithelial Barriers
ISBN 0–12–288030–7

effectively, allowing control of the majority of acute bacterial infections. Perhaps less predictable have been the insights gained into the importance of specific cell-mediated immunity, both systemically and also locally at mucosal surfaces. A range of intracellular bacterial pathogens, viruses, fungi and parasites, which are essentially non-pathogenic or responsible for self-limiting or slowly progressive infections in normal individuals, can produce devastating disease in those individuals who are deficient in T lymphocytes. At mucosal surfaces, uncontrolled infections with cytomegalovirus, *Candida albicans*, *Pneumocystis carinii* or *Cryptosporidium* sp. illustrate graphically the selective forces which have moulded the evolution of local adaptive cell-mediated immunity (Fauci, 1984; Smith, 1992).

During the evolution of the immune system, there has been a close association between mucosal surfaces and lymphoid tissues, illustrated most clearly by the lympho-epithelial relationships in the modern thymus and in the avian bursa of Fabricius. Perhaps more ancient is the association between lymphocytes and the intestinal epithelium. Recent evidence suggests that the intestinal epithelium may be a site for extrathymic maturation of a specific subset of mucosa-associated lymphocytes that expresses T cell receptors (γ/δ and α/β) and a homodimeric form of CD8 (Croitoru *et al.*, 1990; Cerf-Bensussan and Guy-Grand, 1991; Guy-Grand *et al.*, 1991). In modern mammals, a large proportion of the total lymphoid mass is concentrated around mucosal surfaces, which also have specialized structures and functions that are associated with the induction and delivery of local immune responses (Mayrhofer, 1984). This presumably reflects the importance of mucosal surfaces as major sites of antigen exposure, both in health and in disease. The surfaces involved are large, some support a rich commensal microflora, all are sites that are colonized or invaded periodically by pathogens and some are exposed to environmental and dietary antigens. Local immunization of these surfaces is the goal of vaccination against diarrhoeal diseases, respiratory infections and sexually transmitted diseases, and the oral route is also an attractive one for the delivery of a variety of new generation vaccines.

However, defence against infection is not the only important immunological function of epithelial surfaces. The evolution of the immune system as an orchestrator of the inflammatory response to infection carries with it the risk, manifest in allergic disease, of inappropriate inflammatory responses to innocuous antigens. As sites of contact with the environment, antigen concentration will be highest at epithelial surfaces and, in general, these are the areas most affected by hypersensitivity to extrinsic antigens. An important function of epithelial surfaces is therefore to limit absorption of antigens and to deviate immune responses against those which are absorbed into a non-inflammatory mode. Local immunity at mucosal surfaces therefore imposes competing demands – on the

one hand adaptations for antigen sampling and protection against pathogens and toxins, while on the other hand the exclusion of antigens and down regulation of those responses that do occur. This balance must be sufficiently robust to accommodate the inevitable periods of changed permeability which occur at epithelial surfaces during infections or following chemical or physical insults.

An understanding of how immune responses are regulated at mucosal surfaces requires knowledge of how antigens cross epithelial barriers, where antigen uptake occurs, how much antigen is absorbed, what forms of antigens are taken up most readily, and how the antigen is presented to the immune system. Despite a considerable published literature (reviewed by Mayrhofer, 1984; Gardener, 1988) and a pervasive dogma with respect to antigen absorption in the gastrointestinal tract (Walker, 1987), there is in fact very little unambiguous evidence about any of these important aspects of antigen handling. The reasons are simple – the extensive surfaces of mucosal organs make measurements difficult and their integrity is easily damaged by experimental manipulations. There has been some naivety in the use of mucosal organ preparations both *in vivo* and *ex vivo*. The sensitivity of epithelial integrity to mechanical (Rhodes and Karnovsky, 1971) and anoxic (Gardener and Plumb, 1979; Plumb *et al.*, 1987) insults and the effects of this on antigen absorption have not been understood adequately. Another source of artefacts has been the tracer systems used to monitor absorption of antigen.

Of a more fundamental nature are the difficulties associated with attempting to assess the immunological significance of that antigen which is absorbed. It is not known yet which anatomical compartments are significant, particularly in the afferent phase of the immune response. Taking the example of the small intestine, it is not known definitively whether Peyer's patches, the villous mucosa, the epithelium or the draining lymph nodes are the principle sites at which lymphocytes are primed against lumenal antigens; whether priming at these various sites might produce qualitatively and quantitatively different immune responses; or the relative importance of these sites in secondary exposure to antigens or in induction of effector functions. Furthermore, it is difficult to assess the relative amounts of antigen that might be immunogenic locally, in comparison with the amounts that would have to be absorbed in order to be measurable in lymph or blood, or alternatively to be detected in histological preparations by histochemical or immunohistochemical techniques. Finally, there have been major advances in the understanding of the nature of antigen recognized by T cells. It is now clear that peptides containing as few as eight amino acids can be immunogenic for T lymphocytes (Rötzschke *et al.*, 1990; Van Bleek and Nathenson, 1990; Schumacher *et al.*, 1991; Rudensky *et al.*, 1992). Such peptides are unlikely to be detected by antibodies in immunoassays and call for a major revision of the techniques that have been used to assess antigen absorption in the past and reassessment of the conclusions drawn from the earlier work.

The scope of this review is to examine critically the proposition that epithelia are involved in antigen absorption, antigen processing and antigen presentation. The discussion will also extend to the nature and distribution of classic antigen-presenting cells at mucosal surfaces and the immunological outcomes of antigen presentation at these sites. Attention will be focused mainly on the gastrointestinal tract, with examples drawn from other systems where appropriate. However, epithelia have other functions associated with, or complementary to, the immune system. One of these, the active transport of secretory antibodies, is not directly relevant to this review, although IgA antibodies may have a function in assisting exclusion of antigen (Heremans, 1974; Walker *et al.*, 1975a). More recently, epithelium has come to be viewed as more than just the protective, absorptive or secretory interface at mucosal or dermal sites. It may have important functions in directing the activities of the supporting lamina propria tissues and immigrant cells within the lamina propria. There is growing evidence that, on the one hand, epithelial cells orchestrate the functions of immune cells by the secretion of cytokines, while, on the other hand, lymphocyte-derived cytokines can influence epithelial cell behaviour. This is an exciting new development in mucosal immunology and one in which antigen can be expected to play a key role in both normal physiology and in pathology involving mucous membranes.

2. *Antigens at Epithelial Surfaces*

There exists a large body of information (both clinical and experimental) about immune responses to individual antigens, allergens and infectious agents at various epithelial and mucosal sites. The aim of this section is not to review this extensive and often contradictory literature in detail, but rather to look for general patterns of response. There are very practical benefits (medical and economic) to be obtained from finding methods by which antigens can be delivered to mucosal surfaces in doses and forms which can induce reproducible immunogenic or tolerogenic responses. The discussion in this section will centre on the properties of antigens, while absorption and downstream events will be examined in later sections. Research in this area should be directed towards identifying the properties shared by those antigens that are immunogenic at particular mucosal sites, those which are associated commonly with induction of tolerance and those which are associated with significant deviations towards hypersensitivity.

2.1   ANTIGEN ACCESS

To induce an immune response, antigen must reach the epithelial surface in an immunogenic form. For infectious agents, this may mean that the organism must survive, colonize, replicate and possibly invade the epithelium. In the case of the gastrointestinal tract, the organism must resist the extreme acidity of the stomach and the actions of proteolytic enzymes and bile salts (Newby, 1984; Walker, 1984). Colonization of the gut and other epithelial surfaces (e.g. lung, genitourinary tract, conjunctiva) also requires attachment via specific adhesins (reviewed by Isaacson, 1985). However, discussion of these virulence factors is beyond the scope of this chapter.

With respect to non-living antigens, contact with the exposed epithelial surfaces of the skin and the conjunctivae is uncomplicated and can involve a wide range of environmental, industrial, domestic and cosmetic substances. Contact between antigens and either the respiratory tract or the gastrointestinal tract is indirect, and depends on the nature of the antigen and its survival in the secretions of these systems. Nevertheless, the range of potential antigens is large. On the other hand, non-living antigens that reach the urinary tract are those which are of sufficiently small size to pass into the glomerular filtrate from the circulation. In the case of the male and female reproductive tracts, the most significant antigens may be those of sperm and the components of the seminal fluid. Access of antigen to each of these latter areas will be dealt with only briefly.

2.1.1   The Respiratory Tract

Non-infectious antigens of importance in the respiratory tract are those which are inhaled. Sites of deposition are determined chiefly by the sizes of the inhaled particles (Morrow, 1973; Stuart, 1973). Most aeroallergens are inhaled either as aerosols, dried solutes from aerosols or as preformed particles such as spores, pollen grains or the faecal balls of house dust mites. Antigens (allergens) are released from these particles into the secretions of the respiratory tract. Most of a dose of antigen inhaled experimentally as aerosol is removed from the lung and delivered to the gastrointestinal tract (Sedgwick and Holt, 1983), either as free antigen or as material endocytosed by the macrophages which are abundant in the respiratory secretions. Clearance of antigen from the lung is accomplished by the action of the "mucociliary escalator" and by alveolar macrophages, which are delivered to the gut in large numbers each day (Spritzer *et al.*, 1968; Brain, 1970). It is the remaining antigen that is significant immunologically, although a contribution from antigens delivered from the lung to the gut cannot be excluded.

The majority of inhalant allergens are relatively small polypeptides (Marsh, 1975), generally of less than 40 kD. Recently described major allergens from house dust mites fall within this range (Stewart, 1982; Chua *et al.*, 1988; Chua *et al.*, 1990; Dilworth *et al.*, 1991), as do a number of proteolytic enzymes (papain, trypsin, bromelains and subtilisin) associated with industrial respiratory allergies. Molecular size and ability to diffuse through epithelia (Tagesson *et al.*, 1981) may therefore be major factors contributing to allergenicity.

The soluble antigens released from inhaled particles are probably minimally degraded in the relatively bland secretions of the normal respiratory tract. They may be absorbed intact and it is therefore significant that some natural (Stewart *et al.*, 1986; Chua *et al.*, 1988) and some industrial (Milne and Brand, 1975; Flindt, 1978; Baur and Fruhmann, 1979; Liss *et al.*, 1984; Dor *et al.*, 1986) aeroallergens are enzymes. The activities (most are proteases) of these enzymes may facilitate their own permeation through the epithelium of the respiratory tract. On the other hand, some allergens are high molecular weight proteins. An example is Dpt4 (244 kD), a major allergen from house dust mites (Stewart *et al.*, 1983). This allergen is a lipoprotein, suggesting that uptake may be facilitated by its lipophilic properties.

2.1.2   The Gastrointestinal Tract

In contrast to the respiratory system, where annual exposure to aeroallergens may amount to only a few micrograms (Marsh, 1975), the gut is exposed potentially to several hundred grams of protein per day in the diet (Stephen, 1985). A further difference is that the integrity of antigen is generally preserved in the respiratory tract, while the gut is specialized for the digestive destruction of proteins. There is no doubt that the bulk of ingested protein is digested intralumenally to low molecular weight components (Nixon and Mawer, 1970a and b), although a significant amount is absorbed by enterocytes in the small intestine as small peptides (two to four amino acids) and possibly as even larger fragments (Matthews and Payne, 1980). The potential relevance of peptide absorption will be discussed later (see Sections 2.2.1.2 and 3.2.1). Some protein reaches the distal ileum, but it is not known whether this is of dietary or endogenous origin (Stephen, 1985).

However, it is inescapable that small quantities of protein are absorbed intact from the small intestine and that some of this material enters the circulation (reviewed by Mayrhofer, 1984; Gardener, 1988). Past controversy about the magnitude of intact protein absorption has been resolved (Hemmings, 1980; Udall *et al.*, 1981). However, as discussed by Mayrhofer (1984) and Gardener (1988), most studies on antigen absorption have involved the use of tracer quantities of pure material which is administered in solution. There have been no quantitative studies on the absorption of dietary antigens from foodstuffs in normal meals. Protection of antigens is more likely to occur in a non-homogeneous bolus of food, where there may be some buffering capacity and an excess of protein (Jakobsson *et al.*, 1982).

It is difficult to assess the real exposure to food antigens at the absorptive surface of the gut. Of the foods responsible for the majority of intestinal allergies (eggs, peanuts, milk, fish, soya, wheat), most are major or common sources of dietary proteins (Sampson, 1983;

Sampson and McCaskill, 1985). Quantity may itself ensure that immunogenic doses of a protein survives the digestive process, although allergenicity and the concentration of a protein in a dietary substance are not necessarily related (Aas, 1988). Proteolysis is the other major determinant of the exposure of the absorptive surface to antigens. This may vary physiologically (e.g. neonates in comparison with older individuals) or pathologically (e.g. in cystic fibrosis, achlorhydria or chronic pancreatitis) and also depends on the sensitivities of individual proteins to digestive proteases. Although most attention has focused on the absorption of surviving native proteins, it is noteworthy that in coeliac disease, toxicity of gluten is retained even after extensive digestion with pepsin and trypsin (Frazer *et al.*, 1959; Cornell and Townley, 1973) to peptides of 5–8 kD (Cornell and Maxwell, 1982; Jos *et al.*, 1982) and can be reproduced with defined peptides containing as few as 21 amino acids (De Ritas *et al.*, 1988). Similarly, a cod fish allergen that can be absorbed from the diet and transferred from mother to infant in breast milk (Aas, 1988) yields allergenic peptides with as few as 20 amino acids when subjected to tryptic digestion (Elsayed and Apold, 1977; Elsayed and Apold, 1983). The potential importance of peptides as antigens is highlighted by recent clarification of the nature of T cell epitopes and the processing events in antigen-presenting cells (Braciale T and Braciale V, 1991). Digestion in the small intestine has parallels with the processing of exogenous antigens in the endocytic pathways of antigen-presenting cells and it has enormous potential to produce antigenic peptides that could be recognized by T lymphocytes. The fact that most individuals, under normal circumstances, do not experience chronic cell-mediated mucosal inflammation suggests that peptides are excluded from the mucosa or that responses to them are suppressed effectively. Either mechanism (or both) would appear to have been a vital evolutionary adaptation to protect the gut from an inevitable encounter with dietary proteins that contain protease-resistant peptides. The studies of Nixon and Mawer (1970a,b) indicate that partially digested protein can be recovered from the free intestinal juices for a considerable time after leaving the stomach and, unless further digested or trapped in the unstirred mucous layer, the resulting peptides would be available for absorption. Proteolysis may expose cryptic epitopes (Spies *et al.*, 1970) and alter the immunogenicity or tolerogenicity of the parent protein (Muckerheide *et al.*, 1977; Turkin and Sercarz, 1977; Yowell *et al.*, 1979).

Protein digestion has two main phases – peptic digestion in the stomach and pancreatic digestion in the upper small intestine. These processes lead to the production of a complex mixture of peptides and free amino acids. Hydrolysis is completed by peptidases of the brush border (larger peptides) or within enterocytes (mainly dipeptides) (Matthews and Payne, 1980). Of the two, peptic digestion is believed to be less significant in yielding nutritionally important amounts of absorbable products (Hoffman-Goetz, 1984), although it could yield potentially immunogenic fragments. With respect to the generation of humoral responses, peptic digestion may reduce the load of dietary antigens, as indicated by a higher incidence of antibodies to bovine serum albumen in individuals with achlorhydria (Kraft *et al.*, 1967). There is also circumstantial support for this view from the observation of increased antigen absorption in premature infants and neonates, in whom peptic activity (Agnod *et al.*, 1969) and acid secretion (Euler *et al.*, 1979; Hyman *et al.*, 1985) are below levels found in older children and adults (Walker, 1985). However, these are likely to be only two of a number of factors that could contribute to food intolerance in babies. Peptic digestion can reduce the immunogenicity of experimental oral immunogens and this effect is inhibited by prior treatment with cimetidine (Klipstein and Engert, 1979).

Pancreatic digestion is more significant nutritionally (Hoffman-Goetz, 1984), and must therefore also have a major impact on reducing the load of dietary antigen. However, the evidence is mainly circumstantial. Pancreatic enzyme levels are physiologically low in premature and term infants (Hadorn, 1975; Lebenthal and Lee, 1980) and reduced proteolysis may allow greater antigen absorption in this age group (Walker, 1985). Children with cystic fibrosis frequently have severe pancreatic insufficiency (Hadorn, 1975), and it is possible that the resultant antigen absorption is partly responsible for the raised IgE levels and allergy observed in a proportion of patients (Allan *et al.*, 1975; Warren *et al.*, 1975; Turner *et al.*, 1978) and for the common occurrence of autoantibodies (Hodson and Turner-Warwick, 1981). However, other factors such as infection, poor ciliary clearance of inhaled antigens and epithelial defects must also be considered. Experimental studies have shown that inhibitors of proteases can increase the absorption of biologically active molecules such as insulin (Danforth and Moore, 1959), as can ligation of the pancreatic duct (Alpers and Isselbacher, 1967). Furthermore, administration of protease inhibitor can increase the immunogenicity of oral doses of antigen (Walker, 1988), although whether this result should be interpreted in terms of preservation of intact antigen or reduced production of tolerogenic peptides is not clear.

Finally, prior immunization may also affect the accessibility of antigen to the absorptive surface of the gut. Walker *et al.* (1975a) observed extra cellular breakdown of antigen–antibody complexes adsorbed to the mucus layer, glycocalyx or epithelial cell surface of the intestine in immunized rats. Later studies (Walker *et al.*, 1975b) suggested that the enzymes responsible were of pancreatic origin. Walker and associates suggested that immune exclusion of antigens from the gastrointestinal tract may in part be due to complexing with specific antibody, followed by digestion of the complexes by enzymes in the unstirred layer of the gut.

2.1.3 The Genitourinary Tract

With the exception of infectious agents, the nature of antigens within the genitourinary tract is relatively unexplored. Polypeptides with molecular weights comparable to horseradish peroxidase (40 kD) enter the glomerular filtrate and are absorbed by the epithelium of the renal proximal tubules (Ericsson, 1965; Graham and Karnovsky, 1966). It is therefore feasible for small proteins and peptides absorbed from other mucosal surfaces, or arising during infection, to enter the proximal nephrons. The immunological significance of such antigens is unknown but the expression of class II MHC molecules by proximal tubule epithelium is thought-provoking (Mayrhofer and Schon-Hegrad, 1983). Rats are able to respond to protein antigens introduced directly into the urinary bladder (Husband and Clifton, 1989), but the only antigens likely to be encountered by this route physiologically are those from infectious agents.

The female genital tract has a physiological microflora in the vagina and it can be infected by either ascending or haematogenous routes. A physiological source of foreign antigen is sperm and seminal fluid. The amounts of antigen may be quite large – in rabbits and humans, sperm can be found at the ovary after coitus within 1 and 5 min, respectively (Overstreet and Cooper, 1978; Harper, 1988), suggesting a rapid mass movement of semen through the uterus, Fallopian tubes and oviducts. The upper female genital tract receives secretory immunoglobulin (reviewed by Parr and Parr, 1986) and it is not an immunologically privileged site (Beer and Billingham, 1974). The normal lack of response to sperm and seminal fluid components may be related to the lack of specialized antigen-sampling structures in the female upper reproductive tract, although the vaginal epithelium contains abundant Langerhans cells (Parr *et al.*, 1991b). However, this and the complex issue of the fetomaternal relationship will not be discussed further.

Even less is known about the immunology of the male reproductive tract. The only significant sources of foreign antigens are infectious agents. However, the spermatozoa are themselves antigenic, at least in part due to their sequestration within the tubules of the testis and the epididymis. These tubules constitute a "blood–testis" barrier (Johnson, 1970; Setchell *et al.*, 1990), where the developing and mature spermatozoa are segregated from the interstitial tissues by elaborate tight junctions between the epithelial cells of the seminiferous tubules (Sertoli cells) and the epididymis, respectively (Dym and Fawcett, 1970; Russell, 1977). Although endocytosis of seminal fluid proteins by epithelial cells of the vas deferens holds a special place in the history of modern cell biology (Friend and Farquhar, 1967), and a similar phenomenon also occurs in the epididymis (Flickinger *et al.*, 1988), it has not yet been attributed any immunological significance. Readers interested in immunological aspects of the testis are referred to a special issue of the *Journal of Reproductive Immunology* (1990).

2.2   The Nature of the Antigen

Some of the features of antigens whose immunogenicity at mucosal surfaces have been documented, or whose absorption has been studied, have been discussed in Section 2.1. Mechanisms of absorption will be dealt with in Section 3. It has been clear from early studies (Crabbé et al., 1969) that antigens of considerable size (e.g. horse spleen ferritin) can stimulate immune responses in the gastrointestinal tract. However, many allergens are relatively low molecular weight proteins, suggesting that absorption may be related to molecular size. Until recently, most attention has been directed towards the uptake of intact antigens. This may be appropriate at some epithelial surfaces but it is unrealistic to insist, in the case of the gastrointestinal tract, that native proteins are the only relevant antigens. While not denying that some uptake of native protein occurs, it is important to recognize that absorption of peptides may suffice to generate immune responses in the gut (and perhaps elsewhere).

2.2.1   Fluid Phase Antigens

Most potential antigens have no particular affinity for components of the lumenal cytoplasmic membranes of epithelial cells. If they are to make contact with immunologically competent cells, lumenal antigens of this type must pass through the epithelial barrier by either a paracellular route (passive diffusion between the cells) a transcellular route (some form of active transport across the cells) or by lateral diffusion in the cell membrane (for those lipophilic molecules that can insert into the membrane). On the other hand, it is possible theoretically that some antigens (foreign or self) might enter epithelial cells in a retrograde direction from the ablumenal surface (see Section 3.1.3). Diffusion through "pores" in normal intercellular regions with intact tight junctions will be limited by molecular size, while diffusion into cell membranes will be governed by hydrophobicity. In contrast, pinocytosis and vesicular transport of fluid phase molecules will not be limited by molecular size and the process should be non-competitive and non-saturable.

2.2.1.1   *Skin and Squamous Mucosae*

The surfaces which have squamous epithelia are the skin, the oral cavity, the oropharynx, the oesophagus and the lower female genital tract. Of these, only the skin is keratinized in humans, although some areas of the buccal cavity are also keratinized in other species. The membranous portion of the stomach (rodents) and the rumen (ruminants) are evolutionary adaptations of the oesophagus and they also have a keratinized squamous epithelium.

The permeability of the skin has received considerable study because it is a barrier to dehydration and penetration by inflammatory substances and by infectious agents. It is also a route with potential importance for absorption of pharmaceutical products (Ebling, 1992). Normal skin is highly impermeable to water, polar solutes and macromolecules. On the other hand, it is relatively permeable to lipid solvents and lipid-soluble compounds. The permeability of the barrier does not depend on the viability of the integument, and regional variations in permeability are related to lipid content rather than to local thickness (Squier et al., 1991). The permeability of the buccal mucosa is generally greater than that of skin, but it has similar characteristics. Consistent with the permeability of squamous epithelia to lipids, the rumen mucosa is the major site of absorption of nutritionally important volatile fatty acids (Wolin, 1981), while it is a potential barrier to the enormous antigen load of the rumen microflora and digesta. The vaginal epithelium in those animals that have an oestrus cycle may represent a special case. The cyclical changes in the epithelium in the mouse vagina are associated with periods of increased permeability to macromolecules (Parr et al., 1991b).

The skin and buccal mucosa present some interesting comparisons with other epithelia, where the main extracellular permeability barrier is the tight junctions between adjacent cells. In the skin, as demonstrated first by Blank (1953, 1965), the barrier lies in the corium, which is composed of non-viable squames. There can be no active transport of materials through this outer layer of dead cells and the squames are themselves highly impermeable. With the possible exception of very limited diffusion of water and permeation by polar solvents such as dimethyl-sulphoxide, which may pass through the squames, most permeation of the skin involves diffusion of lipid-soluble substances between the cells (Ebling, 1992). The intercellular spaces in the stratum corneum are filled with lipid lamellae formed from the contents of lamellar bodies in the epithelial cells (Landman, 1986). These lipid-filled spaces are highly impermeable to water and polar solutes but they provide a route for diffusion of lipid solvents and lipid-soluble materials and a significant depot for such substances (Stroughton, 1966). Another route by which charged solutes can enter the skin by iontophoresis is the adnexa such as hair follicles and sweat ducts (Grimnes, 1984; Cullander and Guy, 1991), and this may also allow diffusion of some electrolytes.

These considerations indicate that antigen absorption through intact skin (and probably other squamous epithelia) will be limited mainly to lipid-soluble haptenic substances. Most contact sensitizing agents belong to this category (Rycroft and Wilkinson, 1992), although dermatitis caused by contact with metals suggests that some ions must permeate the stratum corneum, possibly through the epithelium of hair follicles and sweat glands. In contrast, permeation of macromolecules through intact skin is unlikely. Histochemical studies using tracer proteins such as horseradish peroxidase indicate that while the deeper layers of the epidermis are readily permeable to the enzyme *in vivo* and *in vitro*, the stratum corneum is completely impermeable (Elias and Friend, 1975; Squier and Hopps, 1976). Later studies using isotopically labelled horseradish peroxidase have indicated that skin may allow some passage of the enzyme (Squier and Hall, 1985), but uncertainties in this study about the stability of the tracer make interpretation of the results difficult.

The permeability of the buccal epithelium to water appears to be greater than that of skin (Squier and Hall, 1985). Most studies have indicated that it is impermeable to diffusion of horseradish peroxidase (McDougall, 1970, 1971; Squier, 1973; Elias and Friend, 1975) but Squier and Hall (1985) have reported that horseradish peroxidase can permeate the buccal mucosa of pigs, especially the non-keratinized sublingual mucosa. This is of some immunological interest, because sublingual desensitization therapy has a long tradition among allergists and its effectiveness has been demonstrated recently in house dust mite allergy (Scadding and Brostoff, 1986). The buccal mucosa is permeable to lipid-soluble therapeutic agents (Moffat, 1971) and is therefore permeable presumably to many contact sensitizing agents. The low incidence of contact sensitivity in the mouth is probably due to low exposure to potential sensitizers, but immunoregulatory mechanisms similar to those in the gut (Chase, 1946; Asherson et al., 1977) may also operate in the oral cavity and deserve exploration.

It is unlikely that the extensive system of antigen-sampling cells in the skin (see Section 4.3) has evolved only to facilitate immune responses against low molecular weight haptens, the majority of which are man-made. Hypersensitivity reactions against these compounds has no obvious protective value. It is more likely that the protective role of cutaneous cell-mediated immune responses is directed against infective agents (bacteria, viruses and fungi) which establish superficial or invasive infections of the skin. A less obvious role, involving recognition of abnormal tissue antigens, may be one of surveillance against cutaneous neoplasia (Halliday et al., 1990a). Antigens such as these are components of invasive microorganisms or have an endogenous origin in the epithelial cells themselves.

2.2.1.2   *Intestinal Mucosa*

The majority of soluble proteins, polysaccharides and free colloidal materials in the gut lumen can be expected to have no specific affinity for the epithelial brush border. There is no doubt that a variety of particulate materials such as erythrocytes (André et al., 1973; Kagnoff, 1982), lymphocytes (Dunkley and Husband, 1986), killed bacteria (Waldmann et al., 1972; Porter et al., 1974; Michalek et al., 1978; Challacombe and Tomasi, 1980; Stokes et al., 1979; Stokes, 1984), polyacrylamide-linked hapten carrier (Cox and Muench, 1984), poly(butyl-2–cyanacrylate)-adsorbed ovalbumen (O'Hagan and Donachie, 1991), liposome-encapsulated antigens (Wachsmann et al., 1986; Childers et al., 1990), protein antigens in ISCOMs (Mowatt and Davies, 1990) and antigens in biodegradable microspheres (Eldridge et al., 1991) are immunogenic via the oral route. Orally administered soluble fluid phase proteins are also immuno-

genic, but usually induce tolerance (Thomas and Parrot, 1974; Stokes, 1984). Absorption of fluid phase proteins and colloids via the specialized Peyer's patch epithelium will be discussed in Section 2.2.2.3. The more controversial issue of absorption of fluid phase materials via the villous mucosa will be addressed here.

Over 100 years ago, Herbst (1844) described starch granules in the lymph and circulation of a dog that had been fed a suspension of starch. This phenomenon, now called persorption (Volkheimer, 1977), has been observed experimentally in a number of species (including humans) using both synthetic and natural particles (reviewed by Baintner, 1986) and it is probably a pathophysiological occurrence, at least in herbivores (Baker et al, 1961; Nottle, 1977). It has been assumed that persorbed particles in lymph, blood and tissues originate from the insinuation of the materials into the lamina propria through areas of weakness in the epithelium overlying villous mucosa. This occurs most commonly at the extrusion zone at the tips of villi in the small intestine (Volkheimer et al., 1968). The appearance of persorbed particles in the circulation and in the gut mucosa both appear to depend on the kneading action of peristalsis, by which the particles are driven into the mucosa by mechanical microtrauma. Persorption of particles in the size range <1–150 μm has been described, but the process is most efficient for those in the range 7–70 μm (Baintner, 1986a).

Two main questions arise with respect to persorption. First, have the persorbed particles recovered from lymph and blood arisen from entry through the villous mucosa? It is tempting to think so, because the histological evidence is persuasive. However, there have been no quantitative studies to measure how frequently particles are persorbed into villi, while some studies have failed to observe the phenomenon (Bockman and Cooper, 1973; Le Fevre et al., 1978; Pappo et al., 1991). The work of Le Fevre et al. (1978) suggests a different explanation for the presence of absorbed particles in villi. These workers found absorbed latex particles only in villi surrounding Peyer's patches. They suggest that these particles originated from the adjacent Peyer's patch and that they were transported to the villi either by connecting lymphatics or by the direct migration of macrophages. If this explanation is correct, then it is conceivable that all "persorption" originates from absorption through the specialized follicle-associated epithelium of Peyer's patches and that the presence of particles in the gut mucosa is simply part of a wider picture of deposition in many tissues of the body. There is no doubt that particles such as 30 nm Percoll microspheres (Matsuno et al., 1983) 20–50 nm India ink particles (Bockman and Cooper, 1973), 2 μm latex microspheres (Le Fevre et al., 1978) and 1–10 μm biodegradable microspheres (Eldridge et al, 1991) are absorbed by cells in the follicle-associated epithelium. Furthermore, infectious reovirus virions (60–80 nm) are taken up exclusively through M cells

(Wolf *et al.*, 1983) and independent evidence suggests that they are redistributed via the lymphatic and blood circulation to the crypt epithelium as part of the general viraemic spread from the Peyer's patches to other sites (Rubin *et al.*, 1985). However, as the size limit for this route appears to be about 10 μm (Eldridge *et al.*, 1991) and persorption of particles up to 150 μm has been reported (Baintner, 1986a), it is possible that the route advocated by Volkheimer (1972) applies to materials in that size range.

The second question about persorption relates to its immunological significance. If persorption (in particular, uptake of particles of size <10 μm) is really a reflection of M cell absorptive activity in follicle-associated epithelium, then it has fundamental importance (Neutra and Kraehenbuhl, 1992). Particles absorbed by this route (Le Fevre *et al.*, 1978), like bacteria that invade Peyer's patches (Mayrhofer *et al.*, 1986), enter not only Peyer's patches but are also carried to the regional lymph nodes by phagocytic cells. The immunological consequences will depend on the nature and antigenicity of the particles. However, the immunological consequences of such persorption as occurs through the villous mucosa is unknown. If the classical persorptive route applies only to larger, rigid particles that enter the mucosa by microtrauma, the immunological significance is doubtful. Such materials are likely to be inorganic and include asbestos fibres and silica, or microscopic residues of plant fibres (Baker *et al.*, 1961; Nottle, 1977; Carter and Taylor, 1980). Furthermore, it should not be assumed that mechanical entry of large particles via the extrusion zone indicates that this is a point of sustained intercellular weakness or increased permeability. Recent evidence indicates that intercellular defects are closed rapidly, first by contraction of subepithelial myofibroblasts (Moore *et al*, 1989a), and followed quickly by repair (Moore *et al.*, 1989b).

Persorption may, therefore, be limited to the accidental uptake of relatively large particles from the digesta. The uptake of small particles, some of which may enter the circulation, is probably by a distinct process limited to the follicle-associated epithelium of Peyer's patches (see Section 2.2.2.3) and similar structures. The same sort of analysis should be applied to the absorption of soluble fluid phase antigens. On the one hand, there is a considerable literature on measurements of small intestinal permeability and on the uptake of fluid phase marker macromolecules through the epithelium of the villous mucosa. On the other hand, the absorption of immunologically significant antigen is attributed by others to the activities of M cells in the Peyer's patch follicle-associated epithelium (Neutra and Kraehenbuhl, 1992). Entry of macromolecules and/or peptides via the villous mucosa will be dealt with here, while discussion of follicle-associated epithelium is deferred until Section 2.2.2.3.

Antigens absorbed through the villous mucosa can engage potentially with lymphoid compartments that are different from those that are taken up into Peyer's patches. The anatomical lymphoid compartment engaged by orally administered antigen could affect the nature of the immune response, as well as its pathological consequences. Although there is general agreement that the normal epithelium is an effective barrier against large-scale absorption, uptake of antigenic molecules via the villous mucosa forms part of the accepted dogma of mucosal immunology. To some extent, this view has arisen from observations on the small intestine in diseases involving intolerance of food antigens. For instance, in conditions such as coeliac disease and food allergies, there is generalized involvement of the areas of villous mucosa that are exposed to the causative antigens. This sort of observation invites the conclusion that these diseases are initiated by antigen permeating the epithelium, leading to local inflammation in the subjacent lamina propria. This apparently straightforward conclusion is nevertheless subject to alternative explanations.

In the case of coeliac disease, it is by no means certain that gluten (or even large proteolytic fragments) is absorbed intact, or that gluten can be considered as a typical fluid phase antigen. The protein may have lectin-like properties (see Section 2.2.2.2), allowing it to bind to brush border glycoproteins (Douglas, 1976; Köttgen *et al.*, 1982). Furthermore, coeliac disease is associated strongly with certain HLA-D haptotypes (Cole and Kagnoff, 1985), and MHC class II molecules are known to be expressed on the brush borders of enterocytes in human small intestine (Mayrhofer *et al.*, 1990). This suggests that gluten peptides could bind directly to class II molecules, perhaps facilitating uptake and intracellular transport by a "receptor-mediated" pathway (see Section 3.2.1).

In food allergies with a pathogenesis involving immediate-type hypersensitivity, there is agreement that local anaphylaxis (Bloch *et al.*, 1979; Roberts *et al.*, 1981) or immune complex-mediated mucosal damage (Tolo *et al.*, 1977) is followed by increased permeability of the gut to soluble antigens. However, the permeability of the mucosa between challenges is less certain. On existing information, it cannot be excluded that the sequence of events following allergen challenge in a sensitive individual is initial absorption via the specialized Peyer's patch epithelium, followed by entry of allergen into the systemic circulation. Clinical evidence (Scadding and Brostoff, 1985) and experimental observations in humans (Prausnitz and Küstner, 1921; Wilson and Walzer, 1935) and animals (Hettwer and Kriz, 1925; Hettwer and Kriz-Hettwer, 1926) indicate that anaphylactic reactions can occur at extra-intestinal sites following absorption of allergens from the gut and that absorbed allergens can even be transferred via breast milk to suckling infants (Kilshaw and Cant, 1984; Jakobsson *et al.*, 1985; Aas, 1988). These findings indicate that sufficient antigen is absorbed into the circulation to activate mast cell degranulation in distant tissues. Therefore, the initial intestinal response to food allergens could be induced by allergens delivered to the lamina propria via the circulation, rather than by direct uptake through the villous epithelium. The resulting inflammatory reaction might be subclinical but it could be responsible for loss of barrier integrity in the overlying epithelium, allowing local diffusion of allergen from the lumen into the mucosa. This later influx of antigen might intensify inflammation, produce symptomatic illness, and maintain the lesion for as long as the allergen persists.

It follows that observations on diseased intestine do not necessarily provide direct clues about where absorption of antigen occurs or about how the absorptive process is mediated. Furthermore, they give no information about how much antigen is absorbed. Despite this, and perhaps because sufficient antigen is absorbed for it to be measured in lymph and portal blood in experimental animals (Warshaw *et al.*, 1971) and in the circulation in humans (Paganelli *et al.*, 1979), the attention of many workers has been focused on the large absorptive area of the villous mucosa. In fact, the amount absorbed from an experimental dose of antigen is very small, estimated at 0.02% in humans (Gruskay and Cooke, 1955), 0.01–0.02% in rats (Danforth and Moore, 1959; Warshaw *et al* 1971, 1974) and 0.01% in rabbits (Rothberg *et al.*, 1970). These measurements, based on absorption of several proteins (ovalbumen, insulin, bovine serum albumin and horseradish peroxidase), are probably overestimates, both for technical reasons (Udall *et al.*, 1981; Skogh, 1982) and because the experiments have involved direct injection of the proteins into the intestinal lumen (thus avoiding normal gastric digestion). Furthermore, there is no reason to suppose that absorption of this magnitude could not occur through Peyer's patches.

Only two studies have addressed directly the relative importance of the villous mucosa and Peyer's patches in the absorption of soluble fluid phase macromolecules (Duroc *et al.*, 1983; Keljo and Hamilton, 1983). As discussed below, both are flawed potentially by the use of *in vitro* organ preparations and neither provides a balance sheet based on the relative areas of the two epithelial surfaces. Furthermore, the results in the two studies were conflicting. Keljo and Hamilton (1983) used Ussing chambers to measure horseradish peroxidase absorption by pig intestinal mucosa. They found that uptake was several-fold greater per unit area through mucosa which contained Peyer's patches than through villous mucosa alone. However, Duroc *et al.* (1983), who studied rabbit small intestine mucosa by a similar technique, reported that transport of horseradish peroxidase was greater through areas of villous mucosa that did not contain Peyer's patches. It should be emphasized that only gross permeability to horseradish peroxidase was measured in these experiments, not epithelial permeability specifically. Furthermore, other studies have raised concerns that physiological barriers between epithelial cells are compromised in *ex vivo* preparations of intestinal mucosa (Rhodes and Karnovsky, 1971; Gardner, 1978; Gardner and Plumb, 1979; Plumb *et al.*, 1987). In particular, it is doubtful that *ex vivo* preparations requiring eversion (everted sacs) or muscle stripping (Ussing chamber preparations) can ever be free from such concerns.

Whether fluid phase antigens are absorbed in significant amounts by the villous mucosa is a complex question, as indicated in an earlier review (Mayrhofer, 1984). One aspect of this complexity is the effect of maturity on the uptake of macromolecules from the gut (reviewed by Baintner, 1986a). It is uncontroversial that immunoglobulin is absorbed from the proximal small intestine in many species, although not in significant amounts by normal full-term human infants (Ammann and Steihm, 1966). In some species, such as mice and rats, absorption of IgG is receptor mediated (Baintner, 1986b) and its uptake is not an example of fluid-phase transport. However, in ruminants and ungulates, immunoglobulin transport from colostrum is non-selective for a period of 24–48 h after birth, and macromolecules are absorbed from the gut indiscriminately and transported into the blood. This occurs by pinocytosis, with transport of the protein into large supranuclear vacuoles by pinocytosis, followed by discharge of the contents of the vacuoles into the extracellular fluid at the basolateral surfaces of the enterocytes (Comline *et al.*, 1951; Payne and Marsh, 1962; Kraehenbuhl and Campiche, 1969). Transport through the enterocytes in ruminant and ungulate intestine appears to circumvent fusion with lysosomes (Kraehenbuhl and Campiche, 1969), although the presence of lysosomal enzymes in the intestinal lymph of lambs during absorption of colostrum suggests that some lysosomal fusion may occur (Dinsdale and Healy, 1982).

In contrast to the massive uptake of non-immunoglobulin macromolecules by ruminant neonates, sufficient to cause proteinuria (Pierce, 1959, 1961), there is only a low level of non-selective uptake of macromolecules by rodent proximal small intestine in the preclosure period (Jordan and Morgan, 1968; Jones, 1974). This appears to result from incorporation of fluid phase proteins within the coated vesicles that transport IgG, as discussed succinctly by Morris (1978) and by Bainter (1986b). Unlike receptor-bound IgG, most of the fluid phase proteins absorbed non-selectively are directed to the lysosomal compartments of enterocytes and are then degraded, and the same may be true for some particulate antigens, such as virions (Worthington and Graney, 1973). In humans and primates, there is evidence of non-selective uptake of macromolecules by enterocytes *in utero* (Lev and Orlic, 1973; Moxey and Trier, 1975), although it is not clear whether significant amounts are transported across the epithelium. Studies in premature babies and in human neonates indicate that these infants absorb small, but nevertheless detectable, amounts of antigenically intact breast milk (Jakobsson *et al.*, 1986; Jakobsson, 1988) and formula (Roberton *et al.*, 1982) proteins. These findings have obvious implications for the development of food intolerance, in view of the early development of immuno-competence in the human fetus and the evidence from mouse studies that oral tolerance may develop comparatively late in the maturation of the mucosal immune system (Strobel, 1988; Strobel and Ferguson, 1984). However, it should be noted that transport across the epithelium is not a prerequisite for presentation of absorbed antigens – antigen that is absorbed non-selectively and delivered to the lysosomal compartment of the enterocytes could be equally important immuno-logically if processed proteins can be presented by the epithelial cells (see Section 3.3).

Receptors for the Fc portion of IgG are confined to the proximal half of the small intestines of mice and rats during the preclosure period (MacKenzie, 1972; Borthistle *et al.*, 1977; Rodewald, 1980). This part of the intestine is whitish in colour and has non-vacuolated enterocytes, in contrast to the brown colour of the distal small intestine. The latter lacks Fc receptors and the enterocytes, which are vacuolated, contain absorbed pig-mented material (Baintner, 1986b). The ileum in human (Bierring *et al.*, 1964) and primate (Ruebner *et al.*, 1974) neonates has a comparable epithelium. Macromolecular absorption occurs in the distal small intestine but, as reviewed elsewhere (Mayrhofer, 1984), there is no compelling evidence that this is accompanied by transport of material across the epithelium. Nevertheless, ileal uptake of macromolecules, for the reasons discussed above, could provide a mechanism for presentation of processed lumenal fluid phase antigens by the epithelium.

The main question is whether non-selective uptake and/ or transport of antigens occurs in the villous mucosa beyond the period of closure, which commences *in utero* in some species (e.g. humans), in the first few days of life (e.g. ruminants) or after the loss of IgG Fc receptors at weaning (e.g. rats and mice). The issues to be addressed are: (1) the extent to which fluid phase macromolecules are absorbed; (2) whether they are transported across the epithelial cells; (3) whether there are regional differences in the magnitude of uptake and the fate of the absorbed antigen; (4) the relative importance of paracellular and transcellular pathways of transport; (5) the influence of artefacts on the observations made in experimental models; and (6) the immunological significance of absorbed antigen.

It is useful to commence by considering the notion of "intestinal permeability". Strictly, it is a measure of the rate of passage of a substance through the epithelial barrier per unit area, whether by transcellular transport or by diffusion through nominal pores in the tight junctions between the enterocytes (reviewed by Tagesson *et al.*, 1981; Gardner, 1984; Peters and Bjarnason, 1985; Cooper, 1986). In discussions of antigen absorption and mucosal immune responses, the term is often used loosely without adequate regard for the relative sizes of permeability probes or nominal antigens. Absorption by

diffusion should be limited by the sizes of the pores and by a function related to the size of the molecule in question. It appears that for monosaccharides, permeability is inversely proportional to molecular volume (Hamilton *et al.*, 1987). Most work in this area has involved studies of intestinal permeability to relatively low molecular weight substances such as non-metabolizable monosaccharides (e.g. xylose and mannitol), non-hydrolysable oligosaccharides (Menzies, 1974, 1983), polyethylene glycol (Chadwick, 1977a,b) and ^{51}Cr-labelled EDTA (Bjarnason *et al.*, 1983). Hamilton *et al.* (1987) have shown a clear cut-off in absorption through an aqueous channel in rat intestine, with an upper limit of molecular volume for monosaccharides of approximately 220×10^{-3} nm^3 (~ 0.18 kD). This pore is believed to be a membrane channel involved in transcellular transport (reviewed by Cooper, 1986). Clearly, the results of studies on intestinal permeability using monosaccharides and other probes of similar size have no relevance to antigen absorption (macromolecular or peptide).

Hamilton *et al.* (1987) have further shown that absorption of oligosaccharides is relatively unaffected by changes in molecular volume in the range from 362×10^{-3} nm^3 (0.342 kD) to the largest tested, 1128×10^{-3} nm^3 (0.666 kD). Their interpretation is that extracellular aqueous pores exist in the intestinal epithelial barrier, that these are less numerous than the small transmembrane channels and that their size is large in comparison with the molecular radii of the largest oligosaccharides tested. It was suggested that these larger pores were located in the intercellular tight junctions between enterocytes, but the study did not exclude uptake via a pinocytotic mechanism. It is noteworthy that this study was performed *in vivo*, under conditions that would not be expected to compromise the physiological permeability of the mucosa. Artefacts due to breaching of tight junction integrity are not, therefore, a valid criticism of these findings (see below).

Other probes with molecular weights in the range 0.24–0.60 kD are PEG 400 and Cr-EDTA. The use of PEG as a probe for measuring intestinal permeability has been criticised on the grounds that it is lipid-soluble (Peters and Bjarnason, 1985; Cooper, 1986) and may therefore cross the epithelium by membrane diffu-sion. Although recent studies have highlighted the hydrophilicity of PEG 400 (Ma *et al.*, 1990), it is nevertheless the case that PEG interacts strongly with cell membranes (witnessed by its use in the production of somatic cell hybrids) and may therefore not qualify as a fluid phase molecule. Measurements of intestinal permeability using PEG 400 indicate an inverse relation-ship between absorption of its component polymers and their molecular weights (Chadwick *et al.*, 1977b; Ma *et al*, 1990). Cr-EDTA is a hydrophilic marker, and measurements on the permeability of human small bowel mucosa *in vitro* (subject to criticism about integrity of tight junctions) also indicate an inverse relationship between uptake and molecular weight (Bjarnason and Peters, 1984). On the basis of the molecular weights of these probes, it is likely that they diffuse through the larger pore described by Hamilton *et al.* (1987). Uptake (which is increased in coeliac disease) appears to be limited to fluid phase molecules of less than 1 kD (Bjarnason and Peters, 1984), indicating an upper limit to the sizes of molecules that can diffuse through this particular pore.

It is uncertain what relevance paracellular diffusion by fluid phase molecules in this size range has to mucosal immune responses. It is clear that results from studies using relatively low molecular weight permeability probes such as Cr-EDTA are not appropriate for predicting uptake of intact polypeptide antigens through the gut epithelium. Nevertheless, this route of uptake might accommodate the absorption of antigenic peptides consisting of up to 10 amino acids, although evidence for transport of intact peptides across the intestinal epithelium is sparse. Di- and tripeptides are taken up by enterocytes but, with the exception of a few peptidase-resistant examples (Matthews and Payne, 1980), absorption is followed by rapid intracellular hydrolysis to amino acids. In contrast, with one reported exception (Chung *et al.*, 1979), there is no evidence that enterocytes take up tetrapeptides (reviewed by Matthews and Payne, 1980; Kerchner and Geary, 1983), even when the normally rapid hydrolysis by brush border enzymes is prevented by peptidase inhibitors. Transcellular transport of higher oligopeptides is therefore extremely unlikely.

However, some evidence has been obtained by Tome *et al.* (1987) that oligopeptide casomorphins can cross the rabbit ileal epithelium *in vitro*, provided that they are protected from intralumenal brush border hydrolysis. This result conflicts with the findings of Woodhouse *et al.* (1987) that the chemotactic tripeptide F-Met-Leu-Phe was not transported measurably across intact rat small intestine, even in the presence of a carboxypeptidase inhibitor that prevented intralumenal digestion. The latter findings suggest that under physiological conditions the intestine is impermeable to oligopeptides, which are known to be produced by enteric bacteria (Chadwick *et al*, 1988). In contrast, the permeability barrier is lost as a result of injury (Anderson *et al.*, 1987) and this may explain the observations of Tome *et al.* (1987) on Ussing chamber preparations. The experimental findings appear to support the argument advanced by Woodhouse *et al.* (1987) that the rarity of neutrophils in the normal intestinal lamina propria reflects the exclusion of lumenal bacterial products (e.g. chemotactic peptides) by the intact epithelium.

Clearly, this is an important issue. As suggested by Peters and Bjarnason (1985), the cut-off for absorption of small molecules via the paracellular route (<1 kD) could include immunologically significant peptides, such as toxic fragments of gluten. Furthermore, antigen-presenting cells can present peptides containing as few as eight to nine residues to T lymphocytes (Rötzschke *et al.*,

1990; Van Bleek and Nathenson, 1990; Schumacker *et al.*, 1991) and stimulation of T cells by peptides is now a commonplace observation (Braciale and Braciale, 1991). In addition, monovalent peptides can be endocytosed and processed by B lymphocytes (Davidson and Watts, 1989; Davidson *et al.*, 1990). It is therefore possible for small peptides, in principle, to contravene the general rule that only proteins over 10 kD can elicit antibody responses. While peptides may not be absorbed significantly through normal villous epithelium, a primary insult to tight junction integrity could nevertheless allow ingress of peptide antigen and thus initiate a vicious circle of local inflammation and increased permeability.

The preceding discussion deals with diffusion of fluid phase molecules through "pores" that might perhaps accommodate antigenic peptides but not native antigens. However, most discussion of antigen absorption in the past has focused on the uptake of larger polypeptides and native proteins (or other macromolecules of comparable size). Loehry *et al.* (1970) examined the relationship between molecular weight and small intestine permeability by measuring clearance of markers from blood into lumenal perfusate in anaesthetized rabbits. In this study, an inverse relationship was observed between clearance and molecular weight for molecules ranging from urea (0.06 kD) to polyvinylpyrrolidone (33 kD). Furthermore, by analysis of bidirectional permeability to various molecular weight fractions of polyvinyl-pyrrolidone, this relationship was extended to include the range 8–80 kD. On the basis of these observations, it was proposed that absorption involves passive diffusion through a graded series of pores, where there are many of the smaller sizes and progressively fewer of the larger sizes.

This study should be free from artefactual changes in permeability caused by experimental manipulation of the intestine (Rhodes and Karnovsky, 1971). However, it is difficult to conceive how a barrier such as the epithelium could present a continuum of pores with selectivity over such a wide range of molecular weights. Furthermore, the tissue barrier from the gut lumen to the circulation is complex, consisting of the lumenal unstirred layer (Anderson *et al.*, 1988), the epithelium, the basement membrane and the vascular endothelium. Permeability for each marker is therefore a summation of the permeabilities of the individual barriers. The so-called "pores" which limit absorption of molecules of different molecular weights are, therefore, not necessarily all located in the epithelium and the smooth inverse relationship between molecular weight and permeability observed by Loehry *et al.* (1970) represents the overall characteristics of the lumen to blood barrier. While some components of this barrier may be diffusional and selective, other components (e.g. M cells or villous epithelium) could be non-selective and utilize an active process such as pinocytosis. In this respect, it is note-worthy that the selectivity of the mucosa appears to be determined by the capillary wall and the epithelial basement membrane, rather than by the epithelium itself (Kingham and Loehry, 1976; Kingham *et al.*, 1976). Furthermore, the bidirectional nature of the permeability barrier (Loehry *et al.*, 1970) does not necessarily argue in favour of passive diffusion across the epithelium. As mentioned later, uptake of marker proteins is possible at the basolateral membranes of enterocytes (Hugon and Borgers, 1968; Mayrhofer *et al.*, 1990), perhaps illustrating that with respect to antigen movements across the mucosa, the intestine has a secretory as well as an absorptive function.

Neither the studies of Loehry *et al.* (1970) nor others using fluorescein-conjugated dextrans (Tagesson *et al.*, 1978 and 1981), insulin (Danforth and Moore, 1959) nor horseradish peroxidase (Warshaw *et al.*, 1971 and 1974) have shown specifically that absorption from the gut lumen into the circulation occurs via the villous mucosa. However, it is of interest that absorption of the non-degradable fluorescent dextrans was limited to molecules of less than 20 kD. Mucolytic agents increased permeability, perhaps by their effects on the unstirred layer (Anderson *et al.*, 1988) but did not alter selectivity. The selectivity for dextrans, which are in some respects ideal fluid phase markers (Steinman *et al.*, 1980), may provide an estimate of the upper limit of the size of pores available for diffusion of macromolecules. Regional differences in the uptake of polypeptides have been suggested (Hemmings *et al.*, 1977; Hemmings and Williams 1978), with the ileum most active. However, uptake of Percoll microspheres by the ileum in suckling mice and macromolecule uptake by the ileum in suckling rats does not lead to transport across the epithelium (Clarke and Hardy, 1969a,b; Matsuno *et al.*, 1983) and ceases after closure (Morris and Morris, 1976). Artefacts in measurement are believed to explain the reported large-scale uptake of polypeptides by the ileum (Udall *et al.*, 1981).

If macromolecules are indeed absorbed via the villous mucosa, the most direct demonstration would be by an *in situ* histological technique, employing either enzyme activity (light and electron microscopy) or an electron-dense marker (electron microscopy) to detect the marker molecule. The marker used most commonly for this purpose has been the enzyme horseradish peroxidase (40 kD), although its validity as a fluid phase molecule is uncertain (Straus, 1983). Horse spleen ferritin (750 kD) has also been used as a marker in ultrastructural studies. A picture has emerged from studies using horseradish peroxidase (Cornell *et al.*, 1971; Walker *et al.*, 1972) and ferritin (Bockman and Winborn, 1966; Worthington and Syrotuck, 1976; Williams, 1978) where active transport of macromolecules through mature enterocytes occurs by a process of pinocytosis and vesicular transport (reviewed by Walker, 1984). Absorption by this process is non-selective and appears to involve uptake of marker molecules through tubular invaginations at the bases of the microvilli that constitute the brush borders of enterocytes. It has been concluded in a number of studies that this process delivers even high molecular weight molecules such as ferritin to the lateral intercellular spaces between the epithelial cells (Bockman and Winborn, 1966; Williams, 1978). However, direct connection of brush border invaginations with vesicles or tubules in the terminal web zone of enterocytes has never been illustrated convincingly, except in the preclosure period.

In contrast to the above studies, other work in adult (Rhodes and Karnovsky, 1971; Kao *et al.*, 1972; Owen, 1977), suckling (Kraehenbuhl and Campiche, 1969; Hugon, 1971; Lecce, 1972; Rodewald, 1973; Abrahamson and Rodewald, 1981) and fetal (Lev and Orlic, 1973) animals have found no evidence that horseradish peroxidase is transported across the villus epithelium. Furthermore, in adult mice and rats there was no uptake even across the enterocyte brush border (Rhodes and Karnovsky, 1971; Kao *et al.*, 1972; Owen, 1977).

The reasons for the qualitatively different results between these two groups of studies appear to lie in the experimental design. In those studies where either uptake or transport of marker molecules by enterocytes was observed, the marker was introduced into short ligated segments of intestine. In contrast, no transport was observed when marker was injected into either unligated intestine (Kraehenbuhl and Campiche, 1969; Rhodes and Karnovsky, 1971; Owen, 1977) or long segments of ligated intestine, or when marker was administered orally (Hugon, 1971; Lecce, 1972; Lev and Orlic, 1973). Moreover, even initial uptake across the brush border was not evident in adult animals (Rhodes and Karnovsky, 1971; Kao *et al.*, 1972; Owen, 1977).

The work of Rhodes and Karnovsky (1971) offers an explanation for why manipulation of the intestine, such as ligation, alters the handling of marker macromolecules. Surgical trauma to the intestine leads to permeation of horseradish peroxidase through tight junctions for a considerable distance around a site of injury. The nature of this propagated damage to tight junctions is unknown, but its patchy nature and spread from the site of original trauma suggests that it may be mediated by neural or neurovascular mechanisms. Such effects might be mediated by cholinergic pathways (Phillips *et al.*, 1987) or neuropeptide release from unmyelinated sensory nerve endings (Bienenstock *et al.*, 1988). Ligation might be expected to induce a similar propagating injury. The earlier work of Hettwer and Kriz (1925) and Hettwer and Kriz-Hettwer (1926) support this hypothesis – they were able to induce systemic anaphylaxis in sensitized guinea-pigs by injection of antigen into ligated gut loops but not by injection into unligated gut.

In addition to loss of tight function integrity, Rhodes and Karnovsky (1971) also observed that injury caused increased binding of horseradish peroxidase to the enterocyte brush borders and the appearance within the cells of vesicles containing the enzyme. The simultaneous localization of horseradish peroxidase to the brush border and within apical cytoplasmic vesicles in enterocytes in injured gut is very similar to the appearance described as "absorption" in experiments using ligated gut segments. Rhodes and Karnovsky suggested an alternative explanation – that damage to tight junctions allows horseradish peroxidase to diffuse from the lumen into the lateral spaces between enterocytes, where it is then absorbed across the basolateral cell membranes by pinocytosis. This explanation is supported by the observation that horseradish peroxidase administered intravenously is absorbed from the extracellular fluid by enterocytes and that it accumulates in endosomes and lysosomes in the apical cytoplasm (Hugon and Borgers, 1968). The subtle and patchy damage to tight junction integrity may have been overlooked in the studies that employed ligated gut segments.

We have re-examined recently the absorption of horseradish peroxidase from the rat small intestine (Mayrhofer *et al.*, 1990). The aim of the experiments was to eliminate all handling of the gut during the course of mucosal exposure to the enzyme. This was achieved by atraumatic perfusion of the upper small intestine in the conscious rat via an indwelling soft silastic cannula, introduced via the pylorus 24 h earlier. After perfusion for 4 h (5 ml/h, 5 mg horseradish peroxidase type I/ml) the gut was fixed, cryopreserved and frozen sections examined for enzyme activity. The results confirmed that in normal gut, tight junction integrity excludes horseradish peroxidase, the enzyme does not associate with the brush border and nor is there evidence of absorption into villous enterocytes. In contrast, there was clear evidence of uptake through the follicle-associated epithelium of Peyer's patches.

In contrast, when horseradish peroxidase was introduced into 3–4 cm acutely ligated gut segments, the enzyme was detected bound to the lumenal brush border, as described by Rhodes and Karnovsky (1971). Production of a controlled ischaemic lesion by occlusion of the superior mesenteric artery for periods of 5, 10 or 15 min was also followed by early binding of horseradish peroxidase to the brush border of the jejunal mucosa, followed by progressive evidence of loss of tight junction integrity and diffusion of the enzyme into the lateral spaces and the lamina propria. Furthermore, intravenous injection of either horseradish peroxidase, fluorescein-conjugated dextran (70 kD) or fluorescein-conjugated ovalbumen was followed by appearance of the enzyme or the conjugates in the extracellular fluid of the mucosa and vesicular uptake by enterocytes of the crypts and villi of the villous mucosa. Finally, injection of horseradish peroxidase into unligated gut of young rats (Bui and Mayrhofer, unpublished findings) confirmed that uptake occurs into enterocytes on the villi of the jejunum and ileum before

closure, but that no uptake was observed at either site after closure.

These findings lay to rest finally the notion that uptake of macromolecules occurs across the brush border of the normal small intestine at least in amounts detectable by immunohistochemistry. They confirm the earlier morphological studies of macromolecule uptake by normal mature small intestine. Furthermore, they are in accord with the studies of Jones (1974), which showed that uptake and transport of both immunoglobulin and serum albumen from the intestine of rats *in vivo* ceases after closure. It must be concluded that the macromolecular antigen detected in body fluids after oral exposure is taken up from the gut via the specialized follicle-associated epithelium of Peyer's patches. The question of uptake across damaged epithelium and the significance of retrograde absorption of antigen by epithelial cells from their basolateral surfaces will be discussed later (see Section 3.1.3).

2.2.1.3 *Respiratory Mucosa*
Antigen absorption is understood less well in the respiratory tract than in the gut and it presents different problems of measurement. In addition to the anatomical difficulties of measuring absorption by the mucosa of even the larger airways, in the respiratory system, antigen is partitioned between the air and the fluid-phase overlying the epithelium. Thus, in addition to considering the molecular weights or the sizes of particles in relation to uptake, attention must be paid to the probability of an antigen making contact at all with the fluid phase in a particular part of the respiratory tract before it can encounter the barrier of the respiratory epithelium. This probability may be affected by factors such as airway size, air flow velocity, humidity and the hygroscopicity of the aerosol, but the major determinant of whether it will be deposited on an epithelial surface (and if so, where) is the size of the particles in the aerosol.

As discussed by Stuart (1973), potential antigens range in diameter from viruses (<0.05 µm) to pollens ($10-100$ µm), although small particles will usually be aggregated or contained in droplets (e.g. viruses), while larger particles (e.g. pollens) may be fragmented to the $1-5$ µm range under environmental conditions (Busse *et al.*, 1972). Larger particles are impacted by their inertia into the mucous layer in the upper respiratory tract or at bifurcations of the lower airways, with a probability for particles 7 µm in diameter of about 33% and for particles of 1 µm diameter of about 1% (Landahl *et al.*, 1951). Particles of $10-20$ µm diameter are removed essentially completely from inspired air in the nasopharynx, while submicrometre particles largely escape deposition in the upper respiratory tract. In contrast, particles of $1-5$ µm diameter have a high probability of reaching deep into the lung parenchyma, where because of their mass a significant deposition occurs by sedimentation in alveolar ducts and alveoli (Gerrity *et al.*, 1979). Submicrometre particles also reach the alveoli but sedimentation is less important in this case and thermal diffusion becomes increasingly important. As a result, only 10% of particles less than 0.5 μm in diameter are deposited in the lung parenchyma (Stuart, 1973).

These considerations affect not only the sites of contact and absorption of airborne antigens, but also the pattern of clinical response to them (Novey et al., 1970; Wilson et al., 1971). Furthermore, the site of antigen deposition within the distal airways has a profound influence on the mechanisms available for clearance. A watershed exists at the distal terminal bronchioles, which have ciliated epithelium and can deliver fluid, cells and solutes to the pharynx via the mucociliary escalator. Beyond the terminal bronchioles, the alveolar ducts and alveoli have non-ciliated epithelium. Antigen clearance there depends more on phagocytosis by macrophages, transepithelial transport and lymphatic drainage, and less on fluid transport across the watershed to link up with the mucociliary escalator (Green, 1973; Morrow, 1973; Kilburn, 1974; Cohen and Gold, 1975). Aspects of particle size in relation to deposition of viruses and bacteria from aerosols within the lung are discussed by Cohen and Gold (1975).

Most of the antigens that have been studied in the respiratory tract can be considered as fluid phase molecules or particles. However, viral and some bacterial pathogens adhere to respiratory mucosa by specific receptors (e.g. ICAM-1 for rhinoviruses, sialic acid for influenza viruses) or adhesins (e.g. *Bordetella pertussis*, *Mycoplasma* sp). This could influence access of the microbial ligands to the immune system. While penetration of the cell is vital to establish infection by the viruses, little is known about whether receptor-mediated transport has an immunological significance in respiratory epithelium. Only the behaviour of fluid phase antigens will be discussed in this review.

In the upper respiratory tract, there have been excellent studies on the handling of particulate antigens (reviewed by Proctor et al., 1973). Particles of size greater than 5 μm diameter are deposited mainly as air is accelerated past the anterior surfaces of the nasal turbinates and to a lesser extent as the air flow changes in direction to the posterior nasal cavity. There is also evidence for significant deposition on nasal mucosa of particles in the 1–5 μm diameter range. Particulate material is deposited on to the thick mucous blanket, which is cleared every 10–15 min by ciliary action and replaced by fresh mucus secreted by the mucosa of the nasal cavity and its associated paranasal air sinuses. Anteriorly, some mucus is moved forwards and deposited in the anterior nares, but most is directed posteriorly to the oropharynx. It is noteworthy that the posterior flow of the mucus is directed across the palatine tonsils (adenoids), allowing antigen to become trapped in the overlying epithelial crypts. The adenoids are undoubtedly involved in the presentation of aeroallergens to the mucosal immune system but the absorption of antigens by these structures has not been studied in detail.

Less is known about the handling of soluble antigens in the nasal cavity, although the rapidity of onset of nasal symptoms in allergic subjects after inhalation of specific antigen-containing aerosols suggests that absorption occurs directly across the nasal epithelium. However, the clinical observations must be interpreted cautiously, because the nasal mucosa may be abnormally leaky in individuals who suffer from allergic rhinitis. Furthermore, leucocytes including basophils are present within the nasal secretions and contact with specific antigen could trigger intralumenal release of mediators, which could in turn alter mucosal permeability.

In contrast to the relative lack of information about antigen absorption in the upper respiratory tract, there has been considerable interest in recent years in the handling of antigens within the bronchial tree, stimulated by the rediscovery of BALT by Bienenstock and colleagues (Bienenstock et al., 1973a, b; Bienenstock and Johnston, 1976). This tissue is more evident in some animal species (Sminia et al., 1989) and although there is some doubt that there is a true homologue of BALT in normal human lung (Pabst, 1992), lymphoid tissue is associated closely with the airways under some circumstances (Emery and Dinsdale, 1973 and 1974; Meuwissen and Hussain, 1982). In view of the preceding discussion about the pattern of deposition of inhaled aerosols, it is significant that in those species where BALT is prominent it is placed specifically where particle impact can be expected at bifurcations in the bronchial tree.

The epithelial barrier of the bronchial tree therefore consists of a large area of ciliated columnar epithelium and discrete areas of specialized lymphoepithelium overlying the BALT. Early studies did not suggest a specialized antigen uptake by the BALT lymphoepithelium (Bienenstock et al., 1973b) but later work has shown that it is the major site of antigen absorption in the bronchial trees of some animals (Richardson et al., 1976; Fournier et al., 1977; Rácz et al., 1977; Tenner-Rácz et al., 1979; Van der Brugge-Gamelkoorn et al., 1985 and 1986). These later studies have demonstrated at light microscope and ultrastructural levels the uptake of fluid phase particles such as India ink, latex and ferritin, and of fluid phase soluble proteins such as horseradish peroxidase. Uptake into pinocytotic vesicles and transport across the epithelium have both been observed. Although obvious comparisons have been made with the follicle-associated epithelium of Peyer's patches and the antigen-absorbing cells have been called "M" cells by analogy (Pankow and von Wichert, 1988), the similarities appear more functional than anatomical (Sminia et al., 1989).

Little is known about the permeability of the bronchial epithelium in the normal lung. Like epithelial cells in the gut, the respiratory epithelium is joined by circumferential tight junctions at the lumenal border which do not allow paracellular passage of horseradish peroxidase (Simani et *al.*, 1974; Richardson *et al.*, 1976; Tenner-Rácz *et al.*, 1979; Myrvick and Ockers, 1982). However, the integrity of the tight junctions can be compromised by irritants such as tobacco smoke (Simani *et al.*, 1974) or anaesthetic ether (Richardson *et al.*, 1976). Some authors believe that ciliated epithelium is able to take up fluid phase markers such as horseradish peroxidase (Van der Brugge-Garmelkoorn *et al.*, 1985; Sminia *et al.*, 1989). Supporting evidence for this view comes from the earlier work of Richardson *et al.* (1976). Using non-inhalational anaesthetic and also unanaesthetized guinea-pigs (thus avoiding irritation or manipulation of the mucosa), these workers showed uptake by tracheal and bronchial epithelium of both ferritin and horseradish peroxidase from aerosols of solutions of the proteins. Although uptake of horseradish peroxidase by ciliated epithelium was not observed by Tenner-Rácz *et al.* (1979) or Fournier *et al.* (1977), it seems likely that normal respiratory epithelium, unlike normal intestinal epithelium (see Section 2.2.1.2), takes up fluid phase molecules from the airways by pinocytosis. Like intestinal epithelium, bronchial epithelium is also able to take up horseradish peroxidase from the ablumenal intercellular fluid when the enzyme is injected intravenously (Myrvick and Ockers, 1982) and, by inference, would also sample antigens that reach the ablumenal surfaces of the cells via leaky tight junctions. Simani *et al.* (1974) did not find evidence that horseradish peroxidase that was taken up by the epithelial cells from the lumen of the bronchus was transported across the cell and released at the mucosal surface.

Beyond the terminal bronchioles, air enters the alveolar ducts and alveoli, where the epithelium is non-ciliated and adapted for gaseous exchange (type I pneumocytes) and for production of surfactant (type II pneumocytes). Type I pneumocytes cover about 95% of the surface area in this compartment (Gail and Lenfant, 1983). The permeability barrier between the alveolar lumen and the blood is constituted by the alveolar epithelium, the alveolar capillary endothelium and the fused basement membrane between these two layers. The permeability of this barrier and its individual components has been of interest to physiologists, while the mechanisms by which particulate matter and soluble macromolecules are cleared from the alveoli has been studied mainly from the point of view of environmental safety and occupational lung disease.

Ultrastructural studies indicate that tight junctions exist between adjacent pneumocytes in the alveolar epithelium. These limit the penetration of horseradish peroxidase both from lumen to mucosa and in the reverse direction (Schneeberger-Keeley and Karnovsky, 1968; Bensch and Dominiquez, 1971; Schneeberger and Karnovsky, 1971; Schneeberger, 1974; Lauweryns and Baert, 1977). It was concluded from these studies that the normal alveolar epithelium is impermeable to soluble macromolecules and to particles such as ferritin, while the endothelium of the alveolar capillaries is freely permeable to horse-

radish peroxidase (40 kD) but retained haemoglobulin (64.5 kD), lactoperoxidase (80 kD), catalase (240 kD) and ferritin (600 kD) (Schneeberger and Karnovsky, 1971; Schneeberger, 1974). Formal measurements of the reflection coefficients of a number of small organic molecules indicate that the equivalent pore radius of the alveolar epithelial cells is 0.6–1.0 nm, while that of the capillary endothelium is 4.0–5.8 nm (Taylor and Garr, 1970). These results confirm the epithelium as a barrier to absorption of fluid phase macromolecules by the paracellular route.

However, these conclusions do not preclude the alveoli as possible sites of antigen absorption. While discussion here will be limited to direct uptake of antigen across the alveolar epithelium, uptake of both particles and soluble proteins by alveolar macrophages is quantitatively the most important route of clearance. The reader is referred to reviews of the role of macrophages in clearance of alveoli (Green, 1973; Lauweryns and Baert, 1977). Although most phagocytosed antigen is believed to be delivered to the pharynx by the mucociliary escalator, other routes are also possible. These include migration from the alveoli or alveolar ducts via the "sumps" described by Macklin (1955) and migration from the bronchi via either the bronchial epithelium or the specialized lymphoepithelium of the BALT. While there is excellent evidence that macrophages that have taken up particles in the lungs can be recovered from regional lymph nodes (Harmsen *et al.*, 1985), it is by no means clear where these cells exited from the airways or, indeed, whether they acquired their phagocytosed load within the lumen of the airways or within the interstitium. Studies using radiolabelled alveolar macrophages have not differentiated these possibilities (Corry *et al.*, 1984) and there is ultrastructural evidence that particles are transported from the alveoli and are phagocytosed by macrophages in the interstitium (Lauweryns and Baert, 1974). The observations are intriguing, particularly in view of the prominence given to BALT in discussions of local immune responses in the lung (Pankow and von Wichert, 1988; Sminia *et al.*, 1989) and the inconclusive evidence that BALT is a significant component of human lung tissue (Pabst, 1992). This points, perhaps, to an important role of antigen-transporting cells (cf. Morris, 1978) and regional lymph nodes in pulmonary immune responses.

A neglected but potentially important role of antigen transport in the lung is the pinocytic activity of the type I pneumocytes (Lauweryns and Baert, 1977). It is clear from several studies that these cells can transport both soluble and particulate fluid phase markers from the alveolar lumen to the interstitium and then to the lymphatics (Lauweryns and Baert, 1974, 1977; Bensch and Dominguez, 1971) and the blood (Bensch and Dominguez, 1971). Rapid transport by a pinocytotic mechanism has been described for particles such as carbon and ferritin (Lauweryns and Baert, 1974) and for soluble horseradish peroxidase (Schneeberger-Keeley and Karnovsky, 1968; Bensch and Dominguez, 1971; Schneeberger and Karnovsky, 1971; Schneeberger, 1974), although part of the pinocytosed material may be degraded in the lysosomes of the type I pneumocytes (Lauweryns and Baert, 1974). It is also noteworthy that this transport process is bidirectional (Schneeberger, 1974) and type I pneumocytes can therefore transport antigens (<65 kD) from the circulation to the alveoli. The immunological significance of antigen transport in either direction is unknown but it is a neglected factor that should enter into any consideration of the immunology of the lung.

2.2.1.4 Miscellaneous Epithelia

The handling and significance of fluid phase antigens at a variety of other epithelia has not been studied in detail. In the case of renal tubular epithelium, it is known that polypeptides which pass into the glomerular filtrate are absorbed avidly by the pinocytotic activities of the cells lining the proximal convoluted tubules (Ericsson, 1965; Graham and Karnovsky, 1966). It can be assumed that soluble antigenic materials in the circulation of less than about 60 kD will be handled in the same way. The physiological role of this mechanism is probably in scavenging protein nitrogen that would otherwise be excreted in the urine. Much of the absorbed material is delivered to lysosomes in the epithelial cells for degradation. However, some antigen could be transported across the epithelium or it could be processed and presented to T lymphocytes at the cell surface. In this context, it is noteworthy that the proximal tubules express class II MHC molecules constitutively in some species (Mayrhofer and Schon-Hegrad, 1983).

An area of greater immunological significance, although relatively unexplored, is the salivary tissues associated with the oral cavity. The ducts of these glands have been shown to provide access to bacteria (Schroeder et al., 1983; Nair and Schroeder, 1985) and to soluble fluid phase markers (Nair and Schroeder, 1983). Recent histological evidence suggests that the minor salivary glands in the oral cavity have specialized lymphoepithelial structures that might be involved in antigen uptake and presentation to lymphocytes (reviewed by Nair and Schroeder, 1986). Reliable induction of local secretory immunity in the oral cavity has been difficult to achieve and these studies offer interesting possibilities for future studies.

2.2.2 Antigens with Affinities for Epithelial Membranes

The preceding section examined the handling of a class of antigens (fluid phase) which has neither specific nor casual affinity for epithelial surfaces. However, there is also a range of potential antigens, such as micro-organisms and some soluble substances, that do have affinity for surface molecules on epithelial cells. In some cases, binding of such antigens to surface molecules

may trigger their uptake by receptor-mediated endo-cytosis and this in turn may influence profoundly their handling within the epithelial cells. In contrast to the uncertainty surrounding the handling of fluid phase molecules by enterocytes (see Section 2.2.1.2), uptake of antigens by receptor-mediated processes is less controversial. Nevertheless, as with fluid phase antigens, the relative importance of the villous mucosa and the follicle-associated epithelium remains uncertain because specific binding of antigenic ligands appears to enhance absorption by both routes.

The examples used in this section are drawn from studies on the small intestine. However, similar mechanisms are probably involved at other mucosal sites. They could, for instance, influence the uptake and presentation of microorganisms and toxins in the respiratory tract e.g. *B. pertussis* (Tuomanen and Hendley, 1983) and *Mycloplasma pneumoniae* (Powell et al., 1976), the urinary tract (e.g. *Escherichia coli* (Svanborg-Eden et al., 1976)) and the urogenital tract (e.g. *Neisseria gonorrhoeae* (Johnson, 1981; Virji and Heckels, 1984)). In some cases, as with *N. gonorrhoeae*, pathogens exploit receptor-mediated endocytosis to gain entry through cells, and this ability constitutes an important virulence determinant.

2.2.2.1 Contribution of Adherence to Immunogenicity of Mucosal Antigens

It is a general rule that soluble antigens which have affinities for components of the epithelial surface are more efficient in generating local secretory immunity in the small intestine (Mayrhofer, 1984; Andrew and MacDonald, 1987). The studies of deAizpurua and Russell-Jones (1988) support this thesis. These workers examined in mice the immunogenicity of a number of lectins and toxins that have affinities for the enterocyte brush border. Antigens such as these and cholera toxin (Pierce, 1978) can generate primary immune responses following a single exposure to antigen. In contrast, various fluid phase antigens including cholera toxoid are weak oral immunogens and they may either generate tolerance (Thomas and Parrot, 1974), or slowly evolving local antibody responses only after prolonged oral exposure (Crabbé et al., 1969; Dolezel and Bienenstock, 1971; André et al., 1973).

In the case of infective agents such as viruses (e.g. polio, rotaviruses, adenoviruses) and enteroinvasive bacteria (e.g. enteropathogenic *E. coli*, *Salmonella* spp, *Shigella* spp., *Campylobacter jejuni*, *Yersinia enterocolitica*) it is difficult to assess accurately the relative contributions of adherence, invasion of the mucosa and replication to the observed immunogenicity of the organisms. The relative ineffectiveness of orally administered killed bacteria as protective immunogens (Chuttani et al., 1972) as compared with live vaccines (Levine et al., 1984a; Levine et al., 1987; Ferreccio et al., 1989) emphasizes the importance of invasion and/or replication in some circumstances (Wassef et al., 1989). These factors can also influence local secretory antibody responses, as illustrated by the studies of Hohmann *et al.* (1979), using *E. coli* and *Salmonella* strains of varying virulence in mice. However, experience with *Vibrio cholerae*, a non-invasive bacterium, indicates that invasion is not a necessary prerequisite for the generation of protective immunity (Cash *et al.*, 1974; Levine *et al.*, 1981). Although the interpretation of this example is complicated by the known immunopotentiating effects of cholera toxin (Chen and Strober, 1990; Wilson *et al.*, 1990), studies on a toxin-deficient attenuated El Tor strain (Levine *et al.*, 1984b) indicate that the adjuventicity of cholera toxin is not essential to induce protective immunity. Adherence and colonization appear to be the most important factors.

Another general problem of interpretation is that of where, anatomically, adherent antigens engage with the immune system and exert their immunogenicity (Mayrhofer, 1984). As with fluid phase antigens, absorption could occur either through the enterocytes covering the villous mucosa or through the specialized M cells of Peyer's patches (Owen and Jones, 1974). No receptors that are restricted to M cells have been detected, although there is evidence that particulate antigens have greater access to the surface membranes of the M cells than to the brush borders of enterocytes (Neutra and Kraehenbuhl, 1992). It follows that in those experiments where the antigenicity of soluble ligands has been studied, it is not possible to discern whether antigen is delivered via the villous mucosa or via M cells, because soluble ligands appear to have access to both absorptive enterocytes and to M cells.

Nevertheless, there is a tendency to attribute the enhanced immunogenicity of those antigens that bind to the mucosal surface to their enhanced uptake by M cells (de Aizpurua and Russell-Jones, 1988; Neutra and Kraehenbuhl, 1992). This may not be correct, because binding of some ligands and microorganisms to the brush borders of enterocytes also leads to more effective uptake and transport (see below). The issue is of more than academic importance, because the route of antigen uptake opens the way to alternative modes of presentation, including processing by enterocytes (non-professional antigen-presenting cells), uptake and processing by lamina propria macrophages and/or dendritic cells (professional antigen-presenting cells), delivery to the regional lymph nodes (Mayrhofer *et al.*, 1986; Liu and MacPherson, 1991) or delivery to Peyer's patches. It is difficult in practice to devise experiments which can distinguish clearly the enterocyte (mucosal) route of absorption from the M cell (Peyer's patch) route. There are no carriers that can target antigens selectively to the absorptive enterocytes, while potential carriers such as reoviruses (Wolf *et al.*, 1981 and 1983), bacteria (Inman and Cantey, 1983; Wassef *et al.*, 1989) and antibodies (Pappo *et al.*, 1991) have relative rather than absolute specificity for M cells. Anatomical approaches, such as excision of Peyer's patches, are difficult and mutilating,

and it is impossible to be sure that all follicles have been removed. Nevertheless, within these anatomical limitations, it has been shown that secretory immunity to *Shigella* sp. develops in rabbits after infection of Thiery-Vella loops fashioned from ileum that lacks Peyer's patches (Keren *et al.*, 1978). Taken at face value, this experiment suggests that the Peyer's patch route of immunization is not obligatory for immunization with adherent/invasive microorganisms, in contrast to its importance in the response to fluid phase antigens (Husband and Gowans, 1978).

In summary, there is evidence that specific binding or attachment to the gut epithelium enhances the immunogenicity of antigens. However, although this effect is believed widely to be due to facilitated absorption of the antigen via M cells, the evidence is still not conclusive. Conjugation of antigens to ligands that bind selectively to either the brush borders of enterocytes or to the surface membranes of M cells could conceivably have contrasting effects on their immunogenicity or on the modality of the resulting immune response. Development of cell-specific ligands therefore has considerable importance for vaccine development (Kraehenbuhl and Neutra, 1992).

2.2.2.2 Antigens that Bind or Adhere to the Enterocyte Brush Border

As noted above, a considerable body of investigation has failed to identify any surface antigens, lectin-binding sites, toxin receptors or adhesin receptors that are unique to either the enterocyte brush border or the lumenal surfaces of M cells (Neutra and Kraehenbuhl, 1992). Some particulate antigens such as bacteria (Inman and Cantey, 1983; Wassef *et al.*, 1989), viruses (Wolf *et al.*, 1983), antibody-coated microspheres (Pappo *et al.*, 1991) and cholera toxin-coated colloidal gold (Neutra and Kraehenbuhl, 1992) have *de facto* specificity for M cell membranes. However, this appears to reflect exclusion of particulate materials by the thick fuzz coat on the brush borders of enterocytes rather than the absence of the relevant receptor.

While there may be few antigens that bind exclusively to absorptive enterocytes, there are a host of soluble peptides and microorganisms that adhere to the brush border of the small intestine. Most are microbial products such as toxins, adhesins or virus capsid proteins, which are primarily ligands involved in microbial pathogenesis. However, some may also be important protective antigens that are recognized by the host. The receptors for a few microbial adhesins are host proteins (e.g. the fibronectin receptor of *Staphylococcus aureus* (Höök *et al.*, 1989) and the laminin receptors on *S. aureus* and *Trichomonas vaginalis* (Carneiro *et al.*, 1989)), but most have carbohydrate specificities and are lectins which presumably confer on the microorganisms greater versatility to colonize a range of habitats. These include the TCP pilus of *V. cholerae* (Taylor *et al.*, 1987); the K88, K99 and 987P pili of enterotoxigenic *E. coli* from animals (Isaacson *et al.*, 1980; Moon and Runnels, 1984; Sellwood, 1984) and the CFA (colonization factor A) pili of enterotoxigenic *E. coli* from humans (Evans *et al.*, 1984). However, the adhesive mechanisms involved in the unique attachment and invasion mechanisms of organisms such as enteropathogenic *E. coli* (Staley *et al.*, 1969; Helie *et al.*, 1991), *Shigella* sp. (Sansonetti, 1991), *Salmonella* sp. (Finlay *et al.*, 1988), *Yersinia* sp. (Isberg *et al.*, 1987), *Eimeria* sp. (Augustine and Danforth, 1984) and *Cryptosporidium* sp. (Current and Reese, 1986) are unknown, although much progress is being made in this area (Leong *et al.*, 1990; Wick *et al.*, 1991). Some bacterial toxins, such as *Clostridium perfringens* enterotoxin (McDonel, 1980), cholera toxin and a number of other bacterial enterotoxins also have carbohydrate specificities (reviewed by Donowitz and Welsh, 1987) but the only known receptor for the enterotoxins of *S. aureus* is the class II MHC molecule (Wood *et al.*, 1991). The receptors for many viruses appear to be carbohydrate components of glycoproteins (reviewed by Wick *et al.*, 1991).

In addition to microbial lectins, lectins of plant origin may also have importance in the gut because of their potential for direct toxicity to the mucosa (King *et al.*, 1982; Köttgen *et al.*, 1982; Lorenzsonn and Olsen, 1982), their intrinsic immunogenicity (de Aizpurua and Russell-Jones, 1988) and the possibility that they could be used to direct the uptake of antigen–lectin conjugates (de Aizpurua and Russell-Jones, 1988). The lectins concanavalin A and pokeweed mitogen and the lectins from *Lens culinaris*, *Helix pomatia*, *Glycine max*, *Arachis hypogea* and *Ulex europus* have all been shown to be immunogenic by the oral route (de Aizpurua and Russell-Jones, 1988). However, the precise routes of absorption and presentation responsible for their immunogenicity remain uncertain (see Section 3.1.2). The possibility that gluten has lectin-like properties has been discussed earlier (see Section 2.2.1.1).

The problem of how plant and microbial lectins are handled by the intestinal epithelium has been addressed in a number of studies. Cholera toxin has been shown to bind to the brush border (Peterson and Verwey, 1974) and to enter enterocytes by endocytosis in simple vesicles (Kao *et al.*, 1972; Strombeck and Harrold, 1975; Montesano *et al.*, 1982; Hansson *et al.*, 1984). Some of the internalized toxin is probably degraded in lysosomes (Strombeck and Harrold, 1975), but there seems little doubt that some is also transported across the enterocyte. Similar observations have been made on the B subunit of the *E. coli* LT enterotoxin (Linder and Russell-Jones, 1990), although the evidence for transcytosis across the enterocyte was indirect. Wilson *et al.* (1990) have suggested that fluid phase antigens may also obtain a pick-a-back ride with the cholera toxin B subunit, perhaps by inclusion into pinocytotic vesicles induced by the binding of the lectin. An ultrastructural impression of how this might occur is provided by the effects on the brush border of lectins such as wheat germ agglutinin and concanavalin

A. At high concentrations, these lectins induce the formation of microvillus pits and vesicles in the apical cytoplasm of enterocytes (Lorenzsonn and Olsen, 1982).

Neutra *et al.* (1987) have investigated by ultrastructural visualization the handling of several plant lectins conjugated to ferritin. In comparison with a bovine serum albumin–ferritin conjugate (which neither binds to the brush border nor is endocytosed or transported by the enterocyte), conjugates of concanavalin A, *Ricinus communis* agglutinin (ricin) and wheat germ agglutinin were all bound to the brush border and were endocytosed by rabbit enterocytes. However, absorptive enterocytes did not appear to transport any of the lectins to the basolateral surfaces for secretion and the conjugates were instead delivered to the lysosomal compartment. This contrasts with the transcellular transport of lectin–antigen conjugates by M cells (see Section 2.2.2.3) and the handling of cholera toxin (described above). The contrasting fates of cholera toxin and these plant lectins suggests that the nature of the target molecule may determine the intracellular routing of the bound ligand. There are interesting parallels in the variety of interactions between various microorganisms and the gut epithelium.

Microorganisms display a fascinating diversity in their modes of interaction with the epithelium (Wick *et al.*, 1991). The interactions range from no association in the case of some commensal bacteria, adherence without invasion in the cases of *V. cholerae* and enterotoxigenic *E. coli*, to adherence plus invasion in the cases of a number of enteropathogenic species. Understanding the molecular pathogenesis of infections of the intestine is certain to illuminate the cell biology of the enterocyte and the functions of some of its surface molecules. For instance, it is of interest to know why some bacteria such as *V. cholerae* adhere to the brush border without inducing either endocytosis or observable structural changes, while other adherent bacteria such as enteropathogenic *E. coli* induce characteristic effacement of microvilli (Staley *et al.*, 1969; Helie *et al.*, 1991) and polymerization of actin in the zone of the terminal web (Knutton *et al.*, 1988). It will also be fascinating to know how protozoa such as *Cryptosporidium* induce enterocytes to envelop them with cytoplasmic processes and how a close association is formed between the effaced brush border and the feeder organelle of the parasite (Takeuchi, 1971; Current and Reese, 1986). In the case of invasive bacteria such as *Salmonella* sp. (Takeuchi, 1966, 1971; Finlay *et al.*, 1988), *Shigella* sp. (Takeuchi, 1971; Sansonetti, 1991) and *Yersinia* sp. (Isberg *et al.*, 1987), and protozoa such as *Eimeria* sp. (Augustine and Danforth, 1984), there are important questions to be answered about how these organisms induce endocytosis by the enterocyte, translocation within the cells and movement from cell to cell.

These exciting areas of recent research are beyond the scope of this review (Sansonetti, 1991; Wick *et al.*, 1991; Sanger and Sanger, 1992). However, the studies of soluble ligands and of microbial behaviour highlight the different ways by which enterocytes respond to engagement of different cell surface molecules (or groups of molecules). The immunological relevance of these various microbial strategies is not known and although it may be confined to their role in disease pathogenesis, they illustrate the potential for manipulating the enterocyte for the purposes of delivering experimental vaccines. In a similar way, investigations on ligands such as the B subunits of either cholera toxin or the LT toxin of *E. coli* may reveal routes through the epithelium that can be exploited for delivery of covalently linked vaccine antigens or peptides. The possible importance of enterocytes as non-professional antigen-presenting cells will be discussed in Section 3.3.

2.2.2.3 *Antigen Binding and Uptake by Membranous Cells in Follicle-Associated Epithelia*

"Membranous" or "microfold" cells (Owen and Jones, 1974) have been described in the follicle-associated epithelium of Peyer's patches in a variety of mammalian species. They are the subjects of recent reviews (Sneller and Strober, 1986; Trier, 1991; Neutra and Kraehenbuhl, 1992) and, because their function as antigen-transporting cells is established, discussion of them in this section will be brief. The discovery of M cells and of similar cells in the epithelia overlying lymphoid aggregates in the appendix (Owen and Nemanic, 1978), caecum (Owen and Nemanic, 1978), large intestine (Bland and Britton, 1984; O'Leary and Sweeney, 1986) and the respiratory tract (Sminia *et al.*, 1989) has been one of the major advances in mucosal immunology in recent years. However, the overall importance of M cells in the absorption of fluid phase antigens (see Section 2.2.2) and antigens that bind to the epithelial surface is still uncertain, although as mentioned earlier, there is a tendency to attribute all immunological effects of oral immunization to transport via this route (see Section 2.2.2.1).

Historically, clinical observations in typhoid fever indicated that Peyer's patches were vulnerable to invasion by *Salmonella typhi* and the experimental studies of Kumagai (Kumagai, 1922) showed that intestinal infection by mycobacteria also followed the same route. Later experimental studies indicated that a number of intracellular bacterial pathogens, including *Salmonella* spp. (Sprinz *et al.*, 1956; Carter and Collins, 1974; Hohmann *et al.*, 1979; Kohbata *et al.*, 1986; Gahring *et al.*, 1990), *Mycobacterium* spp. (Fujimura, 1986) and *Yersinia* spp. (Neutra and Kraehenbuhl, 1992), enter Peyer's patches to establish a primary nidus of infection. As with many interactions between microorganisms and host cells (Wick *et al.*, 1991), transport of the pathogen into the Peyer's patch via M cells is an example of exploitation of a physiological process by a parasite. In this case, uptake does not appear to involve active participation by the microorganisms, except insofar as

they have the ability to adhere. Adhesion appears merely to increase the efficiency of an otherwise constitutive transport mechanism (see below). Other bacteria such as *Shigella* spp. (Wassef *et al.*, 1989), *V. cholerae* (Owen *et al.*, 1986b), avirulent *Salmonella typhimurium* (Hohmann *et al.*, 1979) and probably a range of commensal bacteria (Shimizu and Andrew, 1967; Hohmann *et al.*, 1979) are also transported more or less efficiently. However, these organisms, and others such as *Giardia* spp. (Owen *et al.*, 1981), are destroyed by macrophages and inflammatory cells in the Peyer's patches of immunocompetent animals (Berg and Garlington, 1979; Berg, 1981; Owens and Berg, 1982) and are unable to establish infections. The Peyer's patch is also a portal of entry to the mucosa for a number of viruses, including polio (Bodian, 1955; Sicinski, 1990), reovirus (Wolf *et al.*, 1981 and 1983), mouse mammary tumour virus and perhaps also the human immunodeficiency virus (Neutra and Kraehenbuhl, 1992).

In the case of non-microbial antigens, many studies have shown that a range of fluid phase soluble and particulate markers are transported constitutively across M cells. They include carbon particles (Bockman and Cooper, 1973; Joel *et al.*, 1978), ferritin (Bockman and Cooper, 1973), latex beads (LeFevre *et al.*, 1978; Pappo *et al.*, 1991) and horseradish peroxidase (Owen, 1977). The questions to be explored next are how adherence affects the rate of uptake and transport of antigens and the extent to which M cells are accessible to antigens.

M cells appear to be more accessible than are enterocytes to particulate antigens. For instance, microspheres coated with an antibody against an antigen expressed by both cell types bind only to M cells (Pappo *et al.*, 1991). Similar observations have been made using cholera toxin-coated colloidal gold particles (cited by Neutra and Kraehenbuhl, 1992), although soluble cholera toxin binds to enterocytes (Kao *et al.*, 1972) as well as to M cells (Owen *et al.*, 1986d). The differences in binding of particulate ligands to the two cells may be related to the density of the glycocalyx (or "fuzz") that coats the lumenal surface. In comparison with the enterocyte brush border, the M cell surface membrane is deficient both in glycocalyx and integral membrane proteins (Madara *et al.*, 1984). This may reflect the absence or reduced expression (Owen *et al.*, 1986c) in M cells of the hydrolases characteristic of the brush borders of villus (Madara and Trier, 1987) and follicle-associated enterocytes (Smith, 1985). Moreover, M cells probably do not express certain molecules which have a restricted distribution in the intestine, such as the receptor for the transferrin–cobalamin complex (Levine *et al.*, 1982, 1984; Donaldson, 1987). These differences between M cells and enterocytes may affect access and binding of ligands to cell surface molecules and could explain the preference of some bacteria to adhere to M cells (Inman and Cantey, 1983; Kohbata *et al.*, 1986; Wassef *et al.*, 1989).

Another factor affecting access of antigen to M cells

is the thickness of the mucous blanket. This may be thinner over Peyer's patches, where the follicle-associated epithelium contains fewer goblet cells (Owen and Nemanic, 1978). Goblet cells are more plentiful over the caecal Peyer's patch (Bhalla and Owen, 1982) but their abundance in the epithelium of the colonic lymphoid aggregates has not been described. Differences in thickness of the mucous blanket could be responsible for the selective binding of colloidal carbon to colonic lymphoid follicles in rat large intestine (Bland and Britton, 1984).

There appears little doubt that binding to the M cell surface enhances uptake of both soluble and particulate materials. Several studies have indicated that both specific and non-specific binding of antigens to M cells facilitates their uptake. Cationization of ferritin enhances its binding, uptake and transport by M cells (Bye *et al.*, 1984; Neutra *et al.*, 1987). Furthermore, in a well-controlled quantitative study (Pappo *et al.*, 1991), latex microspheres coated with a monoclonal antibody specific for an M cell surface component were taken up three to four times more efficiently than microspheres coated with an irrelevant antibody of the same isotype. Neutra *et al.* (1987) also observed enhanced uptake of ferritin when it was conjugated to plant lectins that bound to M cell surface glycoconjugates. In this case, uptake of a wheat germ agglutinin–ferritin conjugate was 50-fold greater than that of a bovine serum albumen – colloidal gold conjugate. Unfortunately, in this otherwise excellent study, there were no controls for the possibility that the different electron-dense carrier particles may have influenced the apparent absorption rates of the attached proteins.

A further important observation by Neutra *et al.* (1987) was that bound ligands were transported in coated vesicles. This raises important questions about whether multivalent ligands cause clustering of receptors into coated pits on the surfaces of M cells and whether fluid phase antigens are also endocytosed in clathrin-coated vesicles. This form of uptake and transport may have important consequences for the subsequent fate of absorbed antigens. M cells have few lysosomes in comparison with enterocytes and absorbed materials appear largely to escape delivery to the lysosomal compartment. In contrast, the fate of much of the material absorbed by enterocytes appears to be degredation in lysosomes (Neutra *et al.*, 1987).

Nevertheless, enterocytes may also transport some intact antigens (see Section 2.2.2.2). Because of the large area of the villous mucosa, and its lymphatic drainage to the regional lymph nodes, enterocyte uptake could be an immunologically significant route of antigen uptake. This section has emphasized the transcellular transport of intact antigens. The next section deals with the possible relevance of antigen absorption that culminates in the lysosomal compartments of M cells and enterocytes.

3. Processing and Presentation of Antigens by Epithelial Cells

The uptake of fluid phase and surface-bound antigens by a number of epithelial cell-types has been discussed in the preceding sections. In the case of some bound antigens, the endocytosed material (or some part of it) appears to be transported intact across the cells and released at the basal surface. In addition, some specialized cells (e.g. M cells) also absorb and transport both fluid phase and surface-bound antigens. However, in most cases where fluid phase antigens are endocytosed, most (if not all) of the material is directed towards the lysosomal compartment and is degraded. This may also be the fate of a variable proportion of surface-bound antigens, even in the case of specialized antigen-transporting cells such as M cells (Neutra and Kraehenbuhl, 1992).

It is therefore relevant to question whether those antigens that are degraded within epithelial cells are relevant immunologically. If not, in the cases of antigens that are endocytosed in the antegrade direction (apex to base in polarized epithelia), their destruction can be viewed simply in the context of the general antigen exclusion function of epithelia (see Section 2.1). For those antigens which may be endocytosed in the retrograde direction (i.e. from the basal aspect), uptake might reflect an antigen elimination function (see below). However, the recent advances in understanding of how antigens are processed and presented (Braciale and Braciale, 1991) raise the theoretical possibility that partially degraded peptides in epithelial cells might be presented to T lymphocytes. Antigenic peptides, if released by the epithelial cells, could bind to MHC molecules on adjacent "professional" antigen-presenting cells in the mucosa (macrophages and/or dendritic cells). Alternatively, epithelial cells might themselves function as "non-professional" antigen-presenting cells. While there is no doubt that epithelial cells can present viral (i.e. endogenous) antigens to cytotoxic T cells, it is necessary to consider whether or not they can also process and present exogenous antigens.

The latter possibility was raised by the observation that a number of epithelia express class II MHC molecules either constitutively (reviewed by Mayrhofer, 1984; Forsum *et al.*, 1985; Bland, 1988) or in response to certain physiological (Klareskog *et al.*, 1980; Newman *et al.*, 1980) or local immunological (Mason *et al.*, 1981a; Barclay and Mason, 1982; MacDonald *et al.*, 1988) stimuli. Prior to this, expression of class II MHC molecules (Ia antigens) had been observed only by B lymphocytes and cells of the macrophage/dendritic cell lineage (Hämmerling, 1976) – cells now recognized to be vital antigen-presenting cells in the induction of responses to exogenous antigens by CD4–positive T cells (Unanue *et al.*, 1989). The only previous exception had been the thymic epithelium, where there were intuitive reasons for believing that expression of class II MHC molecules would have a role in T cell development and/ or T cell tolerance. However, other epithelia are exposed to food, environmental or microbial antigens, suggesting that the function of class II MHC molecules at these sites might be to present absorbed antigens to class II-restricted T cells (reviewed by Mayrhofer, 1984; Bland, 1988) and the ability of some other cell types (e.g. endothelium) to present exogenous antigens to primed T cells in a class II-restricted manner (Hirschberg *et al.*, 1980) or produce an IL-1-like factor (e.g. skin keratinocytes (Luger *et al.*, 1981)) appeared to strengthen this possibility. Also influential was the view at this time that epithelia, exemplified by the enterocytes of the small intestine, were active in the uptake and transepithelial transport of antigens (Walker, 1984).

Since the time of these earlier observations, there have been a number of significant advances. It is now recognized that exogenous antigens are usually presented to CD4-positive T cells in association with class II MHC molecules (Swain, 1983; Germain, 1986). Furthermore, it is clear that the majority of T cells that are in direct contact with epithelia, such as those of the gut and the skin, are not of the CD4-positive subset. In the gut, the majority express either a homodimeric form of CD8 (Guy-Grand *et al.*, 1991) or express neither CD4 nor CD8, and a variable proportion express the γ/δ TCR heterodimer (Brandtzaeg *et al.*, 1990; Cerf-Bensussan and Guy-Grand, 1991). Although CD4-positive T cells are not precluded from the gut epithelium, the resident population appears unequipped to respond to antigens presented in association with class II MHC molecules. Indeed, if antigen recognition by T cells expressing the γ/δ TCR is restricted in a manner similar to that which governs responses by α/β TCR-positive T cells, then the nature of the restricting element may not be a classic MHC molecule but rather, a class I-like molecule such as CD1 (reviewed by Calabi and Bradbury, 1991). These are important considerations in interpreting the results of experiments designed to examine *in vitro* the antigen-presenting activities of epithelial cells.

Another area of significant advance is the identification of populations of dendritic cells and related cell types in a variety of epithelia and mucosae. This is the subject of Section 4 of this review. However, it should be noted here that squamous epithelia (skin, oesophagus, buccal cavity, vagina) and a variety of mucosae (gut, airways, urogenital tract, occular) have a rich complement of these cells, which in some situations occupy an intraepithelial position. It is therefore questionable whether there is a need for non-professional antigen presentation by epithelial cells unless there are qualitative differences in the responses by T cells to antigens presented by epithelial cells and by the more classic sorts of antigen-presenting cells. This is an area of current research interest (Bland, 1988).

Finally, the production of co-stimulatory cytokines such as IL-1 or IL-6 by epithelial cells (Luger *et al.*, 1981; Bland, 1987; Beagley *et al.*, 1990; Mayer *et al.*, 1990; Santos *et al.*, 1990; Shirota *et al.*, 1990) has been conducive to the idea that they have a physiological function in antigen presentation. In fact, these cytokines are produced by a variety of cell-types and have pleiotropic effects in addition to co-stimulatory effects on T cells. It can be argued that epithelia in general regulate, and are regulated by, the tissues of the lamina propria. The orchestration of the supporting tissues to the needs of the vital epithelial component of the mucosa may occur, at least in part, through an interplay of cytokine messengers as suggested recently for the uterine endometrium (Guilbert *et al.*, 1993). In this orchestra, local T lymphocytes may be only one group of players among the other important resident cells in the mucosa.

3.1 STRUCTURE OF EPITHELIAL CELLS AND THE FATE OF ABSORBED MACROMOLECULES

Discussion in this section will be limited to the epithelial cells of the small intestine (enterocytes), which are the most extensively studied with respect to the handling of macromolecules. Some aspects of antigen handling have been dealt with in Sections 2.1.2, 2.2.1.2 and 2.2.2.2. Attention will be focused here on the intracellular fate of endocytosed macromolecules. As discussed in Section 2.2.1.2, there are two routes by which macromolecules can enter enterocytes – an antegrade route, whereby potential antigens can be absorbed via the lumenal brush border surface, and a retrograde route, by which macromolecules are absorbed via the ablumenal basolateral cell membrane.

3.1.1 Enterocyte Structure

For the purposes of this discussion, the important structural features of enterocytes are the surface membranes (brush border and basolateral) and the organelles associated with the endocytic pathway. The general ultrastructural features of the enterocyte are reviewed elsewhere (Trier, 1968; Madara and Trier, 1987). The lumenal brush border consists of closely packed microvilli, approximately 0.5–1.0 μm in length and 0.1 μm in diameter. Between the microvilli are scattered pits or invaginations, referred to as microvillus pits. In suckling rats, the microvillus pits are prominent and connect with tubular and vesicular structures extending into the terminal web region of the apical cytoplasm (Kraehenbuhl and Campiche, 1969; Rodewald, 1973, 1980; Abrahamson and Rodewald, 1981). During the preclosure period, this tubulovesicular system is involved with immunoglobulin absorption in the proximal small intestine, while in the distal small intestine it delivers absorbed proteins to the lysosomal compartment (Kraehenbuhl and Campiche, 1969; Rodewald, 1973). Although reviewers often suggest that the microvillus pits and tubules of preclosure enterocytes are clathrin coated (e.g. Baintner, 1986b; McKenzie, 1990), this is not clear from the original descriptions by Rodewald (1973) and Abrahamson and Rodewald (1981). However, in the case of another epithelium involved in immunoglobulin transport, microvillus surface pits in yolk sac epithelial cells appear to have clathrin coats (Wilson and King, 1986).

In older rats, the number of microvillus pits is reduced and connections with smooth vesicles and tubules in the terminal web area are not observed (Rodewald, 1973). The physiological role of the microvillus pits in normal mature enterocytes is uncertain. However, as discussed earlier (see Section 2.2.1.2), it is unlikely that there is significant uptake of fluid phase macromolecules via the brush borders of enterocytes. Receptors for the intrinsic factor–cobalamin complex have been localized immuno-cytochemically to these structures (Levine *et al.*, 1982, 1984), and recent observations on the effects of plant lectins on enterocytes suggest that cross-linking of certain brush border glycoconjugates may stimulate endocytic activity in microvillus pits (Lorenzsonn and Olsen, 1982). It is not known what relevance these observations have to the uptake of bacterial lectins such as cholera toxin (Kao *et al.*, 1972) and the labile toxin of enterotoxigenic *E. coli* (Linder and Russell-Jones, 1990). Available evidence suggests microvillus pits in mature enterocytes are not clathrin coated (Levine *et al.*, 1984), unlike apical pits in a number of other absorptive epithelia (Friend and Farquhar, 1967; Herzog, 1984; Rodman *et al.*, 1986) and this also may be a distinguishing point between enterocytes and M cells, which do exhibit clathrin-coated surface pits (Neutra *et al.*, 1987).

The basolateral surface of the enterocyte is a classic cytoplasmic membrane which is functionally separated from the brush border domain at the zonula occludens. In this respect, enterocytes resemble a variety of other polarized epithelial cells, where the protein and lipid composition of the apical membranes differs dramatically from the composition of the basolateral membranes (Molitoris and Nelson, 1990; Simons and Wandinger-Ness, 1990). In the case of the enterocyte, the zonula occludens marks the boundary between a domain that is highly specialized for digestion and absorption (the brush border) and a domain (the basolateral membrane) which shares much in common with unspecialized cell membranes but which nevertheless has some special functions (Madara and Trier, 1987). For the purposes of this discussion, salient specializations are the expression of MHC molecules (Parr and McKenzie, 1979; Mayrhofer *et al.*, 1983), the expression of the polymeric immunoglobulin receptor (Chintalacharuvu *et al.*, 1991), the absorption of polymeric IgA for transport to the brush border (Brandtzaeg, 1974) and the exocytosis of macromolecules under certain circumstances. Both delivery of transported immunoglobulin to the basolateral membrane in the preclosure period in rats (Rodewald, 1973,

1980; Abrahamson and Rodewald, 1981) and uptake of polymeric IgA in later life (Kraehenbuhl and Neutra, 1992) involve clathrin-coated pits and transport vesicles.

The endocytic pathway in enterocytes will be discussed in more detail later. However, the development of the lysosomal compartment will be dealt with here. At the level of the light microscope, acid phosphatase-containing bodies are abundant in the cytoplasm of mature enterocytes (Barka and Anderson, 1962). They appear histochemically as granules, particularly in the supranuclear cytoplasm but also in the basal parts of the cell, and finer granules are seen in the supranuclear cytoplasm. Ultrastructural studies confirm that the acid phosphatase-containing bodies are lysosomes (Owen *et al.*, 1986a). Developmentally, lysosomes are present in the villus enterocytes of the suckling rodent small intestine and are particularly prominent as supranuclear vacuoles in the ileal epithelium (Rodewald, 1973; Hasegawa *et al.*, 1987). In ruminants and ungulates, increased lysosomal activity (as judged by acid hydrolase content) coincides with the termination of non-selective transport of lumenal proteins in the first 24–48 h of life (Healey, 1984; Baintner, 1986b).

In the context of enterocyte development in the rat intestine preclosure, acid phosphatase is present as granules in the cytoplasm of enterocytes in both crypts and villi (Hasegawa *et al.*, 1987). However, in adult rats, acid phosphatase-containing granules are uncommon in crypt enterocytes (Mayrhofer and Spargo, 1990). Diffuse staining for the enzyme was detected in the supranuclear cytoplasm of cells at the bases of jejunal villi, while cytoplasmic granules appeared in cells on the lower villus. At the ultrastructural level, there are many descriptions of lysosomes in mature enterocytes (Straus, 1983; Madara and Trier, 1987). The appearance varies from multivesicular bodies in the apical cytoplasm (King *et al.*, 1982; Levine *et al.*, 1984; Madara and Trier, 1987; Mayrhofer and Spargo, 1990) to classic dense bodies and large phagolysosomes in the supranuclear cytoplasm of the most mature cells (Mayrhofer and Spargo, 1990).

3.1.2 Antegrade Absorption Pathway

In Section 2.2.1.2 it was argued that fluid phase antigens are not absorbed via the brush border in the mature intestine. Based on this premise, discussion in this section will be focused briefly on the preclosure period and on the handling of antigens whose uptake is facilitated by binding to the brush border.

Various aspects of selective absorption of immuno-globulins by rodent enterocytes (Morris, 1968; Brambell, 1970; Morris, 1978; Baintner, 1986b; McKenzie, 1990) and non-selective absorption of macromolecules in new-born ruminants and ungulates (Baintner, 1986b) have been reviewed elsewhere. In ruminants and ungulates, non-selective uptake of lumenal proteins occurs via tubular extensions of the microvillus pits. No receptor is involved and the resulting endocytic vesicles are not clathrin coated. There does not appear to be any sorting and absorbed macromolecules are delivered to the basolateral cell membranes by coated vesicles, which bud from large apical vesicles (Trahair and Robinson, 1989). Non-selective uptake of proteins ceases within 24–48 h of birth and appears to correlate with the development of lysosomes (see above).

In rodents, IgG absorption is selective and mediated by Fc receptors which are expressed on the brush borders of enterocytes on the lumenal halves of villi in the proximal small intestine (Rodewald, 1980). Endocytosis of Fc receptors via microvillus pits appears to be constitutive and has the potential to pick-a-back fluid phase macromolecules that are included in the endocytic vessicles (Abrahamson and Rodewald, 1981). As in newborn ruminants, absorbed IgG is transported to the basolateral cell membranes of the enterocytes in coated vesicles which bud from the apical endocytic vesicles (Rodewald, 1973, 1980; Abrahamson and Rodewald, 1981). However, an important difference is that, in rodent enterocytes, there is sorting of receptor-bound IgG (which is transported intact) from fluid phase molecules in the endosome. The latter are delivered to multivesicular bodies and classical lysosomes, where they are degraded (Abrahamson and Rodewald, 1981), as are fluid phase proteins and virions that are absorbed by the tubulovesicular system in the ileal enterocytes of neonatal rats (Wolf *et al.*, 1983; Fujita *et al.*, 1990). It is significant to note that enterocytes in the neonatal rat small intestine do not express class II MHC molecules (Mayrhofer *et al.*, 1983). This may preclude presentation of processed peptides derived from degraded exogenous antigens to mucosal T lymphocytes during the preclosure period. It may also be significant that this is a period during which oral exposure to antigens does not lead to development of tolerance (Strobel and Ferguson, 1984).

In rodents, uptake of IgG ceases when expression of Fc receptors is down-regulated at approximately 2–3 weeks of age. At the same time, the tubulovesicular system regresses in the apical cytoplasm of enterocytes throughout the small intestine (Kraehenbuhl and Campiche, 1969; Rodewald, 1973) and a similar process occurs in ruminants at the time of closure (Trahair and Robinson, 1989). It is therefore of interest to enquire whether any vestiges of the process of macromolecule absorption, which is so prominent in young animals, persists into adult life. One possibility is that endocytosis remains possible in mature enterocytes, provided a suitable receptor is available (cf. the Fc receptor in neonates).

Some evidence in support of this idea comes from studies described earlier, concerning the uptake of cholera toxin (Kao *et al.*, 1972), the LT toxin of *E. coli* (Linder and Russell-Jones, 1990) and plant lectins (Lorenzsonn and Olsen, 1982). Although some cholera toxin is transported across the cells, perhaps utilizing the coated vesicles which are observed in mature enterocytes

(Rodewald, 1973), endocytosed lectins are delivered principally to apical multivesicular bodies (Neutra *et al.*, 1987). This process is similar to the handling of macromolecules by ileal epithelial cells in the neonatal ileum (Fujita *et al.*, 1990) and, indeed, the effect of lectin binding is to induce formation of apical tubules in mature enterocytes (Lorenzsonn and Olsen, 1982).

Other ligands which illustrate this principle are the intrinsic factor–cobalamin complex and epidermal growth factor. The intrinsic factor–cobalamin receptor is detected in apical multivesicular bodies, suggesting that the absorbed complex is routed to the lysosomal compartment (Levine *et al.*, 1984). Enterocytes have brush border receptors for epidermal growth factor (Thompson, 1987) and at least part of the absorbed ligand is transported intact into the circulation (Gonnella *et al.*, 1987; Thornburg *et al.*, 1987). Finally, there is evidence that brush border hydrolases are endocytosed constitutively and either recycled or delivered to lysosomes (Matter *et al.*, 1990). These examples suggest that there are a number of surface receptors on the brush border which facilitate endocytosis, although there is variation in the relative amounts of bound ligand that are transported or degraded. They also suggest receptors that could be exploited by pathogens, and interesting experimental systems by which antigen conjugates could be targeted for absorption and presentation.

3.1.3 Retrograde Absorption Pathway

It has been known for a number of years that macromolecules are absorbed by enterocytes via the basolateral membranes (Hugon and Borgers, 1968). The physiological significance of the process may be two-fold.

Firstly, it may account in part for the normal process of plasma protein turnover (Ullberg *et al.*, 1960) and, as such, might have no immunological significance. However, if antigen presentation by the gut epithelium were to give a tolerogenic signal to T cells, it could have a major immunological importance in the maintenance of tolerance to circulating proteins.

Secondly, basolateral absorption could provide an important route of antigen elimination from the mucosa. On the one hand, immune complexes containing locally produced polymeric IgA might be actively transported through the cell and discharged into the gut lumen, as described in the liver (Peppard *et al.*, 1981, 1982). This could provide an important anti-inflammatory role for IgA antibodies at times of increased mucosal permeability, especially in species (like humans) that are less reliant on hepatic transport for delivery of secretory antibody to the gut (Underdown and Schiff, 1986). On the other hand, fluid phase antigens might be endocytosed in a pick-a-back fashion in vesicles with receptor-bound IgA but in this case one would predict that its fate might be delivery to the lysosomal compartment (Underdown and Schiff, 1986).

Enterocytes express the polymeric immunoglobulin

receptor (secretory component) on their basolateral membranes. Expression is greatest in crypt cells and diminishes as the cells ascent the villi (Brandtzaeg, 1974). Transport of IgA into the gut secretions involves binding of polymeric IgA to its receptor on the basolateral memrane of the enterocyte and endocytosis via coated pits into coated vesicles (reviewed by Underdown and Schiff, 1986; Hansson and Brandtzoeg, 1989; Kraehenbuhl and Neutra, 1992). The bound IgA is transported across the enterocyte and discharged at the brush border, avoiding fusion with the lysosomal compartment. IgA transport, like the distribution of the polymeric immuno-globulin receptor, occurs mainly in the crypt epithelium and at the bases of the villi.

It is therefore interesting that the uptake of intravenously administered horseradish peroxidase by enterocytes in the rat is most prominent in the crypt epithelium and less in the epithelium as it ascends the villi (Mayrhofer *et al.*, 1990). This suggests that absorption by the retrograde pathway may be linked to the constitutive endocytosis of the polymeric immunoglobulin receptor. By anology with the hepatocyte (Underdown and Schiff, 1986), it might be expected that material co-endocytosed in vesicles with bound IgA would be sorted in the endosomes and routed to the lysosomal compartment. However, crypt enterocytes contain few lysosomes, and the horseradish peroxidase is contained in tubular apical endosomes, where it retains its enzymatic activity for 12–18 h (Mayrhofer *et al.*, 1990). During this time, it is transported by the moving column of enterocytes onto the villus, where loss of enzyme activity appears to coincide with the development of lysosomes (Mayrhofer and Spargo, 1990). It may be significant that the slow degradation of the enzyme ensures that it survives until it reaches the zone at which enterocytes commence expression of class II MHC molecules (Mayrhofer *et al.*, 1983; Mayrhofer and Spargo, 1989).

3.1.4   Confluence of Antegrade and Retrograde Pathways

The two potential routes of antigen absorption by enterocytes are not necessarily separate. In the ileal enterocytes of suckling rats, Fujita *et al.* (1990) have demonstrated that, when distinct macromolecular markers are administered intralumenally and intravenously, they enter separate endosomes located respectively in the apical and basolateral cytoplasm. However, over a period of minutes, both markers occupy the same late endosomes and multivesicular bodies in the apical cytoplasm. Similar observations have been made in the enterocyte-like Caco-2 human cell line (Hughson and Hopkins, 1990). Basally presented radiolabelled transferrin and apically presented concanavalin A enter a common late endosomal system in the apical cytoplasm.

These findings indicate that antigens taken up at either surface of the enterocyte enter late endosomes and that both routes can culminate in delivery of the antigen to

apical multivesicular bodies and perhaps ultimately to classical lysosomes. If these are steps that lead to immunologically significant processing, antigens from either the lumen or the extracellular fluid of the mucosa would be available for presentation to T cells. The retrograde route of uptake would assume importance during episodes of increased epithelial permeability, when antigens from the lumen could traverse the tight junctions and enter the lateral spaces between enterocytes.

3.2   THE SUBSELLULAR DISTRIBUTION OF CLASS II MAJOR HISTOCOMPATIBILITY COMPLEX MOLECULES IN ENTEROCYTES

As mentioned earlier, a variety of epithelial cell-types express class II MHC molecules, either constitutively or in response to hormonal or immunological stimuli (reviewed by Mayrhofer, 1984; Bland, 1988; Spencer and MacDonald, 1990). In the gut in rodents, the epithelium of the normal stomach and large intestine does not express class II MHC molecules. In contrast, the epithelium covering literally the first villi in the duodenum of rats express class II MHC molecules and the level of expression increases from duodenum to ileum. The reasons for these regional differences in expression of class II MHC molecules is not known. The delay in expression until the postweaning period and the absence of expression in antigen-free grafts in rats (Mayrhofer *et al.*, 1983) suggests a possible influence of local antigen but the presence of class II MHC molecules in the small intestine epithelium in athymic rats (Mayrhofer *et al.*, 1983) and its presence in the small intestine in preterm human fetuses (MacDonald *et al.*, 1988) is an indication that expression is in part a constitutive function of the epithelium.

In contrast to the villus epithelium, crypt enterocytes do not express class II MHC molecules constitutively. However, local immune responses generated by graft-versus-host disease (Bland and Whiting, 1992; Mason *et al.*, 1981b) and infection with either *Trichinella spiralis* (Barclay and Mason, 1982) or *Nippostrongylus brasiliensis* (Mayrhofer, personal observation) induce expression of class II MHC molecules in crypt cells, probably by production locally of γ-interferon (Cerf-Bensussan *et al.*, 1984; Steineger *et al.*, 1989; Hughes *et al.*, 1991). In normal conventional and specific pathogen-free rats, the proportion of crypts expressing class II MHC molecules increases in proportion to the distance from the pylorus, and this may reflect the regional distribution of the comensal microflora (Lee, 1985). Aspects of the ontogeny and induction of class II MHC molecules in human gut are discussed by Bland (1988) and by Spencer and MacDonald (1990).

The main subject of this section is the subcellular distribution of class II MHC molecules in enterocytes, in particular in relation to the endocytic pathway and the distribution of endocytosed macromolecules. At the light microscope level, there is a strong impression that class II MHC molecules are present in the cytoplasm of enterocytes as well as associated with cell membranes (Mayrhofer *et al.*, 1983). Furthermore, the distribution of the cytoplasmic material has a distribution that resembles closely the distribution of acid phosphatase (Barka and Anderson, 1962) and lysosomes (Trier, 1968) in the apical cytoplasm of enterocytes.

3.2.1   Surface Expression of Class II Major Histocompatibility Complex Molecules by Enterocytes

There have been relatively few studies on the ultra-structural localization of class II MHC molecules in either epithelial cells, or cells of any other type. However, the studies of Parr and McKenzie (1979) demonstrated that class II MHC molecules are confined to the basolateral membranes of mouse enterocytes. These studies used an immunoferritin technique on fixed cells and probably could not be expected to demonstrate intracellular antigens (Spargo and Mayrhofer, 1990). The surface distribution of class II molecules was confirmed in rat enterocytes at the light microscope level (Mayrhofer *et al.*, 1983; Bland and Warren, 1986a,b; Mayrhofer and Spargo, 1989) and by immuno-electron microscopy (Mayrhofer and Spargo, 1989, 1990). These findings indicate that class II MHC molecules in rodents cannot have a role in binding antigens or peptides in the gut lumen.

In contrast, the studies of Hirata *et al.* (1986) on human small intestine suggested that HLA-DR antigens were present on the brush borders of human enterocytes. This has been confirmed (Mayrhofer and Spargo, 1989; Mayrhofer *et al.*, 1990), indicating that class II MHC molecules are exposed directly to the lumen contents of the human small intestine, as well as being expressed on the ablumenal membranes of the enterocytes. There are several reasons why this might have importance to the function and pathology of gastrointestinal tract. First, as mentioned in Section 2.1.2, lumenal digestion of proteins in the gut could produce immunogenic peptides. It is therefore possible that potentially immunogenic peptides could bind directly to the peptide-binding groove of the MHC molecules (Germain, 1991). Should these MHC molecules cycle from the surface to apical endosomes (see Section 3.1.2) and thus enter the compartment shared by the antegrade and retrograde pathways of antigen absorption (see Section 3.1.4), it is conceivable that the peptide-laden molecules could be relocated to the basolateral cell membrane, where they would be accessible to mucosal T lymphocytes.

Secondly, as an extension of the peptide-binding hypothesis, the restrictions imposed by specific HLA-D region gene products on the precise peptides which are bound (Babbit *et al.*, 1985; Buus *et al.*, 1987) could have important implications in the pathogenesis of food intolerance and in particular, gluten intolerance. Coeliac disease is linked tightly to the possession of certain HLA-D haplotypes – in particular HLA-DR3 and HLA-DR7 – which are themselves in linkage disequilibrium with HLA-DQw2 (Ciclitira and Hall, 1990). It has been suggested that predisposition to coeliac disease is due to the possession of a specific DQα/DQβ heterodimer, which facilitates binding of a gluten peptide(s) and its presentation to T lymphocytes in the mucosa. This hypothesis begs the question that the peptide is actually absorbed and accessible to antigen-presenting cells and to specifically reactive T cells.

In this respect, expression of HLA-D antigens on the brush border of enterocytes assumes added significance, perhaps explaining why humans are almost alone in their susceptibility to gluten intolerance. Gluten is a major dietary protein and its relative resistance to proteolysis (see Section 2.1.2) may ensure a steady supply of peptides to the mucosal surface. Binding of peptides to MHC molecules on the brush border could predispose to mucosal damage in at least three ways. According to the toxic peptide hypothesis (Cornell *et al.*, 1988), mere binding of the toxic fragment and its transport into the endocytic pathway could lead to enterocyte death. Alternatively, the peptide might be transported through the cell and displayed to T lymphocytes in association with the class II MHC molecule on the basolateral surface. However, it is noteworthy that numbers of CD4-positive intraepithelial lymphocytes are only marginally increased in coeliac disease (Savilahti *et al.*, 1990), although they do show signs of activation (Malizia *et al.*, 1985). A third possibility is that the peptide is transported in association with the class II MHC molecule, released from the enterocyte and bound by classic antigen-presenting cells in the lamina propria. In any event, the primary binding of gluten peptides to surface MHC molecules on the brush border would be pivotal in initiating the disease and in accord with the observed HLA-D associations of gluten intolerance (Ciclitira and Hall, 1990).

A third significance of HLA-D antigen expression on the enterocyte brush border relates to its accessibility to certain enterotoxins. The enterotoxins produced by *Staphylococcus aureus* have an emetic action on humans and primates but not on laboratory rodents (reviewed by Wood *et al.*, 1991). It is therefore of interest that these toxins bind specifically to class II MHC molecules (Frazer, 1989; Mollick *et al.*, 1989). It seems possible that the absorption of the toxins, and their species-specific toxicity, might be related to the expression of HLA-D antigens on the lumenal surface of the human small intestine.

3.2.2   Intracellular Class II Major Histocompatibility Complex Molecules

Despite the many advances in the understanding of how antigens are processed and presented to T lymphocytes (Braciale and Braciale, 1991; Brodsky and Guagliardi,

1991), it is only recently that class II MHC molecules have been demonstrated within the endocytic pathway of antigen-presenting cells. Their presence in the endocytic compartment was an essential prediction of the model proposed by Germain (1986) for processing of exogenous antigens, which at that time was based on circumstantial evidence that newly synthesized class II MHC molecules interacted with transferrin *en route* to the surface membrane (Cresswell, 1985).

Morphological evidence for the presence of class II MHC molecules in early endosomes was obtained by Guagliardi *et al.* (1990). The precise compartment in which processing of antigen and association of peptide with class II MHC molecules occurs is controversial (Davidson *et al.*, 1991; Hackett, 1991), but the molecules have now been described in both early (Guagliardi *et al.*, 1990) and late endosomes (Peters *et al.*, 1991) and also in early lysosomes (Peters *et al.*, 1991). The nature of the antigen and its requirements for greater or lesser amounts of processing (e.g. soluble proteins versus complex microbial organisms) may dictate where in the cell processed peptide engages with class II MHC molecules (Harding *et al.*, 1991). Another area of controversy is the degree to which class II MHC molecules are recycled from the cell surface via the endocytic pathway and the significance of this process in antigen presentation (Reid and Watts, 1990; Davidson *et al.*, 1991; Peters *et al.*, 1991). Overall, it appears that both newly synthesized and recycled class II MHC molecules are significant, the relative importance of each pathway perhaps depending on the nature of the cell involved.

With this background, it is interesting that the first morphological evidence for an intracellular distribution of class II MHC molecules in any cell came from studies on enterocytes (Mayrhofer *et al.*, 1983). The original light microscopic observations were confirmed subsequently by the immunohistochemical and cytochemical demonstration that the intracellular distribution of class II MHC molecules overlaps the distribution of acid phosphatase in rat jejunal epithelium (Mayrhofer and Spargo, 1990). Immuno-electron microscopic studies identified class II MHC molecules associated with multivesicular bodies in the apical cytoplasm of human (Mayrhofer and Spargo, 1989, 1990; Mayrhofer *et al.*, 1990) and rat (Mayrhofer and Spargo, 1989, 1990) enterocytes and also in lysosomes and phagolysosomes in the supranuclear cytoplasm. To date, there has been no simultaneous demonstration that antigens endocytosed by either the antegrade or the retrograde pathways associate with organelles containing class II MHC molecules. However, from examples discussed in Sections 3.1.1 to 3.1.3, it seems certain that endocytosed materials will in fact converge with class II MHC molecules in what are believed to be early (multivesicular bodies) and late (dense bodies and phagolysosomes) lysosomal compartments.

The scene is set, therefore, for enterocytes to process endocytosed exogenous antigens and to present them in association with class II MHC molecules to CD4–positive T cells. The processed antigens could originate from either the lumenal or ablumenal surfaces of the enterocytes and could include viral particles, intracellular bacterial pathogens or soluble antigens. Antigen uptake by enterocytes appears to be different qualitatively to uptake by M cells, which contain few lysosomes (Owen *et al.*, 1986) and do not express class II MHC molecules (Mayrhofer *et al.*, 1983; Neutra and Kraehenbuhl, 1992). The latter appear to have a function that is purely transport of antigen, without processing, although this view may require revision in the light of recent evidence that class II MHC molecules may occur in the endosomes of M cells (Trier, 1991; Neutra and Kraehenbuhl, 1992). The potential importance of M cells in vaccine delivery makes this an important issue to resolve, as is the related question of the significance of CD4-positive T cells in follicle-associated epithelium (Bjerke *et al.*, 1988).

3.3 ANTIGEN PRESENTATION BY EPITHELIAL CELLS

Studies on antigen absorption by mucosal surfaces such as the gut must necessarily be done either *in vivo* or in whole organ preparations. Ideally, they should maintain the normal barrier to antigen diffusion by the paracellular route (see Section 2.2.1.2) and therefore the normal polarity of antigen access to the epithelial cell surface. Similarly, most studies of antigen uptake by intestinal epithelial cells have been conducted *in vivo*, where exposure of the lumenal or ablumenal surfaces of enterocytes to antigens can be controlled by the route of administration. This control is lost if there is a breakdown in mucosal permeability due to mechanical or ischaemic damage (see Section 2.2.1.2) and it is only possible *in vitro* if special care is taken to grow epithelial cells as polarized monolayers on suitable membranes. It is not possible at present to create monolayers of this sort from freshly isolated normal gut epithelial cells.

The anatomical and cellular complexity of mucosae, such as those of the gut, makes design of experiments to investigate the antigen-presenting activities of epithelial cells extremely difficult. All studies of antigen presentation by epithelia have therefore been attempted *in vitro*. Implicit in this approach is the likelihood that such antigen as is processed and presented to T cells has been endocytosed via the basolateral cell membrane (retrograde route). Because this pathway is probably not available to fluid phase antigens under normal conditions *in vivo*, the relevance of antigen presentation by epithelial cells may be limited to lumenal antigens such as cholera toxin or lectins (see Section 2.2.2), or to inflamed mucosa where antigens can diffuse through leaky tight junctions. The notion that epithelial presentation of soluble fluid phase antigens might be responsible for the phenomenon of oral tolerance (Bland and Warren 1986b; Mayer and Shlien, 1987; Bland, 1988) is therefore in question, because oral tolerance can be induced in completely normal animals by primary contact with a soluble antigen – conditions where antigen absorption is likely to occur via M cells rather than through the intact epithelium overlying the villous mucosa (see Section 2.2.2).

Nevertheless, *in vitro* studies have shown conclusively that epithelial cells are capable of processing and presenting macromolecular antigens to T lymphocytes. Bland and Warren (1986a,b) showed that freshly isolated MHC class II-positive epithelial cells from the rat small intestine could present soluble ovalbumin to primed syngeneic T cells. Stimulation was antigen-specific and it was partially inhibited by monoclonal anti-I-A and anti-I-E antisera, indicating that the responses were probably restricted by class II MHC molecules. These studies could be criticised on the grounds of purity of the epithelial cells, because only very small numbers of contaminating dendritic cells are necessary to stimulate mitogen-induced or antigen-induced T cell proliferation (Mayrhofer *et al.*, 1986). However, later studies (Bland and Whiting, 1989) excluded contaminating macrophages, although contamination by a small population of dendritic cells cannot be discounted.

Antigen presentation to primed T lymphocytes by enterocytes has been confirmed in human (Mayer and Eisenhardt, 1990) and mouse (Kaiserlain *et al.*, 1989) systems, and in a system using epithelial cells from the human large intestine (Mayer, 1987; Mayer and Shlien, 1987). The studies of Kaiserlain *et al.* (1989) address the question of antigen specificity and MHC restriction of the response to soluble antigen by the use of a T cell hybridoma as the responding cell. In both mouse and human systems, as in the rat system, the response to antigen was specific and inhibited by antibodies directed against relevant class II MHC molecules. In these studies, the question of epithelial cell purity was addressed rigorously. Kaiserlain *et al.* (1989) detected no contamination of their mouse epithelial cell preparation with cells expressing either macrophage or dendritic cell markers. Mayer and Shlien (1987) also monitored purity of epithelial cell preparations and showed further that a colon carcinoma cell line could also present tetanus toxoid to primed T cells. Recently, Pang *et al.* (1990) have shown in the rat that a clonal small intestinal epithelial cell line can also present antigen to primed T cells.

It is now recognized generally that primed T cells are less dependent for reactivation on classical antigen-presenting cells than are virgin T cells. Therefore, it is of interest that gut epithelial cells can stimulate in a primary allogeneic "mixed leucocyte reaction" (Bland, 1987; Mayer and Shlien, 1987). These findings suggest that epithelial cells can produce the co-stimulators (cytokines and/or surface adhesion molecules) required to activate virgin T cells. The first evidence for cytokine production by epithelial cells came from studies on skin epidermal cells, which produced an IL-1-like factor (Luger *et al.*, 1981). Epidermal cells (Luger and Schwarz, 1991) are now known to produce a large range of cytokines (IL-α, IL-1β, IL-3, IL-6, IL-8, TNF-α, GM-CSF, G-CSF, M-CSF, TGFα and TGFβ) and recent studies indicate that the uterine epithelium also produces a number of cytokines under the regulation of ovarian steroids (Robertson *et al.*, 1992). In the gut, enterocytes have been found to produce IL-1 (Bland, 1987; Mayer *et al.*, 1990; Santos *et al.*, 1990), enterocytes and large intestinal epithelial cells to produce IL-6 (Beagley *et al.*, 1990; Mayer *et al.*, 1990; Shirota *et al.*, 1990), and small intestinal crypt cells to produce TGFβ (Barnard *et al.*, 1989). Furthermore, enterocytes also express the surface adhesion molecules ICAM-1 and LFA-1 (Mayer *et al.*, 1990), and may also share other surface molecules with dendritic cells (Kaiserlain *et al.*, 1990). Epithelial cells, therefore, appear equipped to fulfil many of the functions required of professional antigen-presenting cells. Unfortunately, Bland and colleagues have not yet applied their elegant *in vitro* technique (Williams *et al.*, 1992) for generation of primary T cell responses to the specific question of antigen presentation by epithelial cells.

In the above studies, the responder cells have been either primed lymph node cells (Bland and Warren, 1986a,b), peripheral blood lymphocytes (Mayer and Shlien, 1987) or a T cell hybridoma (Kaiserlain *et al.*, 1989). However, the context of antigen presentation in the gut epithelium *in vivo* involves the entirely different IEL population. The majority of these cells express a homodimeric form of CD8, most are CD5-negative, a variable proportion express the γ/δ T cell receptor heterodimer and very few are CD4-positive (reviewed by Cerf-Bensussan and Guy-Grand, 1991). Recent evidence indicates that many of these cells are not thymus dependent and that they develop in an extrathymic environment which has still to be identified (Bandeira *et al.*, 1991; Lake *et al.*, 1991). There have been no studies on antigen presentation to IELs although, for antigen presentation by epithelial cells to have a physiological significance, it would appear that it should be directed towards the IEL population. It is noteworthy, therefore, that γ/δ T lymphocytes, most of which are either negative for both CD4 and CD8 or express CD8, may not be restricted by classic MHC gene products. Recent evidence suggests that they may be restricted by non-classic MHC class I-like molecules, such as CD1 (reviewed by Calabi and Bradbury, 1991).

These considerations do not sit comfortably with the published data. The groups of Bland (Bland and Warren, 1986a,b; Bland, 1987) and Mayer (Mayer and Shlien, 1987; Mayer and Eisenhardt, 1990) have both observed that antigen presentation by epithelial cells leads to enrichment of CD8-positive cells in the responding population. The evidence that this is due to selective proliferation of CD8-positive cells is not definitive, but it suggests that there are qualitative differences between epithelial cells and adherent mononuclear cells as antigen-presenting cells. Furthermore, the observation that proliferation is inhibited by antibodies directed against class II MHC molecules does not fit with established interactions of either class II MHC molecules or of CD8 molecules. Human enterocytes do not express all class II MHC gene products equally (Pallone *et al.*, 1988; Volk *et al.*, 1989; Mayer *et al.*, 1991). These findings do not resolve the above paradox but they do suggest that enterocytes may have different properties to classic antigen-presenting cells.

The situation is further complicated by the undoubted response of a CD4-restricted T cell hybridoma to antigen presented by MHC class II-positive mouse enterocytes (Kaiserlain *et al.*, 1989), by the failure of MHC class II-positive rat crypt enterocytes to stimulate primed T cells (Bland and Whiting, 1989), and by the observation that antigen presentation by epithelial cells from patients suffering from chronic inflammatory bowel disease (but not from normal bowel) enhances the numbers of CD4-positive cells in the responding population (Mayer and Eisenhardt, 1990). There is clearly much to be learned about antigen processing by immature enterocytes (see Sections 3.1.1 and 3.1.3), about the nature of other MHC and MHC-like molecules expressed by epithelial cells under different conditions, about the nature of the receptor for the homodimeric form of CD8 and about the influence of the source of responder T cells. It must be questioned seriously whether classical class II MHC molecules are actually the restriction molecules in antigen presentation by epithelial cells. Their expression may simply be coordinate with the expression of other MHC-like molecules, which are themselves the definitive restriction elements in responses by T cells expressing the homodimeric CD8 molecule. These are interesting and important areas, with implications for vaccine development and for understanding both immunologically mediated mucosal inflammatory diseases and autoimmunity involving epithelial cells.

Finally, the significance *in vivo* of antigen presentation by epithelial cells is unknown. There are interesting correlations between the onset of expression of class II MHC molecules by small intestinal epithelium in young mice and rats and the ability to induce tolerance by oral administration of antigens (reviewed by Bland and Kambarage, 1991). These observations are consistent with a link between expression of class II MHC molecules and the induction of tolerance. On the other hand, infusion of interferon γ into rats has been observed to upregulate class II MHC molecule expression in the gut and to prevent induction of oral tolerance (Zhang and Michael, 1990). Clearly, the interpretation of the *in vivo* observations is complicated. It is therefore of interest that the CD8-positive cells generated by presentation of antigen by epithelial cells appear to have suppressive effects *in vitro* on T cell proliferation, provision of T cell help and differentiation of B lymphocytes (Bland and Warren, 1986a,b; Mayer and Shlien, 1987; Pang *et al.*,

1990). In the rat system, this suppression appears to be antigen-specific (Bland and Warren 1986b) but in the human system it did not exhibit antigen specificity (Mayer and Shlien, 1987). Furthermore, rat enterocyte preparations are themselves non-specifically suppressive for T lymphocyte proliferation (Bland and Warren, 1986a,b; Pang *et al.*, 1990), but this was not reported for the human system.

These suppressive effects have been interpreted in the context of the phenomenon of oral tolerance, but it must be remembered that antigen entry to the normal mucosa may be limited to Peyer's patches. Furthermore, suppressive phenomena mediated by secretion of inhibitory cytokines such as interferon γ by T lymphocytes are not incompatible with other functions, such as cytotoxicity. There is a clear need for studies on antigen presentation to IELs by epithelial cells and for the development of assays (*in vivo* and/or *in vitro*) with which to examine relevant IEL functions. In particular, the development of genetic (immunogenetic, transgenic) approaches to the study of antigen presentation by epithelial cells *in vivo* appears to offer the only way of unravelling this complex issue.

4. Classical Antigen-Presenting Cells Associated with Epithelia

The main subject of this chapter is the relationship between antigens and the epithelial cells which cover cutaneous and mucosal surfaces of the body. However, having passed the epithelial barrier (by either passive or active mechanisms), antigens then have access to lymphocytes via potential antigen-presenting cells of the classic sort. These are located: either within the epithelium; in the dermis (skin) or lamina propria (mucosae); in mucosal lymphoid tissues; or at more distant sites where antigen-laden cells may be carried by the lymphatic and/or vascular circulations. Such cells are broadly of the macrophage or dendritic cell lineages and their pathophysiology and importance in the induction and regulation of immune responses at epithelial surfaces is the subject of an extensive literature which is beyond the scope of the present discussion. The general subject of dendritic cells has been reviewed recently (Austyn, 1987, 1989; King and Katz, 1990; Steinman, 1991) and Austyn and Larsen (1990) have provided an excellent overview of the migratory properties of dendritic cells. Langerhans cells in squamous epithelia are treated comprehensively in a fine recent multi-author volume edited by Schuler (1991).

This section will not attempt a comprehensive review of all aspects of antigen-presenting cells in tissues associated with epithelial surfaces. It will attempt only to highlight topical issues and to examine some areas of controversy.

4.1 THE GASTROINTESTINAL TRACT

The distribution of potential antigen-presenting cells in the lamina propria of the gut mucosa and those in the organized lymphoid structures are best considered separately. The subject has been reviewed in some detail in recent years (Mayrhofer, 1984; Tanner *et al.*, 1984; Bland and Kambarage, 1991), and the review of Le Fevre *et al.* (1979) is a valuable synopsis of earlier work. Many questions raised by the earlier workers remain unanswered because of the difficulties in obtaining preparations of isolated macrophages/dendritic cells from the gut mucosa. Recent advances stem from the work of Bull and Bookman (Bull and Bookman, 1977, Bookman and Bull, 1979), who pioneered the isolation of viable lymphoid cells from intestinal mucosa and later from the development of monoclonal antibodies which detect specific macrophage and dendritic cell markers.

4.1.1 The Lamina Propria

There is little doubt that the lamina propria of both the small and large bowel contains a population of macrophages (LeFevre *et al.*, 1979). The earlier studies employing enzyme histochemistry demonstrated acid phosphatase-containing cells in the lamina propria of the small intestine in guinea-pigs, rats, humans and mice (Sawicki *et al.*, 1977) and the large intestine of guinea-pigs (Abraham *et al.*, 1974). Phagocytosis of invading bacteria *in situ* by macrophages has been observed in the lamina propria of mice (Takeuchi, 1971) and monkeys (Takeuchi, 1971; Takeuchi *et al.*, 1968). Furthermore, phagocytic adherent mononuclear cells have been isolated from gut mucosa by enzymatic digestion techniques (Bull and Bookman, 1977). Although many other studies report the presence of cells described as macrophages on morphological grounds, it is necessary for reasons discussed below to identify the cells positively by the use of macrophage-specific markers or functional characteristics.

These observations are supported by more recent work, using other markers to identify macrophages. In brief, macrophages have been identified in the lamina propria of small intestine from mice (Hume *et al.*, 1984; Soesatyo *et al.*, 1990), rats (Robinson *et al.*, 1986; Sminia and Jeurissen, 1986; Sminia and Van der Ende, 1989; Sminia *et al.*, 1990) and humans (Bjerke and Brandtzaeg, 1990) by the use of macrophage-specific monoclonal antibodies, and have been purified from mouse small intestine (Pavli *et al.*, 1990). Cells with macrophage morphology and expressing macrophage markers have also been isolated from the mucosa of human large intestine (Pavli *et al.*, 1989; Pavli, 1990).

Large mononuclear cells, identified loosely as either macrophages or dendritic cells, have also been identified in human (Janossy *et al.*, 1980; Selby *et al.*, 1983) and rat (Mayrhofer *et al.*, 1983) small intestine stained with monoclonal antibodies directed against class II MHC antigens. In rats, these cells also express CD4 molecules (Mayrhofer *et al.*, 1983) but their nature is uncertain, because class II MHC molecules and CD4 molecules are shared by macrophages and dendritic cells. As discussed by Mayrhofer (1984) and by Bland and Kambarage (1991), a case can be made on ontogenetic grounds that dendritic cells account for at least some of the class II-positive cells in the lamina propria of rat gut. Cells with dendritic appearance that express class II MHC molecules appear in the fetal gut before it is colonized by lymphocytes (Mayrhofer *et al.*, 1983; Wilders *et al.*, 1983a; Van Rees *et al.*, 1988). These cells express class II MHC molecules constitutively and do not express either rat macrophage markers (Van Rees *et al.*, 1988) or detectable CD4 (Mayrhofer *et al.*, 1983), and it has been suggested for these reasons that they may belong to the dendritic lineage. This phenotype is replaced in the adult rat by a class II-positive population which expresses macrophage markers (Mayrhofer *et al.*, 1983; Sminia and Jeurissen, 1986; Van Rees *et al.*, 1988), and similar changes have been observed in the maturation of human gut (Janossy *et al.*, 1986; Harvey *et al.*, 1990). It is not clear whether the observed changes reflect a maturation process in a single lineage of cells (macrophages) or whether dendritic cells are displaced in large measure by infiltration of the mucosa by macrophages. If the latter is true, it is then of interest to know whether the putative dendritic cell population maintains a presence in the mature intestine.

Recent evidence indicates that dendritic cells are indeed present in the lamina propria of the small intestine. Although outnumbered by class II-positive macrophages, cells expressing both class II MHC molecules and dendritic cell markers have been described in the villous mucosa of adult mice (Soesatyo *et al.*, 1990) and humans (Mahida *et al.*, 1989; Pavli *et al.*, 1989; Bjerke and Brandtzaeg, 1990; Harvey *et al.*, 1990; Pavli, 1990). Pavli *et al.* (1990) have isolated a dendritic cell population from mouse small intestinal mucosa and characterized it morphologically, by surface antigen phenotype and by its functional characteristics. The cells were smaller than lamina propria macrophages, expressed high levels of class II MHC molecules, were positive for the mouse dendritic cell-specific monoclonal antibody 33D1 and also expressed low levels of the antigen detected by monoclonal antibody F4/80, a macrophage-specific marker. The latter observation is interesting, in view of the phagocytic history apparent in many gut-derived veiled cells in rat afferent intestinal lymph (Mayrhofer *et al.*, 1983; Pugh *et al.*, 1983) and the similarity between mucosal dendritic cells and Langerhans cells (see below) with respect to expression of the F4/80 antigen. The relationship of these cells to lymphoid dendritic cells (Phillip and Katz, 1990; Steinman, 1991) is uncertain, as is the relationship between macrophages and dendritic cells.

The function of mouse mucosal dendritic cells as stimulators in the MLR has been examined by Pavli *et al.* (1990). Purified dendritic cells were stimulatory in the MLR, whereas lamina propria macrophages were poor stimulators and were responsible for indomethacin-sensitive suppression of the stimulatory activity of unfractionated lamina propria cells. These findings appear to confirm the suggestion by Bjerke and Brandtzaeg (1990) that the excess of macrophages relative to dendritic cells in the mucosa may function to suppress the level of antigen presentation to local T cells in the lamina propria. A similar function has been proposed for macrophages in the lung parenchyma (Holt *et al.*, 1985; Holt *et al.*, 1988), as discussed in Section 4.2. The relevance of these findings may lie on the one hand in the need for local antigen-dependent expansion of B lymphoblasts in the lamina propria (Husband and Gowans, 1978) and, on the other, the need to control local T lymphocyte-mediated mucosal inflammation. Furthermore, it may delay expression of the full antigen-presenting activities of antigen-laden dendritic cells until after they have been safely released into the afferent lymph, on their way to the regional lymph nodes (Mayrhofer *et al.*, 1986). In a dual-lineage model, macrophages would suppress local activation of T lymphocytes by dendritic cells in the lamina propria. Alternatively, in a single-lineage model, macrophages might be immunosuppressive during their phagocytic phase but they could become less suppressive and more stimulatory as they differentiate towards dendritic cells. In the latter model, the dendritic cells identified in the mucosa would be mature forms, about to enter the lymphatics and travel as veiled cells to the mesenteric lymph nodes.

4.1.2   Peyer's Patches

The history relating to antigen-presenting cells in Peyer's patches is chequered. It was reported by Kagnoff and Campbell (1974) that Peyer's patches were defficient in antigen-presenting cells, but later studies indicated that the non-enzymatic methods used to prepare the Peyer's patch lymphoid cells failed to mobilize an undoubted antigen-presenting population (Richman *et al.*, 1981; MacDonald and Carter, 1982). The cells in the later studies were described as macrophages and other studies have identified macrophages in the dome region of the Peyer's patch by morphological and functional criteria (reviewed by Mayrhofer, 1984). However, in retrospect and in view of recent functional studies by Pavli *et al.* (1990), it must be questioned whether the observed antigen-presenting activities attributed to macrophages were not in fact due to a co-purified population of dendritic cells. The function of the macrophage population may be related more to the antibacterial defences in Peyer's patches, which are at risk because of the ability of M cells to transport potential pathogens (see Section 2.2.2.3).

Peyer's patches certainly contain a rich population of dendritic cells, most of them located in the thymus-dependent interfollicular and dome regions (Parrott and Ferguson, 1974), where they appear as classic class II-positive interdigitating reticulum cells (Mayrhofer *et al.*, 1983; Wilders *et al.*, 1983c). These cells presumably correspond to class II-positive cells with prominent cytoplasmic veils that are released by enzymatic dissociation of Peyer's patches (Wilders *et al.*, 1983b). Another prominent feature of Peyer's patches is the presence of large class II-positive cells in the otherwise class II-negative follicle-associated epithelium (Mayrhofer *et al.*, 1983; Wilders *et al.*, 1983c). These cells probably lie within the pocket formed by the attenuated cytoplasm of M cells and may correspond to macrophages or dendritic cells described in this location in ultrastructural studies (Abe and Ito, 1978; Lause and Bockman, 1981; Owen *et al.*, 1981; Wilders *et al.*, 1983c; Kimura, 1977; Jarry *et al.*, 1989). Wilders *et al.* (1983c) have suggested that these intraepithelial dendritic cells may migrate from the follicle-associated epithelium to settle in the interfolluclar areas and dome as interdigitating cells. They would therefore have an antigen-transporting function analogous to veiled cells in the afferent lymph entering lymph nodes. The intraepithelial dendritic cells/macrophages in Peyer's patch epithelium are unique in the gut – with the single exception of structures in the rat small intestine which we have named lymphocyte-filled villi (Mayrhofer and Bui, unpublished observations). These structures appear to have an antigen-sampling epithelium and, like Peyer's patch follicle-associated epithelium, contain numerous class II-positive dendritic cells. Lymphocyte-filled villi contain an unusual population of lymphocytes, and their function is at present unknown.

The presence of dendritic cells in Peyer's patches of humans (Bjerke and Brandtzaeg, 1990) and mice (Soesatyo *et al.*, 1990) has been confirmed recently by the use of specific monoclonal antibodies. Numerical predominance of dendritic cells over macrophages in the Peyer's patch dome (the opposite to the situation in the mucosa) may favour effective presentation of antigen (Bjerke and Brandtzaeg, 1990). Pavli *et al.* (1990) did not in fact obtain macrophages from dissociated Peyer's patches but they and others (Spalding *et al.*, 1983) have obtained dendritic cells which were effective stimulators in the MLR. Peyer's patch dendritic cells also present exogenous antigens to T cells and appear to stimulate IgA responses selectively (Spalding and Griffin, 1986; Strober, 1990). They may therefore contribute to a unique microenvironment in Peyer's patch which is conducive to generation of secretory antibody responses.

4.1.3   The Afferent Lymph and Regional Lymph Nodes

Non-lymphoid cells described variously as macrophages (Hall and Morris, 1965), mononuclear cells (Kelly *et al.*, 1978), dendritic cells (Mason *et al.*, 1981a; MacPherson and Pugh, 1984) and veiled cells (Drexhage *et al.*, 1979) have been described in afferent lymph, including that draining from the intestine (Steer, 1980). The descriptive name "veiled cell" is now generally applied, although a small proportion of the cells are probably classic macrophages (Drexhage *et al.*, 1979). Veiled cells in non-intestinal afferent lymph will be dealt with in Section 4.3. The development of a technique to collect afferent intestinal lymph from rats and the observation that it contained a non-lymphoid cell component (Pugh *et al.*, 1983) lead to an important series of experiments on the function and characteristics of gut-derived dendritic (veiled) cells.

Thoracic duct lymph (and mesenteric lymph) from mesenteric lymphadenectomized rats contains a population of veiled cells which is essentially absent in normal thoracic duct lymph (MacPherson and Steer, 1979; Pugh *et al.*, 1983) or efferent mesenteric lymph (Mayrhofer, personal observation). The precise mucosal compartment from which these cells arise is unknown. It has been claimed that dendritic cells are only found in Peyer's patches in rats (Wilders *et al.*, 1983a; Wilders *et al.*, 1983b), but direct observations on cells in individual lacteals indicates that veiled cells may also arise from segments of gut that do not contain macroscopically visible organized lymphoid tissue (Steer, 1980). The cells are poorly adherent, non-phagocytic and express a high density of MHC class II molecules (Mason *et al.*, 1981a; Pugh *et al.*, 1983; Mayrhofer *et al.*, 1986). They stimulate strongly in the primary MLR (Mason *et al.*, 1981a), induce strong alloreactive transplantation immunity *in vivo* (Lechler and Batchelor, 1982) and are potent in presenting soluble antigens to primed T cells (Mayrhofer *et al.*, 1986). Recent evidence indicates that veiled cells transport antigen from the gut (as above) and that they can present gut-derived antigens to primed T cells *in vitro* (Liu and MacPherson, 1991). It is presumed that these cells are *en route* to the regional lymph nodes, where it appears that they are cleared from the lymph (Pugh *et al.*, 1983) and take up residence in the paracortex as interdigitating reticulum cells (Fossum, 1988). These antigen-laden cells are different from those described by Bell (Bell and Botham, 1982; Bell, 1984) in thoracic duct lymph after intraperitoneal immunization of rats, both in expression of class II MHC molecules and in tissue-homing preference (Austyn and Larsen, 1990). The latter are probably neither veiled cells nor derived from the gut.

The cellular origin of veiled cells in afferent intestinal lymph is unknown, although the cells have some features in common with veiled cells from cutaneous sites and with cultured Langerhans cells (Romani *et al.*, 1991a). They are non-phagocytic, express Fc receptors and complement receptors weakly, and most are CD4-negative. They are therefore unlike macrophages and the class II MHC-positive lamina propria cells in the gut mucosa which express CD4 (Mayrhofer *et al.*, 1983). However, the occurrence of DNA-containing bodies (Pugh *et al.*, 1983), large phagolysosomes (Mayrhofer *et al.*, 1990) and uptake of invasive enteric bacteria (Mayrhofer *et al.*, 1986) suggests that, like Langerhans cells, these cells have had a phagocytic history prior to entering the lymphatics. In this respect, it is of interest to note that lamina propria "macrophages" in rat small

intestinal mucosa have been observed to contain DNA-containing material thought to derive from ingested effete lamina propria cells (Sawicki *et al.*, 1977).

It appears, therefore, that antigen can be transported from the gut to the regional lymph nodes by antigen-presenting cells. Although the gut has no clear equivalent of the epidermal Langerhans cell (except in the oesophagus), it is possible that lamina propria dendritic cells function as antigen-collecting cells in much the same way. However, unlike resident Langerhans cells (Romani *et al.*, 1991b), which are poor stimulators unless matured *in vitro*, lamina propria dendritic cells are potent stimulators in the MLR (Pavli *et al.*, 1990). It appears, therefore, that intestinal dendritic cells have already developed an antigen-presenting function before entering the lymphatics, although it is difficult to exclude the possibility that lamina propria preparations could have been contaminated with interdigitating cells derived from isolated lymphoid follicles that are scattered throughout the mucosa of the mouse small intestine. The functional significance of antigen presentation by these cells when they arrive in the mesenteric lymph nodes, versus presentation locally in Peyer's patches, is unknown, although it can be argued that the former is a major route for priming against gut antigens (Mayrhofer, 1984).

4.2   THE RESPIRATORY TRACT

Perhaps the obvious candidates for antigen-presenting cells in the lungs are the plentiful alveolar macrophages, which have first access to inhaled antigens and infectious agents. Antigen presentation by alveolar macrophages could be imagined to occur either locally in the alveoli and airways (probably involving memory or effector T cells) or within bronchus-associated lymphoid tissue or draining lymph nodes (perhaps involving virgin T cells). Evidence that alveolar macrophages can function as antigen-presenting cells *in situ* comes from studies with lavaged cells, using both antigen- and mitogen-induced T cell proliferation *in vitro* (Laughter *et al.*, 1977; Lipscomb *et al.*, 1981). It has also been shown definitively that antigen-pulsed adherence-purified alveolar macrophages can induce specific T cell sensitization after intra-tracheal instillation in guinea-pigs (Lyons and Lipscomb, 1983). Although sensitization might have occurred in organized lymphoid tissue rather than in the lung parenchyma in this experiment, further studies indicated that antigen-pulsed macrophages could also induce preferential recruitment of antigen-specific T lymphoblasts by the lung, suggesting that they can indeed have a local antigen-presenting function.

There is also evidence that alveolar macrophages can re-enter the parenchyma and/or bronchial mucosa and reach the draining lymph nodes. Installation of red and green fluorescent latex microspheres into adjacent segments of the lungs of dogs has shown that the microspheres are phagocytosed locally and transported by macrophages to the tracheobronchial lymph nodes (Harmsen *et al.*, 1985). Furthermore, when alveolar macrophages were labelled *in vitro* with microspheres and instilled into the lungs, they were also recovered intact in the draining lymph nodes. Similar results have also been obtained in guinea-pigs by tracing the migration of isotopically labelled alveolar macrophages (Corry *et al.*, 1984).

These studies support a role for alveolar macrophages as local antigen-presenting cells in the airways and as antigen-transporting cells which deliver antigen to the draining lymph nodes. However, the majority of alveolar macrophages, together with ingested antigens, are removed from the lung by the mucociliary escalator (Spritzer *et al.*, 1968; Brain, 1970). Furthermore, it is difficult to exclude the possibility that the stimulatory properties attributed to alveolar macrophages are not due to the presence of small numbers of highly potent dendritic cells, as these cells are difficult to separate by adherence techniques (Mayrhofer *et al.*, 1986). A counter argument can be made that alveolar macrophages in some species (e.g. mice and rats) are highly suppressive of mitogen-induced and antigen-induced T cell activation (Holt and Batty, 1980) and that depletion of the cells *in vivo* enhances immune responses to intratracheal antigens (Thepen *et al.*, 1989).

On the basis of these and other findings, Holt *et al.* (1990a) has argued that an important immunological role of alveolar macrophages under normal circumstances is the suppression of T cell activation in the lung parenchyma, although it has not been excluded that they may have a significant role in antigen transport and initiation of immune responses in draining lymph nodes. It has been further argued that local antigen presentation to T cells in the parenchyma involves local dendritic cells in the alveolar septa of rat (Holt *et al.*, 1985; Holt *et al.*, 1988; Rochester *et al.*, 1988), mouse (Sertl *et al.*, 1986) and human (Nicod *et al.*, 1987) lung, and that cells of this type may be recruited into the alveoli by inflammatory stimuli (Holt *et al.*, 1990a). There is a need for studies on antigen-transporting cells and veiled cells in the pulmonary lymph of large animals, where the afferent lymphatics are accessible surgically.

Within the airways, interdigitating reticulum cells are present within the dome and T-dependent areas of the bronchus-associated lymph nodules (Simecka *et al.*, 1986). A particularly striking feature of the epithelium in the larger airways is the presence of an extensive network of intraepithelial dendritic cells that express large amounts of class II MHC molecules (Holt and Schon-Hegrad, 1988; Holt *et al.*, 1988; Holt *et al.*, 1990b). In appearance, the cells resemble Langerhans cells and they are able to bind antigens administered by aerosol *in vivo*, process them and present them to primed T cells *in vitro* (Holt *et al.*, 1988). It is probable that their function is to carry antigens via the afferent lymph to regional lymph nodes, as observed in the case of Langerhans cells and intestinal dendritic cells (see Section 4.1.3 and 4.3). The

presence of Langerhans-like cells in respiratory epithelium (but not gut epithelium) may be related to the ability of this surface to undergo squamous metaplasia under certain circumstances.

4.3   THE SKIN AND OTHER SQUAMOUS EPITHELIA

The skin, of all epithelial surfaces, contains the best characterized population of classic antigen-presenting cells. The specialized antigen-presenting cells of the epidermis, the Langerhans cells, have been reviewed extensively (Stingl *et al.*, 1980; Rowden, 1981; Wolff and Stingl, 1983; Breathnach, 1988), and the reader is referred to a recent authoratative monograph for an up-to-date account of these important cells (Schuler, 1991). Langerhans cells constitute only approximately 2% of epidermal cells but their density in human skin ranges from 460 to 1000/mm2 (Rowden, 1981). By the nature of their extensive dendritic projections, they form a network which contacts most keratinocytes in the suprabasal stratum spinosum of the epidermis. In this position, they are placed strategically to be exposed to antigens or haptens which diffuse, invade or are introduced through the highly water-impermeable statum corneum and they are also accessible to fluid phase antigens which diffuse from the dermis (see Section 2.2.1.1). Similar cells have also been described in a number of squamous epithelia, including those of the buccal cavity, nasopharynx, tonsil, oesophagus, vagina and conjunctiva (Breathnach, 1991). Other cells with dendritic morphology are found in the dermis but they are less well characterized than Langerhans cells.

There has been considerable debate about the origin of Langerhans cells, particularly with respect to any relationship that they may have with the monocyte/macrophage lineage (reviewed by King and Katz, 1990; Breathnach, 1991; Romani *et al.*, 1991a,b). This has centred on the phagocytic capacity of Langerhans cells, the co-expression of surface markers by macrophages and dendritic cells, the application of dendritic cell-specific surface markers to the two cell types and their functional characteristics as antigen-presenting cells. It is now becoming clear that resident Langerhans cells are but a stage in the maturation of dendritic cells and that the microenvironment of the squamous epithelium may induce certain characteristics which are not seen in dendritic cells in other tissue locations. There is little doubt that Langerhans cells have some endocytic activity (Falck *et al.*, 1985; Wolff and Schreiner, 1970; Wolff and Hönigsmann, 1971; Evans *et al.*, 1984; Parr *et al.*, 1986), and that they can endocytose ligands that bind to certain surface glycoproteins (Hanau *et al.*, 1987; Hanua *et al.*, 1988). Recent evidence suggests that Langerhans cells may exhibit specialized mechanisms of endocytosis, involving formation of the characteristic Birbeck granules (Schuler *et al.*, 1991). Nevertheless, under some circumstances it appears that Langerhans can be frankly phagocytic (Parr *et al.*, 1991a) and that they can engulf apoptotic cells, a property that they may share with the lamina propria precursors of the veiled cells found in afferent intestinal lymph (see Section 4.1.3).

The trend of opinion recently has been in favour of separate macrophage and Langerhans cell lineages. This view has been supported by the constitutive expression of class II molecules by Langerhans cells, their unique Birbeck granules, the expression of specific dendritic cell-associated antigens, lack of expression of many (though not all) macrophage markers, their responsiveness to GM-CSF but not to M-CSF, and their weak phagocytic activities (reviewed by Breathnach, 1991; Romani *et al.*, 1991a,b). However, recent evidence suggests that Langerhans cells (and perhaps all dendritic cells) have their origin in a common progenitor with macrophages. Reid *et al.* (1990) have cultured mixed macrophage–dendritic cell colonies from both bone marrow and peripheral blood leucocytes of humans and have shown that the dendritic cells had the phenotype HLA-DR$^+$, HLA-DQ$^+$, CD4$^+$, CD1a$^+$ and were negative for a macrophage cytoplasmic marker (Y1/82A), while the macrophages in the colonies were HLA-DR$^+$, HLA-DQ$^-$, CD4$^+$, CD1a$^-$ and Y1/82A$^+$. These findings suggest a relatively late separation of the two lineages, with circulating Langerhans cell precursors perhaps still having the potential to differentiate in either direction. It is not clear whether the circulating dendritic cell progenitors (DL-CFU) described by Reid *et al.* (1990) are the CD1$^+$ cells described by others in human blood (Gothelf *et al.*, 1986a; Gothelf *et al.*, 1988) and human bone marrow (Gothelf *et al.*, 1986b).

In the context of this chapter, it is the functional characteristics of the resident Langerhans cells in the epidermis that are important. At this stage in their life history, Langerhans cells are class II-positive and express CD1a in humans, while a small subpopulation also expresses CD1c. They also express Fc receptors and C3bi receptors and in the mouse they stain with the macrophage-specific monoclonal antibody F4/80 (reviewed by Breathnach, 1991; Romani *et al.*, 1991a,b), perhaps reflecting their origin from a circulating cell with dual potential. However, in humans, Langerhans cells do not express the majority of macrophage-specific antigens, and classic macrophages are absent from the epidermis, although large numbers are present in the underlying epidermis (Weber-Matthiesen and Sterry, 1990). This anatomical separation of Langerhans cells and macrophages differs from the situation in the gut mucosa (see Section 4.1.1) and in the lung (see Section 4.2) and may have important implications for the regulation of antigen presentation in the skin (see Section 4.4).

Although resident Langerhans cells are able to take up fluid phase antigens (see above) and bind reactive haptenic molecules, freshly isolated mouse Langerhans cells are poor stimulatory cells in the primary MLR (Schuler and Steinman, 1985), oxidative mitogenesis (as before), concanavalin A mitogenesis (Romani *et al.*, 1989b; Romani *et al.*, 1991b) and anti-CD3 mitogenesis (Romani *et al.*, 1989b) and are also relatively inefficient in priming helper T cells *in vitro* (Inaba *et al.*, 1986). In contrast (Romani *et al.*, 1989c), freshly isolated mouse Langerhans cells present soluble proteins efficiently to antigen-specific T cell clones (i.e. activated T cells), suggesting that resident Langerhans cells are able to process macro-molecular antigens but that they lack the ability at this stage in their differentiation to provide the accessory signals required to activate resting T cells. In humans, the situation may be different, in that freshly isolated Langerhans cells can present processed peptides but not intact protein antigens to T cell clones (Tiegs *et al.*, 1990). This suggests that, unlike mouse resident Langerhans cells, the human counterparts are inefficient at antigen processing.

The rate of turnover of Langerhans cells in the skin is uncertain, but some cells may reside in transplanted skin for many weeks. However, it is clear that the numbers of Langerhans cells and the rate of turnover is subject to local stimuli such as application of contact sensitizers and stripping of the epidermis by repeated applications of adhesive tape (Breathnach, 1991). Langerhans cells are observed in tissue sections crossing the dermoepidermal junction, are present in small numbers in the dermis and have been described in dermal lymphatics. Furthermore, afferent lymph from cutaneous areas contains veiled cells (some of which contain Birbeck granules), and these are thought to enter the paracortex of draining lymph nodes, where they are described as interdigitating reticulum cells (reviewed by Austyn and Larsen, 1990; Breathnach, 1991; Romani *et al.*, 1991a; Schuler *et al.*, 1991; Steinman, 1991). There is little doubt that veiled cells arising in skin can transport contact sensitizers (Macatonia *et al.*, 1987; Søeberg *et al.*, 1978) and protein antigens (Silberberg-Sinakin *et al.*, 1976) to regional lymph nodes, as described for veiled cells elsewhere (Mayrhofer *et al.*, 1986; Bujdoso *et al.*, 1989; Liu and MacPherson, 1991).

Questions of major interest are the immunological significance of the poor antigen-processing/presenting capacities of resident Langerhans cells, the immunological potential of the emigrant cells from the skin and the stimuli which cause Langerhans cells to migrate from the epidermis and thence into the draining lymphatics. The function of resident Langerhans cells will be addressed in Section 4.4. However, it is noteworthy that resident Langerhans cells are not homogeneous, either in the numbers of Birbeck granules that they contain (Schuler *et al.*, 1991), the amounts of class II MHC molecules that they express (Dezutter-Dambuyant *et al.*, 1984; Romani *et al.*, 1985; Romani *et al.*, 1989a; Romani *et al.*, 1991a), level of CD1c expression (Romani *et al.*, 1991a), expression of some macrophage markers (Romani *et al.*, 1991a) including the mouse marker F4/80, and expression of Fc receptors (Romani *et al.*, 1989a; Schmitt *et al.*, 1990). A minority of cells are indeterminate (contain few Birbeck granules), expresses high levels of class II MHC molecules and has the phenotype class II[high], F4/80[−] in the mouse (Romani *et al.*, 1991a). Cells expressing high levels of class II MHC molecules are located mainly near the dermoepidermal junction (Romani *et al.*, 1989a), and appear more active as stimulators in the MLR than the general resident Langerhans cell population (Romani *et al.*, 1991a).

These findings are of particular interest because a case can be made that cells with up-regulated expression of class II MHC molecules and CD1c, down-regulated expression of Fc receptors and F4/80 antigen, and fewer Birbeck granules represent the most mature Langerhans cells. *In vitro* culture of resident Langerhans cells gives rise over several days to a population of cells with a similar phenotype (Romani *et al.*, 1989a; Romani *et al.*, 1991a,b), and these cells in turn resemble the veiled cells and interdigitating reticulum cells found *in vivo*. The evidence is therefore suggestive that the epidermis contains a population of mature Langerhans cells which migrate into the dermis and leave the skin via the afferent lymph. The stimulus responsible for maturation may be GM-CSF (Witmer-Pack *et al.*, 1987; Heufleur *et al.*, 1988) or IL-1 (Heufleur *et al.*, 1988), both of which could be produced locally by keratinocytes (Luger and Schwarz, 1991).

Veiled cells in afferent peripheral lymph (Knight *et al.*,1982) and afferent intestinal lymph (Mason *et al.*, 1981a; Mayrhofer *et al.*, 1986) are extremely potent antigen-presenting cells, both in stimulating a primary MLR or a secondary *in vitro* response to protein antigens. They appear, therefore, capable of both antigen processing and provision of accessory stimulating signals to T cells. However, the preparations of cells that have been used were not highly purified and there must be some reservations with respect to the ability of veiled cells to take up and process protein antigens. This caveat is important, because it has been shown by others that, after "maturation" *in vitro*, both Langerhans cells (Romani *et al.*, 1989a; Romani *et al.*, 1991a) and splenic dendritic cells (Inaba *et al.*, 1990) fail to present intact antigens, while retaining an extremely potent capacity to present preprocessed antigens. At what point in the journey between skin and regional lymph nodes the cells lose their capacity to process and present protein antigens is unknown, although it appears that at least some lymphoid dendritic cells are active when isolated freshly (see Section 4.4 for further discussion).

The stimuli that trigger emigration of Langerhans cells from the epidermis are unknown, although up-regulation of cell interaction molecules such as ICAM-1 (Romani *et al.*, 1991a) and CD11c (Weber-Matthiesen and Sterry, 1990) could facilitate the process. Traffic of veiled cells from skin after exposure to contact sensitizing agents is increased (Macatonia *et al.*, 1986), suggesting that traffic of Langerhans cells is controlled by immuno-

logical mechanisms which recognize altered-self. This in turn implies interaction of Langerhans cells with an antigen-specific cell, although the rapidity of the emigration process on primary exposure argues against a classic primary immune response, and, because the phenomenon occurs in nude mice, it has been viewed as T cell-independent (Hill *et al.*, 1990).

The qualitatively different immunological consequences of exposure to contact sensitizing agents when applied to Langerhans cell-rich as compared with Langerhans cell-poor areas of skin (Bergstresser *et al.*, 1980a,b; Toews *et al.*, 1980) or buccal mucosa (Streilein and Bergstresser, 1981) indicate the importance of the cells in antigen presentation. There is a general consensus of opinion that the sensitization phase of the primary immune response to contact sensitizers applied to the skin occurs in the draining lymph nodes (Romani *et al.*, 1991a,b), and this view is supported by recent studies on the recirculatory pathways of virgin and memory T lymphocytes (Mackay *et al.*, 1990). The migration of antigen-laden Langerhans cells to the regional lymph nodes, and their functional capacity to stimulate resting T cells, is therefore of great importance in the induction of immune responses to dermal antigens. In contrast, activated T blasts or memory T cells may interact directly with antigen presented by Langerhans cells in the epidermis. This is suggested by lymphocyte clustering, which has been observed at sites of contact hypersensitivity after challenge with contact sensitizer in sensitive subjects (Silberberg *et al.*, 1975).

4.4   REGULATION OF ANTIGEN PRESENTATION AT EPITHELIAL SURFACES

There are occasions (e.g. during infection) when it is vital that antigens should induce local protective immune responses. If a cell-mediated immune response is involved, this may require recruitment of activated T cells from the circulation or the local activation of memory T cells in response to antigen presented *in situ* by resident antigen-presenting cells. On the other hand, it is important that the many antigens encountered casually from food, inhaled as aerosols or particles, or contacted via the skin, do not induce the local inflammation observed in those who suffer from the various forms of hypersensitivity diseases. Antigen presentation is one level at which adverse responses to antigens could be controlled, and this might involve a number of mechanisms.

1. Presentation of antigens by non-professional antigen-presenting cells (e.g. epithelial cells) could have tolerizing rather than sensitizing effects. This has been suggested in the case of gut epithelium (see Section 3.3), and it is possible that the absence of dendritic cells from all but the specialized follicle-associated

epithelia in the gut is a contributing factor to the phenomenon of oral tolerance. That is, unless antigens engage with specialized antigen-presenting cells in the organized mucosal lymphoid tissues, primary presentation by epithelial cells could deliver a tolerizing signal. A similar explanation could be applied to the tolerizing effects observed with contact sensitizers on Langerhans cell-deficient skin (Toews *et al.*, 1980) or hamster check pouch (Streilein and Bergstresser, 1981). Down-regulatory effects of keratinocytes have been described (Flood *et al.*, 1991), but other possibilities such as the presence of I-J-positive antigen-presenting cells have also been suggested (Halliday *et al.*, 1990b). It is beyond the scope of this chapter to discuss whether alternative pathways of antigen presentation induce specific suppressor T cells or whether they alter the bias between mutually antagonistic subsets of T helper (T_H1 and T_H2) cells.

2. Suppressive effects of local macrophage populations could inhibit activation of T cells at epithelial surfaces. This has been discussed in relation to the gut mucosa (see Sections 4.11 and 4.12) and the lung (see Section 4.2), although its relevance to the skin is less obvious. In the lung (Holt *et al.*, 1985) and gut (Pavli *et al.*, 1990), dendritic cells have potent antigen-presenting functions when separated from resident macrophages. However, in the skin, Langerhans cells and macrophages normally occupy separate tissue compartments (Weber-Matthiesen and Sterry, 1990). It may therefore be necessary to regulate local antigen presentation in the epidermis in a different way. This may be achieved in the skin by a temporal and spatial disassociation of antigen assimilation in the epidermis and antigen presentation in the regional lymph nodes (Romani *et al.*, 1991b). Antigen-containing Langerhans cells may exit the epidermis rapidly before maturation of full antigen-presenting function.

3. A fascinating aspect of the pathophysiology of Langerhans cells and mucosal dendritic cells is the regulation of their migratory behaviour after antigen exposure. In the case of Langerhans cells (see Section 4.3), increased migration is seen within only a few hours of primary skin exposure to contact sensitizers (Breathnach, 1991). While the migration response has been attributed to antigen, little attention has been given to how it might be mediated. In particular, the involvement of antigen-specific cells has either not been considered, or their involvement has been rejected on the grounds that the migratory response occurs in nude mice (Hill *et al.*, 1990). With the discovery of T cells expressing γ/δ antigen-specific receptors, it is no longer valid to assume that the skin (or other epithelia) in nude mice is deficient in T cells (Yoshikai *et al.*, 1986; Yoshikai *et al.*, 1988).

The precise function of γ/δ T cells remains an enigma. However, one possibility is that these cells are part of a layered immune system, in which they function as a primitive early defence system (Janeway, 1988; Herzenberg and Herzenberg, 1989). It has been suggested that γ/δ T cells have the ability to recognize and respond to processed peptides from certain common stress proteins in association with non-classic MHC molecules. In this context it is of interest to note that candidates for restriction molecules that might present stress proteins are the relatively non-polymorphic MHC class I-like CDI group in humans and molecules of the Tla and Qa groups in mice (reviewed by Calabi and Bradbury, 1991). It may not be coincidental, therefore, that Langerhans cells express CD1 in man and Tla in mice (Romani *et al.*, 1991a). Furthermore, in the gut, where there are no intraepithelial dendritic cells, the enterocytes themselves express CD1 in humans (Blumberg *et al.*, 1991a) and appear able to present antigens to intraepithelial T cells in a CD1-restricted manner (Blumberg *et al.*, 1991b). Non-classic MHC molecules could be recognized by γ/δ T cells and perhaps also by thymus-independent α/β T cells that express the CD8 α homodimer in the gut (Guy-Grand *et al.*, 1991) or by CD4⁻, CD8⁻ α/β T cells that have been described in the skin (Groh *et al.*, 1989; Porcelli *et al.*, 1989).

The function of γ/δ T cells is frequently viewed in terms of cytotoxic activity towards, for instance, virus-infected cells. However, dermal dendritic T cells are stimulated to secrete lymphokines by Langerhans cells (Steiner *et al.*, 1989), and intestinal intraepithelial lymphocytes also secrete a range of lymphokines (Beagley *et al.*, 1990). In view of the known effects of GM-CSF and IL-1 on maturation of Langerhans cells (Witmer-Pack *et al.*, 1987; Heufleur *et al.*, 1988), it seems possible that an important evolutionary nexus exists at the level of the dendritic cell between the primitive, relatively non-specific, γ/δ T cell layer of the immune system and the more specific thymus-dependent α/β T cells. If unprimed γ/δ T cells at epithelial surfaces can recognise altered-self in the context of non-classical MHC molecules, either as stress-related peptides, processed foreign peptides or chemically reactive haptens, they may provide the first step in the adaptive immune response – the maturation of local dendritic cells and their rapid dispatch to the regional lymph nodes. In this way, they would ensure rapid presentation of antigens to the large repertoire of virgin T cells that recirculate through the lymph nodes.

4. It has been mentioned earlier that local presentation of antigen to T cells may be important in the effector phase of the immune response in the gut (see Section 4.1.1) and the skin (see Section 4.2). The work of Mackay (1990) indicates that T cells recirculating through afferent lymphatics have predominantly the memory CD45RO phenotype and specialized homing receptors appear responsible for selective homing of activated T cells to gut (Parrott and Wilkinson, 1981), lung (Yednock and Rosen, 1989) and skin (Picker *et*

al., 1991). In the gut, dendritic cells appear to contain a mature antigen presenting population (Pavli, 1990; Pavli *et al.*, 1990). However, in skin, only a small subpopulation of Langerhans cells (those that express high levels of class II MHC molecules) is active in antigen presentation (Romani *et al.*, 1991a). This small sub-population, or migrant cells in the dermis, may be responsible for the local activation of primed T cells in contact hypersensitivity.

5. *Conclusions*

It is evident from this review that there are still large areas of uncertainty about how antigens are handled at epithelial surfaces and how they are presented to the immune system. The solution to some of the problems, such as the absorption of antigens by epithelia, requires application of relatively classical techniques of physiology and demand that immunologists become much more expert and critical in their use of both *in vivo* and *in vitro* models. Furthermore, there is a place for the use of sophisticated organ perfusion studies, where lumenal, arteriovenous and lymph levels of antigen can be monitored accurately. Such preparations are technically demanding (and, sadly, do not attract funding) but the less exacting methods used to date have reached the limits of their usefulness.

The use of monolayers of differentiated epithelial cells is already being used to investigate some aspects of the cell biology of polarized epithelium. Further advances in the understanding of antigen handling and processing by specialized epithelial cells can be expected to follow from development of techniques to maintain the viability and differentiated characteristics of the cells *in vitro*. These developments will follow the identification of appropriate growth factors, extracellular matrix support materials and essential metabolites for various epithelial cells.

Important advances have also occurred in immuno-electron microscopy and in the development of antibody probes that identify specific subcellular compartments and the intracellular locations of specific molecules. It should become possible to trace the intracellular handling not only of intact antigens but also of specific peptides. Furthermore, it is possible to elute peptide residues of processed antigens from whole cells, to separate them by high-performance liquid chromatography and to identify their origins by sequence analysis (Rötzschke and Falk, 1991). This technology can be extended to epithelial cells, and it has the potential to provide insights into the handling of important dietary and microbial antigens.

Important advances can also be expected in the exploration of immune responses by T lymphocytes *in situ*. Tissue compartments are very much a feature of mucosae and other epithelial surfaces. All techniques which rely on the isolation of cells and study of their properties *in vitro* run the risk of misidentification of the precise compartments from which the cells derive and also the creation of non-physiological microenvironments and cell associations. The recent availability of monoclonal antibodies and nucleic acid probes for T cell products such as cytokines offers enormous promise for investigation of cell functions *in situ*. Genetic approaches, such as transgenesis and site-directed mutagenesis, also offer opportunities to investigate the functions of particular gene products in epithelium and other cells.

Finally, it is clear that understanding antigen handling at epithelial surfaces is important, both for the development of oral vaccines and for devising therapies for those afflictions of epithelial organs that have an immunological pathogenesis. However, there is not a sound theoretical basis at present on which to build vaccine or immuno-therapeutic strategies, and this is evident from the limited successes that has been achieved in either of these areas. The future of attempts to apply immunological techniques to the prevention and care of diseases affecting epithelia will depend on a careful reassessment of accepted wisdom as it relates to the disposition of antigen at the important epithelial interfaces between the body and the external environment.

6. *References*

Aas, K. (1988). The biochemistry of food allergens: what is essential for future research. In "Food Allergy" (eds D. Reinhardt and E. Schmidt), pp 1–14. Raven Press, New York.

Abe, K. and Ito, T. (1978). Fine structure of the dome in Peyer's patches of mice. Arch. Histol. Japn 41, 195–204.

Abraham, R., Fabian, R.J., Golberg, L. and Coulston, F. (1974). Role of lysosomes in carrageenan-induced cecal ulceration. Gastroenterology 67, 1169–1181.

Abrahamson, D.R. and Rodewald, R. (1981). Evidence for the sorting of endocytic vesicle contents during the receptor-mediated transport of IgG across the newborn rat intestine. J. Cell Biol. 91, 270–280.

Agnod, M., Yamaguchi, N., Lopez, R., Luhby, A.L. and Glass, G.B.J. (1969). Correlative study of hydrochloric acid, pepsin, and intrinsic factor secretion in newborns and infants. Am. J. Dig. Dis. 14, 400–414.

Allan, J.D., Moss, A.D., Wallwork, J.C. and McFarlane, H. (1975). Immediate hypersensitivity in patients with cystic fibrosis. Clin. Allergy 5, 255–261.

Alpers, D.H. and Isselbacher, K.J. (1967). Protein synthesis by rat intestinal mucosa. The role of ribonuclease. J. Biol. Chem. 242, 5617–5622.

Ammann, A.J. and Steihm, E.R. (1966). Immune globulin levels in colostrum and breast milk, and serum from formula- and breast-fed newborns. Proc. Soc. Exp. Biol. Med. 122, 1098–1100.

Anderson, B.W., Levine, A.S., Levitt, D.G., Kneip, J.M. and Levitt, M.D. (1988). Physiological measurement of luminal stirring in perfused rat jejunum. Am. J. Physiol. 254, G849–G855.

Anderson, R.P., Woodhouse, A.F., Hobson, C.H., Myers, D.B., Broom, M.F. and Chadwick, V.S. (1987). Hepatobiliary excretion and enterohepatic circulation of bacterial chemotactic peptide (FMLP) in the rat. J. Gastroenterol. Hepatol. 2, 45–52.

André, C., Bazin, H. and Heremans, J.F. (1973) Influence of repeated administration of antigen by the oral route on specific antibody-producing cells in the mouse spleen. Digestion 9, 166–175.

Andrew, E.M. and MacDonald, T.T. (1987). In "Immunology of the Gastrointestinal Tract", Vol. I (eds K. Miller and S. Nicklin), Ch. 3. CRC Press, Boca Raton.

Asherson, G.L., Zembala, M., Perera, M.A.C.C., Mayhew, B. and Thomas, W.R. (1977). Production of immunity and unresponsiveness in the mouse by feeding contact sensitizing agents and the role of suppressor cells in the Peyer's patches, mesenteric lymph nodes and other lymphoid tissues. Cell. Immunol. 33, 145–155.

Augustine, P.C. and Danforth, H.D. (1984). Effects of cationized ferritin and neuraminidase on invasion of cultured cells by *Eimeria meleagrimitis* sporozoites. J. Protozool. 31, 140–144.

Austyn, J.M. (1987). Lymphoid dendritic cells. Immunology 62, 161–170.

Austyn, J.M. (1989). "Antigen-Presenting Cells" (In Focus Series). IRL Press, Oxford.

Austyn, J.M. and Larsen, C.P. (1990). Migration patterns of dendritic leukocytes. Transplantation 49, 1–7.

Babbit, B.P., Allen, P.M., Matsueda, G., Haber, E. and Unanue, E.R. (1985). Binding of immunogenic peptides to Ia histocompatibility molecules. Nature 317, 359–361.

Baintner, K. (1986). "Intestinal Absorption of Macromolecules and Immune Transmission from Mother to Young", pp 137–141. CRC Press, Boca Raton.

Baintner, K. (1986a). "Intestinal Absorption of Macromolecules and Immune Transmission from Mother to Young", Ch. 4. CRC Press, Boca Raton.

Baintner, K. (1986b). "Intestinal Absorption of Macromolecules and Immune Transmission from Mother to Young", Ch. 1. CRC Press, Boca Raton.

Baker, G., Jones, L.H.P. and Milne, A.A. (1961). Opal uroliths from a ram. Aust. J. Agric. Res. 12, 473–481.

Bandeira, A., Itohara, S., Bonneville, M., Burlen-Defranoux, O., Mota-Santos, T., Coutinho, A. and Tonegawa, S. (1991). Extrathymic origin of intestinal intraepithelial lymphocytes bearing T-cell antigen receptor γδ Proc. Natl Acad. Sci. USA 88, 43–47.

Barclay, A.N. and Mason, D.W. (1982). Induction of Ia antigen in rat epidermal cells and gut epithelium by immunological stimuli. J. Exp. Med. 156, 1665–1676.

Barka, T. and Anderson, P.J. (1962). Histochemical methods for acid phosphatase using hexazonium pararosanalin as coupler. J. Histochem. Cytochem. 10, 741–753.

Barnard, J.A., Beauchamp, R.D., Coffey, R.J. and Moses, H.L. (1989). Regulation of intestinal epithelial cell growth by transforming growth factor type β. Proc. Natl Acad. Sci. USA 86, 1578–1582.

Baur, X. and Fruhmann, G. (1979). Papain-induced asthma: diagnosis by skin test, RAST and bronchial provocation test. Clin. Allergy 9, 75–81.

Beagley, K.W., Eldridge, J.H., Aicher, W.K., Fujihashi, K., Taguchi, T., Xu, J., Kiyono, H. and McGhee, J.R. (1990). In "Mucosal Immunology" (ed A.W. Cripps), pp 93–98. Newey and Beath, Newcastle, Australia.

Beer, A.E. and Billingham, R.E. (1974). Host responses to intrauterine tissue, cellular and fetal allografts. J. Reprod. Fert. Suppl. 21, 59–88.

Bell, E.B. (1984). Antigen transport: II. The significance of the localization of antigen-laden cells in the spleen. Cell. Immunol. 88, 273–284.

Bell, E.B. and Botham, J. (1982). Antigen transport: I. Demonstration and characterization of cells laden with antigen in thoracic duct lymph and blood. Immunology 47, 477–487.

Bensch, K.G. and Dominguez, E.A.M. (1971). Studies on the pulmonary air-tissue barrier, part IV: cytochemical tracing of macromolecules during absorption. Yale J. Biol. Med. 43, 236–241.

Berg, R.D. (1981). Promotion of translocation of enteric bacteria from the gastrointestinal tracts of mice by oral treatment with penicillin, clindamycin or metronidazole. Infect. Immun. 33, 854–861.

Berg, R.D. and Garlington, A.W. (1979). Translocation of certain indigenous bacteria from the gastrointestinal tract to the mesenteric lymph nodes and other organs in a gnotobiotic mouse model. Infect. Immunol. 23, 403–411.

Bergstresser, P.R., Toews, G.B., Gilliam, J.N. and Streilein, J.W. (1980). Unusual numbers and distributions of Langerhans cells in skin with unique immunologic properties. J. Invest. Dermatol. 74, 312–314.

Bergstresser, P.R., Toews, G.B. and Streilein, J.W. (1980). Natural and perturbed distributions of Langerhans cells: responses to ultraviolet light, heterotopic skin grafting, and dinitrofluorobenzene sensitization. J. Invest. Dermatol. 75, 73–77.

Bhalla, D.K. and Owen, R.L. (1982). Cell renewal and migration in lymphoid follicles of Peyer's patches and cecum – an autoradiographic study in mice. Gastroenterology 82, 232–242.

Bienenstock, J. and Johnston, N. (1976). A morphologic study of rabbit bronchial lymphoid aggregates and lymphoepithelium. Lab. Invest. 35, 343–348.

Bienenstock, J., Johnston, N. and Perey, D.Y.E. (1973a). Bronchial lymphoid tissue. I. Morphologic characteristics. Lab. Invest. 28, 686–692.

Bienenstock, J., Johnston, N. and Perey, D.Y.E. (1973b). Bronchial lymphoid tissue. II. Functional characteristics. Lab. Invest. 28, 693–698.

Bienenstock, J., Denburg, J., Scicchitano, R., Stead, R., Perdue, M. and Stanisz, A.M. (1988). Role of neuropeptides, nerves and mast cells in intestinal immunity and physioloy. Monogr. Allergy 24, 124–133.

Bierring, F., Anderson, H., Egeberg, J. Bro-Rasmussen, F. and Matthiessen, M. (1964). On the nature of the meconium corpuscles in human foetal intestinal epithelium. I. Electron microscopic studies. Acta Pathol. Microbiol. Scand. 61, 365–376.

Bjarnason, I. and Peters, T.J. (1984). *In vitro* determination of small intestinal permeability: demonstration of a persistent defect in patients with coeliac disease. Gut 25, 145–150.

Bjarnason, I., O'Morain, C., Levi, J.A. and Peters, T.J. (1983). Absorption of ^{51}Cr labelled EDTA in inflammatory bowel disease. Gastroenterology 85, 318–322.

Bjerke, K. and Brandtzaeg, P. (1990). In "Advances in Mucosal Immunology" (eds T.T. MacDonald, S.J. Challacombe, P.W. Bland, C.R. Stokes, R.V. Heatley and A. McI. Mowat), pp 617–618. Kluwer, Dordrecht.

Bjerke, K., Bandtzaeg, P. and Fausa, O. (1988). T cell distribution is different in follicle-associated epithelium of human Peyer's patches and villous epithelium. Clin. Exp. Immunol. 74, 270–275.

Bland, P.W. (1987). Antigen presentation by gut epithelial cells: secretion by rat enterocytes of a factor with IL-1–like activity. Adv. Exp. Med. Biol. 216, 219–225.

Bland, P. (1988). MHC class II expression by the gut epithelium. Immunol. Today 9, 174–178.

Bland, P.W. and Britton, D.C. (1984). Morphological study of antigen-sampling structures in the rat large intestine. Infect. Immun. 43, 693–699.

Bland, P.W. and Kambarage, D.M. (1991). Antigen handling by the epithelium and lamina propria macrophages. Gastroenterol. Clin. North Am. 20, 577–596.

Bland, P.W. and Warren, L.G. (1986a). Antigen presentation by epithelial cells of the rat small intestine. I. Kinetics, antigen specificity and blocking by anti-Ia antisera. Immunology 58, 1–7.

Bland, P.W. and Warren, L.G. (1986b). Antigen presentation by epithelial cells of the rat small intestine. II. Selective induction of suppressor T cells. Immunology 58, 9–14.

Bland, P.W. and Whiting, C.V. (1989). Antigen processing by isolated rat intestinal villus enterocytes. Immunology 68, 497–502.

Bland, P.W. and Whiting, C.V. (1992). Induction of MHC class II gene products in rat intestinal epithelium during graft-versus-host disease and effects on the immune function of the epithelium. Immunology 75, 366–371.

Blank, I.H. (1953). Further observations on factors which influence the water content of the stratum corneum. J. Invest. Dermatol. 21, 259–269.

Blank, I.H. (1965). Cutaneous barriers. J. Invest. Dermatol. 45, 249–256.

Bloch, K.J., Bloch, D.B., Stearns, M. and Walker, W.A. (1979). Intestinal uptake of macromolecules VI. Uptake of protein antigen *in vivo* in normal rats and in rats infected with *Nippostongylus brasiliensis* or subjected to mild systemic anaphylaxis. Gastroenterology 77, 1039–1044.

Blumberg, R.S., Allan, C., Bleicher, P., Landau, S., McDermott, F.V., Trier, J.S. and Balk, S.P. (1991a). Expression of CDID by human gastrointestinal epithelial cells. Gastroenterology 100, A562.

Blumberg, R.S., Landau, S., Ebert, E. and Balk, S.P. (1991b). T cell receptor (TCR) – α expression and recognition of CDI by a human jejunal intraepithelial lymphocyte (IEL) line. Gastroenterology, 100, A562.

Bockman, D.E. and Cooper, M.D. (1973). Pinocytosis by epithelium associated with lymphoid follicles in the bursa of Fabricius, appendix, and Peyer's patches. An electron microscopic study. Am. J. Anat. 136, 455–478.

Bockman, D.E. and Winborn, W.B. (1966). Light and electron microscopy of intestinal ferritin absorption. Observations in sensitized and nonsensitized hamsters (*Mesocricetus auratus*). Anat. Rec. 155, 603–622.

Bodian, D. (1955). Emerging concept of poliomyelitis infection. Science 122, 105–108.

Bookman, M.A. and Bull, D.M. (1979). Characteristics of isolated intestinal mucosal lymphoid cells in inflammatory bowel disease. Gastroenterology 77, 503–510.

Borthistle, B.K., Kubo, R.T., Brown, W.R. and Grey, H.M. (1977). Studies on receptors for IgG on epithelial cells of the rat intestine. J. Immunol. 119, 471–476.

Braciale, T.J. and Braciale, V.L. (1991). Antigen presentation: structural themes and functional variation. Immunol. Today 12, 124–129.

Brambell, F.W.R. (1970). "The Transmission of Passive Immunity from Mother to Young" (North-Holland Research Monographs, Frontiers of Biology, Vol. 18). North-Holland, Amsterdam.

Brain, J.D. (1970). Free cells in the lungs. Some aspects of their role, quantitation, and regulation. Arch. Intern. Med. 126, 477–487.

Brandtzaeg, P. (1974). Mucosal and glandular distribution of immunoglobulin components. Differential localization of free and bound SC in secretory epithelial cells. J. Immunol. 112, 1553–1559.

Brandtzaeg, P., Bjerke, K., Halstensen, T.S., Hvatum, M., Kett, K., Krajci, P., Kvale, D., Müller, F., Wilsson, D., Roynum, T.O., Scott, H., Sollid, L.M., Thrane, P. and Valnes, K. (1990). In "Advances in Mucosal Immunology" (eds T.T. MacDonald, S.J. Challacombe, P.W. Bland, C.R. Stokes, R.V. Heatley and A. McI. Mowat), pp 1–12. Kluwer, Dordrecht.

Breathnach, S.M. (1988). The Langerhans cell. Br. J. Dermatol. 119, 463–469.

Breathnach, S.M. (1991). In "Epidermal Langerhans Cells" (ed G. Shuler), pp 23–47. CRC Press, Boca Raton.

Brodsky, F.M. and Guagliardi, L. (1991). The cell biology of antigen processing and presentation. Ann. Rev. Immunol. 9, 707–744.

Bujdoso, R., Hopkins, J., Dutia, B.M., Young, P. and McConnell, I. (1989). Characterization of sheep afferent lymph dendritic cells and their role in antigen carriage. J. Exp. Med., 170, 1285–1302.

Bull, D.M. and Bookman, M.A. (1977). Isolation and functional characterization of human intestinal mucosal lymphoid cells. J. Clin. Invest. 59, 966–974.

Busse, W.W., Reed, C.E. and Hoehne, J.H. (1972). Where is the allergic reaction in ragweed asthma? II. Demonstration of ragweed antigen in airborne particles smaller than pollen. J. Allergy Clin. Immunol. 50, 289–293.

Buus, S., Sette, A., Colon, S.M., Miles, C. and Grey, H.M. (1987). The relation between major histocompatibility complex (MHC) restriction and the capacity of Ia to bind immunogenic peptides. Science 235, 1353–1358.

Bye, W.A., Allan, C.H. and Trier, J.S. (1984). Structure, distribution and origin of M cells in Peyer's patches of mouse ileum. Gastroenterology 86, 789–801.

Calabi, F. and Bradbury, A. (1991). The CDI system. Tissue Antigens 37, 1–9.

Carneiro, C.R.W., Mota, G.F.A., Sabbaga, J., Marquezini, M., Potocnjak, P. and Bretani, R.R. (1989). In "Molecular Mechanisms of Microbial Adherence" (eds L. Switalski, M. Höök and E. Beachey), pp 118–127. Springer-Verlag, New York.

Carter, P. and Collins, F. (1974). The route of enteric infection in normal mice. J. Exp. Med. 139, 1189–1203.

Carter, R.E. and Taylor, W.F. (1980). Identification of a particular amphibole asbestos fibre in tissues of persons exposed to a high oral intake of the mineral. Environ. Res. 21, 85–93.

Cash, R.A., Music, S.I., Libonati, J.P., Craig, J.P., Pierce, N.F. and Hornick, R.B. (1974). Response of man to infection with *Vibrio cholerae*. II. Protection from illness afforded by previous disease and vaccine. J. Infect. Dis. 30, 325–333.

Cerf-Bensussan, N. and Guy-Grand, D. (1991). Intestinal

intraepithelial lymphocytes. Gastroenterol. Clin. North Am. 20, 549–576.

Cerf-Bensussan, N., Quaroni, A., Kurnick, J.T. and Bhan, A.K. (1984). Intraepithelial lymphocytes modulate Ia expression by intestinal epithelial cells. J. Immunol. 132, 2244–2252.

Chadwick, V.S., Phillips, S.F. and Hofmann, A.F. (1977a). Measurements of intestinal permeability using low molecular weight polyethylene glycols (PEG 400). I. Chemical analysis and biological properties of PEG 400. Gastroenterology 73, 241–246.

Chadwick, V.S., Phillips, S.F. and Hofmann, A.F. (1977b). Measurements of intestinal permeability using low molecular weight polyethylene glycols (PEG 400). II. Application to normal and abnormal permeability states in man and animals. Gastroenterology 73, 247–251.

Chadwick, V.S., Mellor, D.M., Myers, D.B., Selden, A.C., Keshavarzian, A., Broom, M.F. and Hobson, C.H. (1988). Production of peptides inducing chemotaxis and lysosomal enzyme release in human neutrophils by intestinal bacteria *in vitro* and *in vivo*. Scand. J. Gastroenterol. 23, 121–128.

Challacombe, S.J. and Tomasi, T.B. (1980). Systemic tolerance and secretory immunity after oral immunization. J. Exp. Med. 152, 1459–1472.

Chase, M.W. (1946). Inhibition of experimental drug allergy by prior feeding of the sensitizing agent. Proc. Soc. Exp. Biol. Med. 61, 257–259.

Chen, K.-S. and Strober, W. (1990). Cholera holotoxin and its B subunit enhance Peyer's patch B cell responses induced by orally administered influenza virus: disproportionate cholera toxin enhancement of the IgA B cell response. Eur. J. Immunol. 20, 433–436.

Childers, N.K., Michalek, S.M., Pritchard, D.G. and McGhee, J.R. (1990). In "Advances in Mucosal Immunology" (eds T.T. MacDonald, S.J. Challacombe, P.W. Bland, C.R. Stokes, R.V. Heatley and A.Mc.I. Mowatt), pp 345–346. Kluwer, Dordrecht.

Chintalacharuvu, K.R., Piskurich, J.F., Lamm, M.E. and Kaetzel, C.S. (1991). Cell polarity regulates the release of secretory component, the epithelial receptor for polymeric immunoglobulins, from the surface of HT-29 colon carcinoma cells. J. Cell. Physiol. 148, 35–47.

Chua, K.Y., Stewart, G.A., Thomas, W.R., Simpson, R.J., Dilworth, R.J., Plozza, T.M. and Turner, K.J. (1988). Sequence analysis of cDNA coding for a major house dust mite allergen, *Derp* I. J. Exp. Med. 167, 175–182.

Chua, K.Y., Doyle, C.R., Stewart, G.A., Turner, K.J., Simpson, R.J. and Thomas, W.R. (1990). Cloning of the major mite allergen *Derp* II by IgE plaque immunoassay. Int. Arch. Allergy Appl. Immunol. 91, 118–123.

Chung, Y.C., Silk, D.B.A. and Kim, Y.S. (1979). Intestinal transport of a tetrapeptide, L-leucylglycylglycylglycine, in rat small intestine *in vivo*. Clin. Sci. 57, 1–11.

Chuttani, C.S., Prakash, K., Vergese, U., Sharma, P., Singha, B., Ghosh, R. and Agarwal, D.A. (1972). Controlled field trials of oral killed typhoid vaccines in India. Int. J. Epidemiol. 1, 39–43.

Ciclitira, P.J. and Hall, M.A. (1990). In "Clinical Gastroenterology. International Practice and Research", Vol. 4 (ed P.J. Ciclitira), Ch. 3. Baillière Tindall, London.

Clarke, R.M. and Hardy, R.N. (1969a). The use of [^{125}I]polyvinylpyrrolidone K.60 in the quantitative assessment of the uptake of macromolecular substances by the intestine of the young rat. J. Physiol. 204, 113–125.

Clarke, R.M. and Hardy, R.N. (1969b). An analysis of the mechanisms of cessation of uptake of macromolecular substances by the intestine of the young rat ("closure"). J. Physiol. 204, 127–134.

Cohen, A.B. and Gold, W.M. (1975). Defence mechanisms of the lungs. Ann. Rev. Physiol. 37, 325–350.

Cole, S.G. and Kagnoff, M.F. (1985). Celiac disease. Ann. Rev. Nutr. 5, 241–266.

Comline, R.S., Roberts, H.E. and Titchen, D.A. (1951). Route of absorption of colostrum globulin in the newborn animal. Nature 167, 561–562.

Cooper, B.T. (1986). In "Gut Defences in Clinical Practice" (eds M.S. Losowsky and R.V. Heatley), Ch 10. Churchill Livingstone, Edinburgh.

Cornell, H.J. and Maxwell, R.J. (1982). Amino acid composition of gliadin fractions which may be toxic to individuals with coeliac disease. Clin. Chim. Acta 123, 311–319.

Cornell, H.J. and Townley, R.R.W. (1973). Investigation of possible intestinal peptidase deficiency in coeliac disease. Clin. Chim. Acta 43, 113–125.

Cornell, H.J., Auricchio, R.S., De Ritas, G., De Vincenzi, M., Maiuri, L., Raia, V. and Silano, V. (1988). Intestinal mucosa of celiacs in remission is unable to abolish toxicity of gliadin peptides on *in vitro* developing fetal rat intestine and cultured atrophic celiac mucosa. Pediatr. Res. 24, 233–237.

Cornell, R., Walker, W.A. and Isselbacher, K.J. (1971). Small intestinal absorption of horseradish peroxidase. A cytochemical study. Lab. Invest. 25, 42–48.

Corry, D., Kulkarni, P. and Lipscomb, M.F. (1984). The migration of bronchoalveolar macrophages into hilar lymph nodes. Am. J. Pathol. 115, 321–328.

Cox, D.S. and Muench, D. (1984). IgA antibody produced by local presentation of antigen in orally primed rats. Int. Arch. Allergy Appl. Immunol. 74, 249–255.

Crabbé, P.A., Nash, D.R., Bazin, H., Eyssen, H. and Heremans, J.F. (1969). Antibodies of the IgA type in intestinal plasma cells of germfree mice after oral or parenteral immunization with ferritin. J. Exp. Med. 130, 723–744.

Cresswell, P. (1985). Intracellular class II HLA antigens are accessible to transferrin-neuraminidase conjugates internalized by receptor-mediated endocytosis. Proc. Natl Acad. Sci. USA 82, 8188–8192.

Croitoru, K., Stead, R.H., Bienenstock, J., Fulop, G., Harnish, D.G., Schultz, L.D., Jeffrey, P.K. and Ernst, P.B. (1990). Presence of intestinal intraepithelial lymphocytes in mice with severe combined immunodeficiency disease. Eur. J. Immunol. 20, 645–651.

Cullander, C. and Guy, R.H. (1991). Sites of iontophoretic current flow into the skin: Identification and characterization with the vibrating probe electrode. J. Invest. Dermatol. 97, 55–64.

Current, W.L. and Reese, N.C. (1986). A comparison of endogenous development of three isolates of *Cryptosporidium* in suckling mice. J. Protozool. 33, 98–108.

Danforth, E. and Moore, R.D. (1959). Intestinal absorption of insulin in the rat. Endocrinology 65, 118–123.

Davidson, H.W. and Watts, C. (1989). Epitope-directed processing of specific antigen by B lymphocytes. J. Cell Biol. 109, 85–92.

Davidson, H.W., West, M.A. and Watts, C. (1990). Endocytosis, intracellular trafficking and processing of membrane IgG and monovalent antigen/membrane IgG complexes in B lymphocytes. J. Immunol. 144, 4101–4110.

Davidson, H.W., Reid, P.A., Lanzavecchia, A. and Watts, C. (1991). Processed antigen binds to newly synthesized class II molecules in antigen-specific B lymphocytes. Cell 67, 105–116.

de Aizpurua, H.J. and Russell-Jones, G.J. (1988). Oral vaccination: identification of classes of proteins which provide an immune response upon oral feeding. J. Exp. Med. 167, 440–451.

De Ritas, G., Auricchio, S., Jones, H.W., Lew, E.J-L., Bernardin, J.E. and Kasarda, D.D. (1988). *In vitro* (organ culture) studies of the toxicity of specific A-gliadin peptides in celiac disease. Gastroenterology 94, 41–49.

Dezutter-Dambuyant, C., Cordier, G., Schmitt, D., Faure, M., Laquoi, C. and Thivolet, J. (1984). Quantitative evaluation of two distinct cell populations expressing HLA-DR antigens in normal human epidermis. Br. J. Dermatol. 111, 1–11.

Dilworth, R.J., Chua, K.Y. and Thomas, W.R. (1991). Sequence analysis of cDNA coding for a major house dust mite allergen, *Der f* I. Clin. Exp. Allergy 21, 25–32.

Dinsdale, D. and Healy, P.J. (1982). Enzymes involved in protein transmission by the intestine of the newborn lamb. Histochem. J. 14, 811–821.

Dolezel, J. and Bienenstock, J. (1971). γ-A and non γ-A immune response after oral and parenteral immunization of the hamster. Cell. Immunol. 2, 458–468.

Donaldson, R.M. (1987). In "Physiology of the Gastrointestinal Tract", 2nd edn (ed. L.R. Johnson), Ch. 33. Raven Press, New York.

Donowitz, M. and Welsh, M.J. (1987). In "Physiology of the Gastrointestinal Tract", Vol. 2 (ed. L.R. Johnson), Ch. 48. Raven Press, New York.

Dor, P.J., Agarwal, M.K., Gleich, M.C., Welsh, P.W., Dunnette, S.L., Adolphson, C.R. and Gleich, G.J. (1986). Detection of antibodies to proteases used in laundry detergents by the radioallergosorbent test. J. Allergy Clin. Immunol. 78, 877–886.

Douglas, A.P. (1976). The binding of a glycopeptide component of wheat gluten to intestinal mucosa of normal and coeliac human subjects. Clin. Chim. Acta 73, 357–361.

Drexhage, H.A., Mullink, H., deGroot, J., Clarke, J. and Balfour, B.M. (1979). A study of cells present in peripheral lymph of pigs with special reference to a type of cell resembling the Langerhans cell. Cell Tissue Res. 202, 407–430.

Dunkley, M.L. and Husband, A.J. (1986). The induction and migration of antigen specific helper cells for IgA responses in the intestine. Immunology 57, 605–614.

Duroc, R., Heyman, M. Beaufrere, B., Morgat, J.L. and Desjeux, J.F. (1983). Horseradish peroxidase transport across rabbit jejunum and Peyer's patches *in vitro*. Am. J. Physiol. 245, G54–G58.

Dym, M. and Fawcett, D.W. (1970). The blood–testis barrier in the rat and the physiological compartmentation of the seminiferous epithelium. Biol. Reprod. 3, 308–326.

Ebling, F.J.G. (1992). In "Textbook of Dermatology" (eds R.H. Champion, J.L. Burton and F.J.G. Ebling), Ch. 4. Blackwell, Oxford.

Eldridge, J.H., Staas, J.K., Meulbroek, J.A., McGhee, J.R., Tice, T.R. and Gilley, R.M. (1991). Biodegradable microspheres as a vaccine delivery system. Molec. Immunol. 28, 287–294.

Elias, P.M. and Friend, D.S. (1975). The permeability barrier in mammalian epidermis. J. Cell Biol. 65, 180–191.

Elsayed, S. and Apold, J. (1977). Allergenic structure of allergen M from cod. Int. Arch. Allergy Appl. Immunol. 54, 171–175.

Elsayed, S. and Apold, J. (1983). Immunochemical analysis of cod fish allergen M: locations of the immunoglobulin binding sites as demonstrated by the native and synthetic peptides. Allergy 38, 449–459.

Emery, J.L. and Dinsdale, F. (1973). The postnatal development of lymphoreticular aggregates and lymph nodes in infants' lungs. J. Clin. Pathol. 26, 539–545.

Emery, J.L. and Dinsdale, F. (1974). Increased incidence of lymphoreticular aggregates in lungs of children found unexpectedly dead. Arch. Dis. Child. 49, 107–111.

Ericsson, J.L.E. (1965). Transport and digestion of hemoglobin in the proximal tubule. II. Electron microscopy. Lab. Invest. 14, 16–39.

Euler, A.R., Byrne, W.J., Meis, P.J., Leake, R.D. and Ament, M.E. (1979). Basal and pentagastrin-stimulated acid secretion in newborn human infants. Pediatr. Res. 13, 36–37.

Evans, D.G., Evans, D.J., Sack, D.A. and Clegg, S. (1984). In "Attachment of Organisms to the Gut Mucosa", Vol. I (ed. E.C. Boedeker), Ch. 8. CRC Press, Boca Raton.

Falck, B., Andersson, A. and Bartosik, J. (1985). Some new ultrastructural aspects on human epidermis and its Langerhans cells. Scand. J. Immunol. 21, 409–416.

Fauci, F.S. (1984). Aquired immunodeficiency syndrome: epidemiologic, clinical, immunologic and therapeutic considerations. Ann. Intern. Med. 100, 92–106.

Ferreccio, C., Levine, M.M., Rodriguez, H., Contreras, R. and Chilean Typhoid Committee (1989). Comparative efficacy of two, three or four doses of TY21a live oral typhoid vaccine in enteric-coated capsules: a field trial in an endemic area. J. Infect. Dis. 159, 766–769.

Finlay, B.B., Gumbiner, B. and Falkow, S. (1988). Penetration of *Salmonella* through a polarized Madin–Darby canine kidney epithelial cell monolayer. J. Cell Biol. 107, 221–230.

Flickinger, C.J., Herr, J.C. and Klotz, K.L. (1988). Immuno-cytochemical localization of the major glycoprotein of epididymal fluid from the cauda in the epithelium of the mouse epididymus. Cell Tissue Res. 251, 603–610.

Flindt, M. (1978). Health and safety aspects of working with enzymes. Process Biochem. 13, 3–7.

Flood, P.M., Ptak, W. and Granstein, R.D. (1991). In "Epidermal Langerhans Cells" (ed. G. Schuler), pp 273–293. CRC Press, Boca Raton.

Forsum, U., Claesson, K., Hjelm, E., Karlson-Parra, A., Klareskog, L., Scheynius, A. and Tjernlund, U. (1985). Class II transplantation antigens: distribution in tissues and involvement in disease. Scand. J. Immunol. 21, 389–396.

Fossum, S. (1988). Lymph-borne dendritic leukocytes do not recirculate, but enter the lymph node paracortex to become interdigitating cells. Scand. J. Immunol. 27, 97–105.

Fournier, M., Vai, F., Derenne, J.P. and Pariente, R. (1977). Bronchial lymphoepithelial nodules in the rat. Morphologic features and uptake and transport of exogenous proteins. Am. Rev. Resp. Dis. 116, 685–694.

Frazer, A.C., Fletcher, R.F., Ross, C.A.C., Shaw, B., Sammons, H.G. and Schneider, R. (1959). Gluten-induced enteropathy. The effect of partially digested gluten. Lancet ii, 252–255.

Frazer, J.D. (1989). High-affinity binding of staphylococcal enterotoxins A and B to HLA-DR. Nature 339, 221–223.

Friend, D.S. and Farquhar, M.G. (1967). Function of coated vesicles during protein absorption in the rat vas deferens. J. Cell Biol. 35, 357–376.

Fujimura, Y. (1986). Functional morphology of microfold cells (M cells) in Peyer's patches. Phagocytosis, and transport of BCG by M cells into rabbit's Peyer's patch. Gastroenterol. Jpn 21, 325–335.

Fujita, M., Reinhart, F. and Neutra, M. (1990). Convergence of apical and basolateral endocytic pathways at apical late endosomes in absorptive cells of suckling rat ileum *in vivo*. J. Cell Sci. 97, 385–394.

Gahring, L.C., Heffron, F., Finlay, B.B. and Falkow, S. (1990). Invasion and replication of *Salmonella typhimurium* in animal cells. Infect. Immunol. 58, 443–448.

Gail, D.B. and Lenfant, C.J.M. (1983). Cells of the lung: biology and clinical implications. Am. Rev. Resp. Dis. 127, 366–387.

Gardner, M.L.G. (1978). The absorptive viability of isolated intestine prepared from dead animals. Q. J. Exp. Physiol. 63, 93–95.

Gardener, M.L.G. (1984). Intestinal assimilation of intact peptides and proteins from the diet – a neglected field? Biol. Rev. 59, 289–331.

Gardener, M.L.G. (1988). Gastrointestinal absorption of intact proteins. Ann. Rev. Nutr. 8, 329–350.

Gardner, M.L.G. and Plumb, J.A. (1979). Release of dipeptide hydrolase activities from rat small intestine perfused *in vitro* and *in vivo*. Clin. Sci. 57, 529–534.

Germain, R.N. (1986). The ins and outs of antigen processing and presentation. Nature 322, 687–689.

Germain, R.N. (1991). The second class story. Nature 353, 605–607.

Gerrity, T.R., Lee, T.S., Haas, F.J., Marenelli, A., Werner, P. and Lourenco, R.V. (1979). Calculated deposition of inhaled particles in the airway generations of normal subjects. J. Appl. Physiol. 47, 867–873.

Gonnella, P.A., Siminoski, K., Murphy, R.A. and Neutra, M.R. (1987). Transepithelial transport of epidermal growth factor by absorptive cells of suckling rat ileum. J. Clin. Invest. 80, 22–32.

Gothelf, Y., Sharon, N. and Gazit, E. (1986a). A subset of human cord blood mononuclear cells is similar to Langerhans cells of the skin: a study with peanut agglutinin and monoclonal antibodies. Human Immunol. 15, 164–174.

Gothelf, Y., Sharon, N. and Gazit, E. (1986b). Fractionation of human bone marrow mononuclear cells with peanut agglutinin: phenotypic characterization with monoclonal antibodies. Human Immunol. 17, 37–44.

Gothelf, Y., Hanau, D., Tsur, H., Sharon, N. Sahar, E., Cazenave, J.-P. and Gazit, E. (1988). T6 positive cells in the peripheral blood of burn patients: are they Langerhans cells precursors. J. Invest. Dermatol. 90, 142–148.

Graham, R.C. and Karnovsky, M.J. (1966). The early stages of absorption of injected horseradish peroxidase in the proximal tubules of mouse kidney: ultrastructural cytochemistry by a new technique. J. Histochem. Cytochem. 14, 291–302.

Green, G.M. (1973). Alveolobronchiolar transport mechanisms. Arch. Intern. Med. 131, 109–114.

Grimnes, S. (1984). Pathways of ion flow through human skin *in vivo*. Acta. Derm. Venereol. 64, 93–98.

Groh, V., Fabbi, M., Hochstenbach, F., Maziarz, R. and Strominger, J.L. (1989). Double-negative (CD4⁻ CD8⁻) lymphocytes bearing T-cell receptor α and β chains in normal human skin. Proc. Natl Acad. Sci. USA 86, 5059–5063.

Gruskay, F.L. and Cooke, R.E. (1955). The gastrointestinal absorption of unaltered protein in normal infants and in infants recovering from diarrhoea. Pediatrics 16, 763–769.

Guagliardi, L.E., Koppelman, B., Blum, J.S., Marks, M.S., Cresswell, P. and Brodsky, F. (1990). Co-localization of molecules involved in antigen processing and presentation in an early endocytic compartment. Nature 343, 133–139.

Guilbert, L., Robertson, S.A. and Wegmann, T.G. (1993). The trophoblast as an integral component of a macrophage–cytokine network. Immunol. Cell Biol. 71, 49–57.

Guy-Grand, D., Cerf-Bensussan, N., Malissen, B., Melassis-Seris, M., Briottet, C. and Vassalli, P. (1991). Two gut intraepithelial CD8⁺ lymphocyte populations with different T cell receptors: A role for the gut epithelium in T cell differentiation. J. Exp. Med. 173, 471–481.

Hackett, C.J. (1991). Later for the rendevous. Nature 349, 655–656.

Hadorn, B. (1975). In "Paediatric Gastroenterology" (eds C.M. Anderson and V. Burke), Ch. 9. Blackwell, Oxford.

Hall, J.G. and Morris, B. (1965). The immediate effect of antigens on the cell output of a lymph node. Br. J. Exp. Pathol. 46, 450–454.

Halliday, G., Cavanagh, L. and Barnetson, R.S.C. (1990a). Regulation of the skin immune system by local antigen presenting cells. Today's Life Sci. 2, 26–34.

Halliday, G.M., Wood, R.C. and Muller, H.K. (1990b). Presentation of antigen to suppressor cells by a dimethylbenz(*a*)anthracine-resistant, Ia-positive, Thy-1-negative, I-J-restricted epidermal cell. Immunology 69, 97–103.

Hamilton, I., Rothwell, J., Archer, D. and Axon, A.T.R. (1987). Permeability of the rat small intestine to carbohydrate probe molecules. Clin. Sci. 73, 189–196.

Hämmerling, G.J. (1976). Tissue distribution of Ia antigens and their expression on lymphocyte subpopulations. Transplant. Rev. 30, 64–82.

Hanau, D., Fabre, M., Schmitt, D.A., Garaud, J.-C., Pauly, G., Tongio, M.-M., Mayer, S. and Cazenave, J.-P. (1987). Human epidermal Langerhans cells cointernalize by receptor-mediated endocytosis "non-classical" major histocompatibility complex class I molecules (T6 antigens) and class II molecules (HLA-DR antigens). Proc. Natl Acad. Sci. USA 84, 2901–2905.

Hanau, D., Fabre, M., Schmitt, D.A., Garaud, J.-C., Pauly, G. and Cazenave, J.-P. (1988). Appearance of Birbeck granule-like structures in anti-T6 antibody-treated human epidermal Langerhans cells. J. Invest. Dermatol. 90, 298–304.

Hanson, L.Å and Brandtzaeg, P. (1989). In "Immunologic Disorders in Infants and Children" (ed. E.R. Stiehm), pp 116–155. W.B. Saunders, London.

Hansson, H.-A., Lange, S. and Lönnroth, I. (1984). Internalization *in vivo* of cholera toxin in the small intestinal epithelium of the rat. Acta Pathol. Microbiol. Scand. 92, 15–21.

Harding, C.V., Collins, D.S., Slot, J.V., Geuze, H.J. and Unanue, E.R. (1991). Liposome-encapsulated antigens are processed in lysosomes, recycled and presented to T cells. Cell 64, 393–401.

Harmsen, A.G., Muggenburg, B.A., Snipes, M.B. and Bice, D.E. (1985). The role of macrophages in particle translocation from lungs to lymph nodes. Science 230, 1277–1280.

Harper, M.J.K. (1988). In "The Physiology of Reproduction" (eds E. Knobil and J. Neill), Ch. 4. Raven Press, New York.

Harvey, J., Jones, D.B. and Wright, D.H. (1990). Differential expression of MHC- and macrophage-associated antigens in human fetal and postnatal small intestine. Immunology 69, 409–415.

Hasegawa, H., Nakamura, A., Watanabe, K., Brown, W.R. and Nagura, H. (1987). Intestinal uptake of IgG in suckling rats. Distinction between jejunal and ileal epithelial cells demonstrated by simultaneous ultrastructural localization of IgG and acid phosphatase. Gastroenterology 92, 186–191.

Healey, P.J. (1984). Lysosomal enzymes in the intestine of the newborn lamb. Biol. Neonate 46, 186–191.

Helie, P., Morin, M., Jacques, M. and Fairbrother, J.M. (1991). Experimental infection of newborn pigs with an attacking and effacing *Escherichia coli* 045:K "E65" strain. Infect. Immunol. 59, 814–821.

Hemmings, W.A. (1980). Distributed digestion. Med. Hypotheses 6, 1209–1213.

Hemmings, W.A. and Williams, E.W. (1978). Transport of large breakdown products of dietary protein through the gut wall. Gut 19, 715–723.

Hemmings, C., Hemmings, W.A., Patey, A.L. and Wood, C. (1977). The ingestion of dietary protein as large molecular mass degradation products in adult rats. Proc. R. Soc. London B 198, 439–453.

Herbst, G. (1844). "Das Lymphgefässystem und seine Verrichtung". Vandenhoeck und Ruprecht, Göttingen.

Heremans, J.F. (1974). In "The Antigens", Vol. 2 (ed. M. Sela), Ch. 6. Academic Press, New York.

Herzenberg, L.A. and Herzenberg, L.A. (1989). Toward a layered immune system. Cell 59, 953–954.

Herzog, V. (1984). Pathways of endocytosis in thyroid follicle cells. Int. Rev. Cytol. 91, 107–139.

Hettwer, J.P. and Kriz, R.A. (1925). Absorption of undigested protein from the alimentary tract as determined by the direct anaphylaxis test. Am. J. Physiol. 73, 539–546.

Hettwer, J.P. and Kriz-Hettwer, R. (1926). Further observations on the absorption of undigested protein. Am. J. Physiol. 78, 136–149.

Heufler, C., Koch, F. and Schuler, G. (1988). Granulocyte–macrophage colony-stimulating factor and interleukin 1 mediate the maturation of murine epidermal Langerhans cells into potent immunostimulatory dendritic cells. J. Exp. Med. 167, 700–705.

Hill, S., Edwards, A.J., Kimber, I. and Knight, S.C. (1990). Systemic migration of dendritic cells during contact sensitization. Immunology 71, 277–281.

Hirata, I., Austin, L.L., Blackwell, W.H., Weber, J.R. and Dobbins, W.O. (1986). Immunoelectron microscopic localization of HLA-DR antigen in control small intestine and colon and in inflammatory bowel disease. Dig. Dis. Sci. 31, 1317–1330.

Hirschberg, H., Bergh, O.J. and Thorsby, E. (1980). Antigen-presenting properties of human vascular endothelial cells. J. Exp. Med. 152, 249s-255s.

Hodson, M.E. and Turner-Warwick, M. (1981). Autoantibodies in cystic fibrosis. Clin. Allergy 11, 565–570.

Hoffman-Goetz, L. (1984). In "Food Intolerance" (ed. R.K. Chandra), pp 3–16. Elsevier, New York.

Hohmann, A., Schmidt, G. and Rowley, D. (1979). Intestinal and serum antibody responses in mice after oral immunization with *Salmonella, Escherichia coli*, and *Salmonella-Escherichia coli* hybrid strains. Infect. Immunol. 25, 27–33.

Holt, P.G. and Batty, J.E. (1980). Alveolar macrophages. V. Comparative studies on the antigen presentation activity of guinea-pig and rat alveolar macrophages. Immunology 41, 361–366.

Holt, P.G. and Schon-Hegrad, M.A. (1988). Localization of T cells, macrophages and dendritic cells in rat respiratory tract tissue: implications for immune function studies. Immunology 62, 349–356.

Holt, P.G., Degebrodt, A., O'Leary, C., Krska, K. and Plozza, T. (1985). T cell activation by antigen-presenting cells from lung tissue digests: suppression by endogenous macrophages. Clin. Exp. Immunol. 62, 586–593.

Holt, P.G., Schon-Hegrad, M.A. and Oliver, J. (1988). MHC class II antigen-bearing dendritic cells in pulmonary tissues of the rat. Regulation of antigen presentation activity by endogenous macrophage populations. J. Exp. Med. 167, 262–274.

Holt, P.G., Schon-Hegrad, M.A. and McMenamin, P.G. (1990a). Dendritic cells in the respiratory tract. Int. Rev. Immunol. 6, 139–149.

Holt, P.G., Schon-Hegrad, M.A., Oliver, J., Holt, B.J. and McMenamin, P.G. (1990b). A contiguous network of dendritic antigen-presenting cells within the respiratory epithelium. Int. Arch. Allergy Appl. Immunol. 91, 155–159.

Höök, M., Raucci, G., Raja, R., Signäs, C., Jönsson, K. and Lindgren, P.-E. (1989). In "Molecular Mechanisms of Microbial Adherence" (eds L. Switalski, M. Höök and E. Beachey), pp 107–117. Springer-Verlag, New York.

Hughes, A., Bloch, K.J., Bhan, A.K., Gillen, D., Giovino, V.C. and Harmatz, P.R. (1991). Expression of MHC class II (Ia) antigen by the neonatal enterocyte: the effect of treatment with interferon-gamma. Immunology 72, 491–496.

Hughson, E.J. and Hopkins, C.R. (1990). Endocytic pathways in polarized Caco-2 cells: identification of an endosomal compartment accessible from both apical and basolateral surfaces. J. Cell Biol. 110, 337–348.

Hugon, J.S. (1971). Absorption of horseradish peroxidase by the mucosal cells of the duodenum of mouse. II. The newborn mouse. Histochemie 26, 19–27.

Hugon, J.S. and Borgers, M. (1968). Albsorption of horseradish peroxidase by the mucosal cells of the duodenum of mouse. I. The fasting mouse. J. Histochem. Cytochem. 16, 229–236.

Hume, D.A., Perry, V.H. and Gordon, S. (1984). The mononuclear phagocyte system of the mouse defined by immunohistochemical localization of antigen F4/80: macrophages associated with epithelia. Anat. Rec. 210, 503–512.

Husband, A.J. and Clifton, V.L. (1989). Role of intestinal immunization in urinary tract defence. Immunol. Cell Biol. 67, 371–376.

Husband, A.J. and Gowans, J.L. (1978). The origin and antigen-dependent distribution of IgA-containing cells in the intestine. J. Exp. Med. 148, 1146–1160.

Hyman, P.E., Clarke, D.D., Everett, S.L., Sonne, B., Stewart, D., Harada, T., Walsh, J.H. and Taylor, I.L. (1985). Gastric acid secretory function in preterm infants. J. Pediatr. 106, 467–471.

Inaba, K., Schuler, G., Witmer, M.D., Valinsky, J., Atassi, B. and Steinman, R.M. (1986). Immunologic properties of purified epidermal Langerhans cells. Distinct requirements for stimulation of unprimed and sensitized T lymphocytes. J. Exp. Med. 164, 605–613.

Inaba, K., Metlay, J.P., Crowley, M.T. and Steinman, R.M. (1990). Dendritic cells pulsed with protein antigens *in vitro* can prime antigen-specific, MHC-restricted T cells *in situ*. J. Exp. Med. 172, 631–640.

Inman, L.R. and Cantey, J.R. (1983). Specific adherence of *Escherichia coli* (strain RDEC-1) to membranous (M) cells of the Peyer's patch in *Escherichia coli* diarrhea in the rabbit. J. Clin. Invest. 71, 1–8.

Isaacson, R.E. (1985). In "Bacterial Adhesion" (eds D.C. Savage and M. Fletcher), Ch. 11. Plenum Press, New York.

Isaacson, R.E., Dean, E.A., Morgan, R.L. and Moon, H.W. (1980). Immunization of suckling pigs against enterotoxigenic *Escherichia coli*-induced diarrheal disease by vaccinating dams with purified K99 or 987P pili: antibody production in response to vaccination. Infect. Immunol. 29, 824–826.

Isberg, R.R., Voorhis, D.L. and Falkow, S. (1987). Identification of invasin: a protein that allows enteric bacteria to penetrate cultured mammalian cells. Cell 50, 769–778.

Jakobsson, I. (1988). In "Food Allergy" (eds D. Reinhardt and E. Schmidt), pp 234–244. Raven Press, New York.

Jakobsson, I., Lindberg, T. and Benediktsson, B. (1982). *In vitro* digestion of cow's milk proteins by duodenal juice from infants with various gastrointestinal disorders. J. Pediatr. Gastroenterol. Nutr. 1, 183–191.

Jakobsson, I., Lindberg, T. Benediktsson, B. and Hansson, B.G. (1985). Dietary bovine β-lactoglobulin is transferred to human milk. Acta Paediatr. Scand. 74, 342–345.

Jakobsson, I., Lindberg, T., Lothe, L., Axelsson, I. and Benediktsson, B. (1986). Human alpha-lactalbumin as a marker of macromolecular absorption. Gut 27, 1029–1034.

Janeway, C.A. (1988). Frontiers of the immune system. Nature 333, 804–806.

Janossy, G., Tidman, N., Selby, W.S., Thomas, J.A., Granger, S., Kung, P.C. and Goldstein, G. (1980). Human T lymphocytes of inducer and suppressor type occupy different microenvironments. Nature 288, 81–84.

Janossy, G., Bofill, M., Poulter, L.W., Rawlings, E., Burford, G.D., Navarrete, C., Ziegler, A. and Kelemen, E. (1986). Separate ontogeny of two macrophage-like accessory cell populations in the human foetus. J. Immunol. 136, 4354–4361.

Jarry, A., Robaszkiewicz, M., Brousse, N. and Potet, F. (1989). Immune cells associated with M cells in the follicle-associated epithelium of Peyer's patches in the rat. An electron- and immuno-electron-microscopic study. Cell Tissue Res. 255, 293–298.

Joel, D.D., Laissue, J.A. and LeFevre, M.E. (1978). Distribution and fate of ingested carbon particles in mice. J. Reticuloendothel. Soc. 24, 477–487.

Johnson, A.P. (1981). The pathogenesis of gonorrhoea. J. Infect. 3, 299–308.

Johnson, M.H. (1970). An immunological barrier in the guinea-pig testis. J. Pathol. 101, 129–139.

Jones, R.E. (1974). Studies *in vivo* and *in vitro* of the transfer of rat IgG and rat albumen across the intestinal walls of young rats. Biol. Neonate 24, 220–229.

Jordan, S.M. and Morgan, E.H. (1968). The development of selectivity of protein absorption from the intestine during suckling in the rat. Aust. J. Exp. Biol. Med. Sci. 46, 465–472.

Jos, J., Charbonnier, L., Mosse, J., Olives, J.P., deTand, M.F. and Rey, L. (1982). The toxic fraction of gliadin digests in coeliac disease. Isolation by chromatography on Biogel P-10. Clin. Chim. Acta, 119, 263–274.

Journal of reproductive Immunology (1990). Immune privilege in the testis. J. Reprod. Immunol. 18, 1–121 (Special Issue).

Kagnoff, M.F. (1982). Oral tolerance. Ann. N.Y. Acad. Sci. 392, 248–265.

Kagnoff, M.F. and Campbell, S. (1974). Functional characteristics of Peyer's patch lymphoid cells. I. Induction of humoral antibody and cell-mediated allograft reactions. J. Exp. Med. 139, 398–406.

Kaiserlain, D., Vidal, K. and Revillard, J.-P. (1989). Murine enterocytes can present soluble antigen to specific class II-restricted CD4+T cells. Eur. J. Immunol. 19, 1513–1516.

Kaiserlain, D., Vidal, K. and Revillard, J.P. (1990). In "Advances in Mucosal Immunology" (eds T.T. MacDonald, S.J. Challacombe, P.W. Bland, C.R. Stokes, R.V. Heatley and A.McI. Mowat), pp 34–37. Kluwer, Dordrecht.

Kao, V.C.Y., Sprinz, H. and Burrows, W. (1972). Localisation of cholera toxin in rabbit intestine. An immunohistochemical study. Lab. Invest. 26, 148–153.

Keljo, D.J. and Hamilton, J.R. (1983). Quantitative determination of macromolecular transport rate across intestinal Peyer's patches. Am. J. Physiol. 244, G637–G644.

Kelly, R.H., Balfour, B.M., Armstrong, J.A. and Griffiths, S. (1978). Functional anatomy of lymph nodes. II. Peripheral lymph-borne mononuclear cells. Anat. Rec. 190, 5–22.

Kerchner, G.A. and Geary, L.E. (1983). Studies on the transport of enkephalin-like oligopeptides in rat intestinal mucosa. J. Pharmacol. Exp. Ther. 226, 33–38.

Keren, D.F., Holt, P.S., Collins, H.H., Gemski, P. and Formal, S.B. (1978). The role of Peyer's patches in the local immune response of rabbit ileum to live bacteria. J. Immunol. 120, 1892–1896.

Kilburn, K.H. (1974). Clearance zones in the distal lung. Ann. N.Y. Acad. Sci. 221, 276–281.

Kilshaw, P.J. and Cant, A.J. (1984). The passage of maternal dietary proteins into human breast milk. Int. Arch. Allergy Appl. Immunol. 75, 8–15.

Kimura, A. (1977). The epithelial–macrophagic relationship in Peyer's patches; an immunopathological study. Bull. Osaka Med. Sch. 23, 67–91.

King, P.D. and Katz, D.R. (1990). Mechanisms of dendritic cell function. Immunol. Today 11, 206–211.

King, T.P., Pusztai, A. and Clarke, E.M.W. (1982). Kidney bean (*Phaseolus vulgaris*) lectin-induced lesions in rat small intestine. 3. Ultrastructural studies. J. Comp. Pathol. 92, 357–373.

Kingham, J.G.C. and Loehry, C.A. (1976). Permeability of the small intestine after intra-arterial injection of histamine type mediators and irradiation. Gut 17, 517–526.

Kingham, J.G.C., Whorwell, P.J. and Loehry, C.A. (1976). Small intestinal permeability. I. Effects of ischaemia and exposure to acetylsalicylate. Gut 17, 354–361.

Klareskog, L., Forsum, U. and Peterson, P.A. (1980). Hormonal regulation of the expression of Ia antigens on mammary gland epithelium. Eur. J. Immunol. 10, 958–963.

Klipstein, F.A. and Engert, R.F. (1979). Protective effect of active immunization with purified *Escherichia coli* heat-labile enterotoxin in rats. Infect. Immunol. 23, 592–599.

Knight, S.C., Balfour, B.M., O'Brien, J., Buttifant, L., Sumerska, T. and Clark, J. (1982). Role of veiled cells in lymphocyte activation. Eur. J. Immunol. 12, 1057–1060.

Knutton, S., Baldwin, T., Williams, P.H. and McNeish, A.S. (1988). New diagnostic test for enteropathogenic *Escherichia coli*. Lancet i, 1337.

Kohbata, S., Yokayama, H. and Yabuuchi, E. (1986). Cytopathogenic effect of *Salmonella typhi* GIFU 10007 on M cells of murine ileal Peyer's patches in ligated ileal loops: an ultrastructural study. Microbiol. Immunol. 30, 1225–1237.

Köttgen, E., Volk, B., Kluge, F. and Gerok, W. (1982). Gluten, a lectin with oligomannosyl specificity and the causative agent of gluten-sensitive enteropathy. Biochem. Biophys. Res. Commun. 109, 168–173.

Kraehenbuhl, J.P. and Campiche, M.A. (1969). Early stages of intestinal absorption of specific antibodies in the newborn. An ultrastructural, cytochemical, and immunological study in the pig, rat and rabbit. J. Cell Biol. 42, 345–365.

Kraehenbuhl, J.-P. and Neutra, M.R. (1992). Transepithelial transport and mucosal defence II: secretion of IgA. Trends Cell Biol. 2, 170–174.

Kraft, S.C., Rothberg, R.M., Knauer, C.M., Svoboda, A.C., Monroe, L.S. and Farr, R.S. (1967). Gastric acid output and circulating anti-bovine serum albumin in adults. Clin. Exp. Immunol. 2, 321–330.

Kumagai, K. (1922). Über den Resorptions vergang der corpusculären Bestandteile im Darm. Cited by Owen, R.L. (1983). Gastroenterology 85, 468–470.

Lake, J.P., Pierce, C.W. and Kennedy, J.D. (1991). T cell receptor expression by T cells that mature extrathymically in nude mice. Cell. Immunol. 135, 259–265.

Landahl, H.D., Tracewell, T.N. and Lassen, W.H. (1951). On the retention of airborne particles in the human lung. Arch. Industr. Hyg. Occup. Med. 3, 356–366.

Landman, L. (1986). Epidermal permeability barrier: transformation of lamellar granule discs into intercellular sheets by a membrane-fusion process, a freeze-fracture study. J. Invest. Dermatol. 87, 202–209.

Laughter, A.H., Martin, R.R. and Twomey, J.J. (1977). Lymphoproliferative responses to antigens mediated by human pulmonary alveolar macrophages. J. Lab. Clin. Med. 89, 1326–1332.

Lause, D.B. and Bockman, D.E. (1981). Heterogeneity, position, and functional capability of the macrophages in Peyer's patches. Cell Tissue Res. 218, 557–566.

Lauweryns, J.M. and Baert, J.H. (1974). The role of the pulmonary lymphatics in the defences of the distal lung: morphological and experimental studies of the transport mechanisms of intratracheally instilled particles. Ann. N.Y. Acad. Sci. 221, 244–275.

Lauweryns, J.M. and Baert, J.H. (1977). Alveolar clearance and the role of the pulmonary lymphatics. Am. Rev. Resp. Dis. 115, 625–683.

Lebenthal, E. and Lee, P.C. (1980). Development of functional response in human exocrine pancreas. Pediatr. 66, 556–560.

Lecce, J.G. (1972). Selective absorption of macromolecules into intestinal epithelium and blood by neonatal mice. J. Nutr. 102, 69–76.

Lechler, R.I. and Batchelor, J.R. (1982). Restoration of immunogenicity to passenger cell-depleted allografts by the addition of donor strain dendritic cells. J. Exp. Med. 155, 31–41.

Lee, A. (1985). In "Advances in Microbial Ecology", Vol. 8 (ed. K.C. Marshall), pp 115–162. Plenum Press, New York.

LeFevre, M.E., Olivo, R., Vanderhoff, J.W. and Joel, D.D. (1978). Accumulation of latex in Peyer's patches and its subsequent appearance in villi and mesenteric lymph nodes. Proc. Soc. Exp. Biol. Med. 159, 298–302.

LeFevre, M.E., Hammer, R. and Joel, D.D. (1979). Macrophages

of the mammalian small intestine: a review. J. Reticuloendothel. Soc. 26, 553–573.

Leong, J.M., Fournier, R.S. and Isberg, R.R. (1990). Identification of the integrin binding domain of the *Yersinia pseudotuberculosis* invasin protein. EMBO J. 9, 1979–1989.

Lev, R. and Orlic, D. (1973). Uptake of protein in swallowed amniotic fluid by monkey fetal intestine in utero. Gastroenterology 65, 60–68.

Levine, J.S., Nakane, P.K. and Allen, R.H. (1982). Immunocytochemical localization of intrinsic factor-cobalamin bound to the guinea pig ileum *in vivo*. Gastroenterology 82, 284–290.

Levine, J.S., Allen, R.H., Alpers, D.H. and Seetharam, B. (1984). Immunocytochemical localization of the intrinsic factor – cobalamin receptor in dog-ileum: distribution of intracellular receptor during cell maturation. J. Cell Biol. 98, 1111–1118.

Levine, M.M., Black, R.E., Clements, M.L., Cisneros, L., Nalin, D.R. and Young, C.R. (1981). Duration of infection-derived immunity to cholera. J. Infect. Dis. 143, 818–820.

Levine, M.M., Black, R.E., Clements, M.L., Young, C.R., Cheney, C.P., Schad, P., Collins, H. and Boedeker, E.C. (1984a). In "Attachment of Organisms to the Gut Mucosa", Vol. II (ed. E.C. Boedeker), pp 223–244. CRC Press, Boca Raton.

Levine, M.M., Black, R.E., Clements, M.L., Lanata, C., Sears, S., Honda, T. and Finkelstein, R. (1984b). Evaluation in humans of attenuated *Vibrio cholerae* El Tor Ogawa strain Texas Star-SR as a live oral vaccine. Infect. Immunol. 43, 515–522.

Levine, M.M., Herrington, D., Murphy, J.R., Morris, J.G., Losonsky, G., Tall, B., Lindberg, A.A., Svenson, S., Baqar, S., Edwards, M.F. and Stocker, B. (1987). Safety, infectivity, immunogenicity, an *in vivo* stability of two attenuated auxotrophic mutant strians of *Salmonella typhi*, 541 Ty and 543 Ty, as live oral vaccines in humans. J. Clin. Invest. 79, 888–902.

Linder, J. and Russell-Jones, G.J. (1990). In "Mucosal Immunology" (ed. A.W. Cripps), p 168. Newey and Beath, Newcastle, Australia.

Lipscomb, M.F., Toews, G.R., Lyons, C.R. and Uhr, J.W. (1981). Antigen presentation by guinea pig alveolar macrophages. J. Immunol. 126, 286–291.

Liss, G.M., Kominsky, J.R., Gallagher, J.S., Melius, J., Brooks, S.M. and Bernstein, I.L. (1984). Failure of enzyme encapsulation to prevent sensitization of workers in the dry bleach industry. J. Allergy Clin. Immunol. 73, 348–355.

Liu, L.M. and MacPherson, G.G. (1991). Lymph-borne (veiled) dendritic cells can acquire and present intestinally administered antigens. Immunology, 73, 281–286.

Loehry, C.A., Axon, A.T.R., Hilton, P.J., Hider, R.C. and Creamer, B. (1970). Permeability of the small intestine to substances of different molecular weight. Gut, 11, 466–470.

Lorenzsonn, V. and Olsen, W.A. (1982). *In vivo* responses of rat intestinal epithelium to intralumenal dietary lectins. Gastroenterology 82, 838–848.

Luger, T.A. and Schwarz, T. (1991). In "Epidermal Langerhans Cells" (ed. G. Schuler), pp 217–251. CRC Press, Boca Raton.

Luger, T.A., Stadler, B.M., Katz, S.I. and Oppenheim, J.J. (1981). Epidermal cell (keratinocyte)-derived thymocyte-activating factor (ETAF). J. Immunol. 127, 1493–1498.

Lyons, C.R. and Lipscomb, M.F. (1983). Alveolar macrophages in pulmonary immune responses. I. Role in the initiation of primary immune responses and in the selective recruitment of T lymphocytes to the lung. J. Immunol. 130, 1113–1119.

Ma, T.Y., Hollander, D., Krugliak, P. and Katz, K. (1990). PEG 400, a hydrophilic molecular probe for measuring intestinal permeability. Gastroenterology 98, 39–46.

Macatonia, S.E., Edwards, A.J. and Knight, S.C. (1986). Dendritic cells and the initiation of contact sensitivity to fluorescein isothiocyanate. Immunology, 59, 509–514.

Macatonia, S.E., Knight, S.C., Edwards, A.J., Griffiths, S. and Fryer, P. (1987). Localization of antigen on lymph node dendritic cells after exposure to the contact sensitizer fluorescein isothiocyanate. Functional and morphological studies. J. Exp. Med. 166, 1654–1667.

MacDonald, T.T. and Carter, P.B. (1982). Isolation and functional characteristics of adherent phagocytic cells from mouse Peyer's patches. Immunology, 45, 769–774.

MacDonald, T.T., Weinel, A. and Spencer, J. (1988). HLA-DR expression in human fetal intestinal epithelium. Gut 29, 1342–1348.

McDonel, J.L. (1980) Binding of *Clostridium perfringens* (^{125}I) enterotoxin to rabbit intestinal cells. Biochemistry 19, 4801–4807.

McDougall, W.A. (1970). Pathways of penetration and effects of horseradish peroxidase in rat molar gingiva. Arch. Oral Biol. 15, 621–633.

McDougall, W.A. (1971). Penetration pathways of a topically applied foreign protein into rat gingiva. J. Periodont. Res. 6, 89–99.

Mackay, C.R., Marston, W.L. and Dudler, L. (1990). Naive and memory T cells show distinct pathways of lymphocyte recirculation. J. Exp. Med. 171, 801–817.

Mackenzie, D.D.S. (1972). Selective uptake of immunoglobulins by the proximal intestine of suckling rats. Am. J. Physiol. 223, 1286–1295.

McKenzie, N.M. (1990). In "Ontogeny of the Immune System of the Gut" (ed. T.T. MacDonald), pp 69–81. CRC Press, Boca Raton.

Macklin, C.C. (1955). Pulmonary sumps, dust accumulations, alveolar fluid and lymph vessels. Acta Anat. 23, 1–33.

MacPherson, G.G. and Pugh, C.W. (1984). Heterogeneity amongst lymph-borne "dendritic cells". Immunobiology 168, 338–348.

MacPherson, G.G. and Steer, H.W. (1979). Properties of mononuclear phagocytes derived from the small intestinal wall of rats. Adv. Exp. Med. Biol. 114, 433–438.

Madara, J.L. and Trier, J.S. (1987). In "Physiology of the Gastrointestinal Tract", 2nd edn (ed. L.R. Johnson), Ch. 44. Raven Press, New York.

Madara, J.L., Bye, W.A. and Trier, J.S. (1984). Structural features of and cholesterol disribution in M-cell membranes in guinea pig, rat, and mouse Peyer's patches. Gastroenterology 87, 1091–1103.

Mahida, Y.R., Patal, S., Gionchetti, P., Vaux, D. and Jewell, D.P. (1989). Macrophage subpopulations in the lamina propria of normal and inflamed colon and terminal ileum. Gut 30, 826–834.

Malizia, G., Trejdosiewicz, L.K., Wood, G.M., Howdle, P.D., Janossy, G. and Losowsky, M.S. (1985). The microenvironment of coeliac disease: T cell phenotypes and expression of the T2 'T blast' antigen by small bowel lymphocytes. Clin. Exp. Immunol. 60, 437–446.

Marsh, D.G. (1975). In "The Antigens", Vol. 3 (ed. M. Sela), Ch 4. Academic Press, New York.

Mason, D.W., Pugh, C.W. and Webb, M. (1981a). The rat mixed lymphocyte reaction: roles of a dendritic cell in intestinal lymph and T-cell subsets defined by monoclonal antibodies. Immunology 44, 75–87.

Mason, D.W., Dallman, M. and Barclay, A.N. (1981b). Graft-versus-host disease induces expression of Ia antigen in rat epidermal and gut epithelium. Nature 293, 150–151.

Matsuno, K., Schaffner, T., Gerber, H.A., Ruchti, C., Hess, M.W. and Cottier, H. (1983). Uptake by enterocytes and subsequent translocation to internal organs, e.g. the thymus, of Percoll microspheres administered per os to suckling mice. J. Reticuloendothel. Soc. 33, 263–273.

Matter, K., Stieger, B., Klumperman, J., Ginsel, L. and Hauri, H.-P. (1990). Endocytosis, recycling, and lysosomal delivery of brush border hydrolases in cultured human intestinal epithelial cells (Caco-2). J. Biol. Chem. 265, 3503–3512.

Matthews, D.M. and Payne, J.W. (1980). Transmembrane transport of small peptides. Curr. Topics Membrane Transport 14, 331–425.

Mayer, L. (1987). The role of epithelial cells as accessory cells. Adv. Exp. Med. Biol. 216, 209–218.

Mayer, L. and Eisenhardt, D. (1990). Lack of induction of suppressor T cells by intestinal epithelial cells from patients with inflammatory bowel disease. J. Clin. Invest. 86, 1255–1260.

Mayer, L. and Shlien, R. (1987). Evidence for function of Ia molecules on gut epithelial cells in man. J. Exp. Med. 166, 1471–1483.

Mayer, L., Siden, E., Becker, S. and Eisenhardt, D. (1990). "Advances in Mucosal Immunology" (eds T.T. MacDonald, S.J. Challacombe, P.W. Bland, C.R. Stokes, R.V. Heatley and A. McI. Mowat), pp 23–28. Kluwer, Dordrecht.

Mayer, L., Eisenhardt, D., Salomon, P., Bauer, W., Plous, R. and Piccinini, L. (1991). Expression of class II molecules on intestinal epithelial cells in humans. Gastroenterology 100, 3–12.

Mayrhofer, G. (1984). In "Local Immune Responses of the Gut" (eds T.J. Newby and C.R. Stokes), pp 1–96. CRC Press, Boca Raton.

Mayrhofer, G. and Schon-Hegrad, M.A. (1983). Ia antigens in rat kidney, with special reference to their expression in tubular epithelium. J. Exp. Med. 157, 2097–2109.

Mayrhofer, G. and Spargo, L.D.J. (1989). Subcellular distribution of class II major histocompatibility antigens in enterocytes of the human and rat small intestine. Immunol. Cell Biol. 67, 251–260.

Mayrhofer, G. and Spargo, L.D.J. (1990). Distribution of class II histocompatibility antigens in enterocytes of the rat jejunum and their associated organelles of the endocytic pathway. Immunology 70, 11–19.

Mayrhofer, G., Pugh, C.W. and Barclay, A.N. (1983). The distribution, ontogeny and origin in the rat of Ia-positive cells with dendritic morphology and of Ia antigen in epithelia, with special reference to the intestine. Eur. J. Immunol. 13, 112–122.

Mayrhofer, G., Holt, P.G. and Papadimitriou, J.M. (1986). Functional characteristics of the veiled cells in afferent lymph from the rat intestine. Immunology 58, 379–387.

Mayrhofer, G., Spargo, L.D.J., Davidson, G.P. and Huang, Z.-H. (1990). In "Advances in Mucosal Immunology" (eds T.T. MacDonald, S.J. Challacombe, P.W. Bland, C.R. Stokes, R.V. Heatley and A. McI. Mowatt), pp 29–33. Kluwer, Dordrecht.

Menzies, I.S. (1974). Absorption of intact oligosaccharides in health and disease. Biochem. Soc. Trans. 2, 1042–1047.

Menzies, I.S. (1983). In "Intestinal Absorption and Secretion" (eds E. Skadhauge and K. Heintze), pp 527–543. MTP Press, Lancaster.

Meuwissen, H.J. and Hussain, M. (1982). Bronchus-associated lymphoid tissue in human lung: correlation of hyperplasia with chronic pulmonary disease. Clin. Immunol. Immunopathol. 23, 548–561.

Michalek, S.M., McGhee, J.R. and Babb, J.L. (1978). Effective immunity to dental caries: dose-dependent studies on secretory immunity by oral administration of *Streptococcus mutans* to rats. Infect. Immun. 19, 217–224.

Milne, J. and Brand, S. (1975), Occupational asthma after inhalation of dust of the proteolytic enzyme, papain. Br. J. Indust. Med. 32, 302–307.

Moffat, A.C. (1971). Absorption of drugs through the oral mucosa. Top. Med. Chem. 4, 1–26.

Molitoris, B.A. and Nelson, W.J. (1990). Alterations in the establishment and maintenance of epithelial cell polarity as a basis for disease processes. J. Clin. Invest. 85, 3–9.

Mollick, J.A., Cook, R.C. and Rich, R.R. (1989). Class II MHC molecules are specific receptors for staphylococcus enterotoxin A. Science 244, 817–820.

Montesano, R., Roth, J. and Orci, L. (1982). Non-coated membrane invaginations are involved in binding and internalization of cholera and tetanus toxins. Nature 296, 651–653.

Moon, H.W. and Runnels, P.L. (1984). In "Attachment of Organisms to the Gut Mucosa", Vol. I (ed. E.C. Boedeker), Ch. 4. CRC Press, Boca Raton.

Moore, R., Carlson, S. and Madara, J.L. (1989a). Villus contraction aids repair of intestinal epithelium after injury. Am. J. Physiol. 20, G274–G283.

Moore, R.S., Carlson, S. and Madara, J.L. (1989b). Rapid barrier restitution in an *in vitro* model of intestinal injury. Lab. Invest. 60, 237–244.

Morris, B. and Morris, R. (1976). The effects of corticosterone and cortisone on the uptake of polyvinylpyrrolidone and the transmission of immunoglobulin G by the small intestine in young rats. J. Physiol. 254, 389–403.

Morris, I.G. (1968). In "Handbook of Physiology", Sect. 6, Vol. 3 (eds C.F. Code and W. Heidel), Ch. 75. American Physiological Society, Washington, DC.

Morris, I.G. (1978). In "Antigen Absorption by the Gut" (ed. W.A. Hemmings), Ch. 3. MTP Press, Lancaster.

Morrow, P.E. (1973). Alveolar clearance of aerosols. Arch. Intern. Med. 131, 101–108.

Mowatt, A.McI. and Donachie, A.M. (1991). ISCOMS – a novel strategy for mucosal immunization? Immunol. Today 12, 383–385.

Moxey, P.C. and Trier, J.S. (1975). Structural features of the mucosa of human fetal small intestine. Gastroenterology 68, 1002–1003.

Muckerheide, A., Pesce, A.J. and Michael, J.G. (1977). Immunosuppressive properties of a peptic fragment of BSA. J. Immunol. 119, 1340–1345.

Myrvick, Q.N. and Ockers, J.R. (1982). Transport of horseradish peroxidase from the vascular compartment to bronchial-associated lymphoid tissue. J. Reticuloendothel. Soc. 31, 267–278.

Nair, P.N.R. and Schroeder, H.E. (1983). Retrograde access of oral antigens into minor salivary glands in the monkey *Macaca fasicularis*. Arch. Oral Biol. 28, 145–152.

Nair, P.N.R. and Schroeder, H.E. (1985). Architecture of associations of minor salivary gland ducts and lymphoid follicles in *Macaca fasicularis*. An ultrastructural study. Cell Tissue Res. 240, 223–232.

Nair, P.N.R. and Schroeder, H.E. (1986). Duct-associated lymphoid tissue (DALT) of the minor salivary glands and mucosal immunity. Immunology 57, 171–180.

Neutra, M.R. and Kraehenbuhl, J.-P. (1992). Transepithelial transport and mucosal defence I: the role of M cells. Trends Cell Biol. 2, 134–138.

Neutra, M.R., Phillips, T.L., Mayer, E.L. and Fishkind, D.J. (1987). Transport of membrane-bound macromolecules by M cells in follicle-associated epithelium of rabbit Peyer's patch. Cell Tissue Res. 247, 537–546.

Newby, T.J. (1984). In "Local Immune Responses of the Gut" (eds T.J. Newby and C.R. Stokes), pp 143–198. CRC Press, Boca Raton.

Newman, R.A., Ormerod, M.G. and Greaves, M.F. (1980). The presence of HLA-DR antigens on lactating human breast epithelium and milk fat globule membranes. Clin. Exp. Immunol. 41, 478–486.

Nicod, L.P., Lipscomb, M.F., Weissler, J.G., Lyons, C.R., Albertson, J. and Toews, G.B. (1987). Mononuclear cells in human lung parenchyma. Characterization of a potent accessory cell not obtained by bronchoalveolar lavage. Am. Rev. Resp. Dis. 136, 818–823.

Nixon, S.E. and Mawer, G.E. (1970a). The digestion and absorption of protein in man. I. The site of absorption. Br. J. Nutr. 24, 227–240.

Nixon, S.E. and Mawer, G.E. (1970b). The digestion and absorption of protein in man. II. The form in which digested protein is absorbed. Br. J. Nutr. 24, 241–258.

Nottle, M. (1977). Plant particles in kidneys and mesenteric lymph nodes of sheep. Aust. Vet. J. 53, 404–405.

Novey, H.S., Wilson, A.F., Surprenant, E.L. and Bennett, L.R. (1970). Early ventilation – perfusion changes in asthma. J. Allergy 46, 221–230.

O'Hagan, D.T. and Davies, S.S. (1990). In "Advances in Mucosal Immunology" (eds T.T. MacDonald, S.J. Challacombe, P.W. Bland, C.R. Stokes, R.V. Heatley and A. McI. Mowatt), pp 386–387. Kluwer, Dordrecht.

O'Leary, A.D. and Sweeny, E.C. (1986). Lymphoglandular complexes of the colon: structure and distribution. Histopathology 10, 267–283.

Overstreet, J.W. and Cooper, G.W. (1978). Sperm transport in the reproductive tract of the female rabbit. I. The rapid transit phase of transport. Biol. Reprod. 19, 101–114.

Owen, R.L. (1977). Sequential uptake of horseradish peroxidase by lymphoid follicle epithelium of Peyer's patches in the normal unobstructed mouse intestine: an ultrastructural study. Gastroenterology 72, 440–451.

Owen, R.L. and Jones, A.L. (1974). Epithelial cell specialization within human Peyer's patches: an ultrastructural study of intestinal lymphoid follicles. Gastroenterology 66, 189–203.

Owen, R.L. and Nemanic, P. (1978). In "Scanning Electron-microscopy" (eds R.P. Becker and O. Johari), pp 367–378. SEM Inc., Chicago.

Owen, R.L., Allen, C.L. and Stevens, D.P. (1981). Phagocytosis of *Giardia muris* by macrophages in Peyer's patch epithelium in mice. Infect. Immunol. 33, 591–601.

Owen, R.L., Apple, R.T. and Bhalla, D.K. (1986a). Morphometric and cytochemical analysis of lysosomes in rat Peyer's patch follicle epithelium: their reduction in volume fraction and acid phosphatase content in M cells compared to adjacent enterocytes. Anat. Rec. 216, 521–527.

Owen, R.L., Pierce, N.F., Apple, R.T. and Cray, W.C. (1986b). M cell transport of *Vibrio cholerae* from the intestinal lumen into Peyer's patches: a mechanism for antigen sampling and for microbial transepithelial migration. J. Infect. Dis. 153, 1108–1118.

Owen, R.L., Apple, R.T. and Bhalla, D.K. (1986c). Morphometric and cytochemical analysis of lysosomes in rat Peyer's patch follicle epithelium: their reduction in volume fraction and acid phosphatase content in M cells compared to adjacent enterocytes. Anat. Rec. 216, 521–527.

Owen, R.L., Koster, F.T. and Ermak, T.H. (1986d). Autoradiographic localizaion of uptake of cholera toxin in Peyer's patches in non-immune rats. Clin. Res. 34, 101A.

Owens, W.E. and Berg, R.E. (1982). Bacterial translocation from the gastrointestinal tract of athymic (nu/nu) mice. Infect. Immunol. 27, 461–467.

Pabst, R. (1992). Is BALT a major component of the human lung immune system? Immunol. Today 13, 119–122.

Paganelli, R., Levinsky, R.J., Brostoff, J. and Wraith, D.G. (1979). Immune complexes containing food proteins in normal and atopic subjects after oral challenge and effect of sodium cromoglycate on antigen absorption. Lancet i, 1270–1272.

Pallone, F., Fais, S. and Capobianchi, M.R. (1988). HLA-D region antigens on isolated human colonic epithelial cells: enhanced expression in inflammatory bowel disease and *in vitro* induction by different stimuli. Clin. Exp. Immunol. 74, 75–79.

Pang, G., Clancy, R. and Saunders, H. (1990). Dual mechanisms of inhibition of the immune response by enterocytes isolated from the rat small intestine. Immunol. Cell Biol. 68, 387–396.

Pankow, W. and von Wichert, P. (1988). M cell in the immune system of the lung. Respiration 54, 209–219.

Pappo, J., Ermak, T.H. and Steger, H.J. (1991). Monoclonal antibody-directed targeting of fluorescent polystyrene microspheres to Peyer's patch M cells. Immunology 73, 277–280.

Parr, E.L. and McKenzie, I.F.C. (1979). Demonstration of Ia antigens on mouse intestinal epithelial cells by immunoferritin labeling. Immunogenetics 8, 499–508.

Parr, E.L. and Parr, M.B. (1986). Uptake of immunoglobulins and other proteins from serum into epithelial cells of the mouse uterus and oviduct. J. Reprod. Immunol. 9, 339–354.

Parr, M.B., Kepple, L. and Parr, E.L. (1991a). Langerhans cells phagocytose vaginal epithelial cells undergoing apoptosis during the murine estrous cycle. Biol. Reprod. 45, 252–260.

Parr, M.B., Kepple, L. and Parr, E.L. (1991b). Antigen recognition in the female reproductive tract. II. Endocytosis of horseradish peroxidase by Langerhans cells in murine vaginal epithelium. Biol. Reprod. 45, 261–265.

Parrott, D.M.V. and Ferguson, A. (1974). Selective migration of lymphocytes within the mouse small intestine. Immunology 26, 571–588.

Parrott, D.M.V. and Wilkinson, P.C. (1981). Lymphocyte locomotion and migration. Prog. Allergy 28, 193–284.

Pavli, P. (1990). Antigen-presenting cells of the gastrointestinal tract. Ph.D. Thesis, Australian National University, Ch. 5.

Pavli, P., Doe, W.F. and Hume, D.A. (1989). Isolation and enrichment of human colonic antigen-presenting cells (DCs). Gastroenterology 96, A385.

Pavli, P., Woodhams, C.E., Doe, W.F. and Hume, D.A. (1990). Isolation and characterization of antigen-presenting dendritic

cells from the mouse intestinal lamina propria. Immunology 70, 40–47.

Payne, L.C. and Marsh, C.L. (1962). Absorption of gamma globulin by the small intestine. Fed. Proc. 21, 909–912.

Peppard, J., Orlans, E., Payne, A.W.R. and Andrew, E. (1981). The elimination of circulating complexes containing polymeric IgA by excretion in the bile. Immunology 42, 83–89.

Peppard, J.V., Orlans, E., Andrew, E. and Payne, A.W.R. (1982). Elimination into bile of circulating antigen by endogenous IgA antibody in rats. Immunology 45, 467–472.

Peters, P.J., Neefjes, J.J., Oorschot, V., Ploegh, H.L. and Geuze, H.J. (1991). Segregation of MHC class II molecules from MHC class I molecules in the Golgi complex for transport to lysosomal compartments. Nature 349, 669–676.

Peters, T.J. and Bjarnason, I. (1985). In "Food and the Gut" (eds J.O. Hunter and V. Alun Jones), pp 30–44. Baillière Tindall, London.

Peterson, J.W. and Verwey, W.F. (1974). Radiolabelled toxin or studying binding of cholera toxin and toxoids to intestinal mucosal receptor sites. Proc. Soc. Exp. Biol. Med. 145, 1187–1191.

Phillip, P.D. and Katz, D.R. (1990). Mechanisms of dendritic cell function. Immunol. Today 11, 206–211.

Phillips, T.E., Phillips, T.L. and Neutra, M.R. (1987). Macromolecules can pass through occluding junctions of rat ileal epithelium during cholinergic stimulation. Cell. Tissue Res. 247, 547–554.

Picker, L.J., Kishimoto, T.K., Smith, C.W., Warnock, R.A. and Butcher, E.C. (1991). ELAM-1 is an adhesion molecule for skin-homing T cells. Nature 349, 796–799.

Pierce, A.E. (1959). Studies on the proteinuria of the new-born calf. J. Physiol. 148, 469–488.

Pierce, A.E. (1961). Further studies on proteinuria in the newborn calf. J. Physiol. 156, 136–149.

Pierce, N.F. (1978). The role of antigen form and function in the primary and secondary intestinal immune responses to cholera toxin and toxoid in rats. J. Exp. Med. 148, 195–206.

Plumb, J.A., Burston, D., Baker, T.G. and Gardener, M.L.G. (1987). A comparison of the structural integrity of several commonly used preparations of rat small intestine *in vitro*. Clin. Sci. 73, 53–59.

Porcelli, S., Brunner, M.B., Greenstein, J.L., Balk, S.P., Terhorst, C. and Bleicher, P.A. (1989). Recognition of cluster of differentiation 1 antigens by human CD4⁻ CD8⁻ cytolytic T cells. Nature 341, 447–450.

Porter, P., Kenworthy, R., Noakes, D.E. and Allen, W.D. (1974). Intestinal antibody secretion in the young pig in response to oral immunization with *Escherichia coli*. Immunology 27, 841–853.

Powell, D.A., Hu, P.C., Wilson, M., Collier, A.M. and Baseman, J.B. (1976). Attachment of *Mycoplasma pneumoniae* to respiratory epithelium. Infect. Immunol. 13, 959–966.

Prausnitz, C. and Kstner, H. (1921). Studien über die Ueberempfindlichkeit. Zentralbl. Bakteriol. 86, 160.

Proctor, D.F., Andersen, I. and Lundqvist, G. (1973). Clearance of inhaled particles from the human nose. Arch. Intern. Med. 131, 132–139.

Pugh, C.W., MacPherson, G.G. and Steer, H.W. (1983). Characterization of nonlymphoid cells derived from rat peripheral lymph. J. Exp. Med. 157, 1758–1779.

Rácz, P., Tenner-Rcz, K., Myrvik, Q.N. and Fainter, L.K. (1977). Functional architecture of bronchial associated lymphoid tissue and lymphoepithelium in pulmonary cell-mediated reactions in the rabbit. J. Reticuloendothel. Soc. 22, 59–83.

Reid, C.D.L., Fryer, P.R., Clifford, C., Kirk, A., Tikerpae, J. and Knight, S.C. (1990). Identification of haemopoietic progenitors of macrophages and dendritic Langerhans cells (DL-CFU) in human bone marrow and peripheral blood. Blood 76, 1139–1149.

Reid, P.A. and Watts, C. (1990). Cell surface MHC glycoproteins cycle through primaquine sensitive intracellular compartments. Nature 346, 655–657.

Richman, L.K., Graeff, A.S. and Strober, W. (1981). Antigen presentation by macrophage-enriched cells from the mouse Peyer's patch. Cell. Immunol. 62, 110–118.

Rhodes, R.S. and Karnovsky, M.J. (1971). Loss of macromolecular barrier function associated with surgical trauma to the intestine. Lab. Invest. 25, 220–229.

Richardson, J., Bouchard, T. and Ferguson, C.C. (1976). Uptake and transport of exogenous proteins by respiratory epithelium. Lab. Invest. 35, 307–314.

Roberton, D.M., Paganelli, R., Dinwiddie, R. and Levinsky, R.J. (1982). Milk antigen absorption in the preterm and term neonate. Arch. Dis. Child. 57, 369–372.

Roberts, S.A., Reinhardt, M.C., Paganelli, R. and Levinsky, R.J. (1981). Specific antigen exclusion and non-specific facilitation of antigen entry across the gut in rats allergic to food proteins. Clin. Exp. Immunol. 45, 131–136.

Robertson, S.A., Mayrhofer, G. and Seamark, R.F. (1992). Uterine epithelial cells synthesize granulocyte-macrophage colony-stimulating factor (GM-CSF) and interleukin-6 (IL-6) in pregnant and non-pregnant mice. Biol. Reprod. 46, 1069–1079.

Robinson, A.P., White, T.M. and Mason, D.W. (1986). Macrophage heterogeneity in the rat as delineated by two monoclonal antibodies MRC OX-41 and MRC OX-42, the latter recognizing complement receptor type 3. Immunology 57, 239–247.

Rochester, C.L., Goodell, E.M., Stoltenberg, J.K. and Bowers, W.E. (1988). Dendritic cells from rat lung are potent accessory cells. Am. Rev. Resp. Dis. 138, 121–128.

Rodewald, R. (1973). Intestinal transport of antibodies in the newborn rat. J. Cell Biol. 58, 189–211.

Rodewald, R. (1980). Distribution of immunoglobulin G receptors in the small intestine of the young rat. J. Cell Biol. 85, 18–32.

Rodman, J.S., Seidman, L. and Farquhar, M.G. (1986). The membrane composition of coated pits, microvilli, endosomes and lysosomes is distinctive in the rat kidney proximal tubule cell. J. Cell. Biol. 102, 77–87.

Romani, N., Stingl, G., Tschachler, E., Witmer, M.D., Steinman, R.M., Shevach, E.M. and Schuler, G. (1985). The Thy-1 bearing cell of murine epidermis. A distinctive leukocyte perhaps related to natural killer cells. J. Exp. Med. 161, 1368–1383.

Romani, N., Lenz, A., Glassel, H., Stössel, H., Stanzl, U., Majdic, O., Fritsch, P. and Schuler, G. (1989a). Cultured human Langerhans cells resemble lymphoid dendritic cells in phenotype and function. J. Invest. Dermatol. 93, 600–609.

Romani, N., Inaba, K., Puré, E., Crowley, M., Witmer-Pack, M. and Steinman, R.M. (1989b). A small number of anti-CD3 molecules on dendritic cells stimulate DNA synthesis in mouse T lymphocytes. J. Exp. Med. 169, 1153–1168.

Romani, N., Koide, S., Crowley, M., Witmer-Pack, M., Livingstone, A.M., Fathman, C.G., Inaba, K. and Steinman, R.M. (1989c). Presentation of exogenous protein by dendritic cells to T cell clones. Intact protein is presented best by immature, epidermal Langerhans cells. J. Exp. Med. 169, 1169–1178.

Romani, N., Schuler, G. and Fritsch, P. (1991a). In "Epidermal Langerhans Cells" (ed. G. Schuler), pp 49–86. CRC Press, Boca Raton.

Romani, N., Witmer-Pack, M., Crowley, M., Koide, S., Schuler, G., Inaba, K. and Steinman, R.M. (1991b). In "Epidermal Langerhans Cells" (ed. G. Schuler), pp 191–216. CRC Press, Bocca Raton.

Rothberg, R.M., Kraft, S.C., Farr, R.S., Kriebel, G.W. and Goldberg, S.S. (1970). In "The Secretory Immunologic System" (ed. D. Dayton), pp 293–307. US Department of Health, Education and Welfare, Bethesda.

Rötzschke, O. and Falk, K. (1991). Naturally-occurring peptide antigens derived from the MHC class-I-restricted processing pathway. Immunol. Today 12, 447–455.

Rötzschke, O., Falk, K., Deres, K., Schild, H., Norda, M., Metzger, J., Jung, G. and Rammensee, H.-G. (1990). Isolation and analysis of natural processed viral peptides as recognised by cytotoxic T cells. Nature 348, 252–254.

Rowden, G. (1981). The Langerhans cell. CRC Crit. Rev. Immunol. 3, 95–180.

Rubin, D.H., Kornstein, M.J. and Anderson, A.O. (1985). Reovirus serotype I intestinal infection: a novel replicative cycle with ileal disease. J. Virol. 53, 391–398.

Rudensky, A.Y., Preston-Hurlburt, P., Al-Ramadi, B.K., Rothbard, J. and Janeway, C.A. (1992). Truncation variants of peptides isolated from MHC class II molecules suggest sequence motifs. Nature 359, 429–431.

Ruebner, B.H., Kanayama, R., Bronson, R.T. and Blumenthal, S. (1974). Meconium corpuscles in intestinal epithelium of fetal and newborn primates. Arch. Pathol. 98, 396–399.

Russell, L. (1977). Movement of spermatocytes from the basal to the adluminal compartment of the rat testis. Am. J. Anat. 148, 313–328.

Rycroft, R.J.G. and Wilkinson, J.D. (1992). In "Textbook of Dermatology" (eds R.H. Champion, J.L. Burton and F.J.G. Ebling), Ch. 17. Blackwell, Oxford.

Sampson, H.A. (1983). Role of immediate food hypersensitivity in the pathogenesis of atopic dermatitis. J. Allergy Clin. Immunol. 71, 473–480.

Sampson, H.A. and McCaskill, C.M. (1985). Food hypersensitivity in atopic dermatitis: evaluation of 113 patients. J. Pediatr. 107, 669–675.

Sanger, J.W. and Sanger, J.M. (1992). Beads, bacteria and actin. Nature 357, 442.

Sansonetti, P.J. (1991). Genetic and molecular basis of epithelial cell invasion by *Shigella* species. Rev. Infect. Dis. 13, S285–S292.

Santos, L.M.B., Lider, O., Audette, J., Khoury, S.J. and Weiner, H.L. (1990). Characterisation of immunomodulatory properties and accessory cell function of small intestinal epithelial cells. Cell. Immunol. 127, 26–34.

Savilahti, E., Arato, A. and Verkasalo, M. (1990). In "Advances in Mucosal Immunology" (eds T.T. MacDonald, S.J. Challacombe, P.W. Bland, C.R. Stokes, R.V. Heatley and A. McI. Mowat), pp 61–66. Kluwer, Dordrecht.

Sawicki, W., Kucharczyk, K., Szymanska, K. and Kujawa, M. (1977). Lamina propria macrophages of intestine of the guinea pig. Possible role in phagocytosis of migrating cells. Gastroenterology 73, 1340–1344.

Scadding, G.K. and Brostoff, J. (1985). In "Food and the Gut" (eds J.O. Hunter and V. Alun Jones), pp 94–112. Baillière Tindall, London.

Scadding, G.K. and Brostoff, J. (1986). Low dose sublingual therapy in patients with allergic rhinitis to house dust mite. Clin. Allergy 16, 483–491.

Schmitt, D.A., Bieber, T., Cazenave, J.-P. and Hanau, D. (1990). Fc receptors of human Langerhans cells. J. Invest. Dermatol. 94, 15s-21s.

Schneeberger, E.E. (1974). The permeability of the alveolar–capillary membrane to ultrastructural protein tracers. Ann. N.Y. Acad. Sci. 221, 238–243.

Schneeberger, E.E. and Karnovsky, M.J. (1971). The influence of intravascular fluid volume on the permeability of newborn and adult mouse lungs to ultrastructural protein tracers. J. Cell Biol. 49, 319–334.

Schneeberger-Keeley, E.E. and Karnovsky, M.J. (1968). The ultrastructural basis of alveolar-capillary membrane permeability to peroxidase used as a tracer. J. Cell Biol. 37, 781–793.

Schroeder, H.E., Moreillon, M.-C. and Nair, P.N.R. (1983). Architecture of minor salivary gland duct/lymphoid follicle associations and possible antigen-recognition sites in the monkey *Macaca fasicularis*. Arch. Oral Biol. 28, 133–143.

Schuler, G. (ed.) (1991). "Epidermal Langerhans Cells", CRC Press, Boca Raton.

Schuler, G. and Steinman, R.M. (1985). Murine epidermal Langerhans cells mature into potent immunostimulatory dendritic cells *in vitro*. J. Exp. Med. 161, 526–546.

Schuler, G., Romani, N., Stössel, H. and Wolff, K. (1991). In "Epidermal Langerhans Cells" (ed. G. Schuler), pp 87–137. CRC Press, Boca Raton.

Schumacker, T.N., De Bruijin, M.L.H., Vernie, L.N., Kast, W.M., Melief, C.J.M., Neefjes, J.J. and Ploegh, H.L. (1991). Peptide selection by MHC class I molecules. Nature 350, 703–706.

Sedgwick, J.D. and Holt, P.G. (1983). Induction of IgE-isotype specific tolerance by passive antigenic stimulation of the respiratory mucosa. Immunology 50, 625–630.

Selby, W.S., Poulter, L.W. and Hobbs, S., Jewell, D.P. and Janossy, G. (1983). Heterogeneity of HLA-DR-positive histiocytes in human intestinal lamina propria: a combined histochemical and immunohistochemical analysis. J. Clin. Pathol. 36, 379–384.

Sellwood, R. (1984). In "Attachment of Organisms to the Gut Mucosa", Vol. I. (ed. E.C. Boedeker), Ch. 3. CRC Press, Boca Raton.

Sertl, K., Takemura, T., Tschachler, E., Ferrans, V.J., Kaliner, M.A. and Shevach, E.M. (1986). Dendritic cells with antigen-presenting capability reside in airway epithelium, lung parenchyma, and visceral pleura. J. Exp. Med. 163, 436–451.

Setchell, B.P., Uksila, J., Maddocks, S. and Pöllänen, P. (1990). Testis physiology relevant to immunoregulation. J. Reprod. Immunol. 18, 19–32.

Shimizu, Y. and Andrew, W. (1967). Studies on the rabbit appendix. I. Lymphoepithelial relations and the transport of bacteria from lumen to lymphoid follicle. J. Morphol. 123, 231–250.

Shirota, K., LeDuy, L., Yuan, S. and Jothy, S. (1990). Interleukin-6 and its receptor are expressed in human intestinal epithelial cells. Virchows Arch. B 58, 303–308.

Sicinski, P. (1990). Poliovirus type 1 enters the human host through intestinal M cells. Gastroenterology 98, 56–58.

Silberberg, I., Baer, R.L., Rosenthal, S.A., Thorbecke, G.G. and Berezowsky, V. (1975). Dermal and intravascular Langerhans cells at sites of passively induced allergic contact hypersensitivity. Cell. Immunol. 18, 435–453.

Silberberg-Sinakin, I., Thorbecke, G.J., Baer, R.L., Rosenthal, S.A. and Berezowsky, V. (1976). Antigen-bearing Langerhans cells in skin, dermal lymphatics and in lymph nodes. Cell. Immunol. 25, 137–151.

Simani, A.S., Inoue, S. and Hogg, J.C. (1974). Penetration of the respiratory epithelium of guinea pigs following exposure to cigarette smoke. Lab. Invest. 31, 75–81.

Simecka, J.W., Davis, J.K. and Cassell, G.H. (1986). Distribution of Ia antigens and T lymphocyte subpopulations in rat lungs. Immunology 57, 93–98.

Simons, K. and Wandinger-Ness, A. (1990). Polarized sorting in epithelia. Cell 62, 207–210.

Skogh, T. (1982). Overestimate of ^{125}I-protein uptake by the adult mouse gut. Gut 23, 1077–1080.

Sminia, T. and Jeurissen, S.H.M. (1986). The macrophage population of the gastrointestinal tract of the rat. Immunobiology 172, 72–80.

Sminia, T. and Van der Ende, M.B. (1989). Localization, isolation and histochemical characterization of macrophages in the gut and mesenteric lymph nodes of the rat. Adv. Exp. Med. Biol. 237, 621–626.

Sminia, T., van der Brugge-Gamelkoorn, G.J. and Jeurissen, S.H.M. (1989). Structure and function of bronchus-associated lymphoid tissue (BALT). CRC Crit. Rev. Immunol. 9, 119–150.

Sminia, T., Van der Ende, M. and de Groot, C. (1990). Macrophages in rat gut. J. Histochem. Cytochem. 38, 1389.

Smith, M.W. (1985). Selective expression of brush border hydrolases by mouse Peyer's patch and jejunal villus enterocytes. J. Cell Physiol. 124, 219–225.

Smith, P.D. (1992). Gastrointestinal infections in AIDS. Ann. Intern. Med. 116, 63–77.

Sneller, M.C. and Strober, W. (1986). M cells and host defence. J. Infect. Dis. 154, 737–741.

Søeberg, B., Sumerska, T., Binns, R.M. and Balfour, B.M. (1978). Contact sensitivity in the pig. II. Induction by intralymphatic infusion of DNP conjugated cells. Int. Arch. Allergy Appl. Immunol. 57, 114–125.

Soesatyo, M., Biewenga, J., Kraal, G. and Sminia, T. (1990). The localization of macrophage subsets and dendritic cells in the gastrointestinal tract of the mouse with special reference to the presence of high endothelial venules. An immuno- and enzyme-histochemical study. Cell Tissue Res. 259, 587–593.

Spalding, D.M. and Griffin, J.A. (1986). Different pathways of differentiation of pre-B cell lines are induced by dendritic cells and T cells from different lymphoid tissues. Cell, 44, 507–515.

Spalding, D.M., Koopman, W.J., Eldridge, J.H., McGhee, J.R. and Steinman, R.M. (1983). Accessory cells in murine Peyer's patch. I. Identification and enrichment of a functional dendritic cell. J. Exp. Med. 157, 1646–1659.

Spargo, L.D.J. and Mayrhofer, G. (1990). In "Laboratory Methods in Immunology", Vol. II (ed. H. Zola), pp 99–112. CRC Press, Boca Raton.

Spencer, J. and MacDonald, T.T. (1990). In "Ontogeny of the Immune System of the Gut" (ed. T.T. MacDonald), pp 23–49. CRC Press, Boca Raton.

Spies, J.R., Stevan, M.A., Stein, W.J. and Coulson, E.J. (1970). The chemistry of allergens. XX. New antigens generated by pepsin hydrolysis of bovine milk proteins. J. Allergy 45, 208–219.

Sprinz, H., Landy, M., Gaines, S. and Edsall, G. (1956). Experimental typhoid fever in chimpanzees. III. Pathogenesis. Fed. Proc. 15, 614–615.

Spritzer, A.A., Watson, J.A., Auld, J.A. and Guetthoff, M.A. (1968). Pulmonary macrophage clearance. The hourly rates of transfer of pulmonary macrophages to the oropharynx of the rat. Arch. Environ. Health 17, 726–730.

Squier, C.A. (1973). The permeability of keratinized and non-keratinized oral epithelium to horseradish peroxidase. J. Ultrastruct. Res. 43, 160–177.

Squier, C.A. and Hall, B.K. (1985). The permeability of skin and oral mucosa to water and horseradish peroxidase as related to the thickness of the permeability barier. J. Invest. Dermatol. 84, 176–179.

Squier, C.A. and Hopps, R.M. (1976). A study of the permeability barrier in epidermis and oral epithelium using horseradish peroxidase as a tracer *in vitro*. Br. J. Dermatol. 95, 123–129.

Squier, C.A., Cox, P. and Wertz, P.W. (1991). Lipid content and water permeability of skin and oral mucosa. J. Invest. Dermatol. 96, 123–126.

Staley, T.E., Jones, E.N. and Corley, L.D. (1969). Attachment and penetration of *Escherichia coli* into intestinal epithelium of the ileum in newborn pigs. Am. J. Pathol. 56, 371–392.

Steer, H.W. (1980). An analysis of the lymphocyte content of rat lacteals. J. Immunol. 125, 1845–1848.

Steineger, B., Falk, P., Lohmüller, M. and van de Meide, P.H. (1989). Class II MHC antigens in the rat digestive system. Normal distribution and induced expression after interferon-gamma treatment *in vivo*. Immunology 68, 507–513.

Steiner, G., Payer, E., Elbe, A., Wolff, K. and Stingl, G. (1989). Langerhans cells (LC) stimulate lymphokine secretion by dendritic epidermal T cells (DETC). J. Invest. Dermatol. 92, 522.

Steinman, R.M. (1991). The dendritic cell system and its role in immunogenicity. Ann. Rev. Immunol. 9, 271–296.

Steinman, R.M., Mellman, I.S., Muller, W.A. and Cohn, Z.A. (1980). Endocytosis and the recycling of plasma membrane. J. Cell Biol. 96, 1–27.

Stephen, A.M. (1985). In "Food and the Gut" (eds J.O. Hunter and V. Alun-Jones), Ch. 5. Baillière Tindall, London.

Stewart, G.A. (1982). Isolation and characterization of the allergen *Dpt*12 from *Dermatophagoides pteronyssinus* by chromato-focusing. Int. Arch. Allergy Appl. Immunol. 69, 224–230.

Stewart, G.A., Butcher, A. and Turner, K.J. (1983). Characterization of a very high-density lipoprotein allergen *Dpt*4 from the house dust mite *Dermatophagoides pteronyssinus*. Int. Arch. Allergy Appl. Immunol. 71, 340–345.

Stewart, G.A., Butcher, A., Lees, K. and Ackland, J. (1986). Immunochemical and enzymatic analyses of extracts of the house dust mite *Dermatophagoides pteronyssinus*. J. Allergy Clin. Immunol. 77, 14–24.

Stingl, G., Tamaki, K. and Katz, S.I. (1980). Origin and function of epidermal Langerhans cells. Immunol. Rev. 53, 149–174.

Stokes, C.R. (1984). In "Local Immune Responses of the Gut" (eds C.R. Stokes and T.J. Newby), pp 97–141. CRC Press, Boca Raton.

Stokes, C.R., Newby, T.J., Huntley, J.H., Patel, D. and Bourne, F.J. (1979). The immune response of mice to bacterial antigens given by mouth. Immunology 38, 497–502.

Stoughton, R.B. (1966). Hexachlorophene deposition in human stratum corneum. Arch. Dermatol. 94, 646–648.

Straus, W. (1983). Mannose-specific binding sites for horseradish peroxidase in various cells of the rat. J. Histochem. Cytochem. 31, 78–84.

Streilein, J.W. and Bergstresser, P.R. (1981). Langerhans cell function dictates induction of contact hypersensitivity or unresponsiveness to DNFB in Syrian hamsters. J. Invest. Dermatol. 77, 272–277.

Strobel, S. (1988). In "Food Allergy" (eds D. Reinhardt and E. Schmidt), pp 99–115. Raven Press, New York.

Strobel, S. and Ferguson, A. (1984). Immune responses to fed antigen in mice. 3. Systemic tolerance of priming is related to age at which antigen is first encountered. Pediatr. Res. 18, 588–594.

Strober, W. (1990). Regulation of IgA B-cell development in the mucosal immune system. J. Clin. Immunol. 10, 56s–63s.

Strombeck, D.R. and Harrold, D. (1975). Comparison of the rate of absorption and proteolysis of (^{14}C)choleragen and (^{14}C)bovine serum albumin in the rat. Infect. Immun. 12, 1450–1456.

Stuart, B.O. (1973). Deposition of inhaled aerosols. Arch. Intern. Med. 131, 60–73.

Svanborg-Eden, C., Hansson, L.-A., Jodal, U. and Sohl-Akerlund, A. (1976). Variable adherence to normal human urinary tract epithelial cells of *Escherichia coli* strains associated with various forms of urinary tract infections. Lancet i, 490–492.

Swain, S.L. (1983). T cell subsets and the recognition of MHC class. Immunol. Rev. 74, 129–142.

Tagesson, C., Sjdahl, R. and Thoren, B. (1978). Passage of molecules through the wall of the gastrointestinal tract. I. A simple experimental model. Scand. J. Gastroenterol. 13, 519–524.

Tagesson, C., Magnusson, K.-E. and Sundqvist, T. (1981). Intestinal permeability studies and human disease. Monogr. Allergy 17, 250–272.

Takeuchi, A. (1966). Electronmicroscopic studies of experimental Salmonella infection. I. Penetration into the intestinal epithelium by *Salmonella typhimurium*. Am. J. Pathol. 50, 109–136.

Takeuchi, A. (1971). Penetration of the intestinal epithelium by various microorganisms. Curr. Topics. Exp. Pathol. 54, 1–27.

Takeuchi, A., Formal, S.B. and Sprinz, H. (1968). Experimental acute colitis in the Rhesus monkey following peroral infection with *Shigella flexneri*. An electron microscope study. Am. J. Pathol. 52, 503–529.

Tanner, A.R., Arthur, M.J.P. and Wright, R. (1984). Macrophage activation, chronic inflammation and gastrointestinal disease. Gut 25, 760–783.

Taylor, A.E. and Gaar, K.A. (1970). Estimation of equivalent pore radii of pulmonary capillary and alveolar membranes. Am. J. Physiol. 218, 1133–1140.

Taylor, R.K., Miller, V.L., Furlong, D.B. and Mekalanos, J.J. (1987). Use of phoA gene fusions to identify a pilus colonization factor co-ordinately regulated with cholera toxin. Proc. Natl Acad. Sci. USA 84, 2833–2837.

Tenner-Rácz, K., Rácz, P., Myrvick, Q.N., Ockers, J.R. and Geister, R. (1979). Uptake and transport of horseradish peroxidase by lymphoepithelium of the bronchus-associated lymphoid tissue in normal and Bacillus Calmette–Guérin-immunized and challenged rabbits. Lab. Invest. 41, 106–115.

Thepen, T., Van Rooijen, N. and Kraal, G. (1989). Alveolar macrophage elimination *in vivo* is associated with an increase in pulmonary immune response in mice. J. Exp. Med. 170, 499–509.

Thomas, H.C. and Parrot, D.M.V. (1974). The induction of tolerance to a soluble protein antigen by oral administration. Immunology 27, 631–639.

Thompson, J.F. (1987). Specific receptors for epidermal growth factor in rat intestinal microvillus membranes. Am. J. Physiol. 254, G429–G435.

Thornburg, W., Rao, R.K., Matrisian, L.M., Magun, B.E. and Koldovsky, O. (1987). Effect of maturation on gastro-intestinal absorption of epidermal growth factor in rats. Am. J. Physiol. 253, G68–G71.

Tiegs, S.L., Evavold, B.P., Yokoyama, A., Stec, S., Quintans, J. and Rowley, D. (1990). Delayed antigen presentation by epidermal Langerhans cells to cloned Th1 and Th2 cells. J. Invest. Dermatol. 95, 446–449.

Toews, G.B., Bergstresser, P.R., Streilein, J.W. and Sullivan, S. (1980). Epidermal Langerhans cell density determines whether contact hypersensitivity or unresponsiveness follows painting with DNFB. J. Immunol. 124, 445–453.

Tolo, K., Bandtzaeg, P. and Jonsen, J. (1977). Mucosal penetration of antigen in the presence or absence of serum-derived antibody. An *in vitro* study of rabbit oral and intestinal mucosa. Immunology 33, 733–743.

Tome, D., Dumontier, A.-M., Hautefeuille, M. and Desjeux, J.-F. (1987). Opiate activity and transepithelial passage of intact β-casomorphins in rabbit ileum. Am. J. Physiol. 253, G737–G744.

Trahair, J.F. and Robinson, P.M. (1989). Enterocyte ultra-structure and uptake of immunoglobulins in the small intestine of the neonatal lamb. J. Anat. 166, 103–111.

Trier, J.S. (1968). In "Handbook of Physiology", Sect. 6, Vol. 3 (eds C.F. Code and W. Heidel), pp 1125–1175. American Physiological Society, Washington, DC.

Trier, J.S. (1991). Structure and function of intestinal M cells. Gastroenterol. Clin. North Am. 20, 531–545.

Tuomanen, E.I. and Hendley, J.O. (1983). Adherence of *Bordetella pertussis* to human respiratory epithelial cells. J. Infect. Dis. 148, 125–130.

Turkin, D. and Sercarz, E.E. (1977). Key antigenic determinants in regulation of the immune response. Proc. Natl Acad. Sci. USA 74, 3984–3987.

Turner, M.W., Warner, J.O., Stokes, C.R. and Norman, A.P. (1978). Immunological studies in cystic fibrosis. Arch. Dis. Child. 53, 631–638.

Udall, J.N., Bloch, K.J., Fritze, L. and Walker, W.A. (1981). Binding of exogenous protein fragments to native proteins: possible explanation for the overestimation of uptake of extrinsically labelled macromolecules from the gut. Immun-ology 42, 251–257.

Ullberg, S., Birke, G., Ewaldsson, B., Hansson, E., Liljedahl, S.-O., Plantin, L.-O. and Wetterfors, J. (1960). The role of the gastrointestinal tract in the elimination of serum albumin. A preliminary autoradiographic study. Acta Med. Scand. 167, 421–425.

Unanue, E.R., Harding, C.V., Luescher, I.F. and Roof, R.W. (1989). Antigen-binding function of class II MHC molecules. Cold Spring Harbor Symp. Quant. Biol. 54, 383–392.

Underdown, B.J. and Schiff, J.M. (1986). Immunoglobulin A: strategic defence initative at the mucosal surface. Ann. Rev. Immunol. 4, 389–417.

Van Bleek, G.M. and Nathenson, S.G. (1990). Isolation of an endogenously processed immunodominant viral peptide from the class I H-2 K^b molecule. Nature 348, 213–216.

Van der Brugge-Gamelkoorn, G., Van de Ende, M. and Sminia, T. (1985). Uptake of antigens and inert particles by bronchus associated lymphoid tissue (BALT) epithelium in the rat. Cell Biol. Int. Rep. 9, 524.

Van der Brugge-Gamelkoorn, G., Van de Ende, M. and Sminia, T. (1986). Changes occurring in the epithelium covering the bronchus-associated lymphoid tissue of rats after intratracheal challenge with horseradish peroxidase. Cell Tissue Res. 245, 439–444.

Van Rees, E.P., Dijkstra, C.D., Van der Ende, M.B., Janse, E.M. and Sminia, T. (1988). The ontogenetic development of macrophage subpopulations and Ia-positive non-lymphoid cells in gut-associated lymphoid tissue of the rat. Immunology 63, 79–85.

Virji, M. and Heckels, J.E. (1984). The role of common and type-specific pilus antigenic domains in adhesion and virulence of gonococci for human epithelial cells. J. Gen. Microbiol. 130, 1089–1095.

Volk, B.A., Brenner, D.A. and Kagnoff, M.F. (1989). Analysis of RNA transcripts for HLA class II genes in human small intestinal biopsies. Gut 30, 1220–1224.

Volkheimer, G. (1972). "Persorption". Thieme, Stuttgart.

Volkheimer, G. (1977). Persorption of particles: physiology and pharmacology. Adv. Pharm. Chemother. 14, 163–187.

Volkheimer, G., Schultz, F.H., Aurich, I., Strauch, S., Beuthin, K. and Wendlandt, H. (1968). Persorption of particles. Digestion 1, 78–80.

Wachsmann, D., Klein, J.P., Scholler, M., Ogier, J., Ackermans, F. and Frank, R.M. (1986). Serum and salivary antibody responses in rats orally immunized with *Streptococcus mutans* carbohydrate protein conjugate associated with liposomes. Infect. Immunol. 52, 408–413.

Waldmann, R.H., Grunspan, R. and Ganguly, R. (1972). Oral immunization of mice with killed *Salmonella typhimurium* vaccine. Infect. Immunol. 6, 58–61.

Walker, W.A. (1984). In "Food Intolerance" (ed. R.K. Chandra), pp 17–54. Elsevier, New York.

Walker, W.A. (1985). Absorption of protein and protein fragments in the developing intestine: Role of immunologic/allergic reactions. Pediatrics 75(Suppl.), 167–171.

Walker, W.A. (1987). In "Food Allergy and Intolerance" (eds J. Brostoff and S.J. Challacombe), pp 209–222. Baillière Tindall, London.

Walker, W.A. (1988). In "Food Allergy" (eds D. Reinhardt and E. Schmidt), pp 15–32. Raven Press, New York.

Walker, W.A., Cornell, R., Davenport, L.M. and Isselbacher, K.J. (1972). Macromolecular absorption. Mechanism of horseradish peroxidase uptake and transport in adult and neonatal rat intestine. J. Cell Biol. 54, 195–205.

Walker, W.A., Wu, M., Isselbacher, K.J. and Bloch, K.J. (1975a). Intestinal uptake of macromolecules. III. Studies on the mechanism by which immunization interferes with antigen uptake. J. Immunol. 115, 854–861.

Walker, W.A., Wu, M., Isselbacher, K.J. and Bloch, K.J. (1975b). Intestinal uptake of macromolecules. IV. The effect of pancreatic duct ligation on the breakdown of antigen and antigen-antibody complexes on the intestinal surface. Gastro-enterology 69, 1223–1229.

Warren, C.P.W., Tai, E., Batten, J.C., Hutchcroft, B.J. and Pepys, J. (1975). Cystic fibrosis – immunological reactions to *A. fumigatus* and common allergens. Clin. Allergy 5, 1–12.

Warshaw, A.L., Walker, W.A., Cornell, R. and Isselbacher, K.J. (1971). Small intestinal permeability to macromolecules. Transmission of horseradish peroxidase into mesenteric lymph and portal blood. Lab. Invest. 25, 675–684.

Warshaw, A.L., Walker, W.A. and Isselbacher, K.J. (1974). Protein uptake by the intestine: evidence for absorption of inact macromolecules. Gastroenterology 66, 987–992.

Wassef, J.S., Keren, D.F. and Mailloux, J.L. (1989). Role of M cells in initial antigen uptake and in ulcer formation in the rabbit ileal loop model of shigellosis. Infect. Immunol. 57, 858–863.

Weber-Matthiesen, K. and Sterry, W. (1990). Organisation of the monocyte/macrophage system of normal human skin. J. Invest. Dermatol. 95, 83–89.

Wick, M.J., Madara, J.L., Fields, B.N. and Normark, S.J. (1991). Molecular cross talk between epithelial cells and pathogenic microorganisms. Cell 67, 651–659.

Wilders, M.M., Sminia, T. and Janse, E.M. (1983a). Ontogeny of non-lymphoid and lymphoid cells in the rat gut with special reference to large mononuclear Ia-positive dendritic cells. Immunology 50, 303–314.

Wilders, M.M., Drexhage, H.A., Weltevreden, E.F., Mullink, H., Duijvestijn, A. and Meuwissen, S.G.M. (1983b). Large mononucelar-Ia-positive veiled cells in Peyer's patches. I. Isolation and characterization in rat, guinea-pig and pig. Immunology 48, 453–460.

Wilders, M.M., Sminia, T., Plesch, B.E.C., Drexhage, H.A., Weltevreden, E.F. and Meuwissen, S.G.M. (1983c). Large mononuclear Ia-positive veiled cells in Peyer's patches. II. Localization in rat Peyer's patches. Immunology 48, 461–467.

Williams, E.W. (1978). In "Antigen Absorption by the Gut" (ed. W.A. Hemmings), Ch. 6. MTP Press, Lancaster.

Williams, N.A., Wilson, A.D., Bailey, M., Bland, P.W. and Stokes, C.R. (1992). Primary antigen-specific T-cell proliferative responses following presentation of soluble protein antigen by cells from the murine small intestine. Immunology 75, 608–613.

Wilson, A.D., Clarke, C.J. and Stokes, C.R. (1990). Whole cholera toxin and B subunit act synergistically as an adjuvant for the mucosal immune response of mice to keyhole limpet haemocyanin. Scand. J. Immunol. 31, 443–451.

Wilson, A.F., Novey, H.S., Surprenant, E.L. and Bennett, L.R. (1971). Fate of inhaled whole pollen in asthmatics. J. Allergy 47, 107–108.

Wilson, J.M. and King, B.F. (1986). Sorting and transepithelial transport of adsorbed protein tracers: effects of temperature. Anat. Rec. 216, 33–39.

Wilson, S.J. and Walzer, M. (1935). Absorption of undigested proteins in human beings. Am. J. Dis. Child. 50, 49–57.

Witmer-Pack, M.D., Olivier, W., Valinsky, J., Schuler, G. and Steinman, R.M. (1987). Granulocyte/macrophage colony-stimulating factor is essential for the viability and function of cultured murine epidermal Langerhans cells. J. Exp. Med. 166, 1484–1498.

Wolf, J.L., Rubin, D.H., Finberg, R., Kauffman, R.S., Sharpe, A.H., Trier, J.S. and Fields, B.N. (1981). Intestinal M cells: a pathway for entry of reovirus into the host. Science 212, 471–472.

Wolf, J.L., Kauffman, R.S., Finberg, R., Dambrauskas, R., Fields, B.N. and Trier, J.S. (1983). Determinants of reovirus interaction with the intestinal M cells and absorptive cells of murine intestine. Gastroenterology 85, 291–300.

Wolff, K. and Hönigsmann, H. (1971). Permeability of the epidermis and the phagocytic activity of keratinocytes. Ultrastructural studies with thorotrast as a marker. J. Ultrastruct. Res. 36, 176–190.

Wolff, K. and Schreiner, E. (1970). Uptake, intracellular transport and degradation of exogenous protein by Langerhans cells. An electron microscopic–cytochemical study using peroxidase as a tracer substance. J. Invest. Dermatol. 54, 37–47.

Wolff, K. and Stingl, G. (1983). The Langerhans cell. J. Invest. Dermatol. 80, 17s-21s.

Wolin, M.J. (1981). Fermentation in the rumen and human large intestine. Science, 213, 1463–1468.

Wood, A.C., Todd, I., Cockayne, A. and Arbuthnott, J.P. (1991). Staphylococcal enterotoxins and the immune system. FEMS Microbiol. Immunol. 76, 121–134.

Woodhouse, A.F., Anderson, R.P., Myers, D.B., Broom, M.F., Hobson, C.H. and Chadwick, V.S. (1987). Intestinal absorption, metabolism and effects of bacterial chemotactic peptides in rat intestine. J. Gastroenterol. Hepatol. 2, 35–43.

Worthington, B.B. and Graney, D.O. (1973). Uptake of adenovirus by intestinal absorptive cells of the suckling rat. I. The neonatal ileum. Anat. Rec. 175, 37–62.

Worthington, B.S. and Syrotuck, J. (1976). Intestinal permeability to large particles in normal and protein deficient adult rats. J. Nutr. 106, 20–32.

Yednock, T.A. and Rosen, S.D. (1989). Lymphocyte homing. Adv. Immunol. 44, 313–378.

Yoshikai, Y., Reis, M.D. and Mak, T.W. (1986). Athymic mice express a high level of functional γ-chain but greatly reduced levels of α- and β-chain T-cell receptor messages. Nature 324, 482–485.

Yoshikai, Y., Matsuzaki, G., Takeda, Y., Ohga, S., Kishihara, K., Yuuki, H. and Nomoto, K. (1988). Functional T cell receptor δ chain messages in athymic nude mice. Eur. J. Immunol. 18, 1039–1043.

Yowell, R.L., Areneo, B.A., Miller, A. and Sercarz, E.E. (1979). Amputation of a suppressor determinant of lysozyme reveals underlying T-cell reactivity to other determinants. Nature 279, 70–71.

Zhang, Z. and Michael, J.G. (1990). Orally inducible immune unresponsiveness is abrogated by IFN-γ treatment. J. Immunol. 144, 4163–4165.

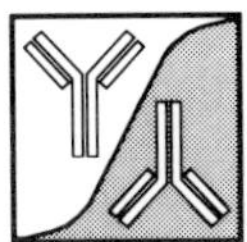

3. Adhesion Molecules and the Modulation of Mucosal Inflammation

Robert H. Gundel *and* L. Gordon Letts

1. Introduction

The existence of a mucosal barrier in the lung is the first layer of a host defence system that the lung possesses. The efficacy of this mucosal barrier is, in part, dependent upon specific cell–matrix and cell–cell adhesion-mediated events. The presence of intact and functional mucosal tissues is therefore essential to normal lung homeostasis. For example, as part of a primary host defence mechanism, the airway epithelium acts as a barrier that inhibits access of substances to the submucosal tissue. In addition, vascular endothelial cells and airway epithelial cells produce and secrete biologically active substances that can modulate their own function as well as that of surrounding cell types. These cells can also serve as a dispensary of enzymes that can modify the function of mediators released by other luminal, mucosal or submucosal cells. In normal lung host defence functions, mucosal tissues contribute to the inflammatory process by the induction of the cell surface expression of adhesion glycoproteins. It is the expression of adhesion molecules (both constitutive and inducible) that directs inflammatory cells to specific tissue sites, allowing clearance of foreign particles, detoxification of released local mediators, and tissue repair processes to occur. However, in diseases such as bronchial asthma, adhesion molecules undoubtably play a major role in orchestrating the excessive recruitment and subsequent activation of inflammatory cells into the lung. It is this infiltration of non-resident cells that is responsible for the maintenance of chronic airway mucosal inflammation which underlies altered airway structure and function.

This chapter reviews the role of adhesion molecules involved in luminal cell–extracellular matrix adhesion and cell–cell adhesion. It will focus on the role of endothelial and epithelial cell adhesive molecules and their importance in normal lung homeostatic functions as well as their possible role in airway disease. The thesis contends that adhesion molecule expression on airway endothelium and airway epithelium alters the basic function of these cells and therefore unveils and projects them as primary critical

Immunopharmacology of Epithelial Barriers
ISBN 0–12–288030–7

steps in the recruitment of inflammatory cells out of the circulation. These cells also transform into an important director of the selectivity of cell recruitment and facilitate the focal adherence, retention and activation at the basolateral surface of the airway epithelium. This selective, focal adherence is causal to airway epithelial cell damage and/or denudation which ultimately results in disruption of the normal functioning of the epithelium as a physical barrier and as a dymanic source of modulatory chemical mediators.

2. Adhesion Molecules

Cell–cell contact and cellular communication are required for the normal functioning of the immune system. For example, in the initiation of an inflammatory response the importance of cell adhesion events is reflected by the fact that specific immunological processes such as antibody production or the generation of antigen-specific T cells depend on cell adhesion for the attachment of lymphocytes to antigen-presenting cells (Dougherty and Hogg, 1987; Dougherty et al., 1988; Smith et al., 1988). Furthermore, the adherence of leucocytes to vascular endothelium is critical for migration out of the vascular space to specific sites of tissue inflammation (Haskard et al., 1986; Siegelman et al., 1989; Stevenson and Paul, 1989). Finally, the effector cell function of leucocytes is optimized by site-oriented and highly specific adherence of leucocytes to selected target cells, thereby promoting the localized release of toxic substances directly onto the target cell (Boyd et al., 1988; Davignon et al., 1981; Dustin et al., 1988; Krensky et al., 1983).

The activation of resident antigen-presenting cells in the lungs following antigen recognition is probably the primary process which results in the cascade of adhesion molecule regulation events in allergic asthma. These events are complex and intricate and have exposed a new paradigm in our understanding of this disease. It is thought that by blocking one or more of the adhesion molecules a new therapy may occur.

For the purposes of review and discussion, the adhesion molecules have been divided into three groups: those which bind luminal cells with the substratum; those which function in mucosal cell–cell adhesion; and those that are expressed on the cells lining the luminal surface and function in cell recruitment, activation and transendothelial migration. In the following sections, particular emphasis will be on epithelial/endothelial expression, function and role in pulmonary homeostasis, disease and pathophysiology.

2.1 CELL–EXTRACELLULAR MATRIX ADHESION MOLECULES

Several cell adhesion molecules, including the integrins, the CD44 hyaluronic acid receptor and proteoglycans, have definitive roles in the binding of luminal mucosal cells to the underlying substratum or extracellular matrix. The extracellular matrix is a complex mixture of proteins such as collagens, elastin, laminin and fibronectin, and functions, in part, as a basement layer to which the luminal cells are fixed to and supported by.

2.1.1 The Integrins

The integrin supergene family of adhesion receptor molecules are transmembrane glycoprotein heterodimers containing non-covalently associated α and β chains (Kishimoto et al., 1989; Larson and Springer, 1990; Sanchez-Madrid et al., 1983). Based on a common β chain, three integrin subfamilies have been designated, each containing distinct α chains that confer ligand specificity (Table 3.1).

The β_1 (VLA) integrins contain the common β chain (CD29) and seven distinct α chains (Hemler, 1990). The β_1 integrins function primarily in cell–matrix adhesion and thus play an important role in the attachment of luminal cells to the underlying substratum. The β_1

Table 3.1 The integrin supergene family

α subunit	Distribution	Ligand
β_1 (CD29)		
CD49a	Lymphocytes Fibroblasts Basement membrane	Laminin, collagen
CD49b	T lymphocytes Platelets Fibroblasts Endothelium Epithelium	Collagen, laminin
CD49c	Epithelium Fibroblasts	Fibronectin, laminin
CD49d	Leucocytes Fibroblasts	Fibronectin, VCAM-1
CD49e	T lymphocytes Fibroblasts Epithelium Endothelium Platelets Thymocytes	
CD49f	T lymphocytes Platelets	Laminin
β_2 (CD18)		
CD11a	Leucocytes	ICAM-1, ICAM-2
CD11b	Macrophages Monocytes Granulocytes	ICAM-1, C3bi, fibrinogen
CD11c	Macrophages Monocytes Granulocytes	
β_3 (CD61)		
CD41	Platelets	Fibronectin, fibrinogen
CD51	B lymphocytes Macrophages	Vitronectin, fibrinogen

integrins can bind to fibronectin, laminin and collagen. Depending on the specific integrin, these molecules may bind to one or more receptors. For example, VLA-5 and VLA-6 (CD49e and CD49f) bind only to fibronectin while VLA-3 (CD49c) binds to laminin, collagen and fibronectin (Buck and Horwitz, 1987; Elices and Hemler, 1990). Interestingly, recent studies have demonstrated that one member of the β_1 integrins, VLA-4 (CD49d/CD29), in addition to being a receptor for fibronectin is involved in leucocyte–endothelial cell adhesion utilizing VCAM-1 as its ligand on vascular endothelium (Elices and Hemler, 1990; Schleimer et al., 1992). The β_2 integrins contain a β chain (CD18) which differs from that of the β_1 integrins, and three distinct α chains designated CD11a (LFA-1), CD11b (Mac-1) and CD11c (p150,95) (Hynes, 1987). In contrast to the β_1 integrins that mainly function in cell–matrix adhesion, the β_2 integrins are expressed on leucocytes and are involved in cell–cell adhesion at the luminal surface (i.e. leucocyte–endothelial cell adhesion). These adhesion molecules will be discussed in detail below.

The β_3 integrins (cytoadhesins) are comprised of the platelet glycoprotein IIb/IIIa complex and the vitronectin receptor (Horton, 1990). The glycoprotein IIb/IIIa complex is found only on platelets. In contrast, the vitronectin receptor is found on vascular endothelial cells and epithelial cells (Horton, 1990). The vitronectin receptor will also bind to fibrinogen and von Willebrand's factor among others.

Thus, the β_1 and β_3 integrins are a group of adhesion molecules that promote binding of luminal cells to the major constituents of the base membrane and to some of the extracellular matrix components present during inflammation and wound healing (Table 3.2).

2.1.2   Other Cell Substratum Adhesion Molecules

Proteoglycans, protein molecules that carry one or more glycosaminoglycan side-chains, are important cell adhesion molecules (Ruoslahti, 1989, 1989). Many of the interactions of proteoglycans with extracellular matrix components are mediated by the glycosaminoglycan component. The

binding action of these molecules may be dependent on the presence of specific sugar sequences present on the glycosaminoglycan or may be purely a charge-dependent mechanism. The binding of proteoglycans to the substratum appears to occur primarily in the charge-dependent manner because the strength of adhesion is determined by the degree of sulphation of the molecule (Ruoslahti, 1989). In addition, it has been shown that larger proteoglycans bind more strongly to extracellular matrix proteins, presumably because larger molecules possess more potential binding sites (i.e. larger molecules have multiple sites of interaction with the matrix proteins).

Syndecan, a cell surface proteoglycan, can bind to collagen and fibronectin through its heparin sulphate chains (Sanders et al., 1989). Syndecan binding to extracellular matrix occurs in concert with binding mediated by the integrin molecules. Proteoglycans such as syndecan are present on the basolateral surface of epithelium, emphasizing their role in the adherence of luminal cells to the substratum.

A hyaluronic acid-binding receptor is present in a lymphocyte-homing receptor responsible for lymphocyte binding to high-venule endothelium in lymphoid tissue (Stamenkovic et al., 1989; Aruffo et al., 1990). This receptor, named Hermes antigen or CD44, is now known to have a wide tissue distribution and exists as a membrane protein on lymphocytes, fibroblasts, smooth muscle and epithelium (Picker et al., 1989). The exact function of CD44 is not clearly understood; however, it clearly plays a role in binding to hyaluronic acid, an important component of the interstitial space.

2.2   CELL–CELL ADHESION MOLECULES

The ability of endothelial and epithelial cells to function as barriers with selective permeability is promoted by cell–cell adhesion. Several adhesion molecules are involved in cell–cell adhesion, including the selectins, cadherins and a member of the immunoglobulin supergene family (Table 3.3). These adhesion molecules promote adhesion between "like" cells and interact with cytoplasmic proteins to form characteristic structures between cells of mucosal tissues such as tight juctions, desmosomes, adherens and gap junctions (Stevenson and Paul, 1989).

PECAM-1 is a member of the immunoglobulin supergene family of adhesion molecules that is localized on the surface of platelets, leucocytes and endothelial cells and is involved in cell–cell adhesion at mucosal barriers (Albelda et al., 1990; Newman et al., 1990). Studies with cultured endothelial cells have demonstrated that PECAM-1 localizes to the cell borders and probably has a major role in cell–cell adhesion and the formation of monolayers (Newman et al., 1990). The importance of the role of PECAM-1 as a cell–cell adhesion molecule is supported by studies in which a monoclonal antibody against PECAM-1 inhibited endothelial cell–endothelial

Table 3.2 Cell–extracellular matrix adhesion molecules

Adhesion molecule	Subunit	Ligand
Integrin		
β_1 (CD29)	CD49a	Laminin/collagen
	CD49b	Laminin/collagen
	CD49c	Laminin/collagen
		Fibronectin
	CD49d	Fibronectin
	CD49e	Fibronectin
	CD49f	Laminin
β_3	α_5	Vitronectin/fibronectin
		Thrombospondin
Proteoglycan (syndecan)	—	Collagen
CD44	—	Hyaluronic acid

Table 3.3 Cell–cell adhesion molecules

Family	Adhesion molecule	Distribution
Immunoglobulin supergene	PECAM-1	Platelets Leucocytes Endothelium
Integrin	CD49b/CD29	Endothelium Epithelium Platelets Lymphocytes
	CD49c/CD29	Epithelium Fibroblasts
	CD49e/CD29	Endothelium Epithelium Platelets Lymphocytes Fibroblasts
Cadherin	E-cadherin	Epithelium
	N-cadherin	Neuronal tissue
	P-cadherin	Placenta
	V-cadherin	Endothelium

cell adhesion and prevent the formation of monolayers (Albelda, 1991). The monoclonal antibody had no effect on existing intact monolayers, again demonstrating its importance in the formation and maintenance of endothelial cell–endothelial cell adhesion. However, the presence of large amounts of PECAM-1 on other cells (e.g. platelets) suggests that this molecule may be involved in many other adhesion events other than that between endothelial cells.

Members of the integrin adhesion molecule family, the β_1 integrins, function in cell–cell adhesion as well as cell–matrix adhesion. Several members of the β_1 integrins including CD49b, CD49c and CD49e have been identified on cell borders of cultured cells *in vitro* (Carter *et al.*, 1990). The role of the β_1 integrins in cell–cell adhesion is supported by studies demonstrating that antibodies directed against certain subunits of the β_1 family will inhibit or disrupt the formation of cultured cell monolayers (Lampugnani *et al.*, 1991). In addition, the arrangement and expression of these molecules in culture indicates that they are expressed all around the cell, and are not concentrated only in the basement membrane region.

The cadherins are a family of molecules involved in cell–cell adhesion. They are similar to the integrins structurally, containing an extracellular component, a single membrane-spanning region and a short cytoplasmic tail capable of interactions with intracellular cytoskeletal components (Takeichi, 1990). In contrast to the integrins, the cadherins contain a single peptide chain (integrins are composed of non-covalently associated α and β chains), and they bind to other cadherins (homophilic binding) on adjacent cells. There are four subclasses of the cadherin molecules: E-cadherins, which are found on mature epithelial cells; N-cadherins, found in neuronal tissue; P-cadherins, found primarily in placental tissue; and

V-cadherins, present on endothelial cells (Takeichi, 1990).

Studies by Behrens and colleagues (1985) have demonstrated the importance of E-cadherins in epithelial cell barrier functioning. In these studies, anti-E-cadherin antibodies altered cell–cell adhesion, resulting in changes in barrier permeability indicated by a decrease in transmembrane electrical resistance. Heimark and co-workers (1990) first identified V-cadherin on cultured endothelial cells and demonstrated their importance in cell–cell adhesion. In these studies, anti-V-cadherin antibodies inhibited endothelial cell aggregation. Other cell–cell adhesion molecules have been described; however, there is still very little information on the functional importance and distribution of these molecules. Further study in this area is warranted.

2.3   LUMINAL CELL ADHESION MOLECULES

2.3.1   The Immunoglobulin Supergene Family

The immunoglobulin supergene family of adhesion molecules comprises a number of related molecules that function primarily in cell–cell adhesion (Table 3.4).

2.3.1.1   *Intercellular Adhesion Molecules*

There are three known ICAMs, which have been designated ICAM-1, ICAM-2 and ICAM-3. ICAM-1 was originally described as the ligand for LFA-1 in hymotypic lymphocyte aggregation (Springer, 1990). ICAM-1 is a 90 kD single-chain glycoprotein with five immunoglobulin-like extracellular domains, a single membrane-spanning region and a short intracellular (cytoplasmic) tail. Each of the extracellular domains appears to be important for ligand-binding specificy. For example, ICAM-1 binding to LFA-1 (CD11a) occurs at the hinge region between domains 2 and 3 while Mac-1 (CD11b) binding occurs at domain 3. Low levels of functionally active ICAM-1 are constitutively expressed on endothelium and epithelium; however, it is also inducible on many cell types. Endothelial and epithelial ICAM-1 is markedly up-regulated by a

Table 3.4 The immunoglobulin supergene family

	Distribution	Molecular mass (kDa)	Ligand
ICAM-1	Endothelium Epithelium Fibroblasts Keratinocytes Lymphoid and myeloid cells	90	CD11a/CD18 CD11b/CD18
ICAM-2	Endothelium Many others	60	CD11a/CD18
VCAM-1	Endothelium Dendritic cells	110	CD49d/CD29

variety of pro-inflammatory cytokines and mediators (Smith *et al.*, 1988; Pober and Cotran 1990). Studies with *in vitro* cell culture systems have demonstrated that induced ICAM-1 expression requires protein synthesis and therefore has a lag phase of approximately 6–12 h before maximum expression occurs (Dustin *et al.*, 1986; Wegner *et al.*, 1991). The presence of ICAM-1 on vascular endothelium is for movement of circulating leucocytes out of the vasculature to specific sites of inflammation (Haskard, 1989). In the normal lung, ICAM-1 expression on airway epithelium is probably involved, via the recruitment and retention of phagocytic leucocytes, in clearance of foreign particles and/or organisms as part of a first line of defence against infection. The leucocyte counter-receptors for ICAM-1 are members of the CD18 integrins LFA-1 (CD11a/CD18) and Mac-1 (CD11b/CD18).

ICAM-2 was first identified in studies using DNA clones of functionally active adhesion molecules in which active clones were identified by binding, in the presence of anti-ICAM-1 monoclonal antibodies, to COS cells transfected with an endothelial cell cDNA library (Staunton *et al.*, 1989). Structurally, ICAM-2 is similar to ICAM-1 but contains only two immunoglobulin-like domains. These domains are most homologous to the first two N-terminal domains of ICAM-1 with 35% amino acid identity. Immunohistochemical staining studies have demonstrated that ICAM-2 is constitutively expressed on a variety of cells but most notably on vascular endothelial cells (Staunton *et al.*, 1989; de Fougerolles *et al.*, 1991). Unlike ICAM-1, ICAM-2 is not readily up-regulated but is constitutively expressed at near-maximal levels. The actual function of ICAM-2 is not fully understood; however, it has been speculated that ICAM-2 may play an important role in the initial adhesion and margination of leucocytes to endothelial cells as part of normal homeostasis in the absence of a strong pro-inflammatory signal. In addition, important differences in both tissue distribution and binding ligands between ICAM-1 and ICAM-2 have been described by de Fougerolles and colleagues (1991). Based on *in vitro* studies of induced expression on a variety of tissues, these authors contend that ICAM-1 is a primary ligand for leucocytes during inflammatory responses while ICAM-2 may be more important in the resting (unstimulated) state or in the early stages of the inflammatory response prior to maximal ICAM-1 expression. In addition, ligand selectivity was suggested by studies demonstrating that ICAM-1 binds to both CD11a (LFA-1) and CD11b (Mac-1) while ICAM-2 apparently binds only to CD11a.

ICAM-3 has recently been described as a glycoprotein ranging from 124 to 87 kD based on the extent of glycosylation. ICAM-3 is expressed at high levels on resting lymphocytes, monocytes and neutrophils while variable amounts are expressed on T and B lymphocytes. Unlike ICAM-1, ICAM-3 appears to be restricted to haematopoetic cells. The primary ligand for ICAM-3 is the β_2 integrin LFA-1.

2.3.1.2 *Vascular Cell Adhesion Molecule*

VCAM-1 was originally described as an inducible endothelial cell adhesion molecule that mediates leucocyte, monocyte and melanoma cell adhesion, in the presence of anti-ICAM-1 antibodies, to activated endothelial cells in culture (Rice *et al.*, 1990). VCAM-1 is a 110 kD single-chain glycoprotein containing six immunoglobulin-like extracellular domains, a single transmembrane region and a short cytoplasmic tail (Rice *et al.*, 1990). The kinetics of VCAM-1 expression in activated cell cultures systems is similar to ICAM-1, with expression beginning 2–4 h post-stimulation, peaking at 8–12 h and persisting for up to 72 h post-stimulation (Osborn *et al.*, 1989). VCAM-1 differs from ICAM-1 in its tissue distribution as it is only expressed on vascular endothelium and not on airway epithelium (Wegner *et al.*, 1991; Montefort *et al.*, 1992). VCAM-1 binds to its β_1 integrin counter-receptor CD49d/CD29 (VLA-4). The expression of VLA-4 on lymphocytes and eosinophils, but not neutrophils, had led to the hypothesis that VCAM-1 may play a dominant role in the development of chronic eosinophilic inflammation that is characteristic of human bronchial asthma. In addition, VCAM-1 expression is also selectively enhanced by IL-4, a product of T_H2 lymphocytes that have been proposed as major players in the development of allergic airway inflammation (Schleimer *et al.*, 1992).

2.3.2 The Selectins

The three members of the selectin supergene family of adhesion molecules are L-selectin, E-selectin and P-selectin (Table 3.5). The selectins are single-chain glycoproteins with an N-terminal lectin-like domain, an epidermal growth factor-like domain and a series of repeat sequences homologous to complement regulatory proteins (two, six and nine consensus repeats in L-selectin, E-selectin and P-selectin, respectively) (Laskey *et al.*, 1989; Stoolman, 1989; Watson *et al.*, 1990). Selectins contain a single transmembrane domain and a C-terminal cytoplasmic tail. The selectins differ in their tissue distribution and are relatively rapidly expressed upon activation. For this reason the selectins are thought to regulate acute, rapidly developing inflammatory processes.

E-selectin is a 115 kD protein expressed exclusively

Table 3.5 The selectin family

	Distribution	Molecular mass (kD)	Ligand
L-selectin	Leucocytes	90	Mannose 6-phosphate Fructose 6-phosphate
E-selectin	Endothelium	115	Sialylated Lewis x
P-selectin	Endothelium	140	Lewis x Sialylated Lewis x CD15

on the surface of endothelial cells (Pober and Cotran 1990; Rice *et al.*, 1990). E-selectin requires *de novo* synthesis for expression, with the onset occurring after 1–2 h, peak levels being expressed at 4 h. Expression of E-selectin is transient and is gone by approximately 16 h after stimulation. E-selectin binds to a carbohydrate group sialylated Lewis x that is found on cell surface glycoproteins and glycolipids of neutrophils and monocytes. L-selectin was originally described and classified as a tissue-specific lymphocyte-homing receptor that mediated lymphocyte binding to specialized high-endothelial venules of peripheral lymph nodes (Laskey *et al.*, 1989). Subsequent studies have demonstrated that L-selectin is present on neutrophils and monocytes and may function in a cell activation capacity as well as a cell adhesive mediator. P-selectin is found on platelets and endothelial cells (Larson *et al.*, 1989; Johnston *et al.*, 1989). P-selectin is stored in cytoplasmic granules and therefore can be rapidly up-regulated and expressed on the cell surface (Lawrence and Springer, 1991). The ligand for P-selectin is a carbohydrate group (non-sialylated Lewis x) found on the surface of leucocytes.

2.3.3   Granulocyte Adhesion Molecules

Adhesion molecules on leucocytes include members from two supergene families: the selectins and the integrins. L-selectin was orginally described and classified as a tissue-specific lymphocyte-homing receptor that mediated lymphocyte binding to specialized high-endothelial venules of peripheral nodes. However, further studies revealed the presence of L-selectin on unstimulated neutrophils and monocytes. Thus, L-selectin is constitutively expressed on leucocytes (Kishimoto *et al.*, 1989a). Integrins present on leucocytes include members of the β_1 (CD29) and β_2 (CD18) family of adhesion molecules. Activation of leucocytes induces a profound change in adhesiveness to endothelial cells (Anderson *et al.*, 1986). After stimulation of leucocytes, L-selectin is rapidly shed from the cell surface via the action of proteases (Kishimoto *et al.*, 1989a). In neutrophils, the shedding of L-selectin occurs concurrently with the rapid up-regulation of Mac-1 (CD11b/CD18) via recruitment from cytoplasmic storage granules. Therefore, activation of neutrophils results in the loss of one adhesion molecule and the up-regulation of another. Based on these observations it has been hypothesized that L-selectin mediates neutrophil rolling

and initial margination to inflamed endothelium. The marginated neutrophil is subsequently exposed to chemotactic stimuli which initiates a switch from L-selectin-mediated adhesion to Mac-1-dependent adhesion. Transendothelial migration is then promoted by Mac-1/ICAM-1 interactions. Thus, a two-step concept of cell recruitment is proposed. The first step is the guiding of cells to the site of inflammation by L-selectin followed by shedding of L-selectin and rapid up-regulation of Mac-1, resulting in transendothelial migration. The second step is a cell-specific signal that causes the cell to marginate and move out of the vascular space. After diapedesis, cell movement through the extravascular space, directed by chemotactic gradients, probably involves the interaction of the β_1 integrins with tissue matrix components (Albelda *et al.*, 1990).

VLA-4 (CD49d/CD29) is present on eosinophils and is the counter-receptor for VCAM-1 on vascular endothelium. It seem probable that the interactions between these adhesion molecules provide an additional (ICAM-1-independent) means of eosinophil recruitment to sites of inflammation.

2.4   REGULATION OF ENDOTHELIAL/ EPITHELIAL ADHESION MOLECULES

Exposure of cultured endothelial or epithelial cells to various cytokines or inflammatory mediators (e.g. IL-1, TNF or LPS) induces the expression of several adhesion ligands (Table 3.6). The inducible expression of adhesion molecules is responsible for the focal migration and retention at specific sites of inflammation. The kinetics of induced expression differs amongst adhesion glycoproteins that can be roughly categorized into those that are rapidly up-regulated and expressed and those that have a delayed onset of expression.

Adhesion molecules that are rapidly up-regulated and expressed on endothelial cells include two members of the selectin family, P-selectin and E-selectin. P-selectin is not constitutively expressed but is rapidly expressed (within 5 min) upon stimulation. The induced expression of P-selectin is rapid but transient, with levels returning to basal levels within 2 h after initial stimulation. The rapid up-regulation of P-selectin is a consequence of its

Table 3.6 Up-regulation of endothelial cell adhesion molecules

Adhesion molecule	Stimulus	Constitutive expression	Maximal expression (h)
E-selectin	IL-1, TNF, LPS	None	4–6 h
ICAM-1	IL-1, TNF, LPS IFNγ	Low level	12–24 h
ICAM-2	??	Moderate levels	
VCAM-1	IL-1, TNF, LPS IL-4	None	12–24 h

storage in intracellular granules called Weibel–Palade bodies, which upon stimulation migrate to the cell surface, fuse with the cell membrane and extrude P-selectin on the cell surface (Johnston *et al.*, 1989). In contrast, E-selectin has a short lag phase before expression begins (approximately 4 h) following stimulation. This lag phase is due to the requirement of new protein synthesis because E-selectin is not preformed and stored in intracellular vessels (Pober *et al.*, 1986; Munro *et al.*, 1989; Pober and Cotran 1990; Rice *et al.*, 1990). Peak levels of E-selectin occur between 6–8 h post-stimulation and return to basal levels by 12–18 h. Because of the rapid transient expression of the selectins, it has been hypothesized that these adhesion molecules are important regulators of acute or rapidly developing inflammatory responses by mediating early margination and initial adherence of leucocytes at specific sites of inflammation.

Adhesion molecules that are delayed in maximal expression include ICAM-1 and VCAM-1. Maximal expression of ICAM-1 and VCAM-1 occurs 12–24 h after stimulation and can be maintained for as long as 72 h. The delayed expression of these adhesion molecules may confer different roles in the inflammatory response. It has been hypothesized that ICAM-1 and VCAM-1, in contrast to the selectins, are involved in the processes of endothelial transmigration, through tissue to specific sites of inflammation. In addition, the prolonged expression of these molecules suggest that they may play important roles in the maintenance of chronic inflammation. In terms of tissue distribution, ICAM-1 differs from VCAM-1 in that it is expressed constitutively and can be up-regulated on both vascular endothelium and airway epithelium (Wegner *et al.*, 1991). As mentioned earlier, there is some evidence that VCAM-1 expression on vascular endothelium, in contrast to ICAM-1, is enhanced by IL-4 (Schleimer *et al.*, 1992). IL-4 is a major product of T_H2 lymphocytes and thus, may have a pivotal role recruitment of eosinophils via VLA-4/VCAM-1 interactions into the lungs following allergen exposure to allergic asthmatics.

3. Adhesion Molecules and Mucosal Inflammation in Asthma

Bronchial asthma is a disease that effects approximately 5–8% of the population (Sheffer, 1991). The acute symptoms of asthma include shortness of breath and wheezy breathing, which are largely the result of the action of mast cell-derived mediators on airway smooth muscle which are released following exposure to specific stimuli. These acute symptoms can be adequately controlled with β_2 agonist bronchodilators (Nelson, 1986); however, it is the chronic symptoms of asthma such as persistent airway obstruction and bronchial hyper-responsiveness that are less well understood and often

unresponsive to available therapies (O'Byrne *et al.*, 1987; Sybert and Weiss, 1985). Current research indicates that chronic symptoms may be a manifestation of the development of airway inflammation involving many different cell types, with eosinophils and T lymphocytes playing prominent roles (Larson, 1987). The development of airway inflammation can be initiated by a large number of stimuli leading to a complex series of events at both the cellular and tissue levels. It is now well recognized that cell adhesion or cell–cell communication is essential to the normal functioning of the immune system and critical to the development of inflammation (Kohl *et al.*, 1984). Cell adhesion molecules direct the initial adherence of inflammatory cells to vascular endothelium and subsequently mediate their transendothelial migration out of the vascular space to specific sites of inflammation. In addition to directing the infiltration of inflammatory cells into the airways, adhesion molecules are also important regulators of inflammatory cell activation (Wright *et al.*, 1988, 1989). Finally, adhesion molecules are responsible for the retention of inflammatory cells at effector cell sites. In these ways, adhesion molecules play crucial roles in disease settings. In asthma, adhesion molecules have a major role in the developing inflammation and alteration in airways function. The focal adherence and activation of leucocytes at mucosal tissues ultimately leads to a disruption of the normal function and integrity of these barriers which we hypothesize is a major contributor to disease morbidity.

Some of the characteristics of asthmatic airways include the presence of activated eosinophils with deposition of eosinophil-derived proteins (e.g. major basic protein) in and around the airways, areas of epithelial cell damage or denudation and increased bronchial responsiveness to physiological, pharmacological or mechanical stimuli (Ryley and Brogan, 1968; Brogan *et al.*, 1975; Filley *et al.*, 1982; Wardlaw *et al.*, 1988). Bronchial hyper-responsiveness is an important characteristic of asthma (Boushey *et al.*, 1980; McFadden, 1987; Wagner *et al.*, 1990). Studies have demonstrated that the degree of airway hyper-responsiveness correlates with the severity of asthma, diurnal variations in peak flow rates and the amount of therapeutic intervention required to control asthmatic symptoms (Hargreave *et al.*, 1981; Chan-Yeung *et al.*, 1982; Ryan *et al.*, 1982; Boulet *et al.*, 1983).

3.1 THE PRIMATE MODEL OF ALLERGIC ASTHMA

We have developed a primate model of allergic asthma to study the role of inflammation in chronic airway dysfunction. Our studies involve the use of adult male cynomolgus monkeys (*Macaca fascicularis*) that have a naturally occurring, IgE-mediated respiratory hyper-sensitivity to protein extracts of the nematode *Ascairs suum*. The methods of assessing airway cellular composition by BAL or bronchial biopsy, airway responsiveness

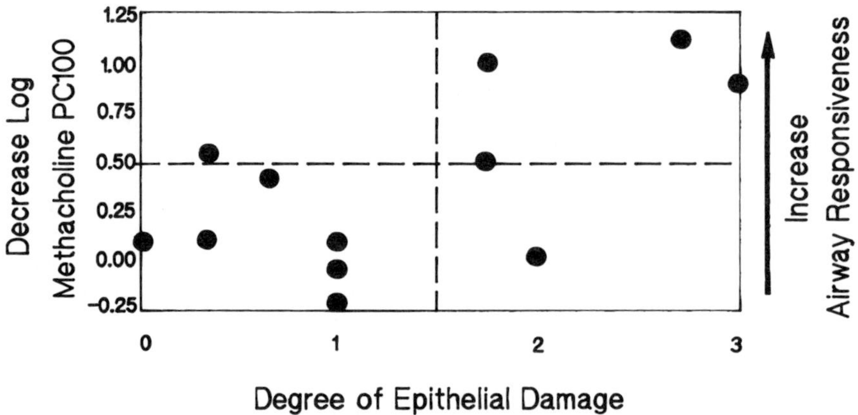

Figure 3.1 The effects of multiple antigen challenges on airway epithelial damage. There is a significant correlation between the degree of airway epithelial cell damage/denudation with the magnitude of the increase in airway responsiveness to inhaled methacholine.

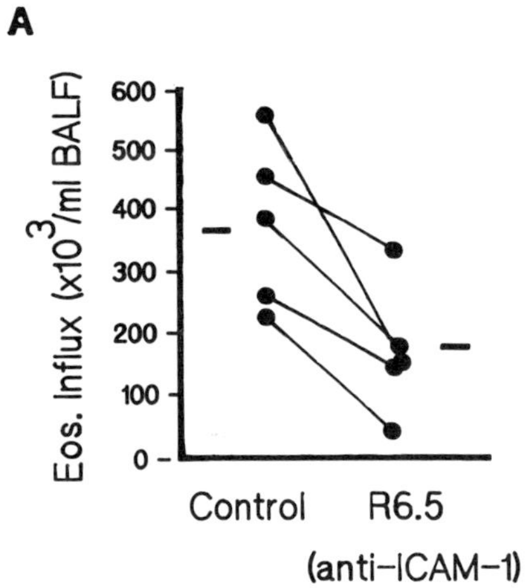

to inhaled methacholine, administration of inhaled antigen and measurement of pulmonary function have been described in detail elsewhere (Gundel *et al.*, 1989, 1990, 1992).

In a series of papers we reported that multiple antigen inhalations induce changes in the airways, including a prolonged and selective airway eosinophilia temporally associated with airway epithelial damage and the onset of airway hyper-responsiveness (Gundel *et al.*, 1989; Wegner *et al.*, 1991; Gundel *et al.*, 1992). Damage to the airway epithelium was assessed by whole lung histology as well as by the presence of creola bodies in BAL fluid. Our results demonstrated a significant correlation between the degree of epithelium damage with the magnitude of the increase in airway hyper-responsiveness (Fig. 3.1). The changes in airway cellular composition, epithelial damage and the magnitude of the increase in airway responsiveness induced in the primate model is similar to that characteristic of human patients with bronchial asthma. Thus, with the use of mAbs to selected adhesion proteins, we have utilized this model to investigate the role of some of the immunoglobulin, selectin and integrin supergene families in the development of airway inflammation and hyperresponsiveness.

Our initial studies examining immunohistological staining of lung biopsy tissue demonstrated that ICAM-1 is up-regulated on both pulmonary vascular endothelium and airway epithelium following multiple antigen inhalations (Wegner *et al.*, 1990). Interestingly, ICAM-1 is markedly up-regulated on the basolateral surface of the airway epithelium and associated with a large accumulation

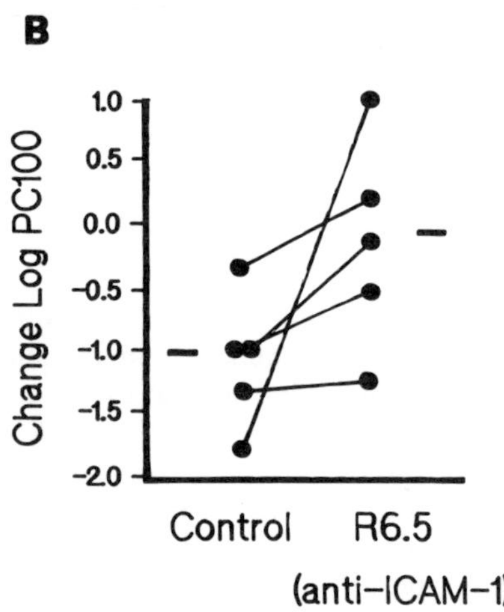

Figure 3.2 The effects of intravenous administration of functionally active anti-ICAM-1 mAb R6.5 (1.76 mg/kg) on (A) antigen-induced eosinophil influx into the airways and (B) increases in airway responsiveness to inhaled methacholine. R6.5 treatment significantly reduced the influx of eosinophils and increases in airway responsiveness induced by multiple antigen inhalation challenges.

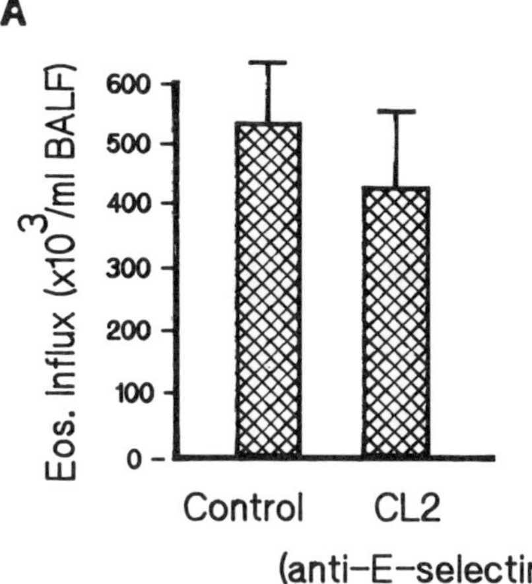

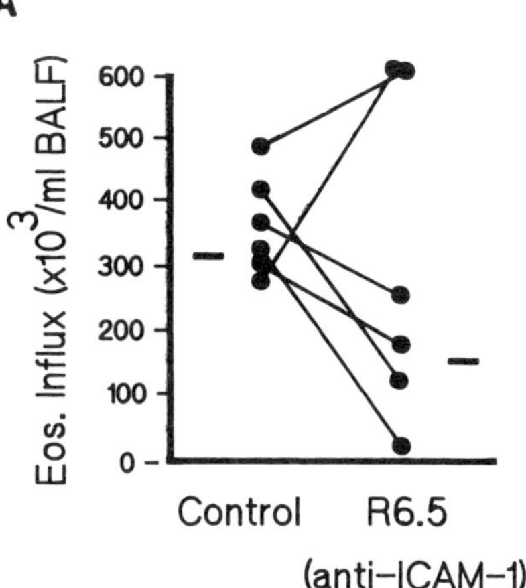

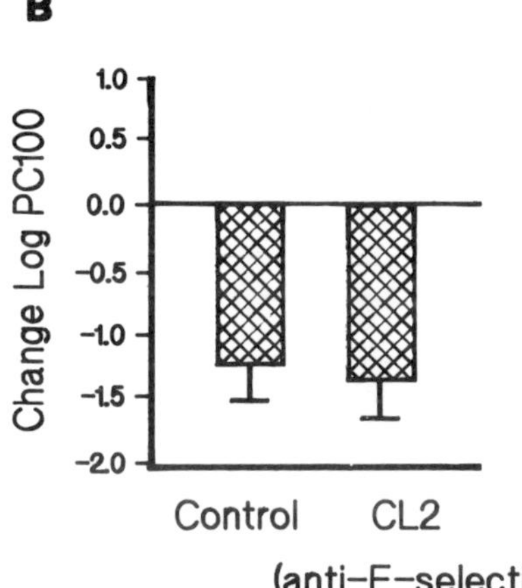

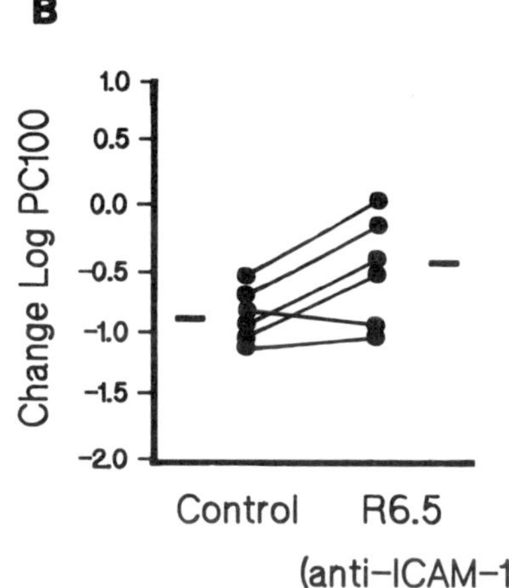

Figure 3.3 Effects of intravenous administration (2 mg/(kg day)) of anti-E-selectin mAb CL2 on (A) antigen-induced airway eosinophilia and (B) hyper-responsiveness induced by multiple antigen inhalation challenges. CL2 did not affect eosinophil influx into the airways nor did it block the increase in airway responsiveness. Values represent the mean ± SEM of five animals. Treatment studies are compared with bracketing control studies.

Figure 3.4 The effects of inhaled anti-ICAM-1 mAb R6.5 on (A) antigen-induced airway eosinophilia and (B) airway hyper-responsiveness. Inhaled R6.5 significantly reduced eosinophil infiltration into the airways and the increase in airway responsiveness.

of granulocytes. Thus, we hypothesized that the up-regulation of ICAM-1 is important in the recruitment of inflammatory cells into the airways. Because of the enhanced expression of ICAM-1 along the basolateral surface of the airway epithelium, it may also mediate epithelial cell damage or denudation leading to changes in airway lung function (e.g. increased airway responsiveness). To further evaluate the role of ICAM-1, in our next series of studies a murine anti-human ICAM-1 mAb (R6.5) was utilized (Wegner *et al.*, 1990). R6.5, administered intravenously, significantly inhibited both the antigen-induced eosinophil influx into the airways and the associated increase in airway responsiveness (Fig. 3.2).

Like ICAM-1, E-selectin expression is also enhanced on inflamed endothelium and has been demonstrated to contribute to granulocyte adherence *in vitro* (Bevilacqua *et al.*, 1987; Luscinskas *et al.*, 1989). However, unlike ICAM-1, E-selectin up-regulation occurs more rapidly and its expression is transient (Bevilacqua *et al.*, 1987). The role of E-selectin in antigen-induced airway eosinophilia and hyper-responsiveness was evaluated using a murine anti-human E-selectin mAb (CL2) following the same protocol as with R6.5 (Wegner *et al.*, 1991). The results of these studies demonstrated that CL2 treatment had no effect on the eosinophil infiltration nor the increase in airway responsiveness induced with multiple antigen inhalations (Fig. 3.3).

Prompted by our results demonstrating efficacy of anti-ICAM-1 in our model and the realization that ICAM-1 and E-selectin differ in their expression on airway epithelium (E-selectin not expressed on airway epithelium),

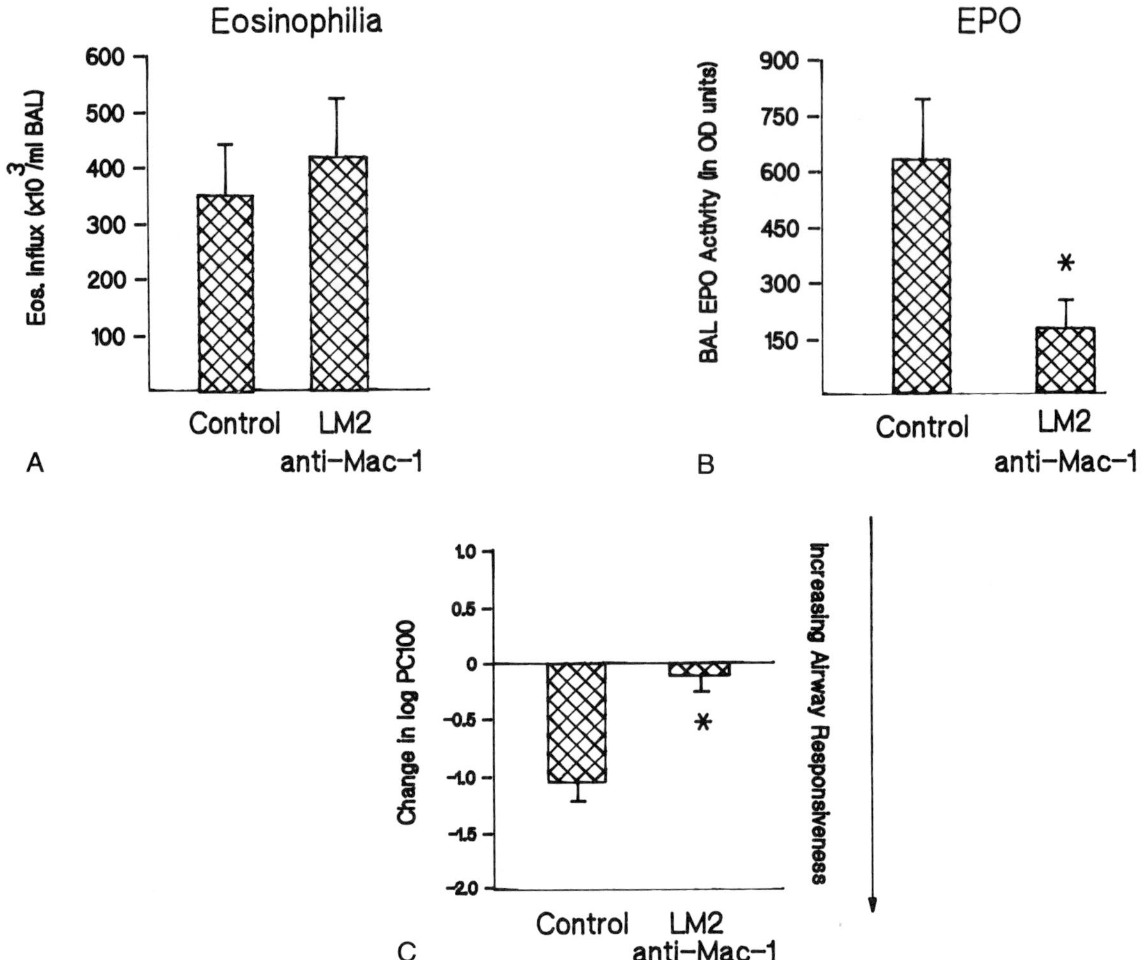

Figure 3.5 Effects of anti-Mac-1 mAb LM2 (2 mg/(kg day) i.v.) on antigen-induced (A) airway eosinophilia; (B) BAL concentrations of eosinophil peroxidase and (C) methacholine PC100 values. LM2 had no effect on eosinophil infiltration into the airways; however, the eosinophil activation/degranulation and airway hyper-responsiveness was significantly inhibited. Values represent the mean ± SEM, n=6.

we postulated that the increase in airway responsiveness may be the result of an epithelial ICAM-1-mediated event. Thus, a study was designed whereby R6.5 (anti-ICAM-1) was administered by inhalation to selectively target the airway epithelial ICAM-1 (Wegner *et al.*, 1991). The results of these studies demonstrated that inhaled R6.5 significantly inhibited the antigen-induced airway epithelial cell damage and airway hyper-responsiveness, supporting our hypothesis (Fig. 3.4).

Further evidence for eosinophil/airway epithelial interactions via ICAM-1 come from a recent study where we examined the role of a murine anti-human Mac-1 (CD11b/CD18) in the primate model (Wegner *et al.*, 1992). Intravenous anti-Mac-1 (LM2) mAb had no effect on antigen-induced eosinophil influx into the airways; however, LM2 treatment effectively inhibited the increase in airway responsiveness. In addition, LM2 treatment

reduced the concentrations of BAL fluid EPO, suggesting that although the eosinophils had infiltrated into the airways they were not "activated" and degranulating in the airways (Fig. 3.5).

We have found that both ICAM-1 and Mac-1 contribute to the airway hyper-responsiveness induced by multiple antigen inhalation in primates. Based on these results, we suggest that eosinophil infiltration into the airways (binding to vascular endothelium and diapedesis) is partially mediated by a CD11a/CD18–ICAM-1-dependent mechanism, and that the activation of eosinophils, once in the airway tissue, and altered airway function associated with the development of airway eosinophilia (increased airway responsiveness) is mediated by a Mac-1 (CD11b/CD18)–ICAM-1-dependent mechanism (Fig. 3.6). We hypothesize that the Mac-1–ICAM-1-mediated binding of eosinophils to the basolateral surface of the airway

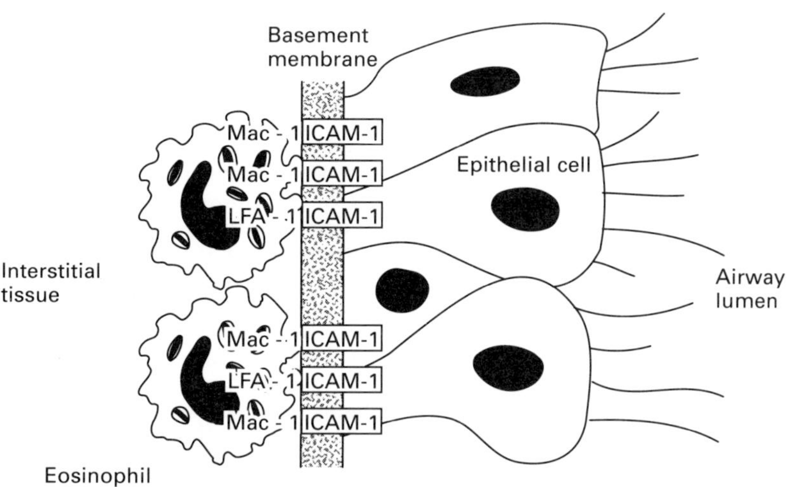

Figure 3.6 A possible mechanism whereby eosinophils adhere to airway epithelium, become activated and release granule-associated mediators that alter epithelial cell function and which leads to epithelial denudation. Adherence of eosinophils to epithelial cells allows the localized, concentrated release of toxic granule contents, thereby optimizing the effects of mediators on airway epithelium.

epithelium results in a further activation and subsequent localized degranulation. This localized degranulation results in a high concentration of toxic eosinophil-derived products (preformed and synthesized *de novo*), including major basic protein, elastase, superoxide anions and others. The action of these mediators function to degrade and breakdown the adhesion between the luminal cells and the substratum. This ultimately results in a dysfunctioning of the mucosal lining via loss of epithelial cell integrity or denudation leading to airway hyper-responsiveness via a number of mechanisms.

3.2 CLINICAL IMPLICATIONS

Whereas experiments leading to the discovery and characterization of cellular adhesion glycoproteins were performed *in vitro* or in animal models of asthma, one must ultimately determine their presence and the potential therapeutic functions of antibodies in human asthma. Evidence suggesting the importance of adhesion molecules in the development of inflammation in humans comes from several lines of evidence. First, using immunocytochemical techniques, Cotran and associates (1986) identified E-selectin on microvascular endothelium during DHR reactions in human volunteers. No E-selectin was identified prior to the initiation of the DHR; however, at 16 and 23 h into the DHR there was a strong signal for E-selectin. In a similar series of experiments, up-regulation of both ICAM-1 and VCAM-1 has been identified on vascular endothelium during DHR in humans (Lewis *et al.*, 1989; Rice *et al.*, 1990). In addition, increased expression of E-selectin and ICAM-1 in skin has been reported after cutaneous injections of recombinant cytokines (Munro *et al.*, 1989). Kyan-Aung and associates

(1991) have reported increased expression of E-selectin and ICAM-1 on vascular endothelium in allergic subjects 6 h after intradermal injection of allergen. In these studies, the kinetics of adhesion molecule up-regulation was similar to the kinetic profile of inflammatory cell infiltrates, suggesting a causal relationship between the two. There was no enhanced expression (up-regulation) of adhesion molecules at adjacent sites of diluent injection in these studies. Finally, several studies have demonstrated expression of both ICAM-1 and E-selectin in bronchial biopsy tissue from allergic asthmatic subjects (Hansel *et al.*, 1992; Montefort *et al.*, 1992; Wardlaw *et al.*, 1992). Thus, there is strong evidence indicating that adhesion molecules are associated and probably involved in the development of allergic inflammation.

The association between adhesion molecule up-regulation and leucocyte infiltration has generated tremendous interest in defining the mechanisms and ligands involved in cell trafficking with hopes of developing a novel and more specific treatment for inflammatory diseases. Data from our laboratory suggest that, in allergic IgE-mediated asthma, the up-regulation of ICAM-1 on vascular endothelium plays a pivotal role in the recruitment of leucocytes (eosinophils) into the lungs. In addition, enhanced expression of ICAM-1 on airway epithelium is an essential component of eosinophil–epithelial interactions leading to eosinophil activation, degranulation, epithelial cell dysfunction/desquamation and increases in airway responsiveness. Since airway hyper-responsiveness is regarded as an important characteristic of the pathophysiology of asthma, its regulation could be of paramount importance in controlling asthmatic symptoms. In addition, many studies suggest that chronic inflammation leads to the development of airway hyper-responsiveness. Therefore, the control or prevention of airway inflammation could ultimately inhibit the onset of airway hyper-responsiveness and expression of asthma symptoms. In this regard, antagonists or regulation of cell surface adhesion molecules may provide an effective prophylactic therapy for bronchial asthma.

4. References

Albelda, S.M. (1991). Endothelial and epithelial cell adhesion molecules. Am. J. Respir. Cell. Mol. Biol. 4, 195–203.

Albelda, S.M., Oliver, PD., Romer, L.H. and Buck, C.A. (1990). Endo-CAM: a novel endothelial cell–cell adhesion molecule. J. Cell Biol. 110, 1227–1237.

Anderson, D.C., Miller, L.J., Schmalstieg, F.C, Rothlein, R. and Springer, T.A. (1986). Contributions of the Mac-1 glycoprotein family to adherence-dependent granulocyte functions: structure–function assessments employing subunit-specific monoclonal antibodies. J. Immunol. 137, 15–27.

Aruffo, A., Stamenkovic, I., Melnick, M., Underhill, C.B. and Seed, B. (1990). CD44 is the principal cell surface receptor for hyaluronate. Cell 61, 1303–1313.

Behrens, J., Birchmeir, W., Goodman, S.L. and Imhof, B.A. (1985). Dissociation of Madin–Darby canine kidney epithelial cells by the monoclonal antibody anti-Arc-1: mechanistic aspects identification of the antigen as a component related to uvomorulin. J. Cell Biol. 101, 1307–1315.

Bevilacqua, M.P., Pober, JS., Mendrick, D.L., Cotran, R.S. and Gimbrone, M.A. (1987). Identification of an inducible endothelial-leukocyte adhesion molecule. Proc. Natl Acad. Sci. USA 84, 9238–9242.

Boulet, L.P., Cartier, A., Thomson, N.C, Roberts, R.S, Dolovich, J. and Hargreave, F.E. (1983). Asthma and increases in nonallergic bronchial responsiveness from seasonal pollen exposure. J. Allergy Clin. Immunol. 71, 399–406.

Boushey, H.A., Holtzman, M.J., Sheller, J.R. and Nadel, J.A (1980). Bronchial hyperreactivity. Am. Rev. Respir. Dis. 121, 389–414.

Boyd, A.W., Wawryk, S.O, Burns, G.F. and Fecondo, J.V. (1988). Intercellular adhesion molecule-1 (ICAM-1) has a central role in cell-cell contact-mediated immune mechanisms. Proc. Natl Acad. Sci. USA. 85, 3095–3099.

Brogan, R.D., Ryley, H.C., Neale, L. and Yassa, L. (1975). Soluble proteins of bronchopulmonary secretions from patients with cystic fibrosis, asthma and bronchitis. Thorax 30, 72–79.

Buck, C.A. and Horwitz, A.F. (1987). Cell surface receptors for extracellular matrix molecules. Annu. Rev. Cell Biol. 3, 179–205.

Carter, W.G., Wayner, E.A., Bouchard, T.S and Kaur, P. (1990). The role of integrins a2b1 and a2b1 in cell–cell and cell–substrate adhesion of human epidermal cells. J. Cell Biol. 110, 1387–1404.

Chan-Yeung, M., Lam, S. and Koener, S. (1982). Clinical features and natural history of occupational asthma due to western cedar (*Thuja plicata*). Am. J. Med. 72, 411–415.

Cotran, R.S., Gimbrone, M.A., Bevilacqua, M.P., Medrick, D.L. and Pober, J.S. (1986). Induction and detection of a human endothelial activation antigen *in vivo*. J. Exp. Med. 164, 661–666.

Davignon, D., Martz, E., Reynolds, T., Kuhrzinger, K. and Springer, T.A. (1981). Monoclonal antibodies to a novel lymphocyte function-associated antigen (LFA-1): mechanism of blockade of T-lymphocyte mediated killing and effects on other T and B lymphocyte functions. J. Immunol. 127, 590–595.

de Fougerolles, A.R., Stacker, S.A., Schwarting, R. and Springer, T.A. (1991). Characterization of ICAM-2 and evidence for a third counter-receptor for LFA-1. J. Exp. Med. 174, 253–267.

Dougherty, G.J. and Hogg, N. (1987). The role of monocyte lymphocyte function-associated antigen 1 (LFA-1) in accessory cell function. Eur. J. Immunol. 17, 943–947.

Dougherty, G.J., Murdoch, S. and Hogg, N. (1988). The function of human intercellular adhesion molecule-1 (ICAM-1) in the generation of an immune response. Eur. J. Immunol. 18, 35–39.

Dustin, M.L., Rothlein, R, Bhan, A.K., Dinarello, C.A. and Springer, T.A. (1986). Induction by IL 1 and interferon-γ: tissue distribution, biochemistry, and function of a natural adherence molecule (ICAM-1). J. Immunol. 137, 245–252.

Dustin, M.L., Singer, K.H, Tuck, D.T. and Springer, T.A. (1988). Adhesion of T lymphocytes to epidermal keratinocytes is regulated by interferon gamma and is mediated by intercellular adhesion molecule I (ICAM-1). J. Exp. Med. 167, 1323–1340.

Elices, M.J. and Hemler, M.E. (1990). The human integrin VLA-2 is a collagen receptor on some cells and a collagen/laminin receptor on others. Proc. Natl Acad. Sci. USA 86, 9906–9910.

Filley, W.V., Holley, K.E., Kephart, G.M., Gleich, G.J. (1982). Identification by immunofluorescence of eosinophil granule major basic protein in lung tissue of patients with bronchial asthma. Lancet ii, 11–16.

Geng, J.-G., Bevilacqua, M.P, Moore, K.L. McIntyre, T.M., Prescott, S.M., Kim, J.M., Bliss, G.A., Zimmerman, G.A. and McEver, R.P. (1990). Rapid neutrophil adhesion to activated endothelium mediated by GMP-140. Nature 343, 757–760.

(1991). Guidelines for the diagnosis and treatment of asthma. I. Definition and diagnosis. J. Allergy Clin. Immunol. 88, 427–438.

Gundel, R.H., Gerritsen, M.E. and Wegner, C.D. (1989). Antigen-coated sepharose beads induce airway eosinophilia and airway responsiveness in cynomolgus monkeys. Am. Rev. Respir. Dis. 140, 629–633.

Gundel, R.H., Gerritsen, M.E, Gleich, G.J. and Wegner, C.D. (1990). Repeated antigen inhalation results in a prolonged airway eosinophilia and airway hyperresponsiveness in primates. J. Appl. Physiol. 68, 779–786.

Gundel, R.H., Wegner, C.D. Torcellini, C.A and Letts, G.L. (1992). Antigen induced acute and late-phase responses in primates. Am. Rev. Respir. Dis. 146, 369–373.

Hansel, T.T., de Vries, J., Carballido, J.M., Blaser, K. and Walker, C. (1992). Eosinophils from the respiratory tissue of asthmatics express functional ICAM-1 and HLA-DR. J. Allergy Clin. Immunol. 89, 165.

Hargreave, F.E., Ryan, G., Thomson, N.C., O'Byrne, P.M., Latimer, K., Juniper, E.F. and Dolovich, J. (1981). Bronchial responsiveness to histamine or methacholine in asthma: Measurement and clinical significance. J. Allergy Clin. Immunol. 68, 347–355.

Haskard, D., Cavender, D., Beatty, P., Springer, T.A. and Ziff, M. (1986). T lymphocyte adhesion to endothelial cells:

mechanisms demonstrated by anti-LFA-1 monoclonal antibodies. J. Immunol. 137, 2901–2906.

Haskard, D.O. (1989). In "Interleukin 1, Inflammation and Disease" (Eds R. Bomford and B. Henderson), pp 123–142. Elsevier, Amsterdam.

Heimark, R.L., Degner, M. and Schwartz, S.M. (1990). Identification of a Ca^{++}-dependent cell–cell adhesion molecule in endothelial cells. J. Cell Biol. 110, 1745–1756.

Hemler, H.E. (1990). VLA proteins in the integrin family: structures, functions, and their role on leukocytes. Annu. Rev. Immunol. 8:365–400.

Horton, M. (1990). Current status review: tronectin receptor: tissue specific expression or adaptation to culture. Int. J. Exp. Pathol. 71, 741–759.

Hynes, R.O. (1987). Integrins: a family of cell surface receptors. Cell 48, 549–554.

Johnston, G.I., Cook, R.G. and McEver, R.P. (1989). Cloning of GMP-140, a granule membrane protein of platelets and endothelium: sequence similarity to proteins involved in cell adhesion and inflammation. Cell 56, 1033–1044.

Kishimoto, T.K., Jutila, M.A., Berg, E.L. and Butcher, E.C. (1989a). Neutrophil Mac-I and MEL-14 adhesion proteins inversely regulated by chenotactic factors. Science 245, 1238–1241.

Kishimoto, T.K., Larson, R.S., Corbi, A.L., Dustin, M.L., Staunton, D.E. and Springer, T.A. (1989b). The leukocyte integrins. Adv. Immunol. 46, 149–182.

Kohl, S., Springer, T.A., Schmalstieg, F.C., Loom, L.S. and Andersonm, D.C. (1984). Defective natural killer cytotoxicity and polymorphomxlear leukocyte antibody-dependent cellular cytotoxicity in patients with LFA-I/OKM-1 deficiency. J. Immunol. 133, 2972–2978.

Krensky, A.M., Sanchez-Madrid, F., Robbins, E., Nagy, J., Springer, T.A. and Burakoff, S. (1983). The functional significance, distribution and structure of LFA-1, LFA-2 and LFA-3: cell surface antigen associated with CTL–target interactions. J. Immunol. 131, 611–616.

Kyan-Aung, U., Harkard, D.O., Poston, R.N., Thornhill, M.H. and Lee, T.H. (1991). Endothelial leukocyte adhesion molecule-1 and intercellular adhesion molecule-1 mediate the adhesion of eosinophils to endothelial cells *in vitro* and are expressed by endothelium in allergic cutaneous inflammation *in vivo*. J. Immunol. 146, 521–528.

Lampugnani, M.C., Resnati, M., Dejana, E. and Marchisio, P.C. (1991). The role of integrins in the maintenance of endothelial monolayer integrity. J. Cell Biol. 112, 479–490.

Larsen, E., Celi, A., Gilbert, G.E, Furie, B.C., Erban, J.K., Bonfanti, R., Wagner, D.D. and Furie, B. (1989). PADGEM protein: a receptor that mediates the interaction of activated platelets with neutrophils and monocytes. Cell 59, 305–312.

Larsen, G.L. (1987). The pulmonary late-phase response. Hosp. Pract. 2, 155–169.

Larson, R.S. and Springer, T.A. (1990). Structure and function of leukocyte integrins. Immunol. Rev. 114, 181–206.

Laskey, L.A., Singer, M.S., Yednock, T.A., Dowbenko, D., Fennie, C., Rodriquez, H., Nguyen, T., Stachel, S. and Rosen, S.D. (1989). Cloning of a lymphocyte homing receptor reveals a lectin domain. Cell 56, 1045–1055.

Lawrence, M.B. and Springer, T.A. (1991). Leukocytes roll on a selectin at physiologic flow rates: distinction from and prerequisite for adhesion through integrins. Cell 65, 859–873.

Lewis, R.E, Buchsbaum, T., Whitaker, D. and Murphy, G.F. (1989). Intercellular adhesion molecule-1 in the evolving human cutaneous delayed hypersensitivity reaction. J. Invest. Dermatol. 93, 672–677.

Luscinskas, F.W., Brock, A.F., Arnaout, M.A. and Gimbrone, M.A. (1989). Endothelial-leukocyte adhesion molecule-1-dependent and leukocyte (CDll/CD18)-dependent mechanisms contribute to polymorphonuclear leukocyte adhesion to cytokine-activated human vascular endothelium. J. Immunol. 142, 2257–2263.

McFadden, E.R. (1987). Exertional dyspnea and cough as preludes to acute attacks of bronchial asthma. N. Engl. J. Med. 292, 555–559.

Montefort, S., Roche, W.R., Howarth, P.H., Djukanovic, R., Gratziou, C., Carroll, M., Smith, L., Britten, K.M., Haskard, D. Lee, T.H. and Holgate, S.T. (1992). Intercellular adhesion molecule-1 (ICAM-1) and endothelial leukocyte adhesion molecule-1 (ELAM-1) expression in the bronchial mucosa of normal and asthmatic subjects. Eur. Respir. J. 5, 815–823.

Munro, J.M., Pober, J.S. and Cotran, R.S. (1989). Tumor necrosis factor and interferon-gamma induce distinct patterns of endothelial activation and associated leukocyte accumulation in skin of *Papio anubis*. Am. J. Pathol. 135, 121–133.

Nelson, H.S. (1986). Adrenergic therapy of bronchial asthma. J. Allergy Clin. Immunol. 77, 771–785.

Newman, P.J., Berndt, M.C., Gorsky, J., White, G.C., Paddock, L.S. and Muller, W.A. (1990). PECAM-1 (CD31): cloning and relation to adhesion molecules of the immunoglobulin gene superfamily. Science 247, 1219–1222.

O'Byrne, P.M., Dolovich, J. and Hargreave, F.E. (1987). Late asthmatic responses. Am. Rev. Respir. Dis. 136, 740–748.

Osborn, L., Hession, C., Tizard, R., Vassallo, C., Luhowskyj, S., Chi-Rosso, G. and Lobb, R. (1989). Direct expression cloning of vascular cell adhesion molecule 1, a cytokine-induced endothelial protein that binds to lymphocytes. Cell 59, 1203–1211.

Picker, L.J., Nakache, M. and Butcher, E.C. (1989). Monoclonal antibodies to human types. J. Cell Biol. 109, 927–937.

Pober, J.S. and Cotran, R.S. (1990). Cytokines and endothelial cell biology. Physiol. Rev. 70, 427–443.

Pober, J.S., Gimbrone, M.A., Lapierre, L.A., Mendrick, D.L., Fiers, W., Rothlein, R. and Springer, T.A. (1986). Overlapping patterns of activation of human endothelial cells by interleukin 1, tumor necrosis factor, and immune interferon. J. Immunol. 137, 1893–1896.

Rice, G.E., Munro, J.M. and Bevilacqua, M.P. (1990). Inducible cell adhesion molecule 110 (INCAN-110) is an endothelial receptor for lymphocytes: a CDll/CD18-independent adhesion mechanism. J. Exp. Med. 171, 1369–1374.

Ruoslahti, P. (1988). Fibronectin and its receptors. Annu. Rev. Biochem. 57, 375–413.

Ruoslahti, P. (1989). Proteoglycans in cell regulation. J. Biol. Chem. 264, 13369–13372.

Ryan, G., Latimer, K., Dolovich, J. and Hargreave, F.E. (1982). Bronchial responsiveness to histamine: relationship to diurnal variation of flow rates and improvement after bronchodilation. Thorax, 37 423–428.

Ryley, H.C. and Brogan, T.D. (1968). Variation in the composition of sputum in chronic chest diseases. Br. J. Exp. Pathol. 49, 25–33.

Sanchez-Madrid, F., Nagy, J., Robbins, E., Simon, P. and Springer, T.A. (1983). A human leukocyte differentiation antigen family with distinct alpha subunits and a common beta subunit. J. Exp. Med. 158, 1785.

Sanders, S., Jalkanen, M., O'Farrell, S. and Bernfield, M. (1989). Molecular cloning of syndecan, an integral membrand proteoglycan. J. Cell Biol. 108, 1547–1556.

Schleimer, R.P., Sterbinsky, S.A., Kaiser, J., Bickel, C.A., Klunk, D.A., Tomioka, K., Newman, W., Luscinskas, F.W., Gimbrone, M.A., McIntyre, B.W. and Bochner, B.S. (1992). IL-4 induces adherence of human eosinophils and basophils but not neutrophils to endothelium. Association with expression of VCAM-1. J. Immunol. 148, 1086–1092.

Siegelman, M.H., Van de Rijn, M. and Weissman, I.L. (1989). Mouse lymph node homing receptor cDNA clone encodes a glycoprotein revealing tandem interaction domains. Science 243, 1165–1172.

Smith, C.W., Rothlein, R., Hughes, B.J., Mariscalco, M.M., Rudloff, H.E. (1988). Determinant for CD18-dependent human neutrophil adherence and transendothelial migration. J. Clin. Invest. 82, 1746–1756.

Springer, T.A. (1990). Adhesion receptors of the immune system. Nature 346, 425–434.

Stamenkovic, E., Amiot, M., Pesando, J.M and Seed, B. (1989). A lymphocyte molecule implicated in lymph node homing is a member of the cartilage link protein family. Cell 56, 1057–1062.

Staunton, D.E., Dustin, M.L. and Springer, T.A. (1989). Functional cloning of ICAM-2, a cell adhesion ligand for LFA-1 homologous to ICAM-1. Nature 339, 61–64.

Stevenson, B.R. and Paul, D.L. (1989). The molecular constituents of intercellular junctions. Curr. Opin. Cell Biol. 1, 884–891.

Stoolman, L.M. (1989). Adhesion molecules controlling lymphocyte migration. Cell 56, 907–910.

Sybert, A. and Weiss, E.B. (1985). In "Bronchial Asthma: Mechanisms and Therapeutics" 2nd edn (eds E.B. Weiss, M.S. Segal and M. Stein), p 808. Little, Brown, Boston.

Takeichi, M. (1990). Cadherins: a molecular family important in selective cell–cell adhesion. Annu. Rev. Biochem. 59, 237–252.

Wagner, E.M., Liu, M.C., Weinmann, G.G., Permutt, S. and Bleecker, E.R. (1990). Peripheral lung resistance in normal and asthmatic subjects. Am. Rev. Respir. Dis. 141, 584–588.

Wardlaw, A.J., Dunnette, S., Gleich, G.J., Collins, J.V. and Kay, A.B. (1988). Eosinophils and mast cells in bronchoalveolar lavage in mild asthma. Relationship to bronchial hyperreactivity. Am. Rev. Respir. Dis. 137, 62–69.

Wardlaw, A.J., Bentley, A.M., Menz, G., Storz, C., Durham, S.R. and Kay, A.B. (1992). Expression of the adhesion molecules ICAM-1 and ELAM-1 in the bronchial mucosa in asthma. J. Allergy Clin. Immunol. 89, 164.

Watson, M.L., Kingsmore, S.F., Johnston, G.I Siegelman, M.H., Le Beau, M.M., Lemons, R.S., Bora, N.S., Howard, T.A., Weisman, I.L., McEver, R.P. and Seldin, M.F. (1990). Genomic organization of the selection family of leukocyte adhesion molecules on human and mouse chromosome 1. J. Exp. Med. 172, 263–272.

Wegner, C.D., Gundel., R.H., Reilly, P., Haynes, N., Letts, L.G. and Rothlein, R. (1990). Intercellular adhesion molecule-1 (ICAM-1) in the pathogenesis of asthma. Science, 247, 456–459.

Wegner, C.D., Rothlein, R. and Gundel, R.H. (1991). Adhesion molecules in the pathogenesis of asthma. Agents Actions 34, 529–544.

Wegner, C.D., Gundel, R.H., Churchill, L., Sousa, D., Stearns, C. and Letts, L.G. (1992). Mac-1 (CD11b/CD18) mediates antigen-induced eosinophil activation and airway hyper-responsiveness in monkeys. Am. Rev. Respir. Dis. (in press).

Wright, S.D., Lo, S.K. and Detmers, P.A. (1989). In "Leukocyte Adhesion Molecules: Structure, Function and Regulation" (eds T.A. Springer, D.C. Anderson, A.S. Rosenthal, and R. Rothlein), pp 190–207. Springer-Verlag, New York.

Wright, S.D., Weitz, J.I., Huang, A.J., Levin, S.M., Silverstein, S.C. and Loike, J.D. (1988). Complement receptor type three (CDllb/CD18) of human polymorphonuclear leukocytes recognizes fibrinogen. Proc. Natl Acad. Sci. USA 85, 7734–7738.

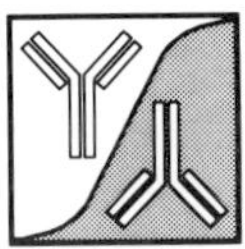

4. Innervation of the Airway Mucosa: Structure, Function and Changes in Airway Disease

Peter K. Jeffery

1. Introduction

The present chapter is complementary to the reviews of Richardson (1979) and Barnes (Barnes, 1989a,b; Barnes *et al.*, 1991a), and extends them by its focus on the distribution, morphology and peptide immunoreactivity of nerves in the conducting airways. In addition, studies which give insight into their functional role in the normal adult are briefly considered and, where known, variations in airway disease are highlighted. The innervation of the respiratory portion of the lung will be dealt with only briefly, as will the innervation of bronchial and pulmonary vessels. For the development of pulmonary innervation the reader is referred to Loosli and Hung (1977). There will be introductory sections on the anatomical aspects, general distribution and classification of pulmonary nerves. Thereafter, the tissue distribution of airway ganglia and nerves supplying tracheobronchial epithelium, submucosal glands and bronchial smooth muscle will be considered. Each major section is broadly subdivided into three: first, a light microscopic section dealing with the results of early silver stains, enzyme histochemistry (e.g. acetylcholinesterase), paraformaldehyde or glyoxylic acid-induced fluorescence (e.g. for catecholamines) and immunohistochemistry, the last used mainly to demonstrate nerves containing neuropeptides; secondly, a section dealing with electron microscopic studies which demonstrate the precise localization of nerve terminals; and, thirdly, one concerned with the salient functional aspects. For a more extensive consideration of the functional aspects of peptidergic innervation the reader is referred to reviews by Barnes *et al.* (1991a,b). The interaction of nerves with the immune system and, where known, alterations to innervation in airway disease are considered in two final sections.

Immunopharmacology of Epithelial Barriers
ISBN 0–12–288030–7

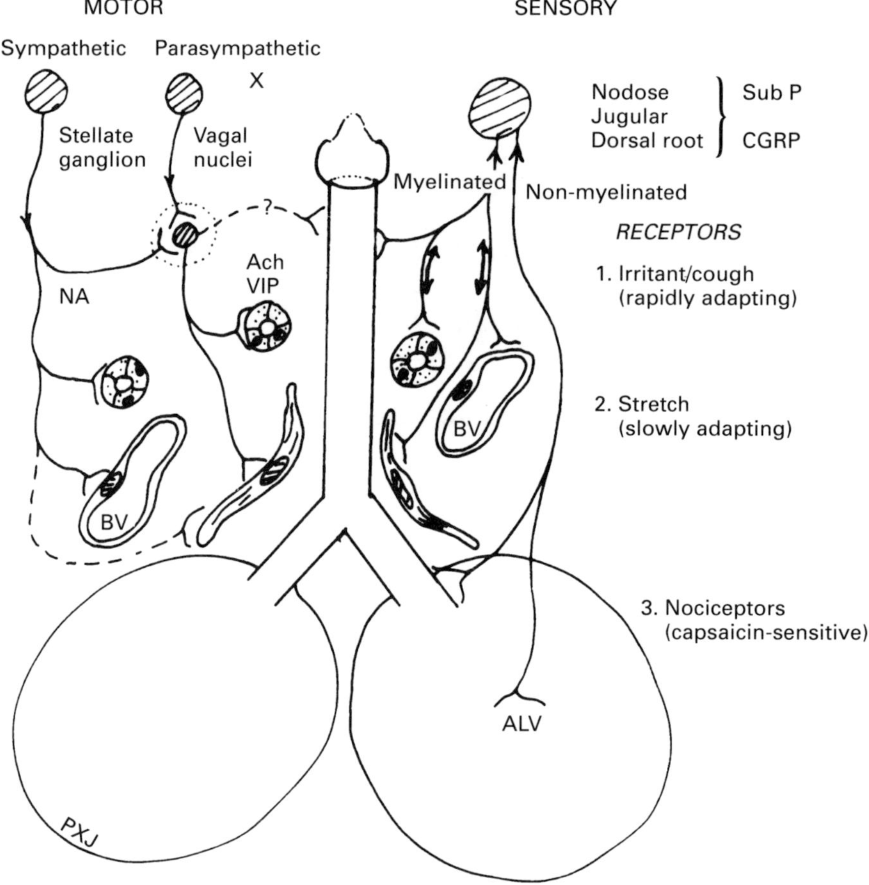

Figure 4.1 Diagrammatic representation of the autonomic and sensory nerve supply to the airways and lung. Autonomic nerve terminals show the coexistence of neuropeptides and "classical" neurotransmitters: sympathetic and peptidergic fibres are likely to innervate parasympathetic ganglia and modulate cholinergic transmission. Sensory fibres are of broadly three types and contain "sensory" neuropeptides which may be released by antidromic stimulation during axonal reflex to induce "neurogenic inflammation". Vagal and sympathetic trunks have both autonomic and sensory nerves.

2. *Anatomy*

The airways of the lung receive both autonomic (efferent) and sensory (afferent) nerve fibres (Fig. 4.1). The efferent innervation has, in general, both excitatory and inhibitory components which together supply and control the functions of effector structures such as mucus-secreting cells, bronchial smooth muscle and vasculature. It appears that during evolution the predominantly excitatory supply to the lung has switched from the sympathetic in lower amphibia to the parasympathetic nervous system in the higher vertebrates (Burnstock, 1969). The predominant neurotransmitters released have also changed and it is now clear that there are many fibres in the mammalian autonomic nervous system that are neither adrenergic nor cholinergic (henceforth refered to as NANC) in type (Burnstock, 1972, 1986). This underlines a major consideration, i.e. that there are major species differences in the predominant type and distribution of autonomic nerves to otherwise similar effector structures.

Sensory endings are of three types based on their position; firing pattern, adaptation to a maintained stimulus and speed of conduction (Widdicombe, 1954a,b, 1964; Widdicombe and Sterling, 1970; Widdicombe, 1974). Many sensory nerves are thought to begin as free endings, either within epithelial structures or interstitial connective tissue, but there may also be specialized associations with specific bronchial cells often referred to as "neuroepithelial bodies" or in avian species as "neurite–receptor cell complexes" (Cook and King, 1969b; Lauweryns *et al.*, 1972). As with autonomic nerves, species differences exist in the type and distribution of sensory nerves.

2.1 AUTONOMIC NERVE SUPPLY

2.1.1 Parasympathetic Nerves

Preganglionic parasympathetic fibres to the airways of the lung arise from the dorsal motor nucleus of the vagus in the floor of the fourth ventricle and descend in the vagus nerve to:

1. supply the larynx and cranial aspects of the trachea via the superior laryngeal nerves;
2. supply the main bulk of the trachea via the recurrent laryngeal nerve; and
3. enter the hila of the lungs via its bronchial branches (Fig. 4.1).

On entering these tissues, preganglionic myelinated fibres synapse with multipolar ganglia located in the airway walls and from these arise non-myelinated post-ganglionic fibres which innervate specific airway effector structures (Larsell, 1921, 1922; Larsell and Dow, 1933; Spencer and Leof, 1964; Widdicombe, 1966; Widdicombe and Sterling, 1970; Nagaishi, 1972; Gabella, 1976).

2.1.2   Sympathetic Nerves

In contrast, sympathetic preganglionic fibres (myelinated) originate from the spinal cord in the first four or five thoracic segments, particularly in the middle and inferior cervical ganglia, and run a short path to synapse in paravertebral ganglia. Three routes to the lung are then distinguished:

1. branches from the middle and inferior cervical ganglion (in animals the latter is often fused with the first thoracic ganglion to form the "stellate ganglion");
2. branches from the cardiac plexus supply the trachea and bronchi;

3. branches from thoracic paravertebral ganglia 1–5 may supply the pulmonary plexuses where they intermingle with vagal fibres.

Depending on species, sympathetic fibres may run an early course together with vagal fibres as in the cat and dog (i.e. the vagosympathetic trunk) or be separate as in rabbits and humans (Widdicombe and Sterling, 1970; Nagaishi, 1972). Fibres arising from the stellate ganglia may also run cranially along the vagosympathetic trunk for variable distances before looping back to the trachea and lung (Cabezas et al., 1971). These factors must be borne in mind when attempting selective stimulation or denervation of the parasympathetic or sympathetic nerve supply to the lung.

2.1.3   Non-adrenergic Non-cholinergic Nerves

The classic view that airways are regulated by cholinergic excitatory and adrenergic inhibitory nerves has been dramatically altered by functional and immunohisto-chemical demonstration of nerves which are NANC in type. NANC nerves have erroneously been referred to as components of a third nervous system comprising nerves with candidate neurotransmitters which include a variety of so-called "regulatory" peptides (Table 4.1) and/or the purines adenosine and ATP (Burnstock, 1972; Zetler, 1976; Polak and Bloom, 1979). In reality, both the classic and NANC systems are superimposed, sharing by co-

Table 4.1. Airway "Regulatory" Peptides: location and airway effect

Peptide	Location	Effect
Calcitonin	NE	?
CGRP	NF, NE	Enhances Sub P-induced bronchoconstriction/ vasodilation and increases blood flow
CCK	?	Brief contraction of tracheal muscle
Enkephalins	NE	Inhibit cholinergic neurotransmission/ stimulates J receptors
Galanin	N, NF	Modulates excitatory NANC transmission
GRP/bombesin, plus effects on lung growth	NE, NF	Bronchoconstriction, gland secretion
NPY	N, NF	Vasoconstriction and relaxes non-vascular smooth
SOM	N, NE	Potentiates cholinergic neurotransmission
Sub P, NkA, NkB, NPK (neurokinins)	N, NF	Bronchoconstriction, gland secretion, mast cell degranulation, increases vascular permeability dilatation
VIP/PHI/PHM[a]	N, NF	Bronchodilatation, vasodilatation, modulates gland secretion

N, neuronal cell body; NF, nerve fibres; NE, neuroendocrine cells.
[a] "PACAP-27" and "helodermin" are VIP-related and have similar effects.

Table 4.2 Co-localization of peptides in airways

Peptide	Site	Reference
VIP + PHI	Cholinergic nerves	Lundberg et al. (1984b)
VIP + Galanin	Cholinergic neurons	Cheung et al. (1985)
Sub P + NkA + CGRP	Sensory nerves	Martling et al. (1987)
Galanin + VIP	Sensory nerves	Luts and Sundler (1989)
NPY + SOM + Galanin + VIP + enkephalin	Adrenergic nerves	Lindh et al. (1989)

localization their distinct putative neurotransmitters within one and the same nerve fibre ending (Fig. 4.1 and Table 4.2). The concept of additional components of the autonomic nervous system is not new. Langley, in 1898, observed that stimulation of the vagus caused gastric relaxation which was unaffected by blockers of the adrenergic system. The observation of inhibitory junction potentials in guinea-pig colonic smooth muscle led to the hypothesis of the existence of a non-adrenergic inhibitory system (Burnstock *et al.*, 1963). The first convincing morphological evidence came from electron microscopic descriptions of three different types of neuronal processes in myenteric ganglia of the large intestines of guinea-pigs, monkeys and humans (Baumgarten *et al.*, 1970). Nerve endings with large, dense-cored granules were found which were similar to those found in the hypothalamic–hypophyseal system (containing the peptides vasopressin and oxytocin). These were referred to as p (for peptidergic)-type nerves and were proposed as responsible for NANC inhibition in the gastrointestinal tract.

The controversies over the role and importance of the sympathetic nervous system in the control of smooth muscle tone took an interesting turn following the results of combined electrophysiological and pharmacological studies of colon, ileum, stomach, vas deferens, and frog lung (Burnstock *et al.*, 1963; Bennett *et al.*, 1966; Robinson *et al.*, 1971; Burnstock, 1986). The techniques of measuring, in organ baths, the effects of field stimulation on the isometric tension of airway muscle strips subsequently led to the discovery of non-adrenergic inhibitory nerves in guinea-pig (Coburn and Tomita, 1973) and human airways (Richardson and Beland, 1976). Electrical field stimulation alone gave rise to muscle constriction while, in the presence of atropine, to relaxation. The relaxation was not blocked by propranolol nor by prior sympathetic depletion by reserpine (the extent of which was checked by the loss of catecholamine fluorescence). In a similar study, Richardson and Bouchard (1973) found that electrical field stimulation in the presence of adrenergic and cholinergic blockade relaxed tracheal rings previously contracted by histamine:

the relaxation was blocked by tetrodotoxin (a ganglionic poison obtained from puffer fish), indicating that the response was neurally mediated. Application of ATP or AMP (i.e. as evidence for a purinergic component) gave similar relaxations though incomplete and less dramatic. *In vivo* confirmation of these findings (in the guinea-pig) has been published (Chesrown *et al.*, 1980). Matsuzaki *et al* (1980) have repeated Richardson's studies of electrical field stimulation and, in addition, measured the release of VIP (assayed by RIA) into the bathing medium: the amount released correlated with the degree of relaxation, its release was blocked by tetrodotoxin and the relaxation was diminished by prior treatment with anti-VIP antisera.

Thus, there is evidence that the peptidergic system plays a major role in the non-adrenergic component of bronchial relaxation, with VIP a strong candidate as a neural mediator. The functional evidence is strong for guinea-pig airways. In human airways there is little or no induced fluorescence (for catecholamines) but much immunoreactivity for VIP in nerves associated with bronchial smooth muscle and mucus-secreting glands, supportive of a functional role for VIP in human airways (Barnes, 1986b). However, more recently nitric oxide has been shown to be an important molecule also mediating bronchial relaxation (Barnes *et al.*, 1991a,b; Belvisi *et al.*, 1992). Whether VIP is a neurotransmitter in the airways will remain unclear until specific VIP antagonists become available.

2.2 SENSORY NERVES

Many sensory fibres are thought to begin as free endings in the lung only to become myelinated, and without a break in continuity, course centrally in the vagus nerves to reach their respective neuronal soma which lie predominantly in the nodose and jugular ganglia (Widdicombe, 1954a,b, 1964; Widdicombe and Sterling, 1970; Nagaishi, 1972; Widdicombe, 1974). Some also have a non-vagal origin, arising from thoracic spinal dorsal root ganglia $T_1–T_6$ (Lundberg *et al.*, 1983a) and

Table 4.3 Sensory innervation (determined by function)

Site	Type (receptor reflex)	Effect
Larynx	I – expiration	Expiratory effort
Extrapulmonary airway	I – cough	Inspiratory and expiratory effort
Intrapulmonary airway	I – "irritant"	Bronchohyperventilation Unpleasant sensation
	II – stretch (Hering–Breuer reflex)	Increasing inhibition of inspiratory centre
	III – airway C fibre	Bronchoconstriction Mucus secretion Rapid, shallow breathing
Lung	III – pulmonary C fibre (J receptor reflex)	Rapid, shallow breathing Bradycardia Pulmonary hypotension

cervical ganglia (Uddman and Sundler, 1987). From here, sensory fibres continue cranially to their termination, most in the dorsal nucleus of the medulla oblongata. Recordings of afferent activity in sympathetic nerves and dorsal roots show that some pulmonary receptors are supplied by afferents coursing along sympathetic routes also (Holmes and Torrance, 1959; Widdicombe, 1964; Kostreva *et al.*, 1975). Recordings of action potentials from single nerve fibres supplying sensory endings located at different sites within the larynx, trachea, bronchi, bronchioli and alveoli have delineated three major functional types of receptor (i.e. types I, II and III, Table 4.3):

- *Type I.*   Rapidly adapting receptors, i.e. (a) expiration; (b) "irritant" or (c) cough receptors supplied by myelinated afferent fibres with conduction velocites of 20–24 m/s. It is probable that most of these are located within or close to surface airway epithelium of the larynx, trachea and bronchus (Widdicombe, 1954a,b; Armstrong and Luck, 1974; Sampson and Vikruk, 1975) (*vide infra*).
- *Type II.*   Slowly adapting ("stretch") receptors. These are also served by myelinated afferents but with slightly faster conduction velocities of 30–50 m/s. Whilst there has been controversy as to the location of their endings, it is likely that they are located deep within the airway way in close association with airway smooth muscle. Their stimulation elicits the Hering–Breuer inflation reflex (particularly strong in animals) and the bronchial dilatation observed following airway distension (Widdicombe, 1954a,b; Bartlett *et al.*, 1979).
- *Type III.*   C fibre endings supplied by non-myelinated fibres with conduction velocities of 1–2 m/s. Two types of C fibre are distinguished in regard to their position and accessibility to chemicals placed into either the bronchial or pulmonary vasculature: (1) airway C fibres (Coleridge *et al.*, 1965; Coleridge and Coleridge, 1977; Davis *et al.*, 1982) or (2) pulmonary C fibres supplying juxta capillary or J receptors (Paintal, 1969).

There is currently intense interest in airway C fibres. Antidromic stimulation of sensory C-fibres is thought to cause the release of mediators (Florey *et al.*, 1932) which produce vasodilatation (in skin and nasal mucosa) and local oedema due to an increase in vascular permeability (in both skin and respiratory tract) (Lundberg *et al.*, 1983a; Saria *et al.*, 1983; Lundberg *et al.*, 1985b). This type of neural response is referred to as "neurogenic inflammation". Several distinct mediators may be involved depending on species and tissue sites: histamine or the regulatory peptides Sub P or NkA either alone or, in skin, potentiated by release of the co-transmitter CGRP (Brain and Williams, 1985; Gamse and Saria, 1985). There is evidence for a similar mechanism in the airways of guinea-pigs and rats following electrical stimulation of the vagus (Lundberg *et al.*, 1982b; Persson *et al.*, 1985) or after irritation by cigarette smoke, ethyl ether or formalin and

also bradykinin or capsaicin (Lundberg and Saria, 1983; Lundberg *et al.*, 1983c; Lundberg and Lundberg, 1984). The neurogenic inflammation caused by these substances is resistant to atropine or ganglionic blockade and reduced in animals pretreated with a Sub P-antagonist (Lundberg *et al.*, 1983c; Lundberg *et al.*, 1984b), supporting the concept of release of Sub P or a closely related tachykinin from sensory endings (Lundberg *et al.*, 1983b; Lundberg *et al.*, 1984a).

The red and green pepper extract capsaicin has proved to be particularly useful, experimentally. It releases Sub P from sensory nerves when given acutely, but chronic administration depletes the lung of Sub P and, if given early enough during neonatal development, destroys sensory C fibres (Lundberg and Saria, 1983). The results of experiments with capsaicin support the concept of a neurogenic inflammation mediated via sensory nerves. In a similar way, selective (i.e. supra- or infra-nodose) denervation by surgical (Das *et al.*, 1979) or by chemical means (e.g. using 6-hydroxydopamine or reserpine to ablate sympathetic fibres) has proved useful in determining the site and extent of sensory innervation of the lung.

2.3   CERVICAL AND THORACIC VAGUS

Several microscopic analyses have been made of the vagus nerve at its different anatomical levels (Evans and Murray, 1954; Agostini *et al.*, 1957; Gabella, 1976; Mei *et al.*, 1980). The study of Mei *et al.* (1980) on the cat vagus is of particular interest as both electron microscopic and experimental techniques were used. By electron microscopy, about 51 000 nerves were estimated for each infranodose vagus (Table 4.4). Myelinated fibres had a diameter of 2–14 μm with 80% between 2 and 4 μm. The number of myelinated nerve fibres was about 7000 for supranodose, whereas 10 000 were found in infranodose vagus. The rise in myelinated fibre number is accounted for by myelinated afferent fibres which enter and terminate in cell bodies located in the nodose ganglia. Two thousand myelinated fibres were found in each of the superior laryngeal nerves. The fall in total fibre number by 10 000 between supra- and infranodose vagus is accounted for by branches to each of the superior laryngeal nerves if a ratio of 4:1 unmyelinated to myelinated fibres is assumed. Unmyelinated fibres showed a range in diameter of 0.3–3 μm, with their unimodal distribution having mean maxima of 0.6 μm, 1.1 μm and 1.4 μm for supranodose, infranodose and diaphragmatic nerves, respectively. A total of about 40 000 (as against 25 000 estimated by light microscopy) non-myelinated fibres were found for the vagus at its cervical level with a ratio of about 4:1 for non-myelinated to myelinated fibres. In diaphragmatic vagus, 99% of nerves were unmyelinated in type, so it is likely that nearly all the 10 000 (i.e. 20% of the total) myelinated fibres found in

Table 4.4 Number of nerve fibres in cat vagi

	Total	Non-myelinated (%)	Myelinated (%)	Ratio
I Entire vagus				
Supranodose	60 700	53 700 (89)	6900 (11)	8:1
Infranodose	51 300	41 200 (80)	10 600 (20)	4:1
Diaphragmatic (posterior trunk)	21 700	21 400 (99)	400 (1)	54:1
II Sensory vagus (i.e. after extracranial supranodose unilateral vagotomy)				
Infranodose	32 000	27 600 (88)	4300 (12)	6:1

Adapted from Mei *et al.* (1980).

Table 4.5 Neuropeptides in the airways: occurrence and tissue distribution[a]

Peptides	Epithelium and subepithelium	Seromucous glands	Non-vascular smooth muscle	Blood vessels
Sub P	++	+	+	++
CGRP	+++	+	+	++
NPY	+	+	++	+++
VIP	+	++	++	++

Modified from Uddman and Sundler (1987)

[a] Data on relative fibre frequency compiled from observations in laboratory animals (mouse, rat, guinea-pig, ferret, cat). Frequency was graded as: +, few fibres; ++, moderate supply of fibres; +++, numerous supply of fibres.

each infranodose vagus will supply the lung leaving its diaphragmatic portion and these consist almost entirely of small-diameter unmyelinated fibres.

Following selective supranodose vagotomy (which in theory causes selective degeneration of motor fibres whilst leaving sensory fibres in infranodose vagus intact) there is a reduction in total fibre number. The fall affects both myelinated and non-myelinated fibres and, in particular, the largest non-myelinated fibres of the thoracic vagus. The ratio of non-myelinated to myelinated fibres is thus altered in "sensory" vagi and is then about 6:1 for infranodose and 260:1 for diaphragmatic vagi, i.e. significantly higher than in the "intact" vagus. Thus, about 63% of fibres in infranodose vagus and 100% in diaphragmatic vagus are sensory (conversely, 37% of fibres in the former may be assumed to be motor). Furthermore, the sensory fibres in infranodose vagus are a mixture of unmyelinated (55% of the total) and myelinated (8% of the total) fibres. Following supranodose section, diaphragmatic vagus showed little change in total fibre number and thus is almost entirely sensory.

3. Tissue Distribution

The general distribution of nerves within tracheobronchial tissue is similar in many species as determined by silver staining methods (Larsell, 1922; Gaylor, 1934; Honjin, 1956; Spencer and Leof, 1964; El-Bermani, 1973; Berkley, 1983). More recently, nerve fibres, ganglia and neuroendocrine cells are all demonstrated to advantage by their immunoreactivity for neurone-specific enolase (NSE) (Wharton *et al.*, 1981), PGP-9.5 (*vide infra*) (Doran *et al.*, 1983) or S-100 (Sheppard *et al.*, 1983). In the trachea and main bronchus, nerve bundles and ganglia are found mainly in the posterior membranous portion of the airway (Fisher, 1964) (Figure 4.2). On entering the lung the nerve bundles divide to form distinct peribronchial and perivascular plexuses (species differences exist in the extent and immediacy with which this happens). The peribronchial plexus then further divides to form extra-and endochondrial (i.e. subepithelial) plexuses. This division depends on the extent of cartilage in succeeding airway generations, there being little in the rat and more in cat and human. By silver stains, both "thick" and "thin" fibres are described in peribronchial nerve bundles, and examination of 1 µm plastic-embedded material has shown them to be myelinated and non-myelinated, respectively (El-Bermani, 1973). It is further estimated that 30–35% of hilar nerves in the rat are myelinated, the percentage decreasing with airway generation. Most of the peribronchial nerves of the rat and human are acetylcholinesterase-positive (i.e. indicative of a cholinergic innervation) compared with only 10–20% of periarterial nerves (Plate 4.1). The tracheobronchial muscular wall is infiltrated by a plexus of fibres ("intermuscular plexus"), most of which are derived from peribronchial nerve bundles. Acetycholinesterase-positive nerve bundles predominate; those found in the bronchiolar region, and beyond, have fewer fibres and ganglia **than**

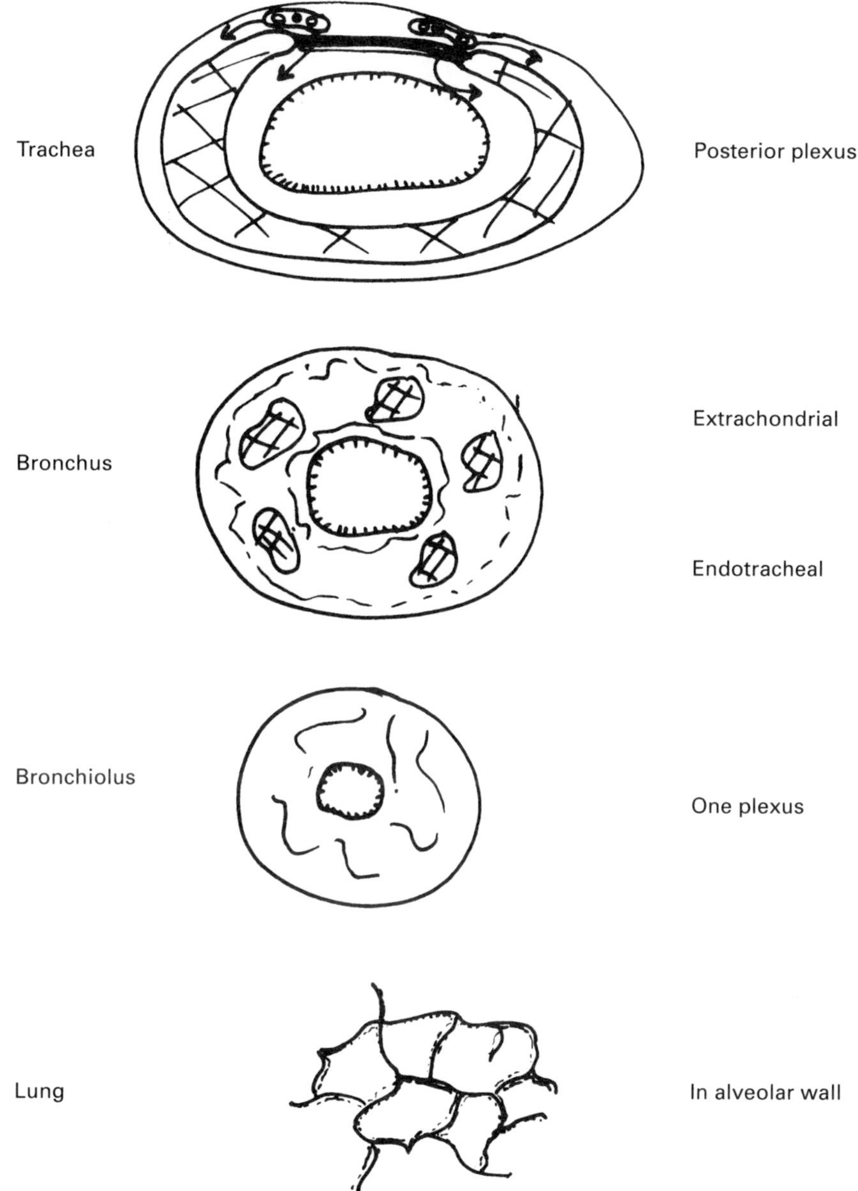

Figure 4.2 Plan diagram to show general tissue distribution of nerves (as seen by silver staining) in cross-sections of airways at distinct levels of branching.

plexuses of airway generations of higher order (Larsell and Dow, 1933; Hirsch *et al.*, 1968a,b; Shipperbottom, 1988). Immunoreactivity for a variety of neuropeptides is abundant, particularly in the larger airways. The tissue distribution and the predominant type varies with site and species (Table 4.5). The early light microscopic reports of an extensive alveolar innervation (Hirsch *et al.*, 1968a) have not been substantiated by later studies. Whilst electron microscopy of the lungs of the rat and human confirms that alveolar structures are innervated (Plate 4.2), such nerves are sparse (Meyrick and Reid, 1971; Fox *et al.*, 1980).

3.1 AIRWAY GANGLIA

Relatively little is known about the airway intramural ganglia first noted by Landois in 1885. Silver impregnation methods demonstrate autonomic ganglia in the airway walls of several species: in general they are concentrated along the posterior wall of the trachea between and close to the free ends of the cartilage plates, at their junction with the trachealis muscle. Their number decreases in more peripheral airway generations but they are present in most species at least as far peripherally as small bronchi (Fisher, 1964; Spencer and Leof, 1964; Loosli and Hung, 1977; Jeffery, 1982). In the monkey

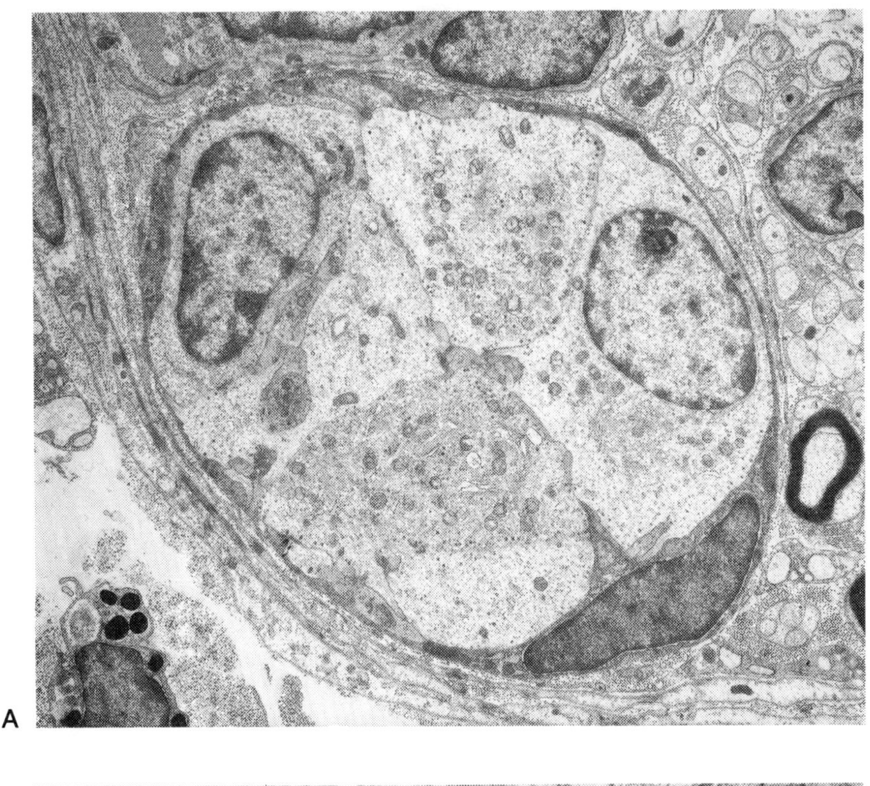

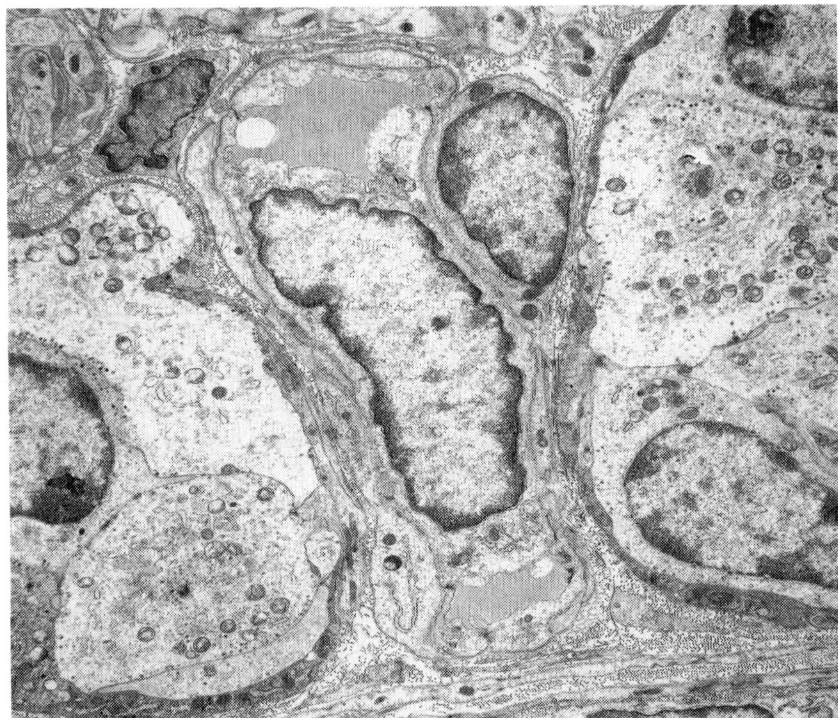

Figure 4.3 Transmission electron micrographs of rat airway wall. (A) An intraganglionic group of cells with dense-cored granules of the peptidergic-type situated at one corner of the ganglion, which also contains myelinated and unmyelinated fibres. Part of a mast cell is shown in the bottom left corner ($\times$ 7500). (B) An intraganglionic blood vessel surrounded by cells containin peptidergic-type granules ($\times$ 10000).

and rat they appear to be particularly abundant in the connective tissue elements about the hilum of each lobe (Zussman, 1966; El-Bermani and McCarthy, 1980). The commonly held view is that ganglia located within the airway wall are parasympathetic in type, and their acetylcholinesterase positivity supports this (Plate 4.3a). However, cell clusters exhibiting catecholamine fluorescence have been found in the dog, cat, rat, calf and human (Blumcke, 1968; Mann, 1971; Jacobowitz et al., 1973; Knight, 1980). Immunoreactivity for VIP has been found in the nodose ganglion, ganglia located along the vagus and in numerous intramural ganglia (Lundberg et al., 1978; Uddman et al., 1980; Shipperbottom, 1988) (Plate 4.3b). Immunoreactivity for CGRP and Sub P have also been demonstrated in nerve fibres present within intramural ganglia (Lundberg et al., 1978; Uddman et al., 1980; Shipperbottom, 1988) (Plate 4.3c). Such peptides may modulate cholinergic ganglionic transmission (vide infra).

Each ganglion contains the cell bodies of about 20–30 cells. But those of the respiratory tract vary greatly in size. The larger ganglia are enclosed by a perineurium and connective tissue capsule. By electron microscopy the neuronal cell bodies are located within a capsule with each cell body up to 60 μm in diameter (Plate 4.4a). The neural soma contains a rounded nucleus with distinct nucleolus, a Golgi zone, neurosecretory vesicles, endoplasmic reticulum and ribosomes (Plate 4.4b). El-Bermani and McCarthy (1980) demonstrated that most preganglionic endings are received in a deep depression or invagination of the neural soma. This arrangement might render a large number of preganglionic terminals inaccesible to the effects of ganglionic blocking drugs. In addition, there are accessory (or satellite) cells, bundles of unmyelinated and myelinated nerves, intraganglionic blood vessels and mast cells.

In electron microscopic studies of rat, cat and human airway ganglia (Richardson and Ferguson, 1976; Knight, 1980; Richardson and Ferguson, 1980; Jeffery, 1982), granulated cells resembling the Kultschitsky (synonyms: neuroendocrine or Feyrter) cells of airway epithelium have been identified in airway ganglia (Fig. 4.3A). Their ultrastructural appearance is similar to cells described in myenteric and mesenteric ganglia (Oosaki, 1970; Elfvin, 1971), superior cervical ganglia (Matthews and Raisman, 1969; Williams and Palay, 1969; Morgan et al., 1976), and abdominal and aortic paraganglia (Morgan et al., 1976). The granulated cells are smaller than adjacent neurones and are usually present as groups of cells, each cell with electron-lucent cytoplasm containing neurosecretory-like vesicles of the large dense-cored type. The peripheral localization of granules within the cell is characteristic. They may also be closely associated with intraganglionic blood vessels (Fig. 4.3B) and release secretion which may alter or modulate neuronal activity in distant parts of the nerve bundle. The intraganglionic granulated cells may be responsible for the catecholamine or immunofluorescence found in ganglia (vide supra) and it has been suggested that they may function as inhibitory catecholamine-secreting interneurones (Jacobowitz et al., 1973). Alternatively, they may represent the cell bodies of the VIPergic nerves found innervating bronchial effectors (Uddman et al., 1978; Shipperbottom, 1988) (see Plate 4.3b). Using a radiolabelled riboprobe for the mRNA encoding for VIP, the presence of mRNA transcripts for VIP in human bronchial intramural ganglia has recently been demonstrated, indicating that VIPergic neurones are probably present and that they synthesize VIP locally in the airway wall (author's unpublished observations).

The earlier concept of mural ganglia being simply relay stations is clearly over-simplistic, and it is probable that these form sites of complex integration. Burnstock and colleagues have studied paratracheal neurones of the rat in cell culture. They observe two types of neurone with distinct patterns of "firing". Their preliminary work on neonatal rat trachealis muscle has demonstrated populations of neurones containing VIP, SOM, NPY and galanin; this is in keeping with the immunoreactive identification of these peptides in situ (Burnstock et al., 1987). Considering their common embryological origin, it is probable that the intrinsic neurones of the gut and airways will have several features in common. It is, therefore, probable that parasympathetic and sympathetic components interact directly at ganglia with presynaptic inhibition of cholinergic transmission by noradrenaline via α-receptors. The NANC system is also likely to have a modulatory effect on ganglionic transmission (Cheung et al., 1985; Dayer et al., 1985; Barnes, 1986b; Cadieux et al., 1986; Barnes, 1992). These structural interrelationships are rendered functionally more complex by the variety of receptor subtypes recently described (Mak and Barnes, 1989, 1990; Barnes, 1992; Mak et al., 1993). Experimental studies also support the concept of ganglionic modulation of cholinergic transmission; for example, postganglionic stimulation has a greater effect on tracheal contraction than preganglionic stimulation (in ferrets) (Skoogh, 1983), and both β_2 and α agonists can modulate ganglionic neurotransmission (Baker et al., 1983; Skoogh, 1986). The airway hyper-responsiveness which results from viral infection may also be abolished experimentally by a ganglionic blocker whereas atropine is relatively ineffective: this suggests that a non-cholinergic postganglionic neural pathway is involved (Buckner et al., 1985; Fryer and Jacoby, 1991). The role of intraganglionic intrinsic peptides in the modulation of ganglionic neurotransmission is only beginning to be studied experimentally. VIP appears to modulate cholinergic neurotransmission in guinea-pig parasympathetic ganglia as well as postganglionic nerves (Sekizawa et al., 1988; Ellis and Farmer, 1989; Stretton et al., 1991), and there is evidence that Sub P, CGRP and galanin can facilitate ganglionic transmission (Barnes, 1991, 1992).

3.2 AIRWAY SURFACE EPITHELIUM

The early use of silver stains showed a rich innervation to airway surface epithelium at all levels of airway in the rat (Berkley, 1893), mouse (Honjin, 1956), cat (Hirsch *et al.*, 1968a; Jabonero and Sabadell, 1972), dog (Elftman, 1943), chicken (Jabonero and Sabadell, 1972), rabbit, monkey and human (Fisher, 1964; Spencer and Leof, 1964; Jabonero and Sabadell, 1972), and in each case this was derived from a subepithelial nerve plexus. All but Fisher (1964) and Hayashi (1937), who suggested a motor supply to epithelial mucous (goblet) cells and to ciliated cells, respectively, believed the supply to be exclusively sensory. The results of the vital stains methylene blue (injected or perfused) gave support to the presence of intraepithelial nerve fibres as far peripherally as the respiratory bronchiole in a number of species (Larsell, 1921; Fillenz and Woods, 1970).

Using histochemical techniques, there was a failure to localize either acetylcholinesterase positivity or catecholamines to terminal fibres of airway surface epithelium of guinea-pigs, rabbits or dogs (Fillenz and Woods, 1970) or humans (Shipperbottom, 1988), implying a lack of motor innervation to this site. However, Ericson *et al.* (1972) described isolated varicose catecholamine-positive nerve fibres penetrating the surface where they ran between epithelial cells and close to the neuroendocrine cells in the upper trachea of mice. In the author's opinion, the catecholamine fluorescence reported within the epithelium of mice is most likely to be due to the elongated processes of single neuroendocrine cells, which also contain amines in granules of the neurosecretory type.

Immunoreactivity for PGP-9.5 in humans demonstrates large numbers of subepithelial nerves which penetrate focally the surface epithelium and then branch and spread throughout the epithelium, along the basement membrane, then running superficially, extending between cells nearly as far as the airway lumen (Fig. 4.4 and Plate 4.5). There are about two nerve fibres per millimetre length of surface epithelium in humans. (Shipperbottom, 1988). In the author's experience, their concentration about the bronchial circumference varies greatly. Depending upon species and site, there is general agreement that Sub P/NkA- and CGRP-containing sensory nerves are found within and immediately below surface epithelium. The respiratory tract of most animals has a moderate supply of Sub P-containing nerve fibres (Lundberg *et al.*, 1984a; Uddman and Sundler, 1987). In humans, in contrast to the numbers of PGP-9.5 immunoreactive nerves, Sub P-positive fibres in surface epithelium are rarely detected by immunofluorescence. Using the more sensitive techniques of IGSS or avidin–biotin complexes, Sub P immunoreactivity in human bronchi obtained postmortem was confined to the surface epithelium of two out of six cases studied (Shipperbottom, 1988) (Plate 4.6). Rapid enzymatic degradation and the effects of smoking and age may be responsible for the

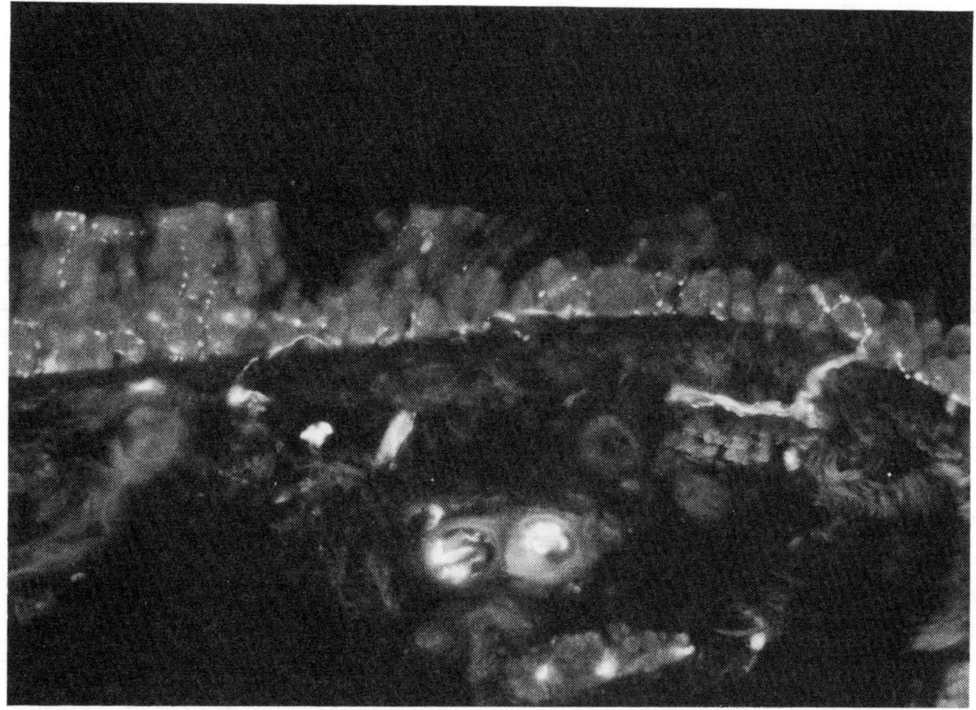

Figure 4.4 Immunoreactivity for PGP-9.5 in human airway wall demonstrating nerve bundles/ganglia and also branching varicose nerves associated with airway surface epithelium (× 300).

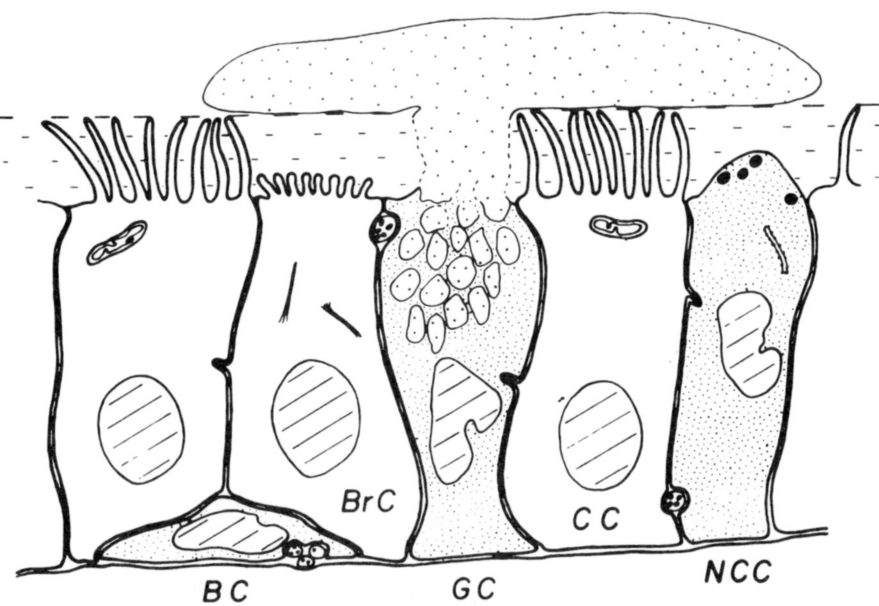

Figure 4.5 Diagrammatic representation of airway epithelium showing ciliated (CC), secretory (GC and NCC), brush (BrC)and basal cells (BC). The majority of intraepithelial nerve profiles lie close to the basement membrane with occasional fibres close to the lumen.

difficulty in its identification in human airways. *In vivo*, Sub P is metabolized rapidly by NEP (synonym: enkephalinase), localized to airway epithelium (Dusser *et al.*, 1988). The tissue activity of NEP is an important determinant of tachykinin effectiveness. Reductions of NEP activity occur after respiratory tract infections, cigarette smoke, ozone and high doses of toluene diisocyanate (Dusser *et al.*, 1989; Barnes, 1991). CGRP is reported to be conspicuous beneath and within surface epithelium in animals, and its distribution closely resembles that of Sub P/NkA, supporting the concept of co-localization of CGRP and Sub P/NkA. In humans, by IGSS, CGRP was rarely found in the nerves of surface epithelium but it was conspicuous in intraepithelial neuroendocrine cells. Both IHC and RIA show that the highest density of Sub P-positive fibres is found in trachea and bronchi, with little in lung: higher concentrations are found in the guinea-pig and rat compared to the cat (Sheppard and Polak, 1986) and there is little or none in human respiratory tract (Laitinen *et al.*, 1983). Consequently the functional role of Sub P, at least in humans, is unclear.

A number of electron microscopic studies have shown unequivocally that nerve fibres do pierce the airway surface epithelial basement membrane and come to lie in apposition to a number of distinct epithelial cell types in the rat (Jeffery and Reid, 1973a), cat (Hung, 1976; Das *et al.*, 1978), rabbit (Fillenz and Woods, 1970), guinea-pig (Boucher *et al.*, 1980; Hulbert *et al.*, 1981), chicken (Cook and King, 1969a,b; King *et al.*, 1974; Walsh and McLelland, 1974), goose (Phipps *et al.*, 1977) and human (Rhodin, 1966; Lauweryns *et al.*, 1972; Jeffery, 1982;

Laitinen, 1985; Jeffery, 1986; Jeffery 1989). However, there appears to be great species variation in their morphology, distribution and number in each succeeding airway generation. As an example, three species will be considered: the rat, cat and human:

Rat. Few studies of epithelial innervation have been quantitative, but in one such study intraepithelial nerve fibres were found in rat extra- but not intrapulmonary airway epithelium (Jeffery and Reid, 1973a). There was a significantly higher concentration of intraepithelial nerve fibres in the upper than in the lower trachea or main bronchus (117, 70 and 23/mm, respectively). In each case, more nerves were found anteriorly than in the membranous posterior wall. Most (87%) of the epithelial nerve fibres were close to the basement membrane and associated particularly with basal cells, 13% were superficial, many within 1 µm of the airway lumen (Fig. 4.5). All were without myelin, Schwann cell sheath or basement membrane and were surrounded by the plasma membranes of epithelial cells and separated from them by a gap of about 15 nm; no specialized synaptic complexes were seen. The mean diameter of fibres was 0.4 µm and each was found either alone or in a group of up to seven fibres (Fig. 4.6A). Thirty-three per cent had dense-cored neurosecretory vesicles (Fig. 4.6B), usually accompanied by a large number of agranular vesicles, 17% had only agranular vesicles, with the remaining 50% without vesicles but containing neurotubules and mitochondria. A few fibres were larger than normal, measuring between 1 and 2 µm and contained accumulations of β-glycogen and vesicles of both types (Fig. 4.7A). Nerve fibres were

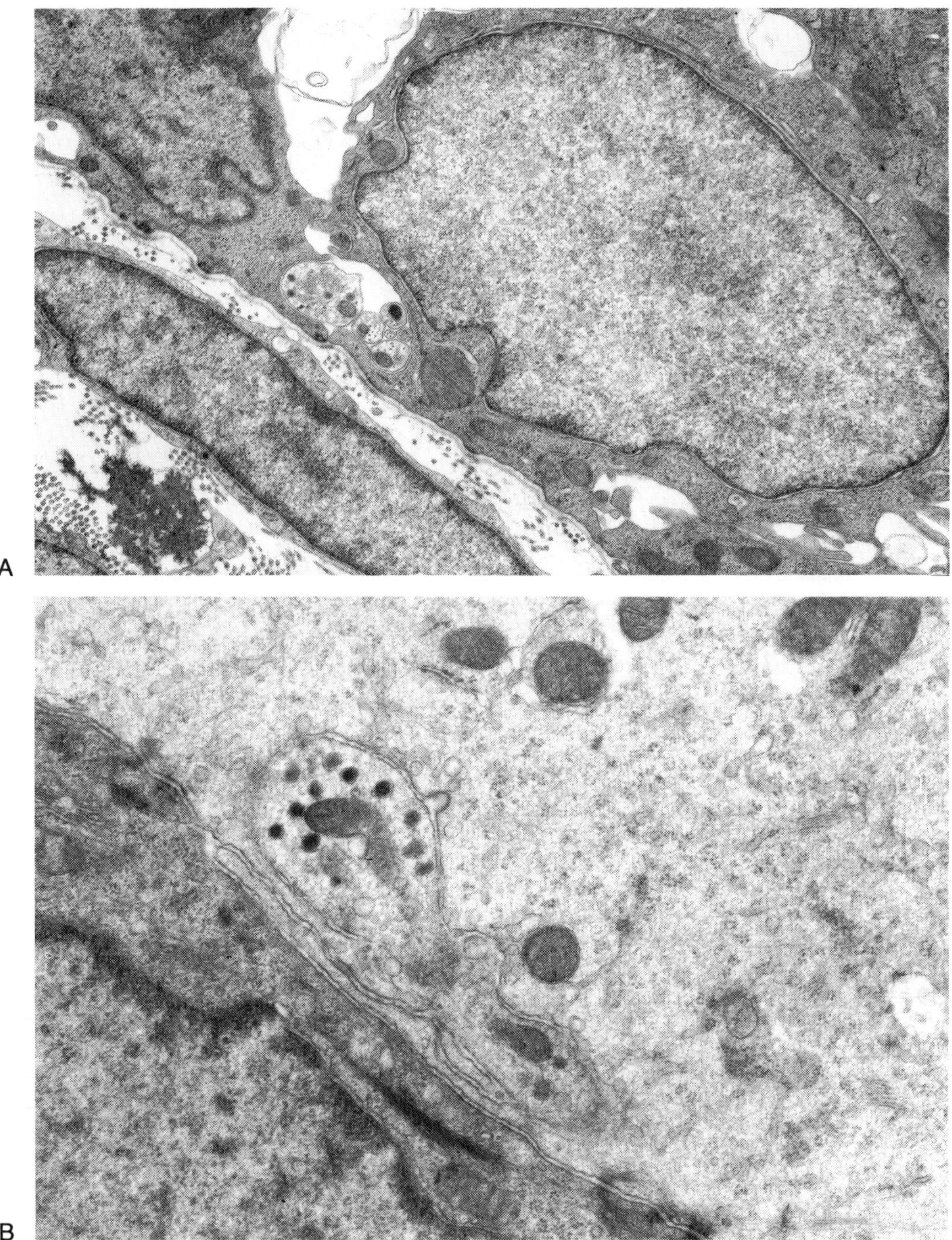

Figure 4.6 Transmission electron micrographs of intraepithelial nerves of rat tracheal surface epithelium: (A) a group of six fibres with the largest containing dense-cored granules of the peptidergic type, together with agranular vesicles (×11 800); (B) a single fibre lying between a basal and superficial ciliated cell containing only peptidergic-type granules (×22 500).

frequently associated with basal (see Fig. 4.6A), ciliated (see Fig. 4.6B) and mucous cells (Fig. 4.7B).

Neuroendocrine cells were infrequently found by Jeffery and Reid (1937b). One showed a multiple innervation of nerve fibres, some of which contained neurosecretory vesicles (Plate 4.7). Neuroendocrine cells are part of the diffuse neuroendocrine system (Polak and Bloom, 1979; Said, 1980) and were previously considered to belong to the APUD (amine precursor uptake and decarboxylation) system (Pearse, 1969). They

epithelial bodies, has also been shown in a number of other species, including humans (Lauweryns *et al.*, 1970).

Cat. A quantitative study of intraepithelial nerves in cat airways showed similarities with the rat (Das *et al.*, 1978). Whilst intraepithelial nerves were identified in a hilar airway, they were absent from the more peripheral intrapulmonary airways. Additionally, the main carina was examined for intraepithelial nerves and was found to have the highest concentration of nerves of all the airways examined. The nerve-to-cell ratios were 1:5, 1:3 and 1:12 for the lower trachea, carina and hilum, respectively. In contrast to the rat, most of the intraepithelial fibres were without neurosecretory vesicles or contained accumulations of mitochondria, suggesting that the majority had a sensory function (Hung, 1976; Das *et al.*, 1978). Experimental studies support this contention (Das *et al.*, 1979) (*vide infra*). As with other species, occasional nerves were found close to granulated Kultschitsky-like cells (Das *et al.*, 1978).

Human. Rhodin's (1966) electron microscopic study of the epithelium of human trachea demonstrated intraepithelial nerve fibres. The author's own observations of bronchial biopsies obtained from normal healthy individuals show that the subepithelial mucosa has a rich innervation with bundles of both myelinated and unmyelinated fibres. Intraepithelial nerve fibres are present in main stem extrapulmonary bronchi and segmental airways (Fig. 4.8) but, in comparison with the rat and cat, they are infrequently found (Laitinen, 1985) (Table 4.6). All but the occasional nerve fibre lack neurosecretory vesicles, when they may be small clear vesicles or, less frequently, large and dense cored. Rarely, a varicosity may be cut in section showing large numbers of mitochondria devoid of mitochondrial cristae (Fig. 4.8C). These are characteristic of receptor endings found in other tissues.

The intraepithelial endings are ideally placed to receive mechanical stimuli or those of irritant gases such as cigarette smoke, ammonia, sulphur dioxide, ethyl ether and carbon dust. The only barrier given to these substances is the tight junction (Figs 4.8C and 4.9), which runs as an apicolateral belt comprised of interlocking strands and complementary grooves joining adjacent cells (Godfrey *et al.*, 1992). Boucher *et al.* (1980) and Hulbert *et al.* (1981) have shown, in the guinea-pig, that the epithelial exclusion of HRP (40 kD) placed in the airway lumen, isolated from the nerve, can be overcome experimentally following 100 puffs of cigarette smoke: the penetration of HRP past the tight junction was observed by its subsequent accumulation around intraepithelial nerve fibres. Whilst circumstantial, it is likely that intraepithelial nerves are the morphological correlate of the cough irritant and airway C fibres described above (see Table 4.3) whose local reflexes give rise to mucus secretion, vasodilatation and the oedema associated with neurogenic inflammation. The position of the "irritant

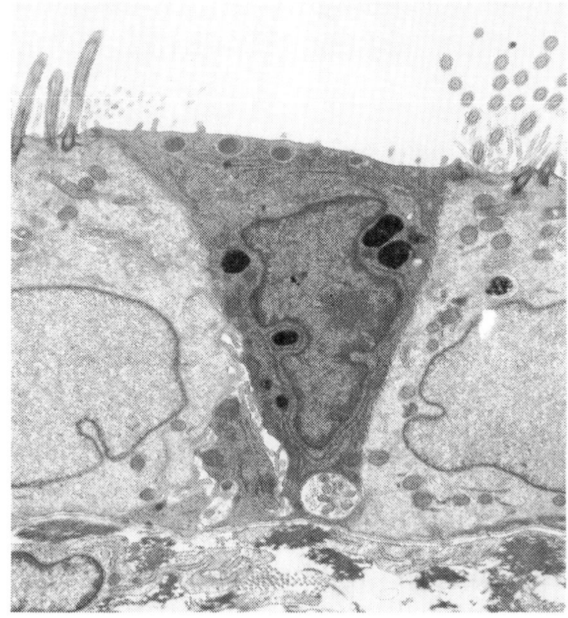

A

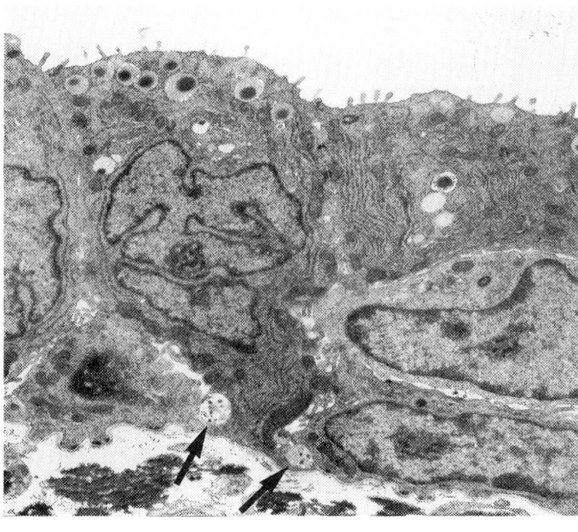

B

Figure 4.7 (A) Transmission electron micrograph of rat tracheal epithelium showing a relatively large (1 μm diameter) ending with mitochondria and β-glycogen (×6300). (B) Intraepithelial nerve fibres with peptidergic-type granules in close apposition (arrows) to mucous cells of rat tracheal surface epithelium (×4800).

are found in airway epithelium of several species, including humans, and are particularly abundant in the fetal and neonatal period (Hage, 1974). Some contain 5-HT (Lauweryns *et al.*, 1982) and others the peptides bombesin, calcitonin, katacalcin and CGRP, localized to them by IHC (Wharton *et al.*, 1978; Becker *et al.*, 1980; Sheppard *et al.*, 1983). The innervation of groups of morphologically similar cells, collectively called neuro-

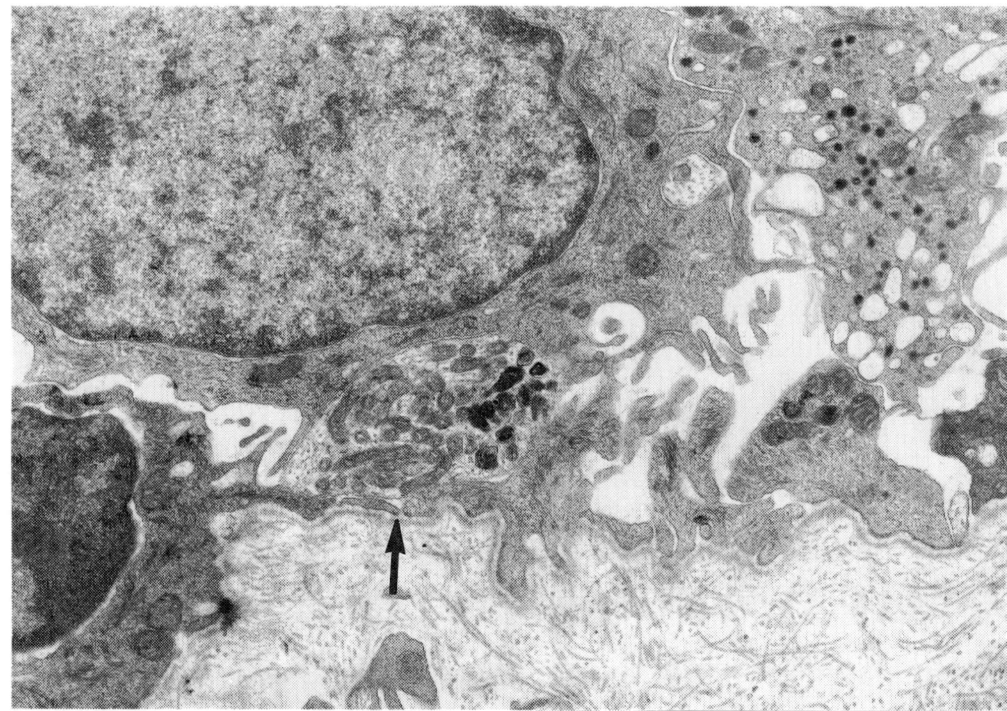

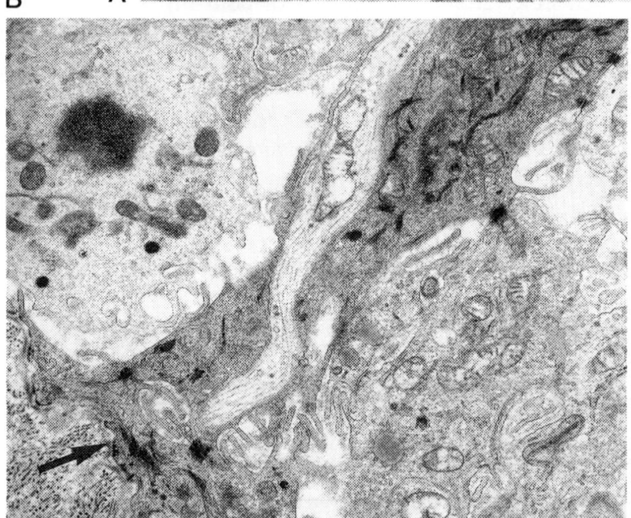

Figure 4.8 Transmission electron micrographs of intraepithelial nerves in human bronchial surface epithelium: (A) cut transversely to show a nerve fibre with mitochondria and neurotubules close to the epithelial basal lamina (arrow) beneath a basal cell, × 10 000; (B) cut in longitudinal section to show a fibre with distinct neurotubules extending towards the airway lumen (top right) away from the basal lamina (arrow), × 7200; (C) a terminal ending packed with mitochondria immediately beneath a tight junction which separates it from the airway lumen (L), × 17 000.

Table 4.6 Intraepithelial nerve fibres per millimetre of surface epithelium: a comparison of three species examined by electron microscopy

Airway	Rat[a]	Cat[a]	Human[b]
Trachea			
Upper/mid	117	41	9
Lower	70	ND	ND
Main bronchus/carina	78	75	ND
Intrapulmonary/lobar	0	17	9
Segmental/distal bronchus	0	0	4

ND, not determined.

[a] Jeffery and Reid (1973) and Das *et al.* (1978). [b] Laitinen *et al.* (1985).

receptor" within the airway wall is thought to be superficial as irritant receptors respond to chemically inert dust blown into the airway lumen and to gentle mechanical probing (Mills *et al.*, 1969; Sellick and Widdicombe, 1971) and can be destroyed by removal of the superficial mucosal lining. The receptive field of such

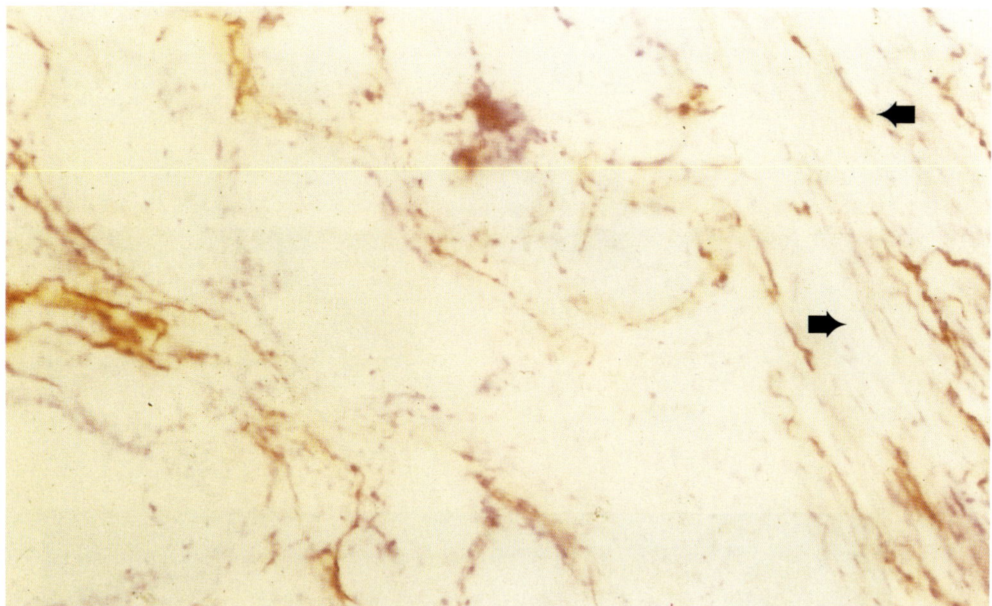

Plate 4.1 Human bronchial wall showing acetylcholinesterase positivity indicative of cholinergic innervation of bronchial smooth muscle (arrows) and, to the left, the acini of submucosal mucus-secreting glands.

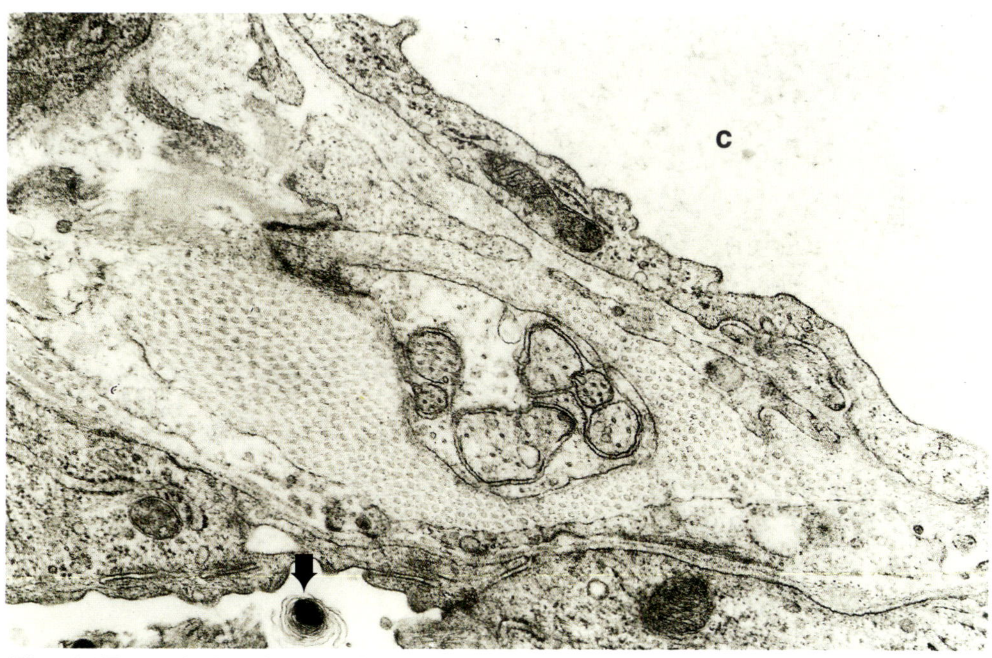

Plate 4.2 Transmission electron micrograph of six nerve fibres cut in cross-section, surrounded by Schwann cell sheath and embedded in collagen in human alveolar wall (x 16 000). The fibres are equally accessible to stimulating agents placed in either the alveolar space (arrow) or pulmonary capillary (c). Reproduced with permission, from Meyrick and Reid (1971).

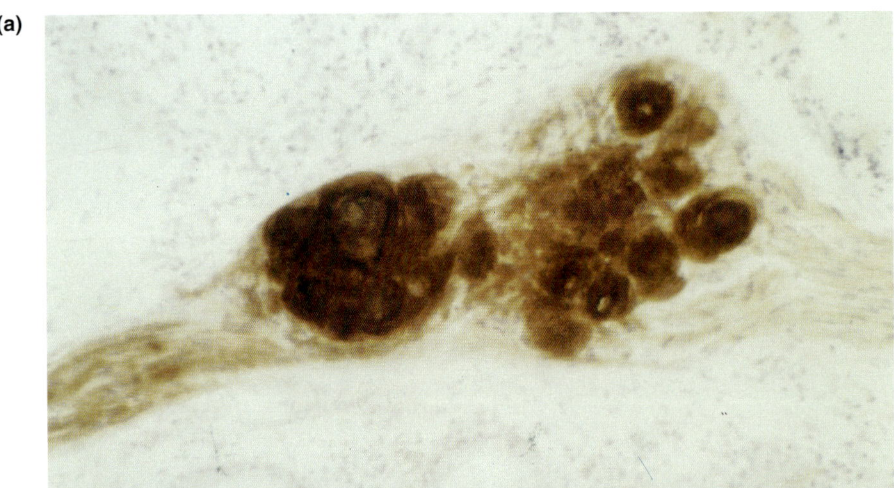

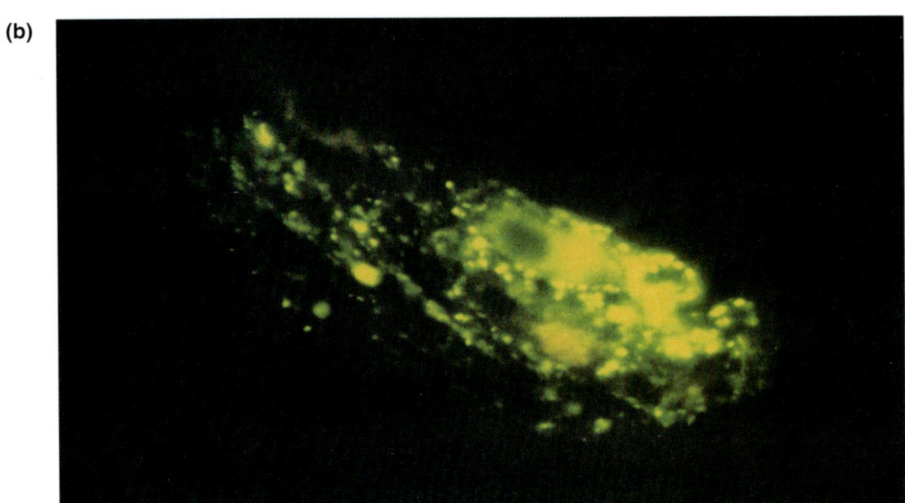

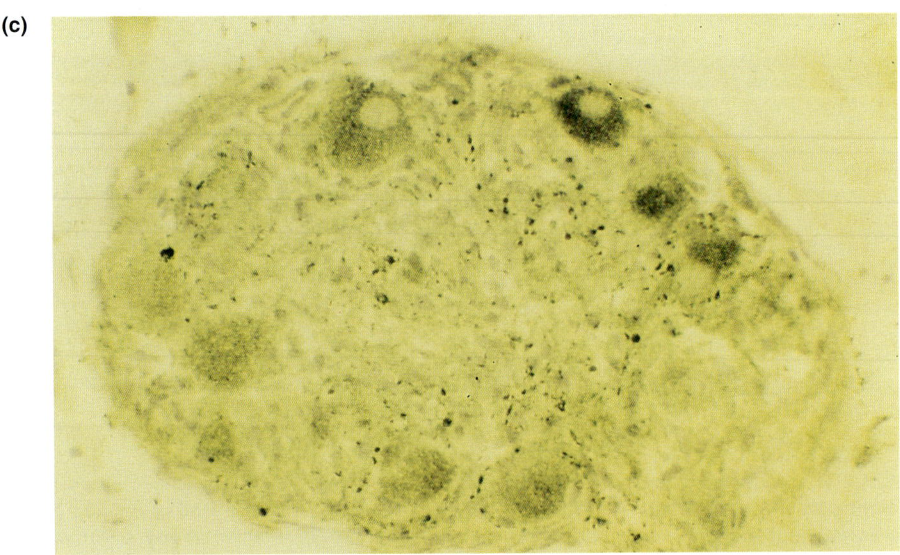

Plate 4.3 (a) Intramural ganglion in human bronchus positive for acetylcholinesterase. (b) A VIP immunoreactive-positive ganglion in human bronchial mucosa demonstrated by immunofluorescence. (c) Human mural ganglion stained with the IGSS technique. There is punctate, dense immuno-gold/silver-positive staining for CGRP of transversely cut nerve fibres lying between the nerve cell bodies.

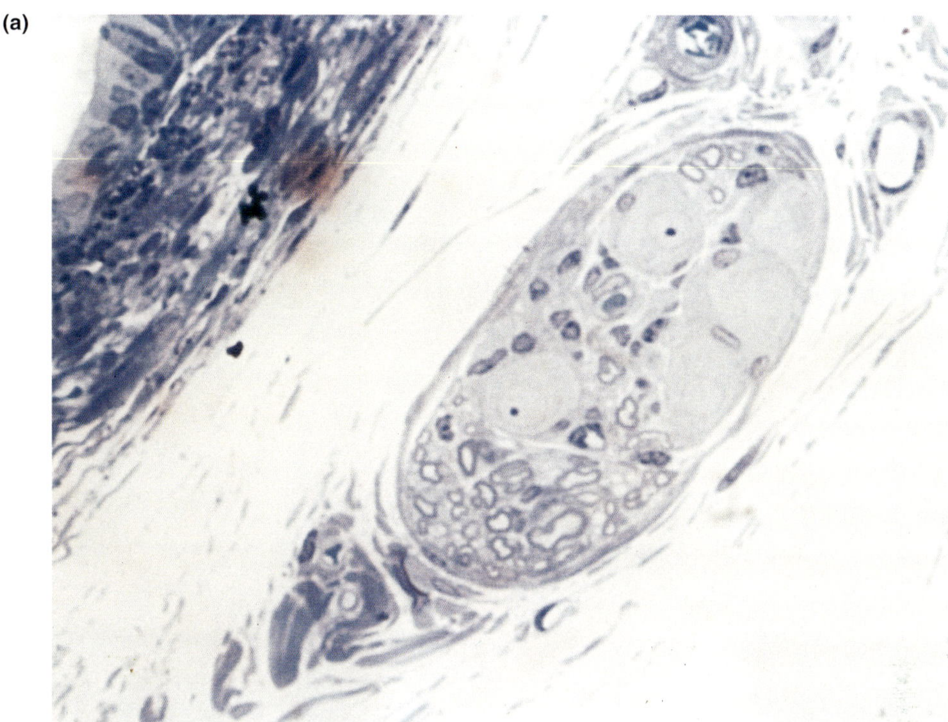

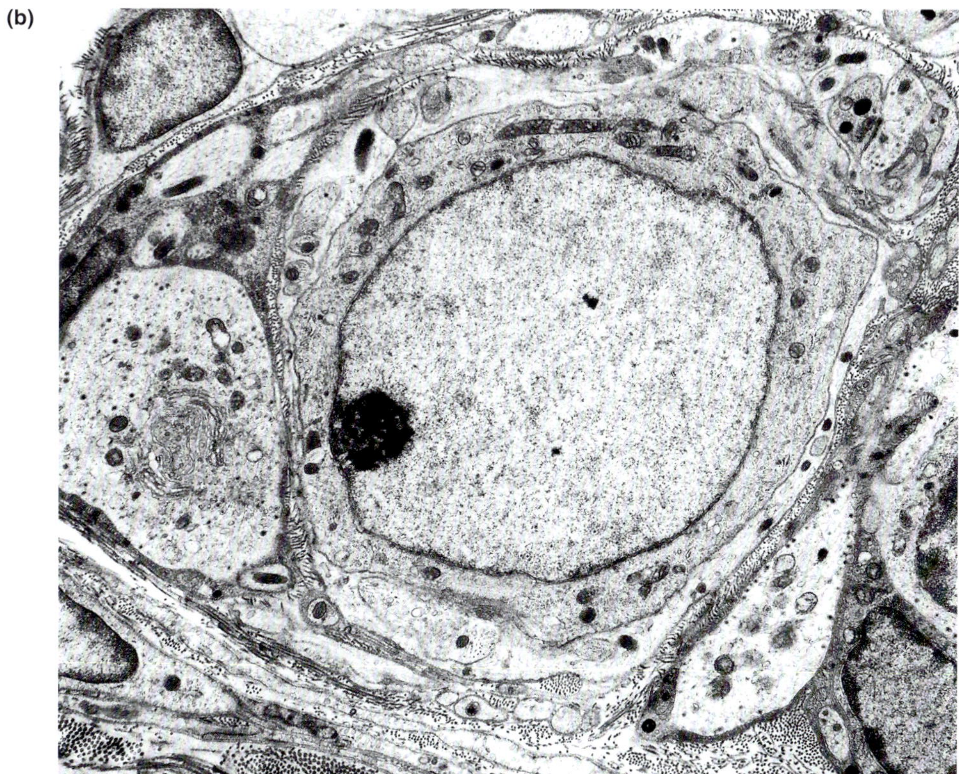

Plate 4.4 Intramural ganglion in rat intrapulmonary airway. (a) Toluidine blue-stained plastic-embedded section showing a collection of five neurones together with myelinated fibres and bronchiolar epithelium (top left) (x 255). (b) Transmission electron micrograph showing a neural cell body containing a large nucleus with distinct nucleolus (x 1500). The neurone is surrounded by many nerve profiles cut in various planes and the processes of cells containing relatively large dense-cored vesicles of the peptidergic type.

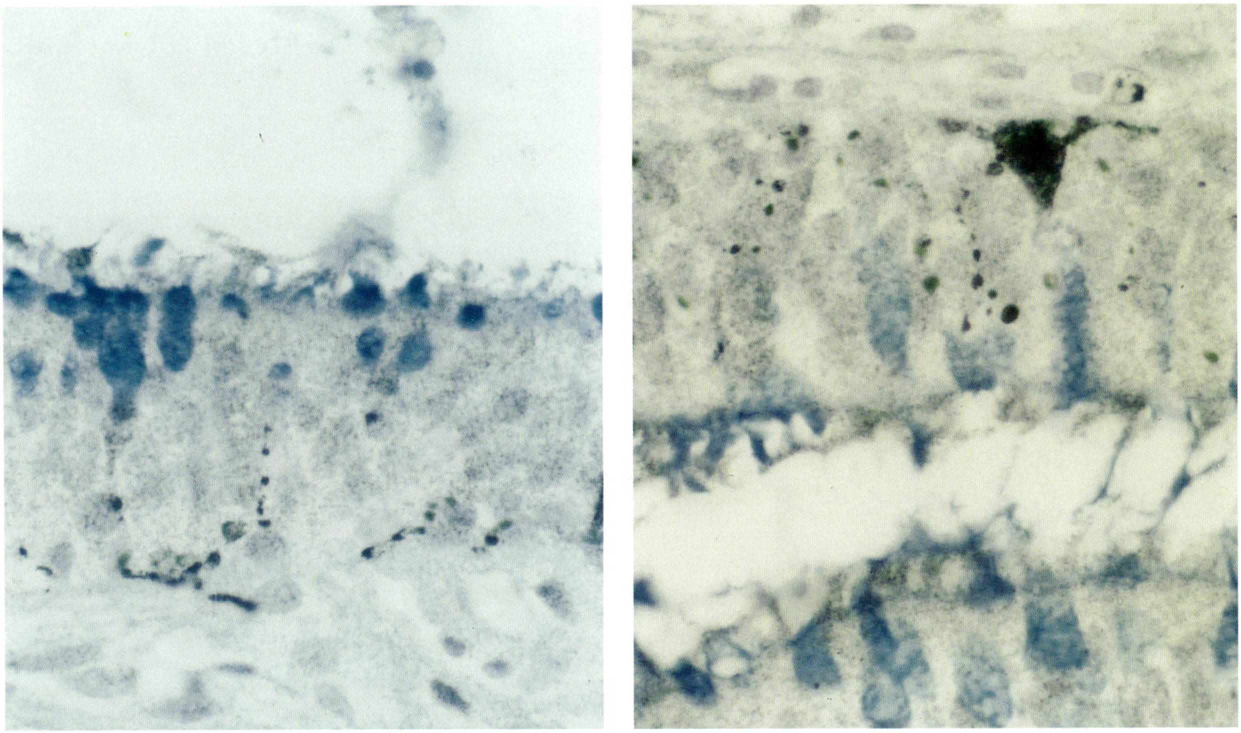

Plate 4.5 IGSS staining of human bronchial epithelium demonstrating: (a) PGP-9.5 immunopositive nerve fibres entering the surface epithelium in which there are alcian blue-positive mucous cells; (b) both intraepithelial nerves and triangular-shaped neuroendocrine cells.

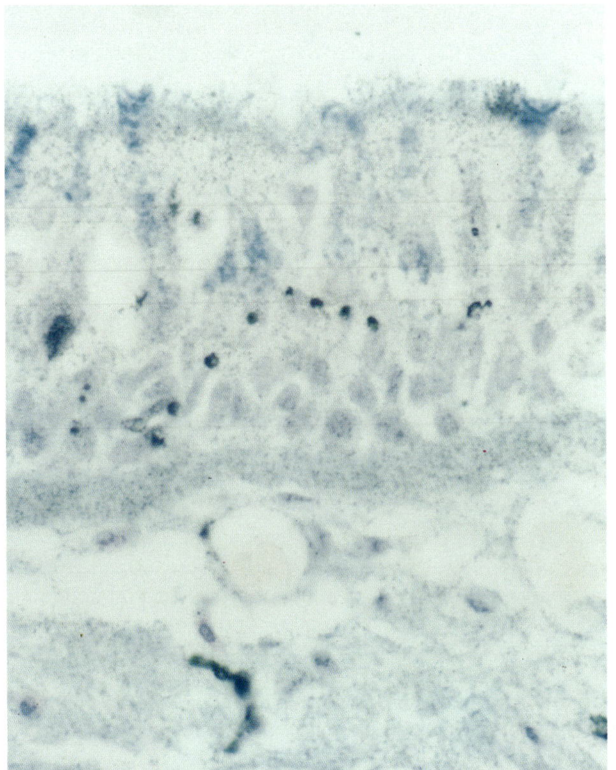

Plate 4.6 Sub P immunoreactivity as shown by the IGSS technique. The presence of Sub P is rare in human mucosa but frequent in other species such as the rat. This immuno-gold/silver-positive fibre is beaded and is shown passing from subepithelium through the basal lamina into the epithelium and extending towards the lumen.

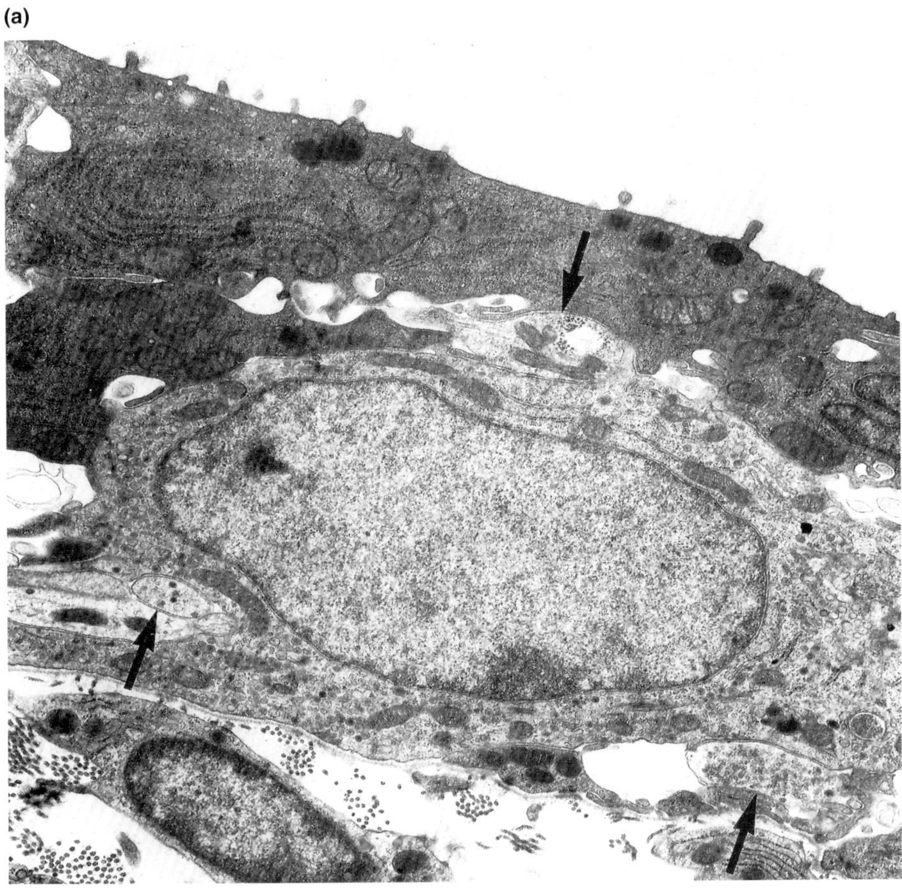

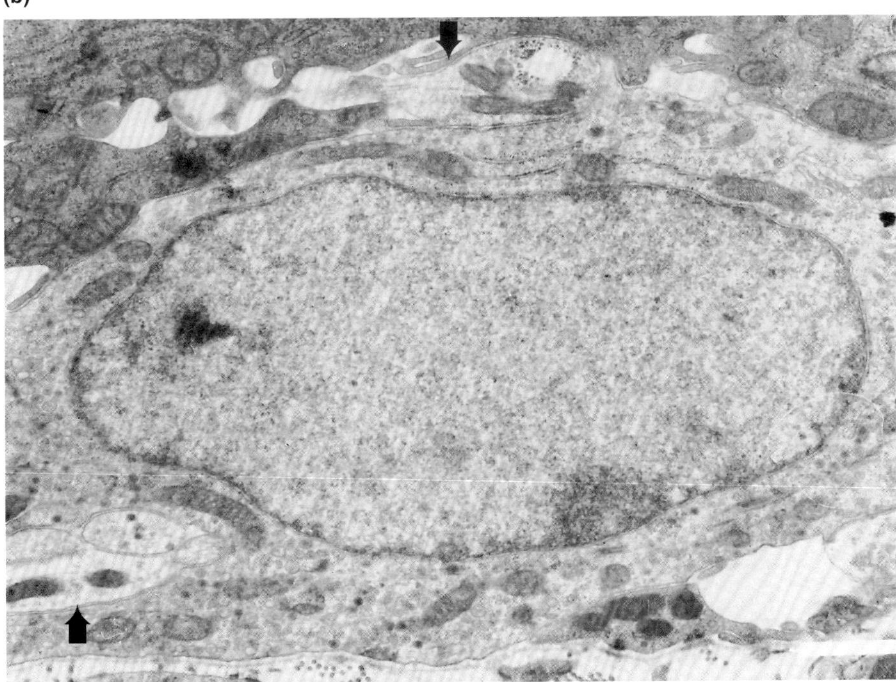

Plate 4.7 Transmission electron micrographs of the rat intraepithelial neuroendocrine cell showing: (a) the relationship to airway lumen and basal lamina (x 11 200); (b) the striking "innervation" of the same neuroendocrine cell by several types of nerve fibres (arrows) (x 15 210).

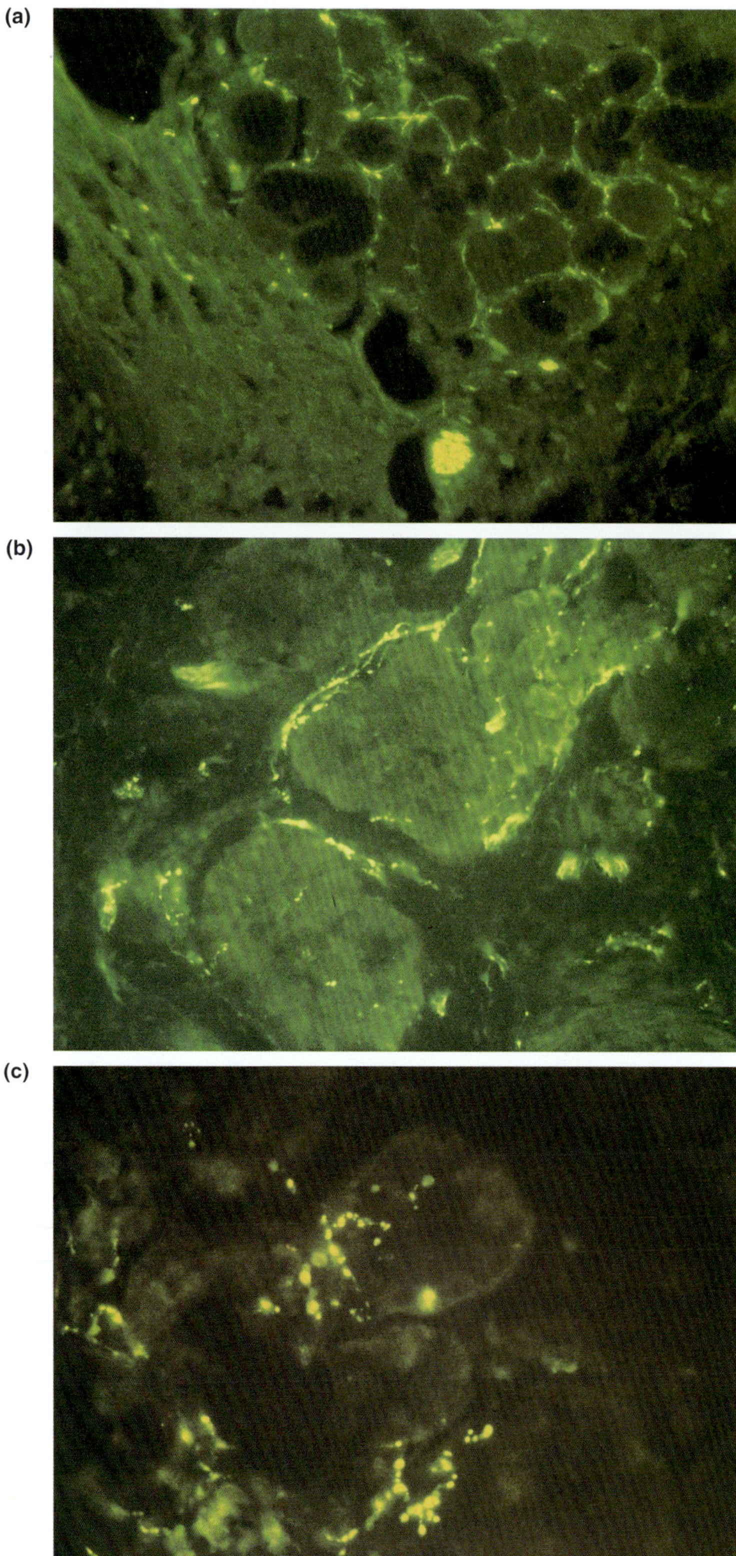

Plate 4.8 Nerve fibres are frequently associated with the acini of human submucosal mucus-secreting glands as demonstrated by: (a) PGP-9.5 immunoreactivity; (b) glyoxylic acid-induced fluorescence; and (c) VIP immunoreactivity.

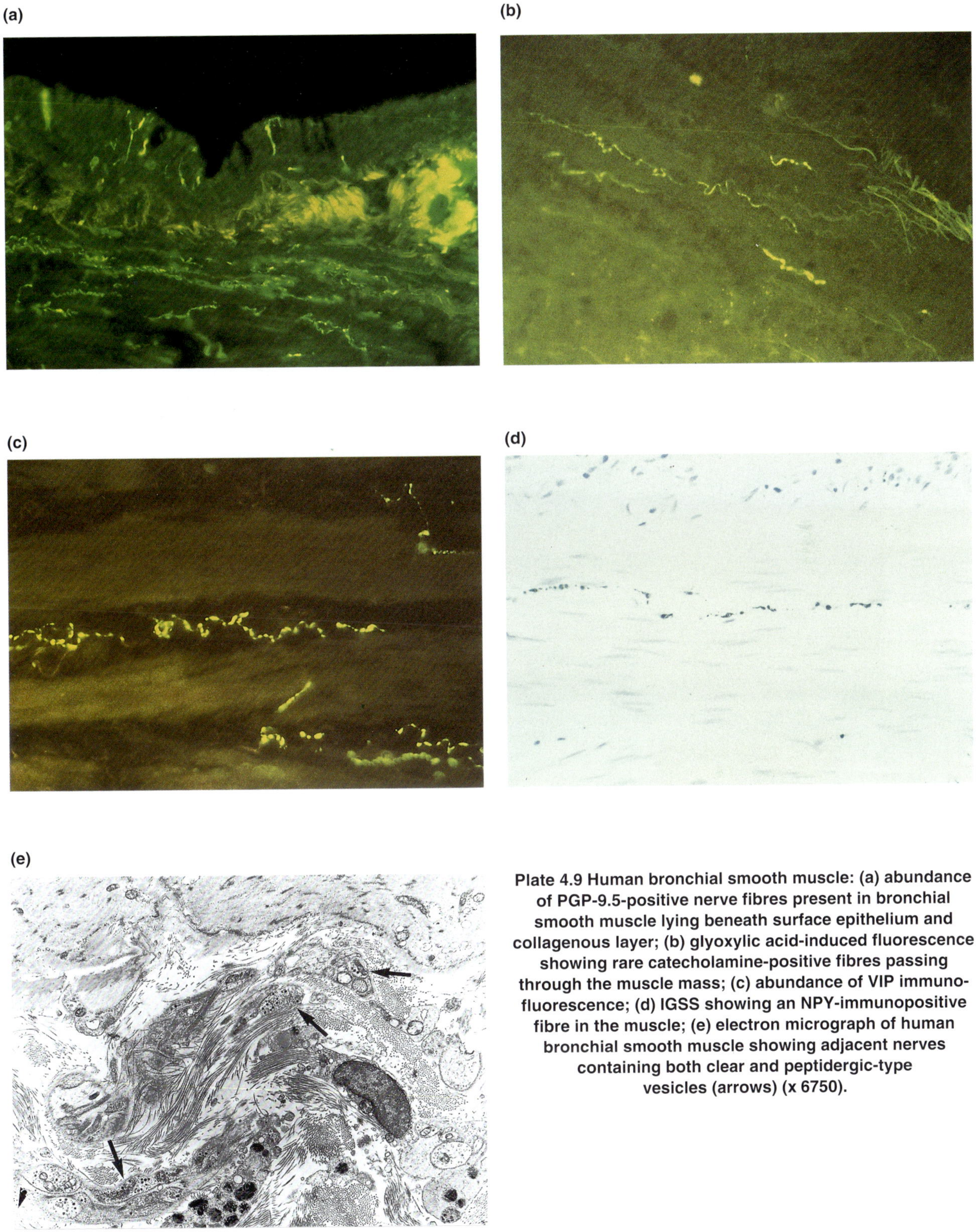

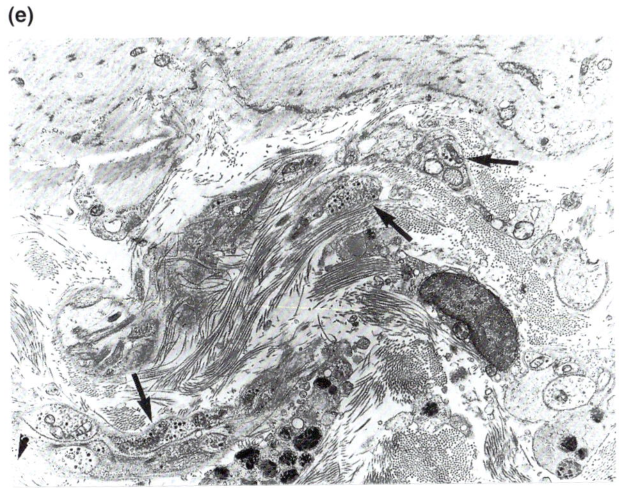

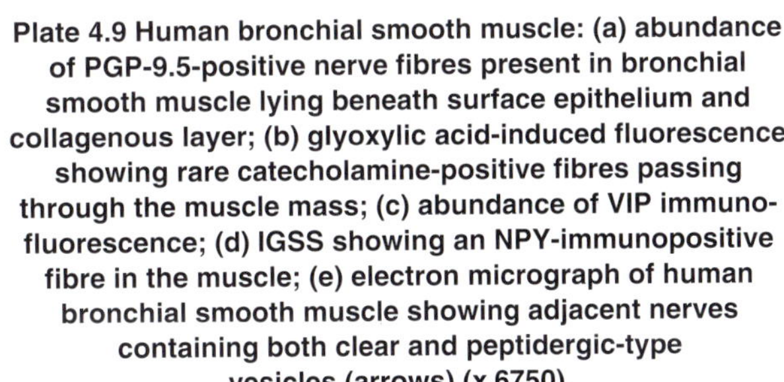

Plate 4.9 Human bronchial smooth muscle: (a) abundance of PGP-9.5-positive nerve fibres present in bronchial smooth muscle lying beneath surface epithelium and collagenous layer; (b) glyoxylic acid-induced fluorescence showing rare catecholamine-positive fibres passing through the muscle mass; (c) abundance of VIP immuno-fluorescence; (d) IGSS showing an NPY-immunopositive fibre in the muscle; (e) electron micrograph of human bronchial smooth muscle showing adjacent nerves containing both clear and peptidergic-type vesicles (arrows) (x 6750).

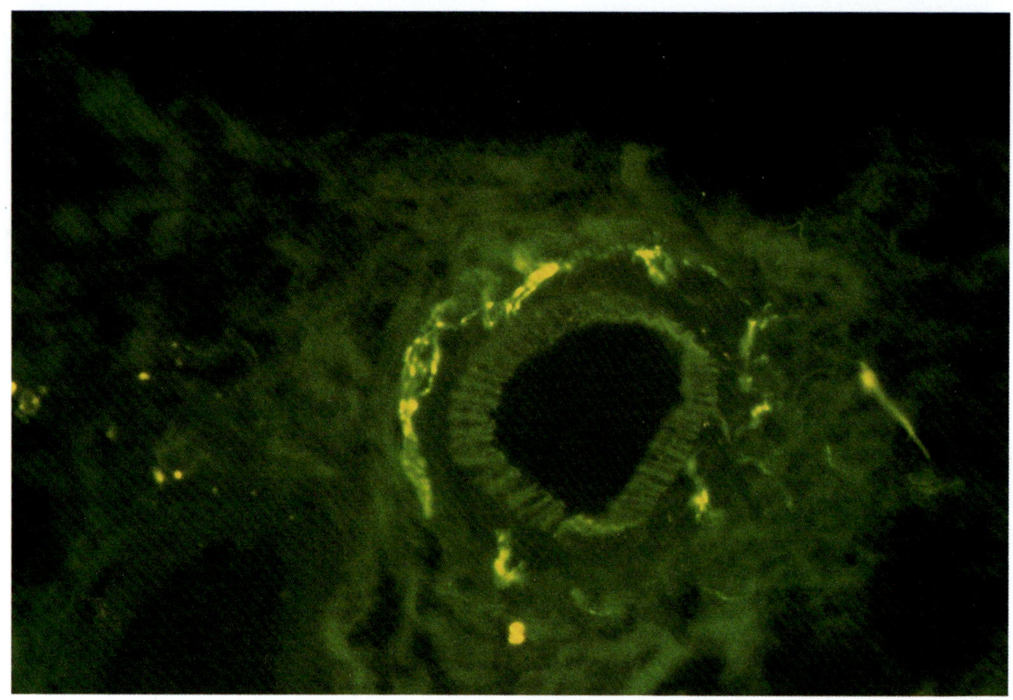

Plate 4.10 Human bronchial artery showing catecholamine-positive fibres. Glyoxylic acid-induced fluorescence.

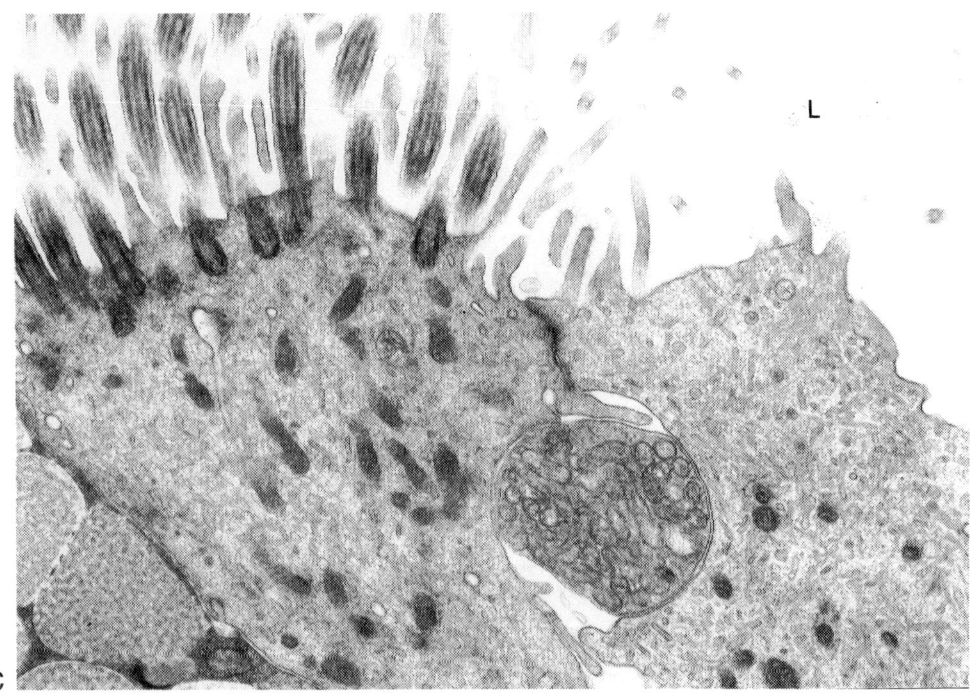

receptors in dog trachea and main stem bronchi is about 1 cm or more in diameter, which is compatible with the morphological studies in that it indicates that the 'ending' supplied by a single fibre is highly branched (Mortola *et al.*, 1975; Sampson and Vikruk, 1975). The reader is referred elsewhere to more detailed discussion of the functional consequences of stimulation of these sensory endings (Jeffery, 1986; McDonald, 1987)

3.2.1   Neuroepithelial Bodies

The physiological role of the clusters of granulated cells which make up the neuroepithelial bodies described in humans and other animals is unclear (Lauweryns *et al.*, 1972; Hung *et al.*, 1973; Lauweryns and Cokelaere, 1973; King *et al.*, 1974; Lauweryns and Goddeeris, 1975). Their presence, close to fenestrated capillaries, and the loss of their serotonin-rich dense-cored vesicles in response to hypoxia, has led to the suggestion that they mediate the hypoxia-induced vasoconstriction which results in a shunt of blood to better ventilated areas of the lung (Lauweryns and Cokelaere, 1973; Moosavi *et al.*, 1973). The function of nerve fibres in apposition to clusters of these groups of cells or single cells (Jeffery and Reid, 1973a; Lauweryns and Cokelaere, 1973) has not yet been elucidated by electrophysiological studies. The speculation is that motor fibres may control degranulation of these cells in response to central nervous or reflex stimulation, but this is unproven.

3.2.2   Cilia

While there is evidence for the neural control of ciliated epithelia in some invertebrates (Takahashi and Murakami,

1969; Aiello, 1974; Sleigh, 1974) and amphibia (Aiello, 1974; Sleigh, 1974), such control has not been shown for the mammalian respiratory system. Indeed, some studies indicate that an intact nerve supply is not required to maintain the normal movement of mucus in dog airways (Lommel, 1908), and evidence from lung transplant patients confirms this (Brody *et al.*, 1972). It is suggested that a local pacemaker determines ciliary autorhythmicity (Laurenzi, 1973), and it has been shown that serotonin and catecholamines and, to a lesser extent, parasympathomimetics are excitatory to cilia while atropine and β-blockers have a cilio-depressive effect; propranolol was, however, less clearly a cilio-depressant (Gosselin, 1966; Laurenzi, 1973; Iravani and Melville, 1975; Foster *et al.*, 1976). Neurally mediated stimulation of mucociliary activity is inhibited by Sub P antagonists, suggesting a role for Sub P in the modulation of mucociliary clearance (Uddman and Sundler, 1987; Wong *et al.*, 1990). Furthermore, in amphibia it has been suggested that parasympathomimetic agents may offer a "protective" effect and maintain ciliary streaming during irritation by, for example, tobacco smoke (Falk *et al.*, 1961). But as yet there is no direct evidence that the autonomic innervation to mammalian airway epithelia has any such effects.

3.2.2.1   *Secretory Cells*

Epithelial secretions derive mainly from three types of surface epithelial cell: the serous and mucous (goblet) cells of the large airways and Clara cells located particularly in terminal bronchioli (Jeffery, 1990). Electrical stimulation of both parasympathetic and/or sympathetic

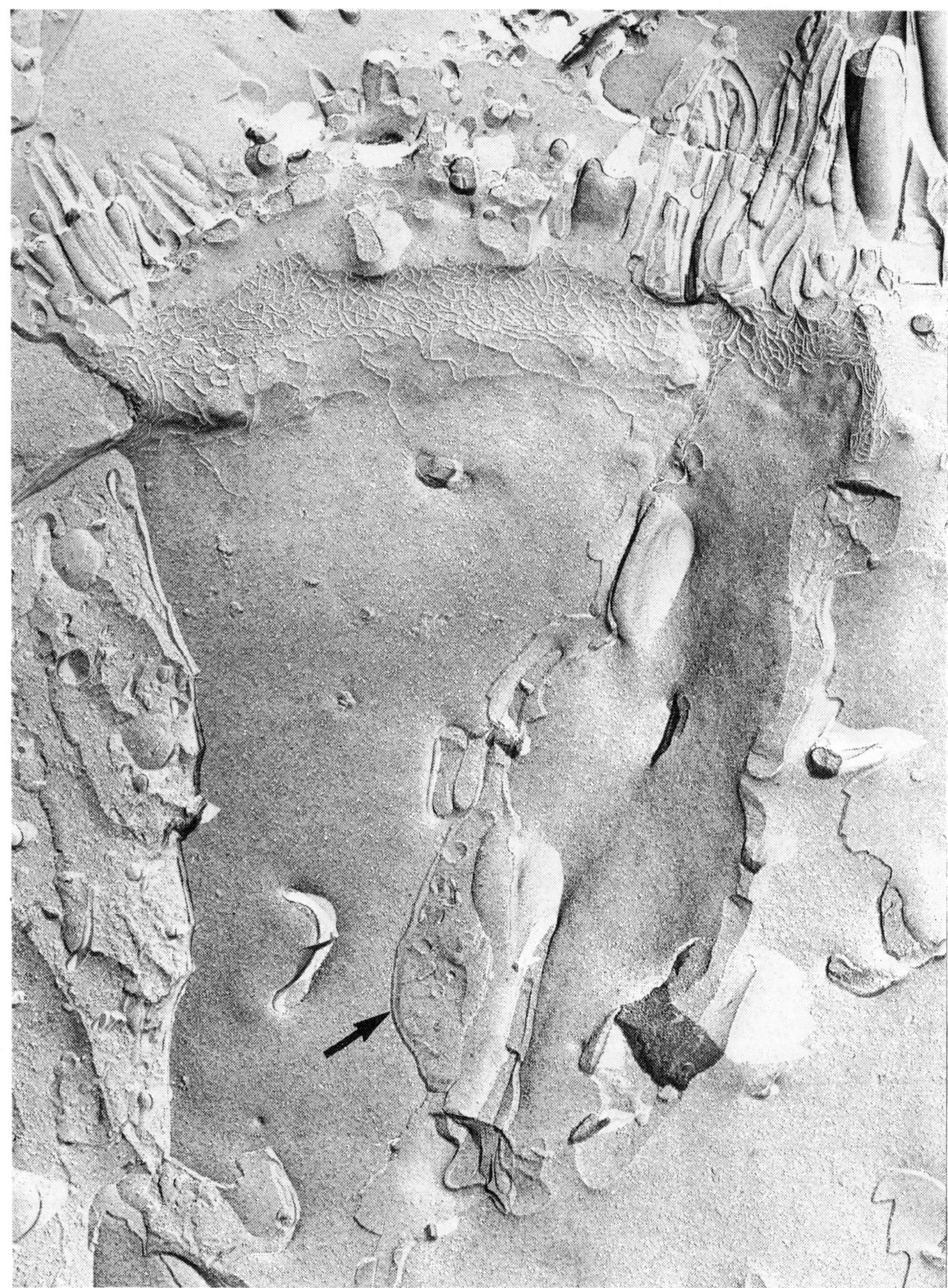

Figure 4.9 Airway epithelial tight junctions as seen by the freeze–fracture technique. The tight junction appears as a network of branching strands and grooves which extend around the apicolateral border of the cell to a depth of about 0.5 μm. The arrow marks a possible fracture of a nerve fibre. (× 21 500.) (Courtesy R. Godfrey)

fibres to airways causes an outpouring of mucus, as does stimulation arising from reflexes initiated in the upper respiratory tract; these effects can be inhibited by autonomic blockers (Phipps and Richardson, 1976; Richardson and Phipps, 1978). In cats, the source of the stimulated mucus is thought to be mainly the submucosal glands with little derived from epithelial mucous cells (Gallagher *et al.*, 1986). In the goose, production of mucus is, however, restricted to the epithelial mucous cells (i.e. intraepithelial glands), and here electrical stimulation of the peripheral cut end of oesophageal nerves causes a marked discharge of mucus, suggesting autonomic control of epithelial mucous cells at least in one avian species (Phipps *et al.*, 1977). In the guinea-pig, Kuo *et al.* (1990) have determined the effect of capsaicin and sensory neuropeptides on tracheal surface epithelial mucus secretion. The capsaicin-induced secretion, which was rapid and sustained to a level 50% above that of controls, was unaffected by prior vagus nerve section or pretreatment with atropine, propanolol and phentolamine. Intravenous Sub P, NkA, NkB and CGRP all produced dose-dependent increases in epithelial mucous cell secretion, and the effects of Sub P were not inhibited by atropine or the histamine receptor antagonists mepyramine or cimetidine. The order of potency for the tachykinins in this effect indicated that the response was probably mediated via tachykinin cell surface receptors of the Nk-1 subtype. These receptors have been localized to guinea-pig surface epithelium by autoradiographic binding and mapping studies (Carstairs and Barnes, 1986). Similar mechanisms are thought to control surface mucous cell discharge in rat (McDonald, 1988). That epithelial mucous cells present at other sites in the body can be induced to discharge their mucin has been shown in culture. Intestinal crypt mucous cells rapidly secreted in response to 10^{-6} M acetylcholine (in the presence of a cholinesterase inhibitor) (Specian and Neutra, 1971). However, those on the villar epithelial surface did not secrete, and a differential response of surface epithelial and deeper-lying mucus-secreting cells appears to occur for surface secretory cells of the gut as well as the airway wall. Pilocarpine induces discharge of Clara cell secretory granules as early as 30 min after a subcutaneous injection (Yoneda, 1977); this effect can be blocked by atropine and, interestingly, also by propranolol (Massaro et al., 1979). Isoproterenol has a similar and equal discharge effect on Clara cell granules in vivo, and this is blocked by propranolol (Massaro et al., 1979); Massaro et al. (1979) suggest that this indicates an "in series cholinergic–adrenergic regulation" of secretion by the Clara cell. Paradoxically, pilocarpine (perfused) does not stimulate secretion in the isolated lung but, rather, it inhibits the secretory effects of isoproterenol (Massaro et al., 1981).

Autoradiography and receptor mapping have shown a high density of β receptors on bronchial, bronchiolar and, particularly, alveolar epithelium (Carstairs et al., 1984; Sharma and Jeffrey, 1990a). The significance of such intense labelling in the absence of a sympathetic innervation is unclear at present. There is also a high density of VIP receptors on human airway epithelium (Carstairs and Barnes, 1986). In the dog, VIP is a potent stimulator of chloride ion transport and, thereby, water secretion (Nathanson et al., 1983); similar mechanisms may operate in humans.

3.3   SUBMUCOSAL GLANDS

The silver staining techniques of Larsell (1922), Elftman (1943) and Spencer and Loef (1964), and the author's more recent studies with PGP-9.5 in humans have shown that the submucosal glands of the the rabbit, dog and human are richly innervated by nerve fibres derived from nearby intramural autonomic ganglia (Plate 4.8a). Using histochemical methods (i.e. acetylcholinesterase), Wardell et al. (1970) have demonstrated that these fibres were cholinergic in the dog; Falk–Hillarp fluorescent techniques indicative of catecholaminergic innervation were negative in this species. In contrast, histochemical and fluorescence techniques, applied to the tracheal airways of the cat and monkey, have indicated the presence of both a cholinergic and a catecholaminergic innervation to both serous and mucous glands (Silva and Ross, 1974; El-Bermani, 1978; Murlas et al., 1980). Similar studies of human resection or postmortem material have shown the presence of cholinergic fibres close to glands (Richardson and Beland, 1976; Shipperbottom, 1988). Glyoxylic acid-induced (catecholamine) fluorescence is present with moderate frequency and is seen as fine varicose endings between gland acini (Pack and Richardson, 1984; Shipperbottom, 1988) (Plate 4.8b). Uddman et al. (1978) have shown the presence of nerve cell bodies and fibres immunoreactive for VIP close to submucosal glands, particularly in the upper, but also in the lower, respiratory tracts of guinea-pigs, rabbits and cats. There are high densities of VIPergic fibres around gland acini in the cat (Sheppard and Polak, 1986) and human (Shipperbottom, 1988) (Plate 4.8c).

The innervation of submucosal glands has also been examined by electron microscopy. The close proximity of nerve fibres to gland acini is not disputed, but the extent to which individual serous, mucous or duct cells are innervated is less known. The author's electron microscopic studies have been concerned with the innervation of cat and human submucosal glands as the former is much used to investigate the neural control of airways secretion. In comparison with the human, cat airway submucosal glands occupy a much larger portion of the airway wall (i.e. 75%) (Jeffery, 1978). Like the human, cat submucosal glands have both serous and mucous acini, but the largest proportion of acini show a mixture of cell types, with the serous cell predominant. In agreement with the studies of Silva and Ross (1974), Murlas et al. (1980), the author has described nerve fibre bundles surrounding cat gland acini or present as single fibres piercing the acinar basement membrane when they become closely invested by individual secretory cells (Fig. 4.10). Murlas et al. (1980) have examined the autonomic innervation of cat submucosal glands following injection of the animal with 5–hydroxydopamine to label specifically the adrenergic fibres. Considering all varicosities within 10 μm of the acini to be within the "effector" sphere of influence, they found that only 10% were adrenergic; the remainder were presumed to be cholinergic. No differential innervation of serous and mucous cells was observed. LeBlond and Jeffery (unpublished observations) found that whilst intra-acinar nerve fibres in the cat were associated with serous, mucous and mixed acini, there were fewer intra-acinar nerves than expected, by random distribution with serous and more with mixed

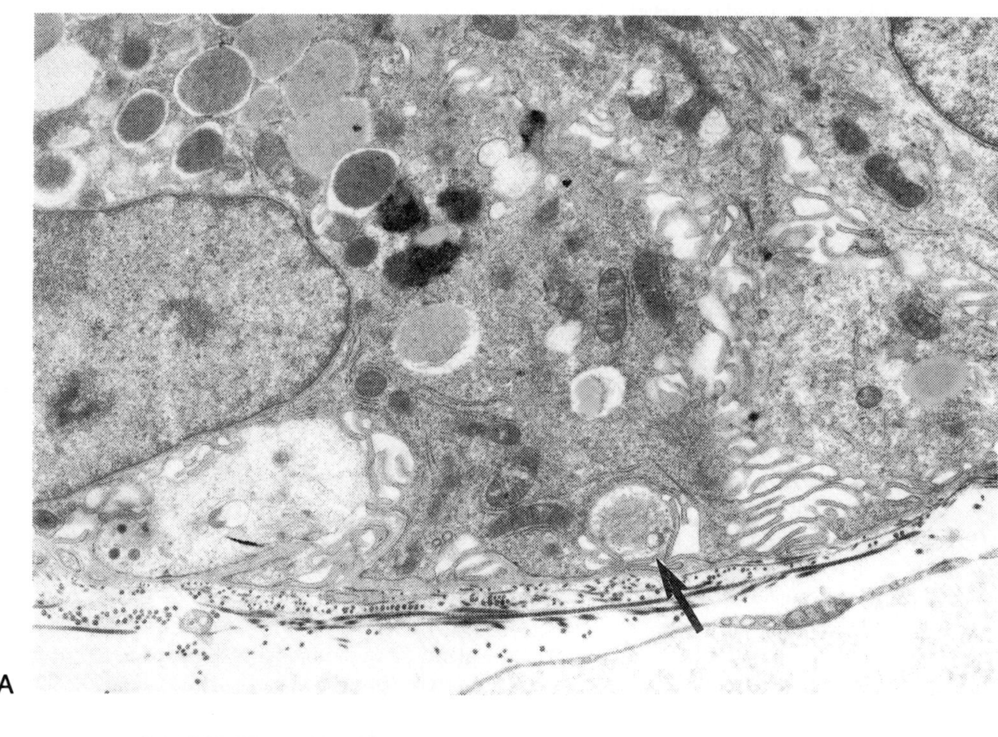

A

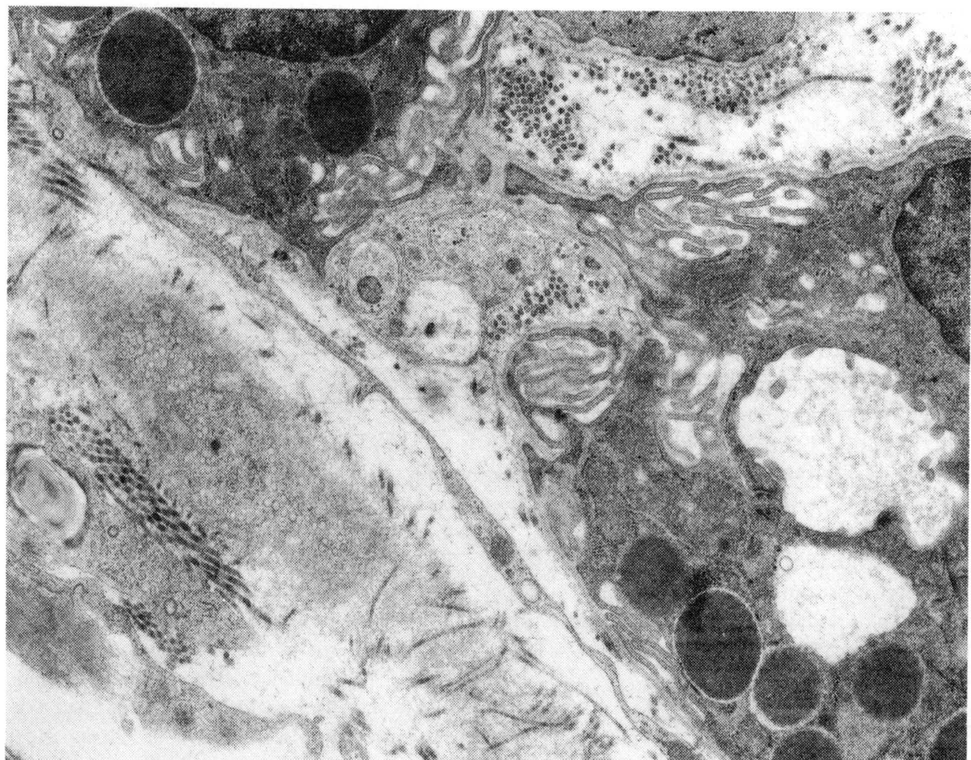

B

Figure 4.10 Transmission electron micrographs of cat airway submucosal gland. Part of a secretory acinus showing: (A) a nerve fibre located within it (arrow), demonstrating the predominance of agranular vesicles ($\times$ 11 300); (B) a group of five nerve fibres immediately adjacent to and lying between two secretory cells with electron-dense secretory granules of the "serous" type ($\times$ 15 300).

Figure 4.11 Transmission electron micrograph of part of a human bronchial submucosal gland acinus of the "mucous" type with a group of nerve fibres lying immediately adjacent, embedded in collagen and containing vesicles of both the agranular and peptidergic types (indicative of the coexistence of acetylcholine and VIP) ($\times$ 28 000).

acini. Conversely, those nerve fibres peripheral but adjacent to acini were particularly associated with serous rather than with mixed acini. Seventy-one per cent of intra-acinar nerves contained neurosecretory vesicles of the agranular type, suggestive of cholinergic function. Nerve fibres with only dense-cored vesicles (5%) were found exclusively outside the acini; the transmitter associated with this latter type of large dense-cored vesicle is not yet known but it may well be VIP (*vide supra*). The possibility of an unequal distribution of nerves to distinct types of acini in the cat may well be important in influencing the volume and type of secretion which is discharged from airway submucosal glands in response to neural stimulation, but this requires confirmatory studies. Preliminary studies by LeBlond and Jeffery (unpublished observations) with human airways show that an intra-acinar innervation is rarely present and that nerve bundles lying outside and adjacent to acini are abundant (Fig. 4.11). Many of the extra-acinar fibres have large dense-cored vesicles compatible with a peptidergic innervation.

In 1932, Florey *et al.* showed that stimulation of the parasympathetic peripheral cut ends of the vagal supply to the cat trachea resulted in an increased volume of intraluminal secretion and a depletion of submucosal gland content while surface epithelial mucous (goblet) cells were not appreciably affected (as judged by histology); this effect could be blocked by atropine. The secretomotor fibres have been shown to run in the recurrent laryngeal and also the superior laryngeal nerves (Johnson, 1935). Cholinergic agonists (acetylcholine, methacholine and pilocarpine), which mimic the effects of parasympathetic nerve stimulation, increase the volume of secreted mucus and mucous glycoproteins both *in vivo* and *in vitro* (Richardson and Phipps, 1978). The physiology of mucus secretion, the bulk of which is derived from submucosal glands, has been reviewed by Richardson and Phipps (1978) and Phipps *et al.* (1982b), to which the reader is referred for a comprehensive coverage. There is a resting secretion of mucus which is independent of extrinsic innervation; this is exemplified by nerve section *in vivo* (Bitensky *et al.*, 1975; Gallagher *et al.*, 1975; Frossard and Barnes, 1987) and by the maintenance of secretion in *in vitro* organ culture of airway mucosa (Meyrick and Reid, 1971; Sturgess and Reid, 1972; Boat *et al.*, 1974; Coles *et al.*, 1982). At least part of this baseline secretion may be due to spontaneous discharge of postganglionic (presumably cholinergic or VIPergic) nerve fibres whose cell bodies lie locally in parasympathetic ganglia; these would be expected to be unaffected by denervation or removal of the tissue for culture. Evidence for cholinergic control is supported by experimental studies which show that atropine reduces the resting secretion seen *in vivo* in the goose (Phipps and

Richardson, 1976; Phipps *et al.*, 1977) and of human airway mucosa *in vitro* (Sturgess and Reid, 1972).

There is also good evidence that parasympathetic efferent pathways are involved in the reflex secretion of airway mucus in response to stimuli originating in nasal or nasopharyngeal epithelium (Phipps and Richardson, 1976), the larynx and tracheal epithelium (Widdicombe, 1954a,b; Richardson and Peatfield, 1987) and also from direct measurements of fluid discharged from individual submucosal gland ducts. The last method has been used to demonstrate that laryngeal stimulation reflexly increases the amount of tracheobronchial fluid discharged from such ducts; the sensory pathway is via the superior laryngeal nerves and the efferent via the cervical vagus (Ueki *et al.*, 1980). The response is atropine-sensitive (German *et al.*, 1980). Electrical stimulation of the central cut end of the vagus or superior laryngeal nerves of one side also causes reflex secretion in the trachea, mediated by secretomotor fibres present in the intact nerves of the other side (Florey *et al.*, 1932; Johnson, 1935). Electrophysiological recordings of such secretomotor fibres have been made in the lungs of cats and dogs, and they are found to be excited by intravenous adrenaline, by stimulation of laryngeal epithelium, by ammonia vapour given into the lung, as well as asphyxia (Widdicombe, 1966).

Sympathetic nerve stimulation (stellate ganglion) has been shown to increase cat tracheal mucin secretion to the same extent as parasympathetic stimulation (Gallagher *et al.*, 1975). The β blocker propranolol blocked the effect but α adrenoceptor antagonists (phentolamine and phenoxybenzamine) did not, showing that the effect is mediated by β adrenoceptors, at least in the cat. There are some indications that this is not the case for all species (Nathanson *et al.*, 1983). The pharmacological results are complementary and both receptor agonists (phenylephrine and methoxamine) and β adrenoceptor agonists (β_1, dobutamine; β_2, salbutamol) increase the volume and quantity of secreted mucous glycoproteins, each response being blocked by the appropriate specific antagonist (Phipps *et al.*, 1982b). Autoradiographic receptor-mapping studies in human show an intense labelling of submucosal gland acini for β receptors, of both subtypes (Carstairs *et al.*, 1985; Sharma and Jeffrey, 1990a). The finding in glands for the β_1 subtype is compatible with the tissue localization of catecholamine-containing nerves. At other tissue sites there is clearly an excess of β_2 receptors, most of which will respond to circulating catecholamine.

Not all the secretory effects of electrical stimulation of nerves are mediated by acetylcholine or catecholamines. In the cat and goose a secretory response remains even after combined cholinergic and receptor blockade (Phipps *et al.*, 1977; Peatfield and Richardson, 1983). This NANC effect is probably due to the release of neuropeptides. The tachykinins Sub P, NkA and NkB have each been shown to stimulate mucin secretion from human bronchial mucosa *in vitro* (Rogers and Barnes, 1989; Rogers *et al.*, 1989). The effects were especially obvious in tissue pretreated with the opioid antagonist naloxone. It was also demonstrated (Rogers and Barnes, 1989; Rogers *et al.*, 1989) that capsaicin-induced secretion of mucus was blocked completely by morphine and that this effect was reversed by naloxone. The studies indicate that sensory nerve stimulation in humans can cause secretion of mucus by an axonal reflex mechanism and that this might be responsible for the hypersecretion of mucus seen after inhalation of irritants such as cigarette smoke. The inhibitory effect of opioid drugs is probably via stimulation of opioid (μ) binding sites on unmyelinated capsaicin-sensitive sensory nerve fibres resulting in an inhibition of Sub P release. This is consistent with the inhibitory effects of opioids on airway vascular leakage and smooth muscle contraction in response to NANC stimulation (Frossard and Barnes, 1987; Belvisi *et al.*, 1988). The order of potency suggests that the secretogogue effect is mediated via the Sub P receptors localized to submucosal glands (Carstairs and Barnes, 1992) and that they are of the Nk-1 subtype.

Sub P is thought to act by causing contraction of myoepithelial cells surrounding gland acini rather than affecting the secretory cells *per se* (Coles *et al.*, 1984; Shimura *et al.*, 1988). In the human nasal mucosa, Sub P stimulates mucous but not serous cell secretion (Baraniuk *et al.*, 1991), whereas the reverse appears to be the case in ferret trachea (Webber, 1989).

The close association of VIPergic nerves and glands in several species including humans suggests a role for VIP also. In ferret airways, VIP stimulates ^{35}S-labelled glycoproteins (Peatfield and Richardson, 1983). But VIP appears to have an inhibitory effect on glycoprotein secretion by human bronchial explants *in vitro* (Coles *et al.*, 1981) and stimulates serous but not mucous glycoprotein secretion by human nasal mucosal explants (Baraniuk *et al.*, 1990). The nature of its secretory effect may depend on the pre-existing basal level of secretion and the nature of baseline neurogenic drive (Shimura *et al.*, 1988).

3.4 AIRWAY SMOOTH MUSCLE

The innervation of tracheobronchial smooth muscle has been recognized since the early days of silver staining (Larsell, 1922; Larsell and Dow, 1933; Gaylor, 1934; Jabonero and Sabadell, 1972; Honjin, 1956). Most studies indicate that fibres from the peribronchial plexus penetrate the muscle and terminate in a simple end-swelling thought to be motor or, in Larsell's (1992) study, a "smooth muscle spindle" thought to be a sensory receptor (Plate 4.9a).

Acetylcholinesterase-positive fibres have been found in most species studied, including humans (El-Bermani, 1970; Mann, 1971; El-Bermani, 1973; Richardson and Ferguson, 1980; Shipperbottom, 1988) (see Plate 4.1).

In contrast, the presence of catecholamine fluorescence in bronchial smooth muscle is species-dependent: abundant in the cat (Dahlstrom *et al.*, 1966), lacking or sparse in many species including humans (Plate 4.9b) or restricted solely to the upper trachea as in the guinea-pig (Coburn and Tomita, 1973; Richardson and Ferguson, 1980). VIP and, in humans, Sub P also have been localized in and around the bronchial smooth muscle of a number of species (Said, 1982, 1987). (Plate 4.9c). PHI, a peptide with striking sequence homology to VIP and its human variant PHM-27, has a tissue distribution indistinguishable from VIP. Their concentrations vary markedly, however, probably due to tissue specific differences in post-translational processing of the precursor (Fahrenkrug *et al.*, 1985). NPY is co-localized with noradrenaline in sympathetic nerve fibres (Polak and Bloom, 1984; Sheppard *et al.*, 1984). NPY immunoreactive nerve fibres are found throughout the respiratory tract of the cat, rat, guinea-pig and human (Plate 4.9d). Fine varicose nerve fibres are found in bronchial smooth muscle at all airway levels, albeit their number decreases in bronchioles.

Electron microscopic studies fail to show synaptic contact of nerve with muscle. Whilst some nerve varicosities lie further than a micrometre away from the muscle, others may share a common investment of basement membrane with muscle yet show no further synaptic specialization (Richardson and Ferguson, 1980). There are also variations in muscle-to-muscle contacts: muscle fibres are separated from each other in guinea-pig and dog airway (and thus act as multiple units) while frequent connections of the nexus (gap junction) type are seen in those of the human. The latter type of junction implies electrical coupling such that the muscle behaves as a single unit, and in this respect is similar to that often found in the gastrointestinal tract (Richardson and Ferguson, 1980). Nerve profiles with 50 nm agranular vesicles (characteristic of cholinergic innervation) are seen in the airways of many species, and vesicles of a similar size but with eccentric dense cores (thought to be adrenergic) can be found in the cat. Endings with relatively large (180–200 nm) dense-cored vesicles of the peptidergic type have been found in cat, guinea pig and human trachealis muscle, and this is supportive evidence for the presence of a peptidergic or purinergic transmitter (Burnstock, 1972; Richardson and Ferguson, 1980) (*vide supra*) (Plate 4.9d).

With regard to sensory endings, the cobalt and HRP diffusion studies of Lacy (1980) convincingly showed sensory innervation of smooth muscle, the efferent fibres of which run in the vagus. In 1970, Fillenz and Woods described, in rabbits, nerve profiles containing numerous mitochondria and glycogen lying close to bronchial muscle. In 1974, Von During *et al.* examined 1 μm serial sections of rat intrapulmonary bronchi, tracing the course of nerve bundles with myelinated fibres and their branches: the branches were found to penetrate the muscle as "leaf-like" expansions, each expansion devoid of myelin or Schwann cell sheath. By electron microscopy they showed the "endings" to be oriented parallel to muscle fibres, embedded in collagen and to contain mitochondria and a so-called "receptor matrix". However, the functional properties of the endings described are not known and their role as stretch receptors is still undetermined. Bartlett *et al.* (1976) localized stretch receptors to the membranous portion of the trachea and main bronchi of dogs by recording the electrical discharge from single nerve fibres supplying the receptors whilst mechanically probing the airway wall. On each occasion the localization was to an area of airway lining of about 3 mm^2. Focal dissection and removal of lining epithelium and lamina did not interrupt receptor discharge whilst the removal of the immediately underlying deep submucosal tissue did. Ten such pieces were examined by electron microscopy and all showed nerve fibre bundles together with bronchial muscle. Serial sections of five such pieces did not, however, reveal a specific receptor structure and the search for a stretch receptor whose structure is known and can be directly linked with physiological function still continues. Bartlett *et al.* (1976) confirmed, however, the localization of the stretch receptor to a region containing muscle. This is of interest as not all studies agree with a submucosal localization: there are data which suggests that stretch receptor activity in rabbits arises from receptors in the surface epithelial lining of airways (Bitensky *et al.*, 1975). An electron microscopic study of trachealis muscle of the mouse following cervical vagotomy indicated that as much as 80% of fibres within this muscle are of the sensory type, and that, by electron microscopy, many of these are filled with mitochondria (Pack and Widdicombe, personal communciation).

The airways of animals and healthy humans are tonically constricted and maintained as such by vagal efferent activity. Thus, cutting or cooling the vagus nerves in animals (Nadel and Widdicombe, 1963; Colebatch and Halmagyi, 1963; Olsen *et al.*, 1965; Karzewski and Widdicombe, 1969) or administration of atropine to animals (Severinghaus and Stupfel, 1955) and humans (Dautrebande *et al.*, 1962; Nadel and Widdicombe, 1962; Sterling and Batten, 1969; Vincent *et al.*, 1970; Widdicombe, 1975; De Troyer *et al.*, 1979) causes bronchodilation. But the degree of tonicity appears to vary between species and is marked in cats, slight in rabbits and not detectable in guinea-pigs (Nadel, 1980). The evidence indicating that vagal nerves play an important role in regulating airway smooth muscle is strong and has been reviewed by Nadel (1980), to which publication the reader is referred. In addition, the localization and distribution of muscarinic receptors and their M$_3$ subtype is entirely consistent with the functional studies and with the localization of cholinergic nerves (Mak and Barnes, 1989; 1990).

The role of the sympathetic nerve supply to bronchial

muscle is less understood; it varies between species and in any event appears to be less important than the parasympathetic supply. The decrease in bronchomotor tone following adrenergic stimulation is found in many animals, but is weak, notably in the guinea-pig and human. The role of the sympathetic nervous supply to the lung in mammals is, therefore, unclear. In agreement, catecholamine fluorescence around or in bronchial smooth muscle is sparse in humans (*vide supra*). The following are pertinent observations:

1. Whilst sympathetic stimulation causes bronchodilation, which is abolished by β adrenoceptor antagonists, there is not an effect (in the dog) if the vagi are first cut to remove initial resting vagal tone. The effect of sympathetic inhibition is greatest in those airways showing greatest initial vagoconstriction.
2. Cutting the sympathetic nerve supply (dogs) results in only a small change in resistance.
3. There is evidence that the efferent sympathetic supply (cats and dogs) shows only a low level of spontaneous activity.
4. Complete inhibition of vagal constriction is never obtained, showing that vagal constrictor pathways predominate.
5. The maximal bronchodilator effects of sympathetic nerve stimulation occur more slowly (30 s) than vagal bronchoconstriction.
6. Bronchoconstriction due to inhaled tobacco smoke is exaggerated in healthy subjects by propranolol premedication, suggesting a modulation of bronchoconstriction by sympathetic fibres. The bronchodilator effect of sympathetic stimulation is, therefore, most probably due to sympathetic inhibition of cholinergic transmission via ganglia (Nadel, 1980; Barnes, 1986b).

The majority of peptides which affect smooth muscle have a bronchoconstrictor effect, the major exception being VIP (Table 4.7). Sub P does not cause bronchoconstriction either by inhalation or by intravenous infusion whereas NkA causes constriction after inhalation in asthmatic subjects and by intravenous infusion in normal individuals (Joos *et al.*, 1987; Evans *et al.*, 1988). NkA, encoded by the same gene as Sub P, is one of the most powerful constrictors of the tachykinin group, probably acting on Nk-2 receptors (Barnes *et al.*, 1991a,b). VIP is about 50 times more potent as a bronchial smooth muscle relaxant than isoprenaline and reduces the bronchoconstriction due to histamine, $PGF_{2\alpha}$, serotonin, LTD_4 and neurokinins (Said, 1987). In human airways, VIP affects distinct airway levels differently: bronchi are potently relaxed whilst bronchioles are unaffected. In contrast, isoprenaline relaxes both airway levels to a similar extent (Palmer *et al.*, 1986; Barnes *et al.*, 1991a,b). CGRP causes constriction of human bronchi *in vitro*, probably by an indirect mechanism (Barnes *et al.*, 1991a,b).

Table 4.7. Effects of peptides on airway smooth muscle

Peptide	Effect
Bombesin/GRP	Contracts
C5a	Contracts
CCK	Brief contraction
Sub P	Constricts
NkA	Powerful constrictor
NkB	Contracts
"Spasmogenic lung peptide"[a]	Contracts
CGRP	Contracts (?)
VIP	Relaxes
PHM/PHI	Relaxes
ANP	Relaxes
Vasopressin/oxytocin	Relaxes

[a] Said (1977).

3.4 VASCULATURE

Relatively little is known about the nerves supplying the bronchial and pulmonary vasculature. Indeed, much still needs to be learnt about the extent and nature of the bronchial vasculature, which appears to vary widely with species (Laitinen *et al.*, 1987a,b). The functions of bronchial vessels include provision of nutrients to airway structures, clearance of chemical mediators from the airways, regulation of interstitial fluid volume (and hence wall thickness) and participation in heat exchange.

Parasympathetic innervation of bronchial and pulmonary vasculature appears to be sparse whereas there is a relatively rich innervation by catecholamine fluorescent and tyrosine hydroxylase immunoreactive fibres (Shipperbottom, 1988) (Plate 4.10). NPY-immunoreactive fibres are also present in bronchial blood vessels of the nose, trachea and bronchus as well as the pulmonary vasculature (Sheppard and Polak, 1986). Fibres containing VIP are localized to the adventitia surrounding vascular smooth muscle, as are those containing PHI, Sub P, CGRP and galanin (Burnstock, 1986; Dey *et al.*, 1988). Functional studies show that the main control of the vascular bed is sympathetic and adrenergic, acting via α-adrenoceptors (Daly and Hebb, 1966; Lung *et al.*, 1976). Non-cholinergic vasodilatation is probably normally due to release of VIP or PHI, particularly in large airways and the nose (Lundberg *et al.*, 1981; Lundberg *et al.*, 1984b; Lundberg *et al.*, 1987; Barnes *et al.*, 1991a,b). In the guinea-pig, Sub P, NkA and NkB, in rank order of potency, increased extravasation of plasma in trachea and main bronchi, with lesser effects in larynx and intrapulmonary airways, the effects are probably mediated by Nk-1 receptors (Rogers *et al.*, 1988). The effects are seen after vagal nerve stimulation and also following inhalation of Sub P to by guinea-pigs (Lotvall *et al.*, 1990). No direct measurements have yet been made in human airways. It is thought that the most important functional effect of tachykinins is on blood flow (Barnes *et al.*, 1991a,b). CGRP, whilst having no direct effect

Table 4.8 Effects of Sub P on immune response and reactions

Immune response/reaction	Effect	Reference
Lymphocyte proliferation	Increased	Payan (1983)
Immunoglobulin synthesis	Increased	Stanisz et al. (1987)
Phagocytosis	Enhanced	Bar-Shavit (1980)
Neutrophils, Chemotaxis and Lysosomal release	Enhanced	Marasco (1981)
Eosinophil	Degranulation	Kroegel et al. (1990)
Monocyte chemotaxis	Promoted	Ruff (1985)
Monocyte cytokines	Released (IL-6)	Lotz et al. (1990)
Macrophage	Activated	Hartung (1983)
Mast cell, histamine and 5-HT	Released[a]	Fewtrell (1982)

[a] Effect varies greatly with site and species.

on airway microvascular leak, potentiates the leakage produced by Sub P in the skin. Unlike its effect in the skin, it does not potentiate the effect of Sub P in guinea-pig airways although this last-mentioned finding is not in agreement with an earlier study (Lundberg et al., 1985a,b). The dog tracheal perfusion studies of Laitinen et al. (1987a,b) demonstrate dose-dependent decreases in vascular resistance by NkA, Sub P, VIP and CGRP. The effects of CGRP and VIP appear to occur later and are more sustained than that by NkA and PHI. It is possible that CGRP may be the predominant mediator of arterial vasodilatation in response to sensory nerve stimulation in the nose and bronchi (Matran et al., 1989). NPY is a tracheal vessel constrictor (at doses above 10^{-11} M). Bombesin also constricts vessels. Capsaicin causes a dual response: low concentrations decrease vasculature resistance whilst high concentrations first increase, then decrease, resistance.

4. Interaction with the Immune System

Nerves, their "transmitters" and factors which regulate their growth may have profound effects on the immune system. Acetylcholine, serotonin, endorphins, SOM, VIP and Sub P have each been shown to modulate immune responses and reactions either directly or by influencing mediator release from mast cells and macrophages with subsequent effects on immunity (Payan et al., 1984; Goetzl et al., 1985; Stanisz et al., 1986). As an example, some of the major effects of Sub P are shown in Table 4.8. These effects are in addition to the vasodilatory effect of Sub P and its effect on inducing plasma extravasation. Sub P binding sites have been shown on T cells, particularly on the T helper subset (Payan et al., 1984). Sub P stimulates lymphocyte proliferation whereas VIP and SOM inhibit it (Stanisz et al., 1986). Sub P induces 5-HT and histamine release from rat peritoneal and pleural mast cells (Fewtrell et al., 1982). The effect in vivo is absent in rats whose C fibres have been destroyed by neonatal capsaicin (Skofitsch et al., 1983). The

secretagogue effect is highly site- and species-specific. Mast cells and Sub P- and CGRP-containing nerves have been shown to be closely apposed in the villi of rat intestine, more than would be expected to be in contact by random association (Stead et al., 1987). VIP has anti-inflammatory effects: it inhibits release of mediators from pulmonary mast cells (Undem et al., 1983) and is thought to interact with T lymphocytes (O'Dorisio et al., 1989). CGRP inhibits the proliferative response of T lymphocytes to mitogens via specific receptors (Umeda and Arisawa, 1989) and inhibits macrophage secretion and the capacity of macrophages to activate T lymphocytes (Nong et al., 1989). Nerve growth factor has been shown to have profound effects on rat mast cell growth and function and may, in the human, exert its effect via T cell products (Stanisz et al., 1986). In the author's experience of electron microscopy of human bronchial biopsies, mast cells are occasionally found in close association with nerves (Fig. 4.12A) but no more so than would be expected by random association, and other immune cell types are similarly "innervated" (Fig. 4.12B,C). The relationships may change, however, in airway disease, and this area of research is currently one in which there is much interest.

5. Alterations in Disease

Autonomic dysfunction has long been implicated as a mechanism contributing to the symptoms of airway diseases such as asthma, cystic fibrosis and COPD. The role of nerves in asthma was suggested almost as soon as their discovery in airways in the 17th century, and this view did not change during the next two centuries (Floyer, 1698; Salter, 1968) and was further supported by clinical and experimental studies, particularly in the dog (Roy and Brown, 1885; Dixon and Brodie, 1903). Subsequently a variety of alternative hypotheses have been considered, including immune dysfunction and alterations of mucosal permeability. The discovery of the NANC "system" and the appreciation of its role in the

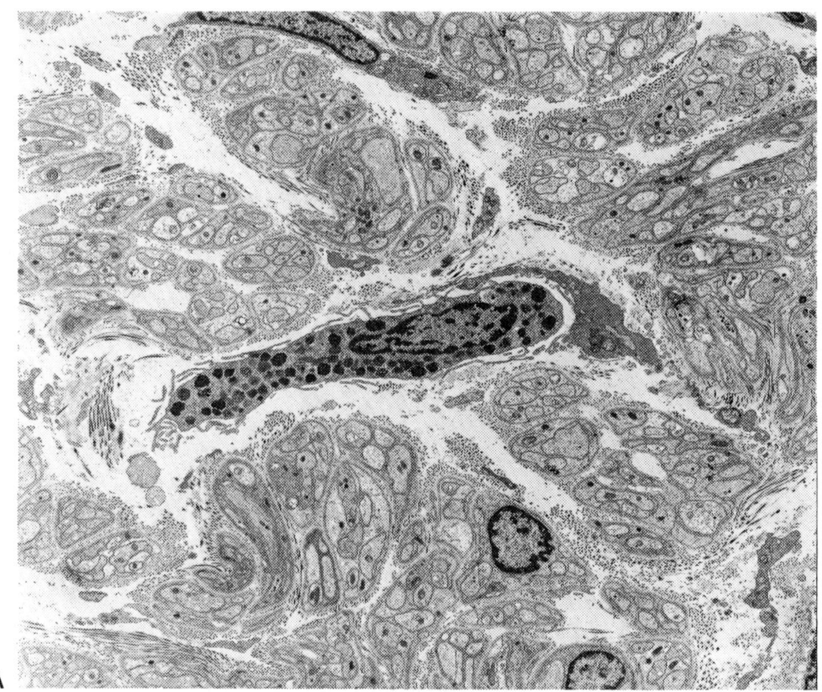

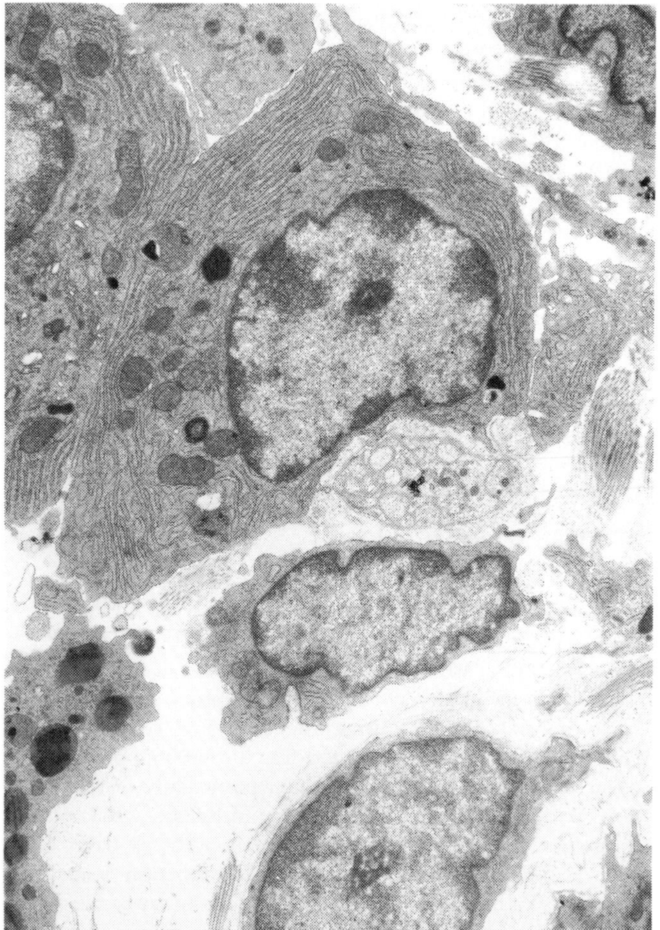

Figure 4.12 Transmission electron micrographs of human bronchial mucosa. (A) The spatial association of a mast cell with a large number of unmyelinated nerve fibres is demonstrated ($\times$ 4000). (B) A plasma cell and lymphomononuclear cell abutting a nerve fibre bundle ($\times$ 7500). (C) A lymphocyte (L) is shown, presumed to be migrating into the surface epithelium through basal lamina and past an adjacent intraepithelial nerve (arrow) ($\times$ 8000). MC, intraepithelial mast cell.

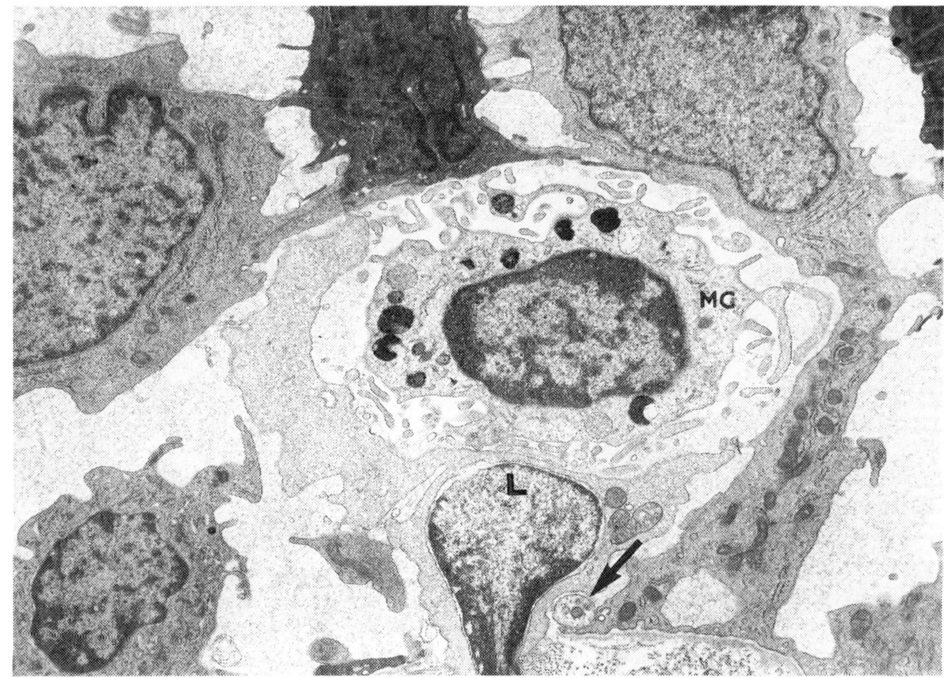

generation of neurogenic inflammation has stimulated a renewed interest in autonomic dysfunction at the level of the ganglion, transmitter and autonomic receptor. In general, the various hypotheses subscribe to the notion that there is an imbalance between excitatory (i.e. muscarinic cholinergic, α-adrenergic and NANC excitatory) and inhibitory (β-adrenergic, NANC inhibitory) components (Barnes *et al.*, 1991a,b). Many of the studies of classical cholinergic and adrenergic neural pathways have failed to confirm abnormalities which can adequately explain the airways hyperresponsiveness and clinical symptoms of asthma (Barnes, 1986b, Barnes *et al.*, 1991a,b). Binding and autoradiographic studies of autonomic receptors have also shown little in the way of significant change in human lung tissue (Goldie *et al.*, 1982; Goldie *et al.*, 1987; Sharma and Jeffery, 1990a; Bai *et al.*, 1992). These studies often do not address receptor function and there may be reduced responsiveness (due to receptor uncoupling) in the presence of a normal number of binding sites. The interest in an imbalance between Sub P (excitatory) and VIP (inhibitory to bronchial smooth muscle) NANC fibres has been stimulated by an immunohistochemical study of Ollerenshaw *et al.* (1989). These workers examined lung tissue obtained mainly at autopsy and lobectomy from patients with asthma ($n = 5$) and subjects without asthma ($n = 9$), respectively. Using immuno-methods applied to sections of intrapulmonary airways, immunoreactive VIP was found in 92% of sections in subjects without asthma but none was found in any of the 468 sections in the asthma cases. In contrast, immunoreactive Sub P was

present in both groups. The conclusion was that the loss of VIP from nerves would diminish neurally mediated bronchodilation, allowing excitatory mechanisms to predominate in asthma. The arguments against this are:

1. the overwhelming evidence in favour of airway inflammation in asthma and the consequent release of peptidases which may rapidly degrade VIP, particularly at the time of death;
2. the absence of pulmonary vascular effects (hypertension) in asthma which would be expected in the absence of VIP innervation of vasculature (Barnes, 1989);
3. the recent observations that there are no differences in VIPergic innervation in bronchial biopsies taken from mild atopic asthmatics (Howarth *et al.*, 1993); and, compatible with this,
4. NANC functional responses appear to be normal in mild asthma (Lammers *et al.*, 1989).

Interestingly, in a recent follow-up publication Ollerenshaw *et al.* (1991) described an increase in the number and "length" of Sub P-containing nerve profiles in airway tissues, obtained from three asthmatics (postmortem) and eight asthmatics in whom bronchial biopsies had been taken for comparison with those of 13 cases without asthma. Sub P-reactive fibres were located in the subepithelium surrounding glands and vessels. These changes would be in accord with those described in inflammatory bowel disease (Holzer, 1988). The results are of great interest and the implications are obvious. In the author's opinion, any firm conclusion should, however, await confirmatory studies.

An increase in vagal sensory activity may also explain the hyper-reactivity of asthma. Both "irritant" and C fibre endings can be stimulated by inflammatory mediators, which include histamine, bradykinin and prostaglandins (Saria *et al.*, 1988). The fragility of surface airway epithelium in asthma described by several workers (Laitinen *et al.*, 1985; Beasley *et al.*, 1989; Jeffery *et al.*, 1989) would be expected to enhance access of irritants and inflammatory mediators to the intraepithelial endings described above. In support of this, damage to epithelium by either viral infection or ozone increases airway responsiveness to histamine, an effect which is blocked by atropine (Empey *et al.*, 1976a; Gold *et al.*, 1978) or abolished by section or cooling of the vagus (Lee *et al.*, 1980). These results support the concept of the involvement of an axon reflex and the initiation of neurogenic inflammation in asthma.

There is almost nothing known about the immuno-histology of autonomic dysfunction in COPD. Interestingly, in the study of Ollerenshaw *et al.* (1991), immunoreactivity for Sub P was compared in airway tissues taken from eight "normal" subjects without respiratory disease and 14 with mild, chronic airflow obstruction: there were no differences in the number and length of Sub P immunoreactive fibres. Alterations to autonomic receptors as determined by radioligand-binding studies have also been described in chronic bronchitis and emphysema (Raaijmakers *et al.*, 1985).

Autonomic dysfunction was also an early suggestion for a prime abnormality in CF (Davis and Kaliner, 1983) whilst the prime gene abnormalities which are now being discovered appear to be unrelated to neural defects, autonomic dysfunction may still be an indirect consequence and contribute to the disease process. In histochemical studies of CF airways obtained at postmortem and freshly, following transplantation, Shipperbottom *et al.* (1987) found no differences in acetylcholinesterase positivity or catecholamine fluorescence but did describe a reduction in VIP-immunoreactive fibres associated with both bronchial smooth muscle and submucosal glands. The finding is of interest, as a previous study of sweat glands (Heinz-Erian *et al.*, 1985) reported similar findings. Autoradiographic and binding studies of autonomic receptors also show a general reduction of both β-adrenergic and VIP-binding sites on effectors in airway tissues from CF patients, but not in asthma (Sharma and Jeffery, 1990a,b).

6. Summary and Conclusion

The neural control of bronchial epithelium, submucosal glands and bronchial and vascular smooth muscle appears to be much more complex than originally proposed. Cholinergic (parasympathetic) innervation seems to be the predominant drive to effectors in mammals, and marked species differences exist in the extent to which this is so. The role of the adrenergic (sympathetic) system is now less clear. There may be (1) direct inhibition of bronchial effectors, (2) indirect inhibition acting on mucosal ganglia to inhibit cholinergic transmission or (3) the main modulation of cholinergic drive may be peptidergic or purinergic with only a minor or no role for adrenergic innervation in humans. The variety of peptides demonstrated in airway tissues by immunohisto-chemistry supports a major role for NANC nerves in several species, including humans, albeit their role in human airway disease requires much further research and confirmation of the interesting discoveries that have recently been made.

7. Acknowledgements

I am particularly grateful to Mrs Jenny Billingham for her patience and care in the preparation of the manuscript and to Mr Andy Rogers for his assistance with preparation of the electron micrographs. I would also like to thank the Cystic Fibrosis Trust and National Asthma Campaign (NAC) (UK) for supporting many aspects of the studies presented herein. Plates 4.1, 4.3b,c, 4.5b, 4.8c, 4.9c, and 4.10 formed part of a submission for a higher degree by Miss A. Shipperbottom during a studentship at the National Heart and Lung Institute, London, supported by the NAC.

8. References

Agostini, E., Chinnock, J.E., Daly, M., de Burgh, M. and Murray, J.G. (1957). Functional and histological studies of the vagus nerve and its branches to the heart, lungs and abdominal viscera in the cat. J. Physiol. 135, 182–205.

Aiello, E. (1974). In "Cilia and Flagella" (ed. M.A. Sleigh), pp 353–376. Academic Press, London.

Armstrong, D.S. and Luck, J.C. (1974). A comparative study of irritant and type J receptors in the cat. Respir. Physiol. 21, 47–60.

Bai, T., Mak, J.C.W. and Barnes, P.J. (1992). A comparison of beta-adrenergic receptors and in vivo relaxant responses to isoproterenol in asthmatic airway smooth muscle. Am. J. Respir. Cell. Mol. Biol. 6, 647–651.

Baker, D.G., Basbaum, C.B., Herbert, D.A. and Mitchell, R.A. (1983). Transmission in airway ganglia of ferrets: inhibition by norepinephrine. Neurosci. Lett. 41, 139–143.

Baraniuk, J.N., Okayama, M., Lundgren, J.D., Okayama, M., Mullol, J., Merida, M., Shelhamer, J.H. and Kaliner, M.A. (1990). Vasoactive intestinal polypeptide (VIP) in human nasal mucosa. J. Clin. Invest. 86, 825–831.

Baraniuk, J.N., Lundgren, J.D., Mullol, J., Okayama, M., Merida, M. and Kaliner, M. (1991). Substance P and neurokinin A in human nasal mucosa. Am. J. Respir. Cell. Mol. Biol. 4, 228–236.

Barnes, P.J. (1986a). Nonadrenergic-noncholinergic neural

control of human airways. Arch. Int. Pharmacodyn. Ther. 280, 208–228.

Barnes, P.J. (1986b). State of art: neural control of human airways in health and disease. Am. Rev. Respir. Dis. 134, 1289–1314.

Barnes, P.J. (1989). Vasoactive intestinal peptide and asthma. N. Eng. J. Med 321, 1128–1129.

Barnes, P.J. (1992). Modulation of neurotransmission in airways. Physiol. Rev. 72, 699–729.

Barnes, P.J., Baraniuk, J.N. and Belvisi, M.G. (1991a). Neuropeptides in the respiratory tract. Am. Rev. Respir. Dis. 144, 1187–1198.

Barnes, P.J., Baraniuk, J.N. and Belvisi, M.G. (1991b). Neuropeptides in the respiratory tract. Pt. II. Am. Rev. Respir. Dis. 144, 1391–1399.

Bar-Shavit, Z., Goldman, R., Stabinsky, Y., Gottlieb, P., Fridkin, M., Teichberg, V.I., and Blumberg, S. (1980). Enhancement of phagocytosis – A newly found activity of substance P residing in its N-terminal tetrapeptide sequence. Bioch. Biophys Res. Commun. 94, 1445–1451.

Bartlett, D. and John, W.M. (1979). Adaption of pulmonary stretch receptors in different mammalian species. Respir. Physiol. 37, 303–312.

Bartlett, D., Jeffery, P.K., Ambrogio, G. and Wise, J.C.M. (1976). Location of stretch receptors in the trachea and bronchi of dog. J. Physiol. 258, 409–420.

Baumgarten, H.G., Holstein, A.F. and Owman, C.H. (1970). Auerbach's plexus of mammals and man: electron microscopic identification of three different types of neuronal processes in myenteric ganglia of large intestine from rhesus monkey, guinea pig and man. Z. Zelforsch. Mikrosk. Anat. 106, 376–397.

Beasley, R., Roche, W., Roberts, J.A. and Holgate, S.T. (1989). Cellular events in the bronchi in mild asthma and after bronchial provocation. Am. Rev. Respir. Dis. 139, 806–817.

Becker, K.L., Monaghan, K.G. and Silva, O.L. (1980). Immunocytochemical localization of calcitonin in kulchitsky cells of human lung. Arch. Path. Lab. Med. 104, 196–198.

Belvisi, M.G., Chung, K.F., Jackson, D.M. and Barnes, P.J. (1988). Opioid modulation on noncholinergic neural bronchoconstriction in guinea pig in vivo. Br. J. Pharmacol. 95, 413–418.

Belvisi, M.G., Stretton, C.D., Yacoub, M. and Barnes, P.J. (1992). Nitric oxide is the endogenous neurotransmitter of bronchodilator nerves in humans. Eur. J. Pharmacol. 210, 221–222.

Bennett, M.R., Burnstock, G. and Holman, M.E. (1966). Transmission from perivascular inhibitory nerves to the smooth muscle of the guinea pig taenia coli. J. Physiol. 182, 527–540.

Berkley, H.J. (1893). The intrinsic pulmonary nerves by the silver method. J. Comp. Neurol. 3, 107–111.

Bitensky, L., Chambers, D.J., Chayen, J., Cross, B.A., Guz, A., Jain, F.K. and Johnston, J.J. (1975). Evidence concerning the site of receptors mediating the Hering–Breuer reflex. J. Physiol. 249, 30–31P.

Blumcke, S. (1968). Experimental and morphological studies on the efferent bronchial innervation I. The peribronchial plexus. Beit. Pathol. Anat. Pathol. 137, 239–286.

Boat, T.F., Kleinerman, J.I., Carlson, D.M., Maloney, W.H. and Matthews, L.W. (1974). Human respiratory tract secretions. I. Mucous glycoproteins secreted by cultured nasal polyp epithelium from subjects with allergic rhinitis and cystic fibrosis. Am. Rev. Respir. Dis. 110, 428–441.

Boucher, R.C., Johnson, J., Inoue, S., Hulbert, W. and Hogg, J.C. (1980). The effect of cigarette smoking on the permeability of guinea pig airways. Lab. Invest. 43, 94–100.

Brain, S.D. and Williams, T.J. (1985). Inflammatory oedema induced by synergism between calcitonin generelated peptide (CGRP) and mediators of increased vascular permeability. Br. J. Pharmacol. 86, 855–860.

Brody, J.S., Klempfner, G., Staum, M.M., Vidyasagar, D., Kuhl, D.E. and Waldhausen, J.A. (1972). Mucociliary clearance after lung denervation and bronchial transection. J. Appl. Physiol. 32, 160–164.

Buckner, C.K., Songsiridej, V., Dick, E.C. and Busse, W.W. (1985). In vivo and in vitro studies on the use of the guinea pig as a model for virusprovoked airway hyperreactivity. Am. Rev. Respir. Dis. 132, 305–310.

Burnstock, G. (1969). Evolution of the autonomic innervation of visceral and cardiovascular systems in vertebrates. Pharmacol. Rev. 21, 247–324.

Burnstock, G. (1972). Purinergic nerves. Pharmacol. Rev. 24, 509–581.

Burnstock, G. (1986). The nonadrenergic noncholinergic nervous system. Arch. Int. Pharmacodyn. 280, 1–15.

Burnstock, G., Campbell, G., Bennett, M. and Holman, M.E. (1963). Inhibition of the smooth muscle of the taenia coli. Nature 200, 581–582.

Burnstock, G., Allen, T.G.J. and Hassel, C.J.S. (1987). The electrophysiologic and neurochemical properties of paratracheal neurons in situ and in dissociated cell culture. Am. Rev. Respir. Dis. 136, S23–S26.

Cabezas, G.A., Graf, P.D. and Nadel, J.A. (1971). Sympathetic versus parasympathetic nervous regulation of airways in dogs. J. Appl. Physiol. 31, 651–655.

Cadieux, A., Springall, D.R., Mulderry, P.K., Rodrigo, J., Chatei, M.A., Terenghi, G., Bloom, S.R. and Polak, J.M. (1986). Occurrence, distribution and ontogeny of CGRP immunoreactivity in the rat lower respiratory tract: effect of capsaicin treatment and surgical denervations. Neuroscience 19, 605–627.

Carstairs, J.R. and Barnes, P.J. (1986). Visualisation of VIP receptors in human and guinea pig lung. J. Pharmacol. Exp. Ther. 239, 249–255.

Carstairs, J.R. and Barnes, P.J. (1992). Autoradiographic mapping of substance P receptors in lung. Eur. J. Pharmacol. 127, 295–296.

Carstairs, J.R., Nimmo, A.J. and Barnes, P.J. (1984). Autoradiographic localisation of beta adrenoreceptor in human lung. Eur. J. Pharmacol. 105, 189–190.

Carstairs, J.R., Nimmo, A.J. and Barnes, P.J. (1985). Autoradiographic visualization of beta-adrenoceptor subtypes in human lung. Am. Rev. Respir. Dis. 132, 541–547.

Chesrown, S.E., Venugopalan, C.S., Gold, W.M. and Drazen, J.M. (1980). In vivo demonstration of nonadrenergic inhibitory innervation of the guinea pig trachea. J. Clin. Invest. 65, 314–320.

Cheung, A., Polak, J.M., Bauer, F.E., Cadieux, A., Christofides, N.D., Springall, D.R. and Bloom, S.R. (1985). Distribution of galanin immunoreactivity in the respiratory tract of pig, guinea-pig, rat and dog. Thorax 40, 889–896.

Coburn, R.F. and Tomita, T. (1973). Evidence for nonadrenergic inhibitory nerves in the guinea pig treachealis muscle. Am. J. Physiol. 224, 1072–1080.

Colebatch, H.J.H. and Halmagyi, D.F.J. (1963). Effect of vagotomy and vagal stimulation on lung mechanics and circulation. J. Appl. Physiol. 18, 881–887.

Coleridge, H.M. and Coleridge, J.C.G. (1977). Impulse activity in afferent vagal C-fibres with endings in the intrapulmonary airways of dogs. Respir. Physiol. 29, 125–142.

Coleridge, H.M., Coleridge, J.C.G. and Luck, J.C. (1965). Pulmonary afferent fibres of small diameter stimulated by capsaicin and by hyperinflation of the lungs. J. Physiol. 179, 248–262.

Coles, S.J., Said, S.I. and Reid, L.M. (1981). Inhibition by vasoactive intestinal peptide of glycosugate and lysozyme secretion by human airways in vitro. Am. Rev. Respir. Dis. 124, 531–536.

Coles, S.J., Bhaskar, K.R., O'Sullivan, D.D. and Reid, L.M. (1982). In "Mucus in Health and Disease II" (eds E.N. Chantler and Elder, J.B.), pp 357–360. Plenum Press, New York.

Coles, S.J., Neill, K.H. and Reid, L.M. (1984). Potent stimulation of glycoprotein secretion in canine trachea by substance P. J. Appl. Physiol. 57, 1323–1327.

Cook, R.D. and King, R.S. (1969a). A neurite–receptor complex in the avian lung: electron microscopical observations. Experientia 25, 1162–1164.

Cook, R.D. and King, R.S. (1969b). Nerves of the avian lung: electron microscopy. J. Anat. 105, 202.

Dahlstrom, A., Fuxe, K., Hokfelt, T. and Norberg, K. (1966). Adrenergic innervation of the bronchial muscle of the cat. Acta Physiol. Scand. 66, 507–508.

Daly, M. de B. and Hebb, C. (1966). "Pulmonary and Bronchial Vascular Systems" (Monographs of the Physiological Society). Edward Arnold, London.

Das, R.M., Jeffery, P.K. and Widdicombe, J.G. (1978). The epithelial innervation of the lower respiratory tract of the cat. J. Anat. 126, 123–131.

Das, R.M., Jeffery, P.K. and Widdicombe, J.G. (1979). Experimental degeneration of intraepithelial nerve fibres in cat airway. J. Anat. 128, 259–267.

Dautrebande, L., Lovejoy, F.W. and McCredie, R.M. (1962). New studies on aerosols XVIII. Effects of atropine aerosols on the airway resistance in man. Arch. Int. Pharmacodyn. Ther. 139, 198–211.

Davis, B., Roberts, A.M., Coleridge, H.M. and Coleridge, J.C.G. (1982). Reflex tracheal gland secretion evoked by stimulation on bronchial C-fibres in dogs. J. Appl. Physiol. 53, 1080–1087.

Davis, P.B. and Kaliner, M. (1983). Autonomic nervous system abnormalities in cystic fibrosis. J. Chron. Dis. 36, 269–278.

Dayer, A.M., Demey, J. and Will, J.A. (1985). Localization of somatostatin, bombesin, and serotoninlike immunoreactivity in the lung of the fetal rhesus monkey. Cell Tissue Res. 239, 621–625.

de Troyer, A., Yernault, J.C. and Rodenstein, D. (1979). Effects of vagal blockage on lung mechanisms in normal man. J. Appl. Physiol. 46, 217–226.

Dey, R.D., Hoffpauir, J. and Said, S.I. (1988). Colocalization of vasoactive intestinal peptide in substance P containing nerves in cat bronchi. Neuroscience 24, 275–281.

Dixon, W.E. and Brodie, T.G. (1903). Contributions to the physiology of the lungs. Part one. The bronchial muscles, their innervation and the actions of drugs upon them. J. Physiol. 23, 97–173.

Doran, J.F., Jackson, P., Kynock, P.A. and Thompson, R.J. (1983). Isolation of PGP9.5, a new human neurone-specific protein detected by high resolution twodimensional electrophoresis. J. Neurochem. 40, 1542–1547.

Dusser, D.J., Umeno, E., Graf, P.D., Djokic, T., Borson, D.B. and Nadel, J.A. (1988). Airway neutral endopeptidase-like enzyme modulates tachykinin-induced bronchoconstriction in vivo. J. Appl. Physiol. 65, 2585–2591.

Dusser, D.J., Djokic, T.D., Borson, D.B. and Nadel, J.A. (1989). Cigarette smoke induces bronchoconstrictor hyperresponsiveness to substance P and inactivates airway neutral endopeptidase in the guinea pig lung. Possible role of free radicals. J. Clin. Invest. 84, 900–906.

El-Bermani, A.-W. (1973). Innervation of the rat lung. Acetylcholmesterase-containing nerves of the bronchial tree. Am. J. Anat. 137, 19–29.

El-Bermani, A.-W. (1978). Pulmonary noradrenergic innervation of rat and monkey: a comparative study. Thorax 33, 167–174.

El-Bermani, A.-W. and McCarthy, L.F. (1980). Synaptic specialization of pulmonary parasympathetic ganglia: a three dimensional study. Acta Anat. (Basel) 107, 361–372.

Elftman, A.G. (1943). The afferent and parasympathetic innervation of the lungs and trachea of the dog. Am. J. Anat. 72, 1–27.

Elfvin, L.-G. (1971). Ultrastructural studies on the synaptology of the inferior mesenteric ganglion of the cat. J. Ultrastruct. Res. 37, 432–448.

Ellis, J.L. and Farmer, S.G. (1989). Modulation of cholinergic neurotransmission by vasoactive intestinal peptide and peptide histidine isoleucine in guinea pig tracheal and smooth muscle. Pulm. Pharmacol. 2, 107–112.

Empey, D.W., Laitinen, L.A., Jacobs, L., Gold, W.M. and Nadel, J.A. (1976a). Mechanisms of bronchial hyperreactivity in normal subjects after upper respiratory tract infection. Am. Rev. Respir. Dis. 113, 131–137.

Ericson, L.E., Hakanson, R., Larson, B., Owman, C. and Sundler, F. (1972). Fluorescence and electron microscopy of amine-storing enterochromaffin-like cells in tracheal epithelium of mouse. Zeitsch. Zellforsch. 124, 532–545.

Evans, D.H.L. and Murray, J.G. (1954). Histological and functional studies on the fibre composition of the vagus nerve of the rabbit. J. Anat. 88, 320–337.

Evans, T.W., Dixon, C.M., Clarke, B., Conradson, T.B. and Barnes, P.J. (1988). Comparison of neurokinin A and substance P on cardiovascular and airway function in man. Br. J. Pharmacol. 25, 273–275.

Fahrenkrug, J., Bek, T., Lungberg, J. and Hokfelt, T. (1985). VIP and PHI in cat neurons: colocalization but variable tissue content possible due to differential processing. Regul. Pept. 12, 21–34.

Falk, H.L., Kotin, P. and Tremer, H.M. (1961). Protective effect of parasympatheticcomminetic agents on ciliated mucus-secreting epithelium. J. Nat. Cancer Inst. 27, 1379–1392.

Fewtrell, C.M.S., Foreman, J.C., Jordan, C.C., Oehme, P., Renner, H. and Stewart, J.M. (1982). The effects of substance P on histamine and 5-hydroxytryptamine release in the rat. J. Physiol. 330, 393–411.

Fillenz, M. and Woods, M.J. (1970). In "Breathing: Hering–Breuer Centenary Symposium" (Ciba Foundation Symposium) (ed. R. Porter), pp 101–109, Churchill Livingstone.

Fisher, A.W.F. (1964). The intrinsic innervation of the trachea. J. Anat. 98, 117–124.

Florey, H., Carleton, H.M. and Wells, A.Q. (1932). Mucus secretion in the trachea. Br. J. Exp. Pathol. 13, 269–284.

Floyer, J. (1698). A Treatise of Asthma. Wilkins and Innis, London.

Foreman, J.C. and Jordan, C. (1984). Neurogenic inflammation. Trends Pharmacol. Sci. 5, 116–119.

Foster, W.M., Bergofsky, E.H., Bohning, D.E., Lippman, M. and Albert, R.E. (1976). Effect of adrenergic agents and their mode of action on mucociliary clearance in man. J. Appl. Physiol. 41, 146–152.

Fox, B., Bull, T.B. and Guz, A. (1980). Innervation of alveolar walls in the human lung: an electron microscopic study. J. Anat. 131, 683–692.

Frossard, N. and Barnes, P.J. (1987). μ-Opioid receptors modulate non-cholinergic constrictor nerves in guinea pig airways. Eur. J. Pharmacol. 141, 519–521.

Fryer, A.D. and Jacoby, D.B. (1991). Parainfluenza virus infection damages inhibitory M2-muscarinic receptors on pulmonary parasympathetic nerves in the guinea pig. Br. J. Pharmacol. 102, 267–271.

Gabella, G. (1976). "Structure of the Autonomic Nervous System". Chapman and Hall, London.

Gallagher, J.J., Hall, R.L., Phipps, R.J. and Jeffery, P.K. et al. (1986). Mucus-glycoproteins (mucins) of the cat trachea – characterization and control of secretion. Biochim. Biophys. Acta 886, 243–255.

Gallagher, J.T., Kent, P.W., Passotore, M., Phippa, R.J. and Richardson, P.S. (1975). The composition of tracheal mucus and the nervous control of its secretion in the cat. Proc. R. Soc. Lond. B 192, 49–76.

Gamse, R. and Saria., A. (1985). Potentiation of tachykinin-induced plasma protein extravasation by calcitonin gene-related peptide. Eur. J. Pharmacol. 114, 61–66.

Gaylor, J.B. (1934). The intrinsic nervous mechanism of the human lung. Brain 57, 143–160.

German, V.F., Ueki, I.E. and Nadel, J.A. (1980). Micropipette measurement of airway submucosal gland secretion: laryngeal reflex. Am. Rev. Respir. Dis. 122, 413–416.

Godfrey, R.W.A., Severs, N.J. and Jeffery, P.K. (1992). Freeze–fracture morphology and quantification of human bronchial epithelial tight junctions. Am. J. Cell. Mol. Biol. 6, 453–458.

Goetzl, E.J., Chernov, T., Renold, F. and Payan, D.G. (1985). Neuropeptide regulation of the expression of immediate hypersentivity. J. Immunol. 135, 802s–805s.

Gold, W.M., Nadel, J.A. and Boushey, H.A. (1978). Bronchial hyperirritability in healthy subjects after exposure to ozone. Am. Rev. Respir. Dis. 118, 287–294.

Goldie, R.G., Paterson, J.W. and Wale, J.L. (1982). A comparative study of beta-adrenoreceptors in human and porcine lung parenchyma strip. Br. J. Pharmacol. 76, 523–526.

Goldie, R.G., Spina, D., Henry, P.J., Lulich, K.M. and Paterson, J.W. (1986). In vitro responsiveness of asthmatic and non-diseased bronchus to carbachol, histamine, beta-adrenoceptor agonists and theophylline. Br. J. Clin. Pharmacol., 22, 669–676.

Goldie, R.G., Spina, D., Rigby, P.J. and Paterson, J.W. (1987). Alpha and beta-adrenoceptors in human non-diseased and asthmatic lung. American Thoracic Society, New Orleans, Louisiana, United States of America. 9th–13th May. Am. Rev. Respir. Dis., 135, A90.

Gosselin, R.E. (1966). Physiologic regulators of ciliary motion. Am. Rev. Respir. Dis. 93, 41–60.

Hage, E. (1974). Histochemistry and fine structure of endocrine cells in fetal lungs, of the rabbit, mouse and guinea pig. Cell Tissue Res. 149, 513–524.

Hartung, H.P. and Toyka, K.V. (1988). Activation of macro-phages by substance P: induction of oxidative burst and thromboxane release. Eur. J. Phamacol. 89, 310–305.

Hayashi, S. (1937). Mikroskopische Studieu zur Innervation der Lungs. J. Orient. Med. 27, 37–79.

Heinz-Erian, P., Rey, R., Flux, M. and Said, S.I. (1985). Deficient vasoactive intestinal peptide innervation in the sweat gland of CF patients. Science, 229, 1407–1408.

Hirsch, E.F., Kaiser, A.C., Borner, H.B., Cooper, T. and Rains, J.J. (1968a). Innervation of the mammalian lung I – the afferent receptors. Arch. Pathol. (Chicago) 85, 51–61.

Hirsch, E.F., Kaiser, G.C., Borner, H.B., Nigro, S.L., Hamonda, F., Cooper, T. and Adams, W.E. (1968b). The innervation of the mammalian lung III. Regression of the intrinsic nerves and of their afferent receptors following thoracic sympathectomy, cervical vagotomy or thoracic stripping of the vagus. Arch. Surg. (Chicago) 96, 149–155.

Holmes, R. and Torrance, R.W. (1959). Afferent fibres of the stellate ganglion. Q. J. Exp. Physiol. 44, 271–280.

Holzer, P. (1988). Local effector functions of capsaicin-sensitive sensory nerve endings: involvement of tachykinins, calcitonin gene related peptide and other neuropeptides. Neuroscience 24, 739–768.

Honjin, R. (1956). On the nerve supply of the lung of the mouse with special reference to the structure of the peripheral vegatative nervous system. J. Comp. Neurol., 105, 587–625.

Hoppin, F.C. Jr., Green, M. and Morgan, M.S. (1978). Relationships of central and peripheral airway resistance to lung volume in dogs. J. Appl. Physiol. 4, 728–737.

Howorth, P.H., Djukanovic, R., Wilson, J.W., Hogabe, S.T., Springall, D.R. and Polak, J.M. (1993). Neuropeptide-containing nerves in endobronchial biopsies from asthmatic and nonasthmatic subjects. Am. J. Resp. Cell. Mol. Biol. (in press).

Hulbert, W.C., Walker, D.C., Jackson, A. and Hogg, J.C. (1981). Airway permeability to horseradish peroxidase in guinea pigs: the repair phase after injury by cigarette smoke. Am. Rev. Respir. Dis. 123, 320–326.

Hung, K.-S., Hertweck, M.S., Hardy, J.D. and Loosli, C.G. (1973). Ultrastructure of nerves and asssociated cells in the bronchiolar epithelium of mouse lung. J. Ultrastruct. Res. 43, 426–437.

Hung, K.S. (1976). Fine structure of tracheobronchial epithelial nerves of the cat. Anat. Rec. 185, 85–91.

Iravani, J. and Melville, G.N. (1975). Effects of drugs and environmental factors on ciliary movement [in German]. Respiration, 32, 157–164.

Jabonero, V. and Sabadell, J. (1972). The sensory innervation of the smooth musculature of the respiratory tract. Z. Mikrosk. Anat. Forsch. 86, 213–243.

Jacobowitz, D., Kent, K.M., Fleisch, J.H. and Cooper, T. (1973). Histofluorescent study of catecholamine-containing elements in cholinergic ganglia from the calf and dog lung. Proc. Soc. Exp. Biol. Med. 144, 464–466.

Jeffery, P.K. (1978). In "Respiratory Tract Mucus, 56th CIBA Foundation Symposium" (ed. R. Porter), pp 5–24. Elsevier/Excerpta Medica, Amsterdam.

Jeffery, P.K. (1982). In "The Lung in its Environment Ettore Majorana" (Life Science Series, Vol. 6) (eds G. Cumming and G. Bonsignore.), pp 57–80. Plenum Press, New York.

Jeffery, P.K. (1986). In "Asthma: Clinical Pharmacology and Therapeutic Progress" (ed. A.B. Kat), pp 376–392. Blackwell, Oxford.

Jeffery, P.K. (1990). In "Respiratory Medicine" (eds R.A.L.

Brewis, G.J. Gibson and G.M. Geddes), pp 57–78. Baillière Tindall, London.

Jeffery, P.K. and Reid, L. (1973a). Intraepithelial nerves in normal rat airways: a quantitative electron microscopic study. J. Anat. 114, 33–45.

Jeffery, P.K. and Reid, L. (1973b). The ultrastructure of normal large bronchi. Bronches 23, 368–380.

Jeffery, P.K., Wardlaw, A.J., Nelson, F.C., Collins, J.V. and Kay, A.B. (1989). Bronchial biopsies in asthma. An ultrastructural, quantitative study and correlation with hyperreactivity. Am. Rev. Respir. Dis. 140, 1745–1753.

Johnson, J. (1935). Effect of superior laryngeal nerves on tracheal mucus. Ann. Surg. 101, 494–499.

Joos, G., Pauwels, R. and van der Straeten, M.E. (1987). Effect of inhaled substance P and neurokinin A in the airways of normal and asthmatic subjects. Thorax 42, 779–783.

Karczewski, W. and Widdicombe, J.G. (1969). The effect of vagotomy, vagal cooling and efferent vagal stimulation on breathing and lung mechanics of rabbits. J. Physiol. 201, 259–270.

King, A.S., McLelland, J., Cook, R.D., King, D.Z. and Walsh, C. (1974). The ultrastructure of afferent nerve endings in the avian lung. Resp. Physiol. 22, 21–40.

Knight, D.S. (1980). A light and electron microscopic study of feline intrapulmonary ganglia. J. Anat. 131, 413–428.

Kroegel, C., Giembycz, M.A. and Barnes, P.J. (1990). Characterization of eosinophil activation by peptides. Differential effects of substance P, melittin, and f-Met-Leu-Phe. J. Immunol 145, 1581–1587.

Kostreva, D.R., Zuperku, E.J., Hess, G.L., Coon, R.L. and Kampine, J. (1975). Pulmonary afferent activity recorded from sympathetic nerves. J. Appl. Physiol. 39, 37–40.

Kuo, H.-P., Rohde, J.A.L., Tokuyama, K., Barnes, P.J. and Rogers, D.F. (1990). Capsaicin and sensory neuropeptide stimulation of goblet cell secretion in guinea-pig trachea. J. Physiol. 431, 629641.

Lacy, M.G. (1980). Pulmonary afferent innervation in rat demonstrated by intra-axonal diffusion of cobalt. Proc. Phys. Soc. 63P

Laitinen, A. (1985). Ultrastructural organization of intraepithelial nerves in the human airway tract. Thorax 40, 488–492.

Laitinen, L.A., Laitinen, A., Panula, P.A., Partanen, M., Tervo, K. and Tervo, T. (1983). Immunohistochemical demonstration of substance P in the lower respiratory tract of the rabbit and not of man. Thorax 38, 531–536.

Laitinen, L.A., Heino, M., Laitinen, A., Kava, T. and Haahtela, T. (1985). Damage of the airway epithelium and bronchial reactivity in patients with asthma. Am. Rev. Respir. Dis. 131, 599–606.

Laitinen, L.A., Laitinen, A.A., Salonen, R.D. and Widdicombe, J.G. (1987a). Vascular actions of airway neuropeptides. Am. Rev. Respir. Dis. 136, 559–564.

Laitinen, L.A., Laitinen, A. and Widdicombe, J. (1987b). Effects of inflammatory and other mediators on airway vascular beds. Am. Rev. Respir. Dis. 135, S67–S70.

Lammers, J.W., Minette, P., McCuster, M.T., Chung, K.F. and Barnes, P.J. (1989). Capsaicin-induced bronchodilatation in mild asthmatic subjects: possible role of nonadrenergic inhibitory system. J. Appl. Physiol. 67, 856–861.

Landois, L. (1885). "A Textbook of Human Physiology", Vol. 1. Charles Griffin, London.

Langley, J.N. (1898). On the inhibitory fibres in the vagus to the end of the oesophagus and stomach. J. Physiol. 23, 407–414.

Larsell, O. (1921). Nerve terminations in the lung of the rabbit. J. Comp. Neurol. 33, 105–131.

Larsell, O. (1922). The ganglia, plexuses, and nerve-terminations of the mammalian lung and pleura pulmonalis. J. Comp. Neurol. 35, 97–132.

Larsell, O. and Dow, R. S. (1933). The innervation of the human lung. Am. J. Anat. 52, 125–146.

Laurenzi, G.A. (1973). The mucociliary stream. J. Occup. Med. 15, 175–176.

Lauweryns, J.M. and Cokelaere, M. (1973). Intrapulmonary neuroepithelial bodies hypoxia-sensitive neuro(chemo-) receptors. Experientia 29, 1384–1386.

Lauweryns, J.M. and Goddeeris, P. (1975). Neuroepithelial bodies in the human child and adult lung. Am. Rev. Respir. Dis. 111, 469–476.

Lauweryns, J.M., Peuskens, J.C. and Cokelaere, M. (1970). Argyrophil, fluorescent and granulated (peptide and amine producing?) AFG cells in human infant bronchial mucosa. Light and electron microscopic studies. Life Sci. 9, 1417–1429.

Lauweryns, J.M., Cokelaere, M. and Theunynck, P. (1972). Neuroepithelial bodies in the respiratory mucosa of various mammals. Zeitsch. Zellforsch. Mikrosc. Anat. 135, 569–592.

Lauweryns, J.M., de Bock, V., Verhofstad, A.A.J. and Steinbusch, H.W.M. (1982). Immunohistochemical localization of serotonin in intrapulmonary neuroepithelial bodies. Cell Tissue Res. 226, 215–223.

Lee, L.Y., Djokic, T.D. and Dumont, C. (1980). Mechanisms of ozone-induced tachypneic response to hypoxia in conscious dogs. J. Appl. Physiol. 48, 163–168.

Lindh, B., Lundberg, J.M. and Hokfelt, T. (1989). NPY-, galanin-, VIP/PHI-, CGRP- and Substance P-immunoreactive neuronal populations in cat autonomic and sensory ganglia and their projections. Cell. Tiss. Res. 259–273.

Lommel, F. (1908). Zur Physiologie und Pathologie desfliunuerepithelo der Atmungsorgane. Dtsch. Arch. Clin. Med. 94, 365.

Loosli, C.G. and Hung, K.-S. (1977). In "Lung Biology in Health and Disease", Vol. 6 (ed. W.A. Hodson), pp 269–306. Marcel Dekker, New York.

Lotz, M., Vaughn, J.H. and Carson, D.M. (1988). Effect of neuropeptides on production of inflammatory cytokines by human monocytes. Science 241, 1218–1221.

Lotvall, J.O., Lemen, R.J., Hui K.P., Barnes, P.J. and Chung, K.F. (1990). Airflow obstruction after substance P aerosol: contribution of airway and pulmonary edema. J. Appl. Physiol. 69, 1473–1478.

Lundberg, J.M., Sand, A. (1982b). Capsaicin-sensitive vagal neurons involved in control of vascular permeability in rat trachea. Acta Physiol. Scand. 115, 521–523.

Lundberg, J.M., Anggard, A., Emson, P., Fahrenkrug, J. and Hokfelt, T. (1981). Vasoactive intestinal polypeptide and cholinergic mechanisms in cat nasal mucosa: studies on choline acetyltransferase and release of vasoactive intestinal polypeptide. Proc. Natl Acad. USA 78, 5255–5259.

Lundberg, J.M., Brodin, E. and Saria, A. (1983a). Effects and distribution of vagal capsaicin-sensitive substance P neurons with special reference to the trachea and lungs. Acta Physiol. Scand. 119, 243–252.

Lundberg, J.M., Franco-cereceda, A., Hua, X., Hokfelt, T. and Fishcher, J.A. (1985a). Coexistence of substance P and calcitonin gene related peptide like immunoreactivities in sensory nerves in relation to cardiovascular and bronchoconstrictor effects of capsaicin. Eur. J. Pharmacol. 108, 315–319.

Lundberg, J.M., Hokfelt, T., Martling, C.-R., Saria, A. and Cuello, C. (1984a). Substance P-immunoreactive sensory nerves in the lower respiratory tract of various mammals including man. Cell Tissue Res. 235, 251–261.

Lundberg, J.M., Hokfelt, T., Nilsson, G., Terenius, L., Rehfeld, J., Elde, R. and Said, S. (1978). Peptide neurons in the vagus, splanchnic and sciatic nerves. Acta Physiol. Scand. 104, 499–501.

Lundberg, J.M., Lundblad, C., Martling, C., Saria, A., St Jarne, P. and Anggard, A. (1987). Cooexistence of multiple peptides and classic transmitters in airway neurons: functional and pathophysiologic aspects. Am. Rev. Respir. Dis. 136, S16–S22.

Lundberg, J.M., Lundbland, L., Saria, A. and Anggard, A. (1984b). Inhibition of cigarette smoke-induced oedema in the nasal mucosa by capsaicin pretreatment and a substance P antagonist. Naunyn-Schmiedeb. Arch. Pharmacol. 326, 181–185.

Lundberg, J.M., Martling, C.-R. and Saria, A. (1983b). Substance P and capsaicin-induced contraction of human bronchi. Acta Physiol. Scand. 119, 49–53.

Lundberg, J.M. and Saria, A. (1983). Capsaicin-induced desensitization of the airway mucosa to cigarette smoke, mechanical and clinical irritants. Nature 302, 251–253.

Lundberg, J.M., Saria, A., Brodin, E., Rosell, S. and Folkers, K. (1983c). A substance P antagonist inhibits vagally induced increase in vascular permeability and bronchial smooth muscle contraction in the guinea pig. Proc. Natl Acad. Sci. USA 80, 1120–1124.

Lundberg, J.M., Saria, A., Theodorsson-Norheim, E., Brodin, E., Hua, X.-Y., Martling, C.-R., Gamse, R. and Hokfelt, T. (1985b). In "Tachykinin Antagonists" (eds R. Hakanson and F. Sundler), pp 159–169. Elsevier, Amsterdam.

Lundbland, L. and Lundberg, J.M. (1984). Capsaicin sensitive sensory neurons mediate the response to nasal irritation induced by the vapour phase of cigarette smoke. Toxicology 33, 1–7.

Lung, M.A., Wang, J.C.C. and Cheng, K.K. (1976). Bronchial circulation: an autoperfusion method for assessing its vaso-motor activity and the study of alpha-adrenoreceptors in the bronchial artery. Life Sci. 19, 577–580.

Luts, A. and Sundler, F. (1989). Peptide containing nerve fibers in the respiratory tract of the ferret. Cell. Tiss. Res; 258: 259–267.

McDonald, D.M. (1987). Neurogenic inflammation in the respiratory tract: actions of sensory nerve mediators on blood vessels and epithelium of the airway mucosa. Am. Rev. Respir. Dis.. 136, S65–S71.

McDonald, D.M. (1988). Neurogenic inflammation in the rat trachea I. Changes in venules, leucocytes and epithelial cells. J Neurocytol. 17, 583–603.

Mak, J.C.W. and Barnes, P.J. (1989). Muscarinic receptor subtypes in human and guineapig lung. Eur. J. Pharmacol. 164, 223–230.

Mak, J.C.W. and Barnes, P.J. (1990). Autoradiographic visual-ization of muscarinic receptor subtypes in human and guinea pig lung. Am. Rev. Respir. Dis. 141, 1559–1568.

Mak, J.C.W., Baraniuk, J. and Barnes, P.J. (1992). Localization of muscarinic receptor subtype messenger RNAs in human lung. Am. J. Respir. Cell. Mol. Biol. 7, 344–348.

McLelland, J. (1969). Observations with the light microscope on the ganglia and nerve plexuses of the intrapulmonary bronchi of the bird. J. Anat. 105, 202.

Mann, S.P. (1971). The innervation of mammalian bronchial smooth muscle: the localization of catecholamines and cholinesteroses. Histochem. J., 3, 319–331.

Marasco, W.A., Showell, H.J. and Becker, E.L. (1981). Substance P binds to the formylpeptide chemotaxis receptor on the rabbit neutrophil. Bioch. Biophys. Res. Commun; 99: 1065–1072.

Martling, C.R., Theodorsson-Norheim, E. and Lundberg, J.M. (1987). Occurence and effects of multiple tachykinins: substance P, neurokinin A, and neuropeptide K in human lower airways. Life Sci 1987; 40: 1633–1634.

Massaro, G., Paris, M. and Thet, L. (1979). *In vivo* regulation of secretion of bronchiolar Clara cells in rats. J. Clin. Invest. 63, 167–172.

Massaro, G., Fischmann, C., Chiang, M.-J., Amado, C. and Masoro, D. (1981). Regulation of secretion in Clara cells. Studies using isolated ventilated perfused lung. J. Clin. Invest. 67, 345–351.

Matran, R., Alving, K., Martling, C.R., Lacroix, J.S. and Lundberg, J.M. (1989). Effects of neuropeptides and capsaicin on tracheobronchial blood flow in the pig. Acta Physiol. Scand. 135, 335–342.

Matsuzaki, Y., Hamasaki, Y. and Said, S.I. (1980). Vasoactive intestinal polypeptide: a possible transmitter of nonadrenergic relaxation of guinea-pig airways. Science 210, 1252–1253.

Matthews, M.R. and Raisman, G. (1969). The ultrastructure and somatic efferent synopses of small granule-containing cells in the superior cervical ganglion. J. Anat. 105, 255–282.

Mei, N., Condamin, M. and Boyer, A. (1980). The composition of the vagus nerve of the cat. Cell Tissue Res. 209, 423–431.

Meyrick, B. and Reid, L. (1971). Nerves in rat intra-acinar alveoli: an electron microscopic study. Respir. Physiol. 11, 367–377.

Mills, J.E., Sellick, H. and Widdicombe, J. A. (1969). Activity of lung irritant receptors in pulmonary microembolism anaphylaxis and drug-induced bronchoconstrictions. J. Physiol. (Lond.) 203, 337–357.

Moosavi, H., Smith, P. and Heath, D. (1973). The Feyrter cell in hypoxia. Thorax 28, 729–741.

Morgan, M., Pack, R.J. and Howe, A. (1976). Structure of cells and nerve endings in abdominal vagal paraganglia of the rat. Cell Tissue Res. 169, 467–484.

Mortola, J.P.G., Sant'ambrogio, G. and Clement, M.G. (1975). Localization of irritant receptors in the airways of the dog. Respir. Physiol. 24, 107–114.

Murlas, C., Nadel, J.A. and Basbaum, C.B. (1980). A morpho-metric analysis of the autonomic innervation of cat tracheal glands. J.Autonom. Nervous Syst. 2, 23–37.

Nadel, J.A. (1980). In "Physiology and Pharmacology of the airways", Vol. 15 (ed. C. Lenfant), pp 217–257. Marcel Dekker, New York.

Nadel, J.A. and Widdicombe, J.G. (1962). Reflex effects of upper airway irritation on total lung resistance and blood pressure. J. Appl. Physiol. 17, 861–865.

Nadel, J.A. and Widdicombe, J.G. (1963). Reflex control of airway size. Ann. N.Y. Acad. Sci. 109, 712–722.

Nagaishi, C. (1972). "Functional Anatomy and Histology of the Lung." University Park Press, Baltimore.

Nathanson, I., Widdicombe, J.H. and Barnes, P.J. (1983). Effect of vasoactive intestinal peptide on ion transport across dog treacheal epithelium. J. Appl. Physiol. 55, 1844–1848.

Nong, Y.H., Titus, R.G., Riberio, J.M. and Remold, H.G.

(1989). Peptides encoded by the calcitonin gene inhibit macrophage function. J. Immunol. 143, 45–49.

O'Dorisio, M.S., Shannaon, B.T., Fleshman, D.J. and Campolito, L.B. (1989). Identification of high affinity receptors for vasoactive intestinal peptide on human lymphocytes of B cell lineage. J. Immunol. 142, 3533–3536.

Ollerenshaw, S., Jarvis, D., Woolcock, A., Sullivan, C. and Scheiber, T. (1989). Absence of immunoreactive vasoactive intestinal polypeptide in tissue from the lungs of patients with asthma. N. Engl. J. Med. 320, 1244–1248.

Ollerenshaw, S.L., Jarvis, D., Sullivan, C.E. and Woolcock, A.J. (1991). Substance P immunoreactive nerves in airways from asthmatics and nonasthmatics. Eur. Respir. J. 4, 673–682.

Olsen, C.R., Colebatch, H.J.H., Mebel, P.E., Nadal, J.A. and Staub, N.C. (1965). Motor control of pulmonary airways studied by nerve stimulation. J. Appl. Physiol. 20, 202–208.

Oosaki, T. (1970). A granular vesicle-containing ganglion cell in the Auerbach's plexus of the rat small intestine. Fukushima J. Med. Sci. 17, 41–50.

Pack, R.J. and Richardson, P.S. (1984). The aminergic innervation of the human bronchus: a light and electron microscopic study. J. Anat. 138, 493–502.

Paintal, A.S. (1969). Mechanisms of stimulation of type J pulmonary receptors. J. Physiol. 203, 511–532.

Palmer, J.B.D., Cuss, F.M.C. and Barnes, P.J. (1986). VIP and PHM and their role in nonadrenergic inhibitory responses in isolated human airways. J. Appl. Physiol. 61, 1322–1328.

Payan, D.G., Brewster, D.R. and Goetzl, E.J. (1983). Specific stimulation of human T-lymphocytes by substance P. J Immunol 131, 1613–1615.

Payan, D.G., Brewster, D.R., Missirian-Bastian, A. and Goetzl, E.S. (1984). Substance P recognition by a subset of human T lymphocytes. J. Clin. Invest. 74, 1532–1539.

Pearse, A.G.E. (1969). The cytochemistry and ultrastructure of polypeptide hormone-producing cells of the APUD series and the embryololgic, physiologic and pathologic implications of the concept. J. Histochem. Cytochem. 17, 303–313.

Peatfield, A.C. and Richardson, P.S. (1983). Evidence for noncholinergic, nonadrenergic nervous control of mucus secretion into the cat trachea. J. Physiol. (Lond.) 342, 335–345.

Persson, C.G.A., Erjefalt, I. and Karlsson, J.-A. (1985). In "Tachykinin Antagonists" (eds R. Hakanson and F. Sundler), pp 171–179. Elsevier, Amsterdam.

Phipps, R.J. and Richardson, P.S. (1976). The effects of irritation at various levels of the airway upon tracheal mucus secretion in the cat. J. Physiol. 261, 563–581.

Phipps, R.J., Richardson, P.S., Corfield, A., Gallagher, J.T., Jeffery, P.K., Kent, P.W. and Passatore, M. (1977). A physiological biochemical and histological study of goose tracheal mucin and its secretion. Phil. Trans. R. Soc. (Lond.) B 279, 513–543.

Phipps, R.J., Williams, I.P., Pack, R.J. and Richardson, P.S. (1982a). Adrenergic stimulation of mucus secretion in human bronchi. Chest 81, 19S.

Phipps, R.J., Williams, I.P., Richardson, P.S., Pell, J., Pack, R.J. and Wright, N. (1982b). Sympathomimetic drugs stimulate the output of secretory glycoproteins from human bronchi in vitro. Clin. Sci. 63, 23–28.

Polak, J.M. and Bloom, S.R. (1979). The diffuse neuroendocrine system. J. Histochem. Cytochem. 27, 1398–1400.

Polak, J.M. and Bloom, S.R. (1984). Regulatory peptides: the distribution of two newly discovered peptides: PHI and NPY. Peptides, 5, 79–89.

Raaijmakers, J.A.M., van Rozen, A.J. and Terpstra, G.K. (1985). Autonomic receptor deficiencies in the pathogensis of chronic obstructive lung disease. Prog. Resp. Res. 19, 153–158.

Rhodin, J. (1966). The ciliated cell. Ultrastructure and function of the human tracheal mucosa. Am. Rev. Respir. Dis. 93, 1–15.

Richardson, J.B. (1979). Nerve supply to the lungs. Am. Rev. Respir. Dis. 119, 785–802.

Richardson, J.B. and Beland, J. (1976). Nonadrenergic inhibitory nerves in human airways. J. Appl. Physiol. 41, 764–771.

Richardson, J.B. and Bouchard, T. (1973). Demonstration of a nonadrenergic inhibitory nervous system in the trachea of the guinea pig. J. Allergy Clin. Immunol. 56, 473–480.

Richardson, J.B. and Ferguson, C.C. (1976). The fine structure of the ganglia in human lung. J. Cell Biol. 70, 48.

Richardson, J.B. and Ferguson, C.C. (1980). In "Lung Biology in Health and Disease" (ed. J.A. NADEL) pp 1–30.

Richardson, P.S. and Peatfield, A.C. (1987). The control of airway mucus secretion. Eur. J. Respir. Dis. 153, 43–51.

Richardson, P.S. and Phipps, R.J. (1978). The anatomy, physiology, pharmacology and pathology of tracheobronchial mucus secretion and the use of expectorant drugs in human disease. Pharmacol. Ther. (B) 3, 441–479.

Robinson, P.M., McLean, J.R. and Burnstock, G. (1971). Ultrastructural identification of nonadrenergic inhibitory nerve fibres. J. Pharmacol. Exp. Ther. 179, 149–160.

Rogers, D.F. and Barnes, P.J. (1989). Opioid inhibition of neurally mediated mucus secretion in human bronchi. Lancet i, 930–932.

Rogers, D.F., Belvisi, M.G., Aursudkij, B., Evans, T.W. and Barnes, P.J. (1988). Effects and interactions of sensory neuropeptides on airway microvascular leakage in guinea-pigs. Br. J. Pharmacol. 95, 1109–1116.

Rogers, D.F., Aursudkij, B. and Barnes, P.J. (1989). Effect of tachykinins on mucus secretion in human bronchi in vitro. Eur. J. Pharmacol. 174, 283–286.

Roy, C.S. and Brown, G. (1885). On bronchial contraction. J. Physiol. 6, 21–25.

Ruff, M.R., Wahl, S.M. and Goetzl, E.J. (1985). Substance P receptor-mediated chemotaxis of human monocytes. Peptides; 6: 107–111.

Said, S.I. (1980). "Neuroendocrinology". Raven Press, New York.

Said, S.I. (1982). Vasoactive peptides in the lung, with special reference to vasoactive intestinal peptide. Exp. Lung Res. 3, 343–348.

Said, S.I. (1987). Influence of neuropeptides on airway smooth muscle. Am. Rev. Respir. Dis. 136, S52–S58.

Salter, H.H. (1968). "On Asthma: Its Pathology and Treatment". Churchill, London.

Sampson, S.R. and Vidruk, E.H. (1975). Properties of "irritant" receptors in canine lung. Respir. Physiol. 25, 9–22.

Saria, A., Lundberg, J.M., Skofitsch, G. and Lembeck, F. (1983). Vascular protein leakage in various tissues induced by substance P, capsaicin, bradykinin, serotonin, histamine and by antigen challenge. Naunyn-Schmiedeb. Archs. Pharmacol. 324, 212–218.

Saria, A., Martling, C.R., Yan, Z., Theodorsson-Norheim, E., Gamse, R. and Lundberg, J.M. (1988). Release of multiple tachykinins from capsaicin-sensitive nerves in the lung by bradykinin, histamine, dimethylphenylpiperainium, and vagal nerve stimulation. Am. Rev. Respir. Dis. 137, 1330–1335.

Sekizawa, K., Tamaoki, J., Graf, P. and Nadel, J.A. (1988).

Modulation of cholinergic transmission by vasoactive intestinal peptide in ferret trachea. J. Appl. Physiol. 69, 2433–2437.

Sellick, H. and Widdicombe, J.G. (1971). Stimulation of lung irritant receptors by cigarette smoke, carbon dust and histamine aerosols. J. Appl. Physiol. 31, 15.

Severinghaus, J.W. and Stupfel, M. (1955). Respiratory dead space increase following atropine in man, and atropine, vagal or ganglionic blockade and hypothermia in dogs. J. Appl. Physiol. 8, 81–87.

Sharma, R. and Jeffery, P.K. (1990a). Airway β-adrenoceptor number in cystic fibrosis and asthma. Clin. Sci. 78, 409–417.

Sharma, R.K. and Jeffery, P.K. (1990b). Airway V.I.P. receptor number is reduced in cystic fibrosis but not asthma. Am. Rev. Respir. Dis. 141, A726.

Sheppard, M.N. and Polak, J.M. (1986). In "Asthma: Clinical Pharmacology and Therapeutic Progress" (ed. A.B. Kay), pp 73–92. Blackwell, London.

Sheppard, M.N., Kurian, S.S., Henzen-Logmans, S.C., Michetti, F., Cocchia, D., Cole, P., Rush, R.A., Marangos, P.J., Bloom, S.R. and Polak, J.M. (1983). Neurone-specific enolase and S100: new markers for delineating the innervation of the respiratory tract in man and other mammals. Thorax 38, 333–340.

Sheppard, M.N., Polak, J.M., Allen, J.M. and Bloom, S.R. (1984). Neuropeptide tyrosine (NPY): a newly discovered peptide is present in the mammalian respiratory tract. Thorax 39, 326–330.

Shimura, S., Sasaki, T., Ekeda, K., Sasaki, H. and Takashima, T. (1988). VIP augments cholinergic-induced glyconjugate secretion in tracheal submucosal glands. J. Appl. Physiol. 65, 2537–2544.

Shipperbottom, C.A. (1988). Histochemical studies of the autonomic innervation of 'normal' and diseased human lung. M.Phil. Thesis University of London.

Shipperbottom, C.A., Jeffery, P.K. and Jones, C.J. (1987). Preliminary studies on pulmonary innervation and alterations in cystic fibrosis. J. Pathol. 151, 61A

Silva, D.G. and Ross, G. (1974). Ultrastructural and fluorescence histochemical studies on the innervation of the tracheo-bronchial muscle of normal cats and cats treated with 6-hydroxydopamine. J. Ultrastruct. Res. 47, 310–328.

Skofitsch, G., Donnerer, J., Petronijevic, S. and Lembeck, F. (1983). Release of histamine by neuropeptides from the perfused rst hindquarter. Naunyn Schmiedeb. Arch Pharmacol. 322, 153–157.

Skoogh, B.-E. (1983). Transmission through airway ganglia. Eur. J. Respir. Dis. 64(suppl 131), 159–170.

Skoogh, B.-E. (1986). Parasympathetic ganglia in the airways. Bull. Eur. Physiopathol. Respir. 22(Suppl.) 7, 143–147.

Sleigh, H.A. (1974). In "Cilia and Flagella" Pergamon Press, Oxford.

Specian, R.D. and Neutra, M.R. (1971). Mechanism of rapid mucus secretion in goblet cells stimulated by acetycholine. Cell Biol. 85, 626–640.

Spencer, H. and Leof, D. (1964). The innervation of the human lung. J. Anat. 98, 599–609.

Stanisz, A.M., Befus, D. and Bienenstock, J. (1986). Differential effects of vasoactive intestine peptide, substance P and somatostatin on immunoglobulin synthesis and proliferation by lymocytes from Peyer's patches, mesenteric lymph nodes, and spleen. J. Immunol. 136, 152–156.

Stanisz, A., Sicchitano, R., Stead, R., Matsuda, H., Tomioka, M., Denburg, J., and Bienenstock, J. (1987). Neuropeptides and immunity. Am. Rev. Respir. Dis. 136, S48–S51.

Stead, R.H., Tomioka, M., Quinonez, G., Simon, G.T., Felten, S.Y. and Bienenstock, J. (1987). Intestinal mucosal mast cells in normal and nematode-infected rat intestines are in intimate contact with peptidergic nerves. Proc. Natl. Acad. Sci. USA 84, 2975–2979.

Sterling, G.M. and Batten, J.C. (1969). Effect of aerosol propellants and surfactants on airway resistance. Thorax 24, 228–231.

Stretton, C.D., Belvisi, M.G. and Barnes, P.J. (1991). Modulation of neural bronchoconstrictor responses in the guinea pig respiratory tract by vasoactive intestinal peptide. Neuropeptides 18, 37–44.

Sturgess, J. and Reid, L. (1972). The organ culture study of the effects of drugs on the secretory activity of the humanbronchial submucosal gland. Clin. Sci. 43, 533–543.

Uddman, R. and Sundler, F. (1987). Neuropeptides in the airways. Am. Rev. Respir. Dis. 136, S3–S8.

Uddman, R., Malm, L. and Sundler, F. (1980). The origin of vasoactive polypeptide (VIP) nerves in the feline nasal mucosa. Acta Otolaryngol. 89, 152–156.

Ueki, I., Germam, V.F. and Nadel, J.A. (1980). Micropipette measurements of airway submucosal gland secretion. Am. Rev. Respir. Dis. 121, 351–357.

Umeda, Y. and Arisawa, H. (1989). Characterization of the calcitonin gene related peptide receptor in mouse T-lymphocytes. Neuropeptides, 14, 237–242.

Undem, B.J., Dick, E.C. and Buckner, C.K. (1983). Inhibition by vasoactive intestinal peptide of antigen-induced histamine release from guinea pig minced lung. Eur. J. Pharmacol. 88, 247–250.

Vincent, N.J., Knudson, R., Leith, D.E., Macklem, P.T. and Mead, J. (1970). Factors influencing pulmonary resistance. J. Appl. Physiol. 29, 236–243.

von During, M., Andres, K.H. and Iravani, J. (1974). The fine structure of the pulmonary stretch receptor in the rat. Z. Anat. Entwicklungsgesch. 143, 215–222.

Walsh, C. and McLelland, J. (1974). Intraepithelial axons in the avian trachea. Z. Zellforsch. Mibrosk. Anat. 147, 209.

Wardell, J.R., Chabrin, L.W. and Payne, B.J. (1970). The canine tracheal pouch: a model for use in respiratory mucus research. Am. Rev. Respir. Dis. 101, 741–754.

Webber, S.E. (1989). Receptors mediating the effect of substance P and neurokinin A on mucous secretion and smooth muscle tone of the ferret trachea: potentiation by enkephalinase inhibition. Br. J. Pharmacol. 98, 1197–1206.

Wharton, J., Polak, J.M., Bloom, S.R. Ghatel, M.A., Solcia, E., Brown, M.R. and Pearse W.G. (1978). Bombesinlike immuno-reactivity in the lung. Nature 273, 769–770.

Wharton, J., Polak, J.M., Cole, G.A., Marangos, P.J. and Pearse, A.G.E. (1981). Neuron-specific enolase as an immunocyto-chemical marker for the diffuse neuroendocrine system in human fetal lung. J. Histochem. Cytochem. 29, 1359–1364.

Widdicombe, J.G. (1954a). Receptors in the trachea and bronchi of the cat. J Physiol. 123, 71–104.

Widdicombe, J.G. (1954b). Respiratory reflexes from the trachea and bronchi of the cat. J. Physiol. 123, 55–90.

Widdicombe, J.G. (1964). In "Handbook of Physiology. American Physiological Society" (eds. W.O. Fenn and H. Rahn) pp 585–630. Williams and Wilkins, Baltimore.

Widdicombe, J.G. (1966). Action potentials in parasympathetic and sympathetic efferent fibres to the trachea and lungs of cats and dogs. J. Physiol. 186, 56–88.

Widdicombe, J.G. (1974). In "The Peripheral Nervous System" (ed. J.I. Hubbord) pp 455–485. Plenum Press, New York.

Widdicombe, J.G. (1975). Reflex control of airways smooth muscle. Postgrad. Med. J. 51, 36–43.

Widdicombe, J.G. and Sterling, G.M. (1970). The autonomic nervous system and breathing. Arch. Intern. Med. 126, 311–329.

Williams, T.H. and Palay, S.L. (1969). Ultrastructure of the small neurons in the superior cervical ganglion. Brain Res. 15, 17–34.

Wong, L.B., Miller, I.F. and Yeates, D.R. (1990). Stimulation of tracheal ciliary beat frequency by capsaicin. J. Appl. Physiol. 68, 2574–2580.

Yoneda, K. (1977). Pilocarpine stimulation of the bronchiolar Clara cell secretion. Lab. Invest. 37, 447.

Zetler, G. (1976). The peptidgeric neuron – a working hypothesis. Biochem. Pharmacol. 25, 1817–1818.

Zussman, W.V. (1966). Fluorescent localization of catecholamine stores in the rat lung. Anat. Rec. 156, 19–29.

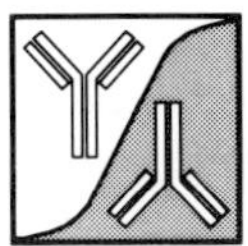

5. Inflammatory Mediators and Modulation of Epithelial/Smooth Muscle Interactions

Douglas W.P. Hay, Stephen G. Farmer *and* Roy G. Goldie

1. Introduction

Until the last decade or so, the prevailing but naive view amongst most scientists was that the sole role of the airway epithelium was as a formidable barrier protecting the respiratory tract from the external environment. However, as a result of increasing and intensive research, it has become apparent that the airway epithelium is a very specialized structure which contains numerous cell types and which possesses an array of important properties and functions which may play critical roles in homeostasis in the pulmonary system. These include the secretion of fluids and mucus, production of many biologically active substances and the provision of sites of drug metabolism.

Immunopharmacology of Epithelial Barriers
ISBN 0–12–288030–7

The crucial role of this cellular layer is perhaps best illustrated by noting that alterations and abnormalities in airway epithelial function may be central to the elaboration of several pulmonary diseases, such as asthma and cystic fibrosis. It will become apparent that, although the epithelium has a significant barrier function, it is also the source of several biologically active substances, is the site of action of many agents and possesses a multifunctional role in lung homeostasis. In this chapter an attempt will be made to review our understanding of the interplay between several inflammatory cells and their constituent mediators and the airway epithelium. A major area of discussion will be interactions between the epithelium and airway smooth muscle. Communication between the airway epithelium and various other cell types, such as nerves, fibroblasts and vascular smooth muscle, will also be considered.

2. Airway Epithelial Cell Types

The respiratory tract is a very complex and diverse configuration of cells and tissues which can be broadly divided into two main areas: the proximal conducting airways and the distal respiratory regions. The latter contains alveolar ducts and air sacs and is the site of exchange of gases. Although it is recognized that the primary role of this system is to permit and regulate the intake of O_2 concomitant with the elimination of CO_2, it is also apparent that the lung serves various other important functions, including clearance of inhaled foreign substances, hormonal regulation, ion transport, and metabolism and excretion of xenobiotics (Penney, 1988; Harkema *et al.*, 1991). Thus, the lung can be considered as a very important metabolic, defence, haemodynamic and also immunological organ (Breeze and Turk, 1984).

Rather than being a homogeneous cellular lining, the epithelium is a remarkably diverse and complicated layer consisting of various populations of cells. There are at least 11 main types of epithelial cells present in mammalian airways, although the function of many of them remains uncertain (Breeze and Wheeldon, 1977; Jeffery, 1983). These cells include ciliated, goblet, brush and K cells (including neuroepithelial body cells), basal and Clara cells, globule leucocytes (derived from subepithelial mast cells), intermediate cells, epithelial serons, oncocytes, special-type non-ciliated columnar and squamous cells (stratified); the latter five cell types are not found in all species. In bronchioles, only six cell types have been noted, namely ciliated, brush, basal, K and Clara cells and globule leucocytes (Breeze and Wheeldon, 1977), whereas, in the alveoli, squamous type V pneumocytes predominate along with cuboidal type II pneumocytes and, in some species, type III pneumocytes (Breeze and Wheeldon, 1977). Not only are there species differences in the composition of the epithelium of the airways, but within

each species there is a striking diversity in the epithelial architecture and the relative distribution of the various cells in different regions (Breeze and Wheeldon, 1977; Breeze and Turk, 1984; Harkema *et al.*, 1991). Grossly, the airway epithelium decreases in thickness down the respiratory tract and varies from a pseudostratified, columnar type in the tracheobronchial lining to a single layer of cuboidal cells in the bronchioles and the highly attenuated squamoid type I cells and cuboidal type IV cells of the alveoli (Breeze and Wheeldon, 1977; Breeze and Turk, 1984; Penney, 1988).

Perhaps not surprisingly, concomitant with significant morphological differences in the various epithelial populations are important differences in their functional characteristics and in the response to injury (Penney, 1988; Harkema *et al.*, 1991). For example, the main function of ciliated cells is to promote the flow of mucus (Harkema *et al.*, 1991), whilst various secretory cells including goblet serous cells and Clara cells play a key role in the generation and maintenance of the mucociliary system. Cytochrome P-450 isozymes are present in only two cell types, namely the Clara cell and the type II pneumocytes, the former possessing the largest rate of cytochrome P-450-mediated metabolism of any cell type (Devereux *et al.*, 1981; Harkema *et al.*, 1991).

Several inflammatory cells including lymphocytes (Breeze and Wheeldon, 1977; Breeze and Turk, 1984), mast cells (Irani *et al.*, 1986) and eosinophils have been localized to the airway epithelium; in particular, in asthma appreciable numbers of eosinophils are often present in the airway epithelium (Jeffery *et al.*, 1989).

In addition, peripheral branches of peptide-containing sensory nerves are detected in the respiratory epithelium (Wharton *et al.*, 1979; Lundberg *et al.*, 1984). The materials released from these fibres include tachykinins such as substance P which have been implicated in neurogenic airway inflammation (McDonald, 1987). Intraepithelial axons are observed associated with all cell types, in particular, basal cells (Jeffrey and Reid, 1973) brush cells (Luciano *et al.*, 1968), K cells (Hung *et al.*, 1973) and neuroepithelial bodies (Lauweryns *et al.*, 1972).

The architecture of the epithelial layer provides an intricate intercellular communication system between the many specialized epithelial cells, nerves and inflammatory cells. The preceding discussion of the anatomy of the airway epithelium highlights the diversity of its constituents and suggests a complex, multifunctional cellular network which plays a critical role in the normal physiology and regulation of homeostasis in the respiratory tract. An alteration in the normal structure and function of these cells is likely to contribute significantly to the pathogenesis of several airway diseases. For example, there is an increased number of mucous goblet cells in chronic bronchitis and cystic fibrosis (Harkema *et al.*, 1991) and the aetiology of the latter disease has been attributed primarily to a defect in the epithelial chloride channel

(Knowles *et al.*, 1986; Anderson *et al.*, 1992). In the remainder of this chapter, various aspects of the functional characteristics of several of the different cell types will be discussed in more detail and, in particular, the impact that the epithelium has on various inflammatory mediators and inflammatory processes.

3. Physiology of Airway Epithelium

3.1 BARRIER FUNCTION

The airway epithelium serves as an effective physical barrier between the underlying tissues and the airway lumen. This is in large part due to the intracellular junctions that connect the apical surfaces of individual cells (Breeze and Wheeldon, 1977; Gumbiner, 1987). This barrier function not only protects the airways from noxious inhaled substances, but also prevents the uncontrolled leakage of water and solutes into the airway lumen. The junction complexes, which are observed between all epithelial cells, are made up of three components: (1) a proximal, continuous zonula occludens or tight junction; (2) an intermediate, continuous, zonula adherens or intermediate junction; and (3) a distally located macula adherens or desmosome (Farquhar and Palade, 1963; Gumbiner, 1987; Schneeberger *et al.*, 1992). In addition, another specialized intercellular junction, the gap junction, which is involved in the transmission of communication between cells, is observed during some stages of fetal development and also between some epithelial cells in adults (Sheridan and Atkinson, 1985; Pitts and Finbow, 1986). It is the tight junctions that provide the barrier function, whereas the intermediate junctions, which possess adhesion molecules, are involved in cell adhesion and recognition. Desmosomes are involved in maintaining the integrity of the epithelium (Rennard *et al.*, 1991; Schneeberger *et al.*, 1992). On the other hand, gap junctions act as channels permitting the non-specific intercellular exchange of small molecules and ions apparently *via* passive diffusion (Revel *et al.*, 1985; Sheridan and Atkinson, 1985) and have been postulated to be involved in epithelial cell–cell communication (Schneeberger *et al.*, 1992).

There has been considerable speculation and interest in the potential impact that dysfunction in the epithelium has on the pathophysiology of pulmonary diseases. It is recognized that a cardinal feature of asthma, even of a mild nature, is damage, shedding or desquamation of the epithelium (Laitinen *et al.*, 1985; Beasley *et al.*, 1989; Jeffery *et al.*, 1989), with a correlation observed between the extent of damage and the degree of airway hyper-responsiveness (Beasley *et al.*, 1989). A study by Lozewicz and co-workers (1990), however, comparing airway biopsies from non-asthmatic and stable asthmatics, did not observe any alteration in the morphology of the epithelium (Lozewicz *et al.*, 1990). The possibility of damage from artefacts to the epithelium as a result of

tissue processing has to be considered (Laitinen *et al.*, 1991). Epithelial damage may result in increased permeability and easier access of allergens, pollutants, inflammatory and contractile agents to intra- and sub-epithelial nerves, inflammatory cells and vascular and airway smooth muscle. Epithelial damage with associated bronchial hyper-responsiveness is observed in patients with farmer's lung (Heino *et al.*, 1982) and viral infections (Hers, 1966). Aerosol administration of substances that cause bronchial epithelial damage, including ozone, nitrogen dioxide or toluene diisocyanate (Golden *et al.*, 1978; Holtzman *et al.*, 1983b; Mapp *et al.*, 1985), results in bronchial hyper-responsiveness. In experimental animals, exposure of the airways to inflammatory stimuli such as cigarette smoke, which increase epithelium permeability, increased the responsiveness of the airways to spasmogens (Boucher *et al.*, 1980; Hulbert *et al.*, 1981; Hulbert *et al.*, 1985). Experiments in humans have provided conflicting data, with some suggesting increased permeability in asthma (Ilowite *et al.*, 1989), and others failing to observe this (Elwood *et al.*, 1983; O'Byrne *et al.*, 1984).

3.2 SECRETORY FUNCTION

As part of the complex defence mechanisms to protect the lungs from harmful substances present in the external environmental, in addition to maintaining the normal homeostatic balance, cells in the airways, including the epithelium, secrete a diverse array of materials including mucus, water and electrolytes, sugars, lipids glycoproteins and proteins, and also immunoglobulin, proteases and bacteriocidal and bacteriostatic agents (Boat and Matthews, 1973; Robinson *et al.*, 1989). Controversy exists as to whether Clara cells also produce surfactant (Penney, 1988). In addition, the epithelium releases a variety of neurohumoral agents and inflammatory mediators. In this section the focus will be on mucus secretion and release of inflammatory mediators and neurohumoral agents.

3.2.1 Mucus

Mucus is an important component of the mobile and protective fluid lining the airways, acting in concert with the motion of cilia and fluid secretions, which form the mucociliary clearance apparatus. This system lubricates the airways and entraps and removes foreign particulate matter and cellular debris from the airways (Marin, 1986; Basbaum and Finkbeiner, 1989; Widdicombe, 1991). The fluid lining the airways has two components, a mucous gel on what is termed the periciliary sol layer and a watery film which bathes the cilia (Lucas and Douglas, 1934; Widdicombe, 1991). There may be limited mucus production in healthy airways, but its amount, composition and viscosity may be altered in airway diseases such as asthma, chronic bronchitis and cystic fibrosis as well as in viral infections (Breeze and Wheeldon, 1977; Marin and Culp, 1986; Bhaskar *et al.*, 1985). In these diseases there may be a reduction in tracheal mucus velocity in conjunction with a decrease in the ciliation of the airway epithelium (Wanner, 1986). In addition, hypertrophy of tracheobronchial submucosal glands and hyperplasia of goblet cells is a frequent observation in asthma, chronic bronchitis and cystic fibrosis (Basbaum and Finkbeiner, 1989). Mucus-secreting cells include epithelial serous, goblet and Clara cells and serous and mucous submucosal gland cells (Marin, 1986; Basbaum and Finkbeiner, 1989). In the larger airways of several species it appears that, by virtue of their larger volume, the submucosal glands are quantitatively the major contributor of mucosubstances. In contrast, in the lower airways, the Clara cell seems to be the main secretory cell, with the product not typically mucoid (Breeze and Wheeldon, 1977; Breeze and Turk, 1984; Penney, 1988).

With regard to the mechanisms involved in the regulation of cellular secretions, much more research has been conducted on submucosal glands than on surface epithelial cells (Marin and Culp, 1986). Data from physiological, histological and anatomical studies provide extensive evidence for neuronal innervation of sub-mucosal gland secretion, including adrenergic, cholinergic and peptidergic inputs (Finkbeiner and Widdicombe, 1989). Autoradiographic and radioligand-binding studies have also indicated the presence of adrenergic and cholinergic receptors on glands (Barnes and Basbaum, 1983; Basbaum *et al.*, 1984; Marin and Culp, 1986). An array of substances elicit effects on submucosal gland secretions, including PGD_2, $PGF_{2\alpha}$ (which increase secretion) and PGE_2 (which decreases secretion), LTC_4 and LTD_4, glucocorticosteroids, cholinoceptor agonists, adrenoceptor agonists, substance P and histamine (Basbaum and Finkbeiner, 1989; Finkbeiner and Widdicombe, 1989). There is evidence that Clara cells are influenced by α and β-adrenoceptor activation and perhaps also by the activation of receptors for prostaglandins (Massaro *et al.*, 1982; Massaro, 1987). Goblet cells are not responsive to pharmacological agents and there is no evidence that they have neuronal inputs (Breeze and Wheeldon, 1977), although they are stimulated by local irritation (Basbaum, 1986). Furthermore, although intraepithelial axons have been associated with several types of surface epithelial cells, there is no convincing evidence that they have a secretory/motor function (Breeze and Wheeldon, 1977).

3.2.2   Release of Mediators and Cytokines

The airway epithelium can synthesize a diverse array of biologically active substances which interact with and influence the activity of several neighbouring cells. To date, probably the most extensively studied epithelium-derived substances are the numerous cyclooxygenase and lipoxygenase products of arachidonic acid metabolism. Their release has been demonstrated from various sources, including native epithelium, freshly dispersed cells, primary and transformed cultures and also cell sonicates (Widdicombe and Pack, 1982; Van Scott *et al.*, 1991).

The cyclooxygenase and lipoxygenase pathways are expressed at high levels in epithelial cells (Holtzman, 1991). The major arachidonic acid metabolite released from primary tracheal epithelial cultures of several species, including humans, may be PGE2 (Leikauf *et al.*, 1988; Churchill *et al.*, 1989; Salari and Chan-Yeung, 1989; Widdicombe *et al.*, 1989). In many instances, however, the profile of released products is significantly different in primary cultures compared with freshly isolated, dispersed cells (Widdicombe, 1991). For example, in human tracheal epithelium, the main products released from freshly isolated, dispersed cells by the calcium ionophore A23187 or Bk is 15-HETE and, to a lesser extent, 12-HETE and 8-HETE, rather than PGE_2 (Churchill *et al.*, 1989).

In human airway epithelial cells, the predominant lipoxygenase pathway appears to be via 15-lipoxygenase (Henke *et al.*, 1988; Holtzman *et al.*, 1988; Salari and Chan-Yeung, 1989). In contrast, 12-lipoxygenase pre-dominates in bovine epithelium (Hansbrough *et al.*, 1989) and 12-lipoxygenase (with low levels of 5-lipoxygenase), in canine and sheep cells (Shannon *et al.*, 1979; Holtzman *et al.*, 1983a; Eling *et al.*, 1986; Holtzman *et al.*, 1988). It was recently reported that PAF stimulated the release of 15-HETE from cultured human bronchial epithelial cells in sufficient quantities to elicit airway smooth muscle contraction (Salari and Schellenberg, 1991). Exposure to ozone stimulated the release of cyclooxygenase and lipoxygenase products from cultured bovine tracheal epithelial cells (Leikauf *et al.*, 1988).

In addition to products of arachidonic acid metabolism, the airway epithelium releases a variety of substances. For example, the release of endothelin, the potent 21-amino acid vasoconstrictor peptide (Yanagisawa *et al.*, 1988) from cultured canine, bovine and human bronchial epithelial cells (Black *et al.*, 1989; Mattoli *et al.*, 1990) and guinea-pig tracheal epithelial cells (Ninomiya *et al.*, 1991), has been demonstrated. Release from guinea-pig tracheal epithelial cells was increased by endotoxin and various cytokines. Release of the reactive oxygen species H_2O_2 from guinea-pig tracheal epithelial cells was also demonstrated. Stimulation was increased with PAF, perhaps via a protein kinase C-dependent mechanism (Kinnula *et al.*, 1992).

Airway epithelial cells express and produce several cytokines including GM-CSF, IL-6 and IL-8 (Smith *et al.*, 1989; Marini *et al.*, 1992). Interestingly, cultured bronchial epithelial cells from asthmatics released more GM-CSF (0.88 ng/0.5 $\times$ 10^6 cells) than cells from non-asthmatic individuals (0.21 ng/0.5 $\times$ 10^6 cells) and they expressed increased levels of mRNA for GM-CSF (Soloperto *et al.*, 1991). Exposure of human bronchial epithelial cell cultures to toluene diisocyanate resulted in a marked increase in the release of IL-1β and IL-6 (Mattoli *et al.*, 1991).

The release of LTB_4, the potent neutrophil chemotactic factor, from freshly isolated canine airway epithelial cells has been demonstrated (Holtzman *et al.*, 1983a). More recent studies have indicated that the airway epithelium is also able to release various non-lipoxygenase products, which are chemotactic for a variety of respiratory cell types. For example, bovine and human bronchial epithelial cells, in culture, release fibronectin, which is chemotactic for lung fibroblasts (Shoji *et al.*, 1989). This product may be important in epithelial–mesenchymal interactions and in the recruitment and function of fibroblasts that occurs in repair processes after injury and airway morphogenesis (Shoji *et al.*, 1989). In addition, bovine bronchial epithelial cells in culture appeared to release at least two different substances that were chemotactic for neutrophils and monocytes: (1) the major component was a low molecular weight, lipid-extractable, protease-resistant substance, released after 24 h (the release was decreased by 5-lipoxygenase or phospholipase A_2 inhibitors); (2) the minor component of activity resided in a high molecular weight, lipid-inextractable, protease-sensitive protein (Koyama *et al.*, 1989). The chemotactic activity released from epithelial cells can be increased by various stimuli, including acetylcholine, cigarette smoke, endotoxin, opsonized zymosan and grain dust extracts (Koyama *et al.*, 1989; Von Essen *et al.*, 1989). Supernatants from cultured bovine bronchial epithelial cells possess chemotactic activity for T helper lymphocytes and β lymphocytes, which was perhaps resident in two factors of different molecudar weights (Robbins *et al.*, 1989).

In addition, bronchial epithelial cells, which are capable of directed migration, release substances, such as fibronectin, which are chemotactic for epithelial cells (Shoji *et al.*, 1990).

3.3 SITE OF METABOLISM

The epithelium acts as a significant site of metabolism, and by virtue of this property is able to modulate the activity of many groups of substances. For example, the airway epithelium is a rich source of endopeptidase – 24.11 (EC 3.4.24.11) (Johnson *et al.*, 1985; Sekizawa *et al.*, 1987). This enzyme is also known as NEP or enkephalinase, a membrane-associated neutral metalloendopeptidase originally identified in rabbit kidney (Kerr and Kenny, 1974). NEP, which is localized to the surface of airway epithelial cells, is capable of hydrolysing a variety of peptides (Burnett, 1987), including the tachykinins, Bk (Dusser *et al.*, 1988), VIP (Said, 1987) and ET (Hay, 1989). The use of potent inhibitors of NEP, such as thiorphan or phosphoramidon, has assisted greatly in determining the physiological role of this enzyme in regulating the function of several peptides. For example, thiorphan or phosphoramidon potentiate the effects of sub P on mucous gland secretion *in vitro* (Borson *et al.*, 1986), ciliary beat frequency (Kondo *et al.*, 1990) and airway smooth muscle contraction (Sekizawa *et al.*, 1987; Devillier *et al.*, 1988; Fine *et al.*, 1989). Inhibition of NEP has been observed to modulate the

relaxant or contractile effects of several other peptides in isolated airway tissues, including Bk (Frossard *et al.*, 1990), ET (Hay, 1989), atrial natriuretic peptides (Fernandes *et al.*, 1992) and VIP (Farmer and Togo, 1990; Rhoden and Barnes, 1990). The actions of thiorphan or phosphoramidon in epithelium-containing isolated airway smooth muscle mimics to a large extent the effects of epithelium removal on agonist-induced responses (Devillier *et al.*, 1988; Fine *et al.*, 1989; Frossard *et al.*, 1989; Hay, 1989; Frossard *et al.*, 1990; Fernandes *et al.*, 1992), providing evidence that the modulatory influence of the epithelium on responses elicited by peptides is in large part due to its acting as a significant site of their metabolism.

Preliminary evidence was recently provided from functional studies using a tracheal tube preparation suggesting that histaminase present in guinea-pig tracheal epithelium contributes to the inhibitory effect of the epithelium on histamine-induced contraction (Lindström *et al.*, 1991). Similarly, in isolated guinea-pig tracheal strips, the epithelium may be a major site of extraneuronal uptake for catecholamines such as isoprenaline (Farmer *et al.*, 1986). Conflicting evidence was also provided that the guinea-pig tracheal epithelium is a locus for the uptake and degradation of adenosine (Farmer *et al.*, 1986; Advenier *et al.*, 1988).

The Clara cell (non-ciliated bronchiolar secretory cell), which is the most abundant non-ciliated cell located predominantly, but not exclusively, in the bronchioles (Jeffrey and Reid, 1973; Breeze and Wheeldon, 1977; Plopper *et al.*, 1980), is highly metabolically active and possesses a variety of enzymes including acid and alkaline phosphatases, non-specific esterase, hydroxylases, transferases, peroxidases and catalase, in addition to cytochrome P-450 monoxygenase (Breeze and Wheeldon, 1977; Goldenberg *et al.*, 1978; Jones *et al.*, 1983). Since Clara cells possess the highest rate of cytochrome P-450-mediated metabolism of any lung cell, one of their main functions is to metabolize xenobiotic cytotoxic agents that enter the respiratory tract via the air or blood (Boyd, 1977; Widdicombe and Pack, 1982; Gail and Lenfant, 1983). Furthermore, the Clara cell appears to be an important target cell for pneumotoxic, mutagenic or carcinogenic substances which need activation via the cytochrome P-450 system (Breeze and Turk, 1984). Evidence from histochemical analysis suggests that hydroxylation of aniline occurs in the bronchial epithelium (Grasso *et al.*, 1971). Type II cells are also very metabolically active (Breeze and Turk, 1984) and, by virtue of their role as the source of pulmonary surfactant, are intimately involved in the metabolism of diverse phospholipids (Voelker and Mason, 1989).

3.4 ANTIGEN EXPRESSION

In humans and animal species, airway epithelial cells have the ability to express class I and class II MHC antigens which are necessary for accessory cell function in antigen presentation (Glanville *et al.*, 1989; Kalb *et al.*, 1989; Spurzem *et al.*, 1990); this expression is influenced by inflammatory and immune stimuli, including interferon γ and TNF. Accordingly, the epithelium may play a role in regulating the immune response subsequent to inhalation of antigens (Kalb *et al.*, 1989). HLA-DR mRNA was present in human ciliated bronchial epithelial cells (Rossi *et al.*, 1990). HLA class IV antigen expression in basal epithelial cells from bronchial biopsies was observed in samples from asthmatic but not from non-asthmatic individuals concomitant with an increase in the numbers of macrophages, T lymphocytes and eosinophils (Poston *et al.*, 1992). Of interest are findings showing that a complex network of dendritic cells, located in the airway epithelium, are involved in the trapping of inhaled antigen and in its presentation to T lymphocytes in the local immune response in airways (Sertl *et al.*, 1986; McMenamin *et al.*, 1991).

In canine lung allografts after single-lung transplantation there was increased expression of MHC class II antigens in the bronchial epithelium, with the level of expression increasing with the progression of rejection and correlating with BAL levels of TNF, IL-2 and interferon γ. The increase in MHC class II antigen expression and cytokine levels was prevented by treatment with cyclosporine, an immunosuppressive agent (Chang *et al.*, 1990).

Increased expression of class II HLA-DR and HLA-DS complex antigens was observed in alveolar epithelium of lung biopsies from patients with idiopathic pulmonary fibrosis and sarcoidosis (Kallenberg *et al.*, 1987). Similarly, equivocal, preliminary evidence suggests enhanced expression of MHC class II antigens on the epithelium of diseased versus normal lungs (Taylor and Rose, 1989; Yousem *et al.*, 1990).

4. *Physiology and Pharmacology of Inflammatory Mediators*

4.1   HISTAMINE

Histamine was first characterized pharmacologically over 80 years ago by Dale and Laidlaw (1910), who noted that it was a potent vasoactive substance. It can be regarded as the first mediator that was considered as playing a role in asthma. In 1919, Dale and Laidlaw demonstrated that administration of histamine to guinea-pigs mimicked anaphylactic bronchoconstriction. In 1927, Lewis reported that the antigen–antibody reaction and stimulation of skin released a material with properties similar to histamine.

The major sites of distribution of histamine in tissues and in blood are mast cells and basophils, respectively, where it is stored preformed in cytoplasmic granules in association with a heparin–protein complex in mast cells

and chrondroitin sulphates in human basophils. Accordingly, histamine concentrations are highest in locations which contain abundant numbers of mast cells such as the skin, gastrointestinal tract and respiratory tract. In the airways, the predominant sites for mast cells are the submucosa and mucosa, near blood vessels and glands, although some are located in the airway epithelium (Brinkman, 1968; Cutz *et al.*, 1978). Histamine exerts several biological effects, the most noteworthy of which are stimulation of gastric acid secretion and its involvement in immediate hypersensitivity reactions and the allergic response. The latter is a consequence of the activation of the mast cell and release of its constituents, including histamine, following the antigen–IgE antibody interaction on the surface of the cells.

Histamine-induced effects are mediated via activation of H_1, H_2 or H_3 receptors. H_1 receptor-mediated effects include contraction of airway and gastrointestinal smooth muscle, increased vascular permeability, prostaglandin generation and tachycardia and activation of vagal afferent nerves. Some H_2 receptor-mediated events include bronchial dilation, stimulation of suppressor T cells and increases in mucus secretion (White *et al.*, 1987). H_3 receptors appear to be predominantly located in the CNS, where histamine is thought to act as a neurotransmitter (Arrang *et al.*, 1983; Schwartz *et al.*, 1986; Arrang *et al.*, 1987). However, peripheral effects of H_3 receptor activation have been observed, including inhibition of cholinergic neurotransmission in the airways (Ishikawa and Sperelakis, 1987; Ichinose and Barnes, 1989; Ichinose *et al.*, 1990).

Histamine elicits H_1-mediated bronchospasm in most species, including humans (Marin *et al.*, 1977), and bronchodilation (mediated via H_2 receptors) in others. Although there is minimal evidence in humans for H_2 receptor mediator bronchodilation, in the presence of an H_1 receptor antagonist, histamine will produce relaxation of airway smooth muscle in all species (Eiser, 1982). When administered alone, H1 receptor antagonists have only minor inhibitory effects against antigen-induced contraction of guinea-pig or human isolated airways, whereas in combination with a leukotriene receptor antagonist, the response is abolished (Hay *et al.*, 1987b).

Histamine-induced increased microvascular permeability in airways is mediated via H_1 receptor activation in endothelial cells, resulting in mucosal oedema. H_2 receptor activation has been demonstrated to stimulate airway mucus secretion in humans (Shelhamer *et al.*, 1980), and it has been suggested that histamine-induced stimulation of epithelial H_1 receptors alters ion and fluid transport and affect the viscosity of the mucus.

4.2   LEUKOTRIENES

The leukotrienes are a family of potent biologically active substances comprising the sulphidopeptidoleukotrienes, LTC_4, LTD_4 and LTE_4, and the chemoattractant LTB_4.

They are formed from the metabolism of membrane-derived arachidonic acid by the catalytic activity of the 5-lipoxygenase enzyme. The mammalian 5-lipoxygenase enzyme has been cloned and its molecular mechanism characterized (Dixon *et al.*, 1988; Rouzer *et al.*, 1988). It is a Ca^{2+}- and ATP-dependent enzyme of approximately 78 kD (Goetze *et al.*, 1985; Rouzer and Samuelsson, 1985; Shimizu *et al.*, 1986).

There are numerous cellular and tissue sources of the leukotrienes, including a variety of inflammatory cells such as eosinophils, mast cells, macrophages, basophils, neutrophils and monocytes; the qualitative and quantitative nature of the release of the leukotrienes depends on the cell type and also the stimulus (Borgeat and Samuelsson, 1979; Fels *et al.*, 1982; Peters *et al.*, 1987). Freshly isolated canine tracheal epithelium is able to convert arachidonic acid to LTB_4 and, in some cases, LTC_4 (Holtzman *et al.*, 1983a; Eling *et al.*, 1986; Holtzman *et al.*, 1988). The peptidoleukotrienes and LTB_4 produce a diverse array of biological effects, which are mediated via an interaction with specific membrane receptors, In human airways, unlike in the guinea-pig, contractions elicited by LTC_4, LTD_4 and LTE_4 appear to be mediated via an interaction with a homogeneous leukotriene receptor population (Buckner *et al.*, 1986; Hay *et al.*, 1987b). The actions of LTB_4 are mediated via two populations of specific receptors.

The leukotrienes have been implicated in the pathogenesis of a variety of disorders. Asthma is the disease for which the evidence is most suggestive of a key role for the leukotrienes, particularly the peptidoleukotrienes. Thus, the leukotrienes mimic many of the classic features of the disease. For example, the peptidoleukotrienes are very potent bronchoconstrictor agonists in human airways *in vitro* and *in vivo*, possessing up to greater than 1000-fold the potency of histamine or cholinoceptor agonists (Griffin *et al.*, 1983; Barnes *et al.*, 1984; Bisgaard *et al.*, 1985; Adelroth *et al.*, 1986; Buckner *et al.*, 1986; Davidson *et al.*, 1987), they potently stimulate mucus secretion from human airways *in vitro* (Marom *et al.*, 1982; Coles *et al.*, 1983) and increase microvascular permeability (Dahlén *et al.*, 1981; Hua *et al.*, 1983; Evans *et al.*, 1989).

LTB_4 has many pro-inflammatory properties, the most notable feature of which is its highly potent chemotactic activity for neutrophils (Ford-Hutchinson *et al.*, 1980; Bray, 1983) and, to a lesser extent, eosinophils (Wardlaw *et al.*, 1986). It also stimulates the release of oxygen radicals and lysosomal enzymes (Feinmark *et al.*, 1981; Serhan *et al.*, 1982; Bray, 1983) from neutrophils, increases C3b receptor expression on human neutrophils and eosinophils (Nagy *et al.*, 1982), promotes neutrophil aggregation (Ford-Hutchinson *et al.*, 1980), stimulates adhesion of leucocytes to the vascular endothelium (Gimbrone *et al.*, 1984; Hoover *et al.*, 1984) and has been demonstrated to enhance vascular permeability (Goetzl, 1980). LTB_4 administration *in vivo* in many

species, including humans, elicits PMN accumulation at many sites in the body (Bray, 1983; Thorsen, 1986; Ford-Hutchinson, 1990).

4.3   PLATELET-ACTIVATING FACTOR

PAF, a choline phosphoglyceride, was discovered and identified in the 1970s independently by different groups of researchers as 1–0-alkyl-2-acetyl-*sn*-glycero-3-phosphocholine (Demopoulos *et al.*, 1979). PAF was originally recognized and named from previous studies demonstrating its ability to elicit aggregation of rabbit platelets following its release by provocation of sensitized leucocytes with antigen (Henson, 1970; Siraganian and Osler, 1971; Benveniste *et al.*, 1972). PAF has been shown subsequently to possess a battery of biological activities not related to platelet activation, including stimulation of various other inflammatory cells, stimulation of glycogenolysis, hypotensive action, contraction of airway and uterine smooth muscle, increased vascular permeability and oedema (Page, 1988).

PAF can be produced by platelets, macrophages, neutrophils, basophils, eosinophils and mast cells, and also endothelial cells (Page, 1988). There is evidence that, unlike for example the eicosanoids, not all the PAF that is synthesized is secreted into the extracellular milieu, but some is retained intracellularly. The relative amounts of release versus retained PAF depends on the cell type (Cluzel *et al.*, 1988; Lynch and Henson, 1986; Prescott *et al.*, 1990; Stewart *et al.*, 1990). These observations led to speculation that PAF may act as an intracellular messenger.

Various groups of investigators have demonstrated saturable stereospecific high-affinity binding sites for PAF in a number of tissues and cell, including platelets (Valone *et al.*, 1982; Kloprogge and Akkerman, 1984), neutrophils (Valone *et al.* 1982), macrophages (Lambrecht and Parnham, 1986) and lung tissue (Goldie *et al.*, 1990b; Hawang *et al.*, 1985). Furthermore, studies have shown a correlation between the binding and functional activities of PAF (Braquet *et al.*, 1987). There is evidence from binding and functional studies examining relative potencies of different PAF receptor antagonists, which suggests the existence of at least two subtypes of PAF receptors (Lambrecht and Parnham, 1986; Hwang, 1987, 1988). The PAF receptor has proven difficult to work with and characterize, in part due to the amphophilic nature and high levels of non-specific binding and also the relatively small number of receptors in most cells (Prescott *et al.*, 1990).

Of the spectrum of biological actions of PAF and potential physiological and pathophysiological relevance, it is probably its activities in the pulmonary system and its putative central role in asthma which have received the most widespread attention. Research suggests strongly that PAF, at least in animals, mimics many of the hallmark features of the disease, including bronchoconstriction, airway hyper-reactivity, enhanced microvascular leakage and mucosal oedema, mucus hypersecretion and inflammatory cell chemotaxis and infiltration (Barnes *et al.*, 1988; Barnes and Henson, 1991).

In humans, PAF is one of the most potent substances identified to elicit bronchoconstriction, enhanced vascular permeability and inflammatory cell chemotaxis (O'Flaherty *et al.*, 1981; Archer *et al.*, 1984; Cuss *et al.*, 1986; Wardlaw *et al.*, 1986). Furthermore, single-aerosol administration of PAF to non-asthmatic volunteers was reported initially to induce airway hyper-reactivity for up to 29 days, although subsequent studies have failed to confirm this finding. Increased levels of PAF were detected in blood and BALs of asthmatics and following antigen exposure (Court *et al.*, 1987; Nakamura *et al.*, 1987). Aerosol administration of PAF in rabbits elicited airway epithelial damage and loss (Cammussi *et al.*, 1983). PAF was proposed to stimulate Cl^- secretion across canine, cultured tracheal but not bronchial epithelium via a prostaglandin-dependent mechanism (Tamaoki *et al.*, 1991). However, it was observed to be without effect on rabbit lung epithelial or endothelial cell permeability (Bolin *et al.*, 1987).

4.4 BRADYKININ

Bk is a potent inflammatory nonapeptide whose production in the body elicits many physiological responses (Proud and Kaplan, 1988; Farmer and Burch, 1991, 1992; Bhoola *et al.*, 1992). Kinins are generated from precursor proteins, the kininogens, by the kallikreins and other serine proteases in tissues and plasma. Many studies in animals as well as a few in humans implicate Bk, kallidin (Lys-Bk, KD) and their hydroxyproline derivates, $[Hyp^3]$-Bk and $[Hyp^4]$-KD, in inflammatory diseases of the upper and lower airways (Proud and Kaplan, 1988; Baraniuk, 1991; Farmer, 1991; Pongracic *et al.*, 1991).

In the airways, kinins, acting on specific cell surface receptors, may produce responses including neurogenic inflammation via: stimulation of sensory nerve fibres; arteriolar dilation, venoconstriction and plasma protein extravasation, leading to mucosal oedema, congestion and swelling; airway smooth muscle contraction, causing bronchoconstriction; and release of many inflammatory mediators such as tachykinins, cytokines and metabolites of arachidonic acid (Proud and Kaplan, 1988; Farmer, 1991; Farmer and Burch, 1991, 1992; Marceau and Regoli, 1991; Pongracic *et al.*, 1991; Bhoola *et al.*, 1992).

Kinin receptors are currently classified as B_1 and B_2 subtypes, based upon the relative potencies of various agonists and antagonists (for reviews, see Farmer and Burch, 1991, 1992). In general, B_2 receptors exhibit much higher affinity for Bk and KD than for their respective carboxypeptidase N products, desArg^9-Bk or desArg^{10}-KD, whereas B_1 receptors are more sensitive to the desarginine metabolites.

B_1 receptors have been characterized largely in rabbit vascular smooth muscle (Bouthillier *et al.*, 1987; DeBlois *et al.*, 1988), although they have also been demonstrated in some other cell types in tissue culture (Straus and Pang, 1984; Lerner and Modéer, 1991). To date, however, there is scant evidence supporting B_1 receptor expression in human airway tissues. The pharmacology of B_1 receptors has recently been reviewed (Farmer and Burch, 1991; Marceau and Regoli, 1991).

Most effects of Bk are mediated by B_2 receptors although there are increasing reports of heterogeneity among this subtype, in addition to the reputed existence of airway "B_3 receptors" (Farmer *et al.*, 1989; Farmer *et al.*, 1991; Field *et al.*, 1992). The evidence for heterogeneity among B_2 receptors, as well as other receptor subtypes, is discussed elsewhere (Farmer and Burch, 1992).

The kinins have a half-life in plasma of only a few seconds, a single passage through the pulmonary vascular bed resulting in 80–90% enzymatic destruction (Steranka *et al.*, 1989; Erdös, 1990; Ward, 1991) by kininases. The principal enzymes distributed within the lungs and vascular beds are kininase I (carboxypeptidase N) and kininase II (ACE), although several other kininases have been identified. The abundance of kininases results in low circulating levels of kinins as a balance between simultaneous production and metabolism. Kinin-metabolizing enzymes can be divided into groups according to their site of action on the peptide sequence (reviewed by Ward, 1991). Although kininase I and II seem to be major contributors to kinin metabolism, there are several other enzymes, particularly NEP, that may have important roles under certain circumstances.

Several studies indicate that Bk may have an important role in asthma. For example, it has been recognized for 30 years that Bk is a very potent bronchoconstrictor in asthmatic patients (Herxheimer and Stresemann, 1961; LeComte *et al.*, 1962; Girard and Moret, 1963). Thus, in asthmatics, but not in non-asthmatic subjects, inhalation of Bk or KD aerosols results in cough, wheezing and bronchoconstriction (Simonsson *et al.*, 1973; Dixon *et al.*, 1987; Fuller *et al.*, 1987a; Dixon and Barnes, 1989; Polosa and Holgate, 1990).

Bk-induced bronchoconstriction in humans is inhibited by muscarinic receptor antagonists such as atropine (Simonsson *et al.*, 1973) and ipratropium bromide (Fuller *et al.*, 1987a). Thus, Bk-induced bronchoconstriction in humans is probably mediated via vagal reflex. That Bk effects in human airways may be largely neuronal is supported further by the fact that the kinin also causes cough (Fuller *et al.*, 1987a; Dixon and Barnes, 1989; Polosa and Holgate, 1990). Further evidence for a role of neuronal mechanisms may be provided by the observation that disodium cromoglycate and nedocromil sodium provide significant protection against Bk-induced bronchoconstriction and cough (Fuller *et al.*, 1987a; Dixon and Barnes, 1989). There are several reports that indicate the effects of these drugs, at least in part, is due to their inhibition of sensory nerve function (Eady and

Jackson, 1989; Dixon and Ind, 1990; Wright *et al.*, 1990; Mansour *et al.*, 1992).

Inhaled Bk-induced bronchoconstriction in asthmatics is unaffected by prior treatment with cyclooxygenase inhibitors (Fuller *et al.*, 1987a; Polosa *et al.*, 1990), suggesting that prostaglandins are not involved. The airway effects of Bk in asthmatics, however, are not affected by the ACE inhibitors (Dixon *et al.*, 1987; Fuller *et al.*, 1987a). Moreover, Bk-induced bronchoconstriction in asthmatics does not involve histamine, acting via H_1 receptors, as terfenadine has no effect (Polosa *et al.*, 1990).

Evidence of a role for kinins in the pathogenesis of asthma is provided by observations that elevated kinin levels were detected in plasma of asthmatic patients (Abe *et al.*, 1967; Christainsen *et al.*, 1987), and correlations between kinin levels and symptom severity were evident (Abe *et al.*, 1967). Bronchoalveolar kinin-generating activity and levels of immunoreactive kinins increase, following allergen challenge, in allergic asthmatics (Christiansen *et al.*, 1987, 1992). Analysis of the bronchoalveolar lavage fluid showed the predominant kinin to be KD. Further, the major kininogenase activity was compatible with tissue kallikreins. Thus, Christiansen and co-workers (1987) speculated that the presence of tissue and plasma kallikrein activities in the asthmatic airways could, via kinin generation, participate in the cascade of events leading to asthma.

Although there are, as yet, no published reports on effects of Bk receptor antagonists in asthmatic patients, several studies in animals suggest that Bk is involved in airway inflammation and hyper-responsiveness. For example, *Ascaris suum* antigen-induced airway hyper-responsiveness to carbachol in sheep was markedly attenuated by prior inhalation of a Bk B_2 receptor antagonist, DArg-[Hyp^3,DPhe^7]-Bk (NPC 567) (Solér *et al.*, 1990). Furthermore, in the same species, allergen-induced late-onset airway obstruction is associated by elevated BAL concentrations of several inflammatory mediators (peptidoleukotrienes, prostaglandins and thromboxane) as well as immunoreactive kinin (Abraham *et al.*, 1991). NPC 567 inhibited both the development of late-phase obstruction and also significantly reduced the elevations in BAL mediators (Abraham *et al.*, 1991).

In a recent study in ovalbumin-sensitized guinea-pigs, chronic exposure to inhaled antigen caused marked infiltration of the airways by eosinophils as well as hyper-responsiveness to histamine and ACh (Farmer *et al.*, 1992). Inhalation of the B_2 receptor antagonists NPC 567 or DArg-[Hyp^3,Thi^5,DTic^7,Tic^8]-Bk (NPC 16731, a recently described, more potent antagonist; Farmer *et al.*, 1991), prior to and during each antigen challenge, inhibited both airway hyper-responsiveness and eosinophil influx (Farmer *et al.*, 1992). Furthermore, NPC 16731 abrogated antigen-induced cyanosis and retarded significantly the onset of dyspnoea (Farmer *et al.*, 1992). Thus, the ability of Bk receptor antagonists to inhibit antigen-induced airway hyper-responsiveness and eosinophilia in animal models of asthma indicates an important role for endogenous kinins. Moreover, the decreased eosinophil infiltration, in animals administered the Bk antagonists, suggests that Bk has a significant function in maintaining allergic inflammation of the airways.

4.5 TACHYKININS

The mammalian tachykinins, also known as neurokinins, are a family of sensory neuropeptides which comprises Sub P, NkA and NkB (also known as neurokinin K) and which possesses a conserved C-terminal sequence of Phe X-Gly-Leu-Met-NH_2. Tachykinin-containing nerves and receptors are widely but selectively distributed in both the central and peripheral nervous systems where they are thought to serve as neurotransmitters. In the periphery the tachykinin-containing nerves are localized to many sites, prominent amongst them are those in the gastrointestinal and respiratory tracts (Nakanishi, 1987). In the pulmonary system, the tachykinins are contained in capsaicin-sensitive peripheral unmyelinated branches of primary afferent neurones, also termed "sensory C fibres". The peptides are synthesized in the nodose and jugular ganglia and then transported to the nerve endings (Lundberg *et al.*, 1983a, 1984; Lundberg and Saria, 1987). Sub P and NkA have been localized histochemically in many species to fibres both beneath and within the airway epithelium, surrounding blood vessels associated with several ganglia and, to a lesser degree, the mucous glands and airway smooth muscle (Lundberg *et al.*, 1984; Uddman and Sundler, 1987). Information regarding the prominence of tachykinin-containing nerves in human airways is conflicting (Laitinen *et al.*, 1983; Lundberg *et al.*, 1984). In addition to capsaicin, a neurotoxin which destroys sensory C fibres (Buck and Burks, 1986), Bk, histamine and vagal nerve stimulation have been shown to release tachykinins from the airways (Saria *et al.*, 1985, 1988).

The tachykinins have been demonstrated to be effective substrates for NEP, which is present in abundance in airway epithelium and which modulates responses to the tachykinins. There is information from studies *in vitro* and *in vivo* that Sub P-induced bronchospasm in guinea-pigs is modulated by ACE (Shore *et al.*, 1988).

Pharmacological analysis of the functional activities of the tachykinins and more-selective related analogues in various tissues, in addition to radioligand-binding studies, provided the initial evidence that there are three distinct tachykinin receptors, designated Nk-1, Nk-2 and Nk-3, which mediated their adverse biological effects (Lee *et al.*, 1986; Maggi *et al.*, 1987; Regoli *et al.*, 1988). There is some evidence from functional studies for subtypes of Nk-2 and Nk-1 receptors, although it is not apparent if they represent merely species differences. The distribution of radiolabelled Sub P-binding sites in guinea-pig airways has been localized by autoradiographic techniques largely to airway smooth muscle (from the large to smaller airway segments), with less dense binding observed on vascular smooth muscle, epithelium and submucosal glands (Carstairs and Barnes, 1986; Goldie, 1990). In human airways, the majority of such sites are seen in association with non-airway smooth muscle sites in the submucosa, including postcapillary venules and mucous glands (Goldie, 1990).

The tachykinins are potent bronchoconstrictors of isolated airways from several species including humans; in humans, bronchospasm appears to be via an Nk2-mediated mechanism (Advenier *et al.*, 1987; Martling *et al.*, 1987; Naline *et al.*, 1989). *In vivo* in humans, administration of Sub P produces little effect on broncho-motor tone but exerts significant cardiovascular effects (Fuller *et al.*, 1987c), while NkA has more pronounced bronchoconstrictor effects and less cardiovascular effects (Joos *et al.*, 1987; Evans *et al.*, 1988). Further, the tachykinins may be responsible for non-adrenergic, non-cholinergic bronchoconstriction (Andersson and Grundstrom, 1983; Lundberg *et al.*, 1983a; Maggi, 1990). Sub P is a potent stimulant of mucus secretion in animal and human airways *in vitro* (Barnes *et al.*, 1986; Borson *et al.*, 1986; Shimura *et al.*, 1987). Sub P produces a rapid, transient increase in short-circuit current and conductance in canine isolated epithelium concomitant with an increase in Cl^- secretion, but does not effect Na^+ absorption (Rangachari and McWade, 1985). The tachykinins are potent vasodilators and hypotensive agents in animals and humans (Fuller *et al.*, 1987b; Laitinen *et al.*, 1987).

In addition, they increase microvascular leakage in rat and guinea-pig airways and may be responsible for that elicited by vagus nerve stimulation (McDonald, 1987; Brokaw *et al.*, 1990). These effects appear to be mediated via Nk-1 receptor activation. Sub P has been reported to affect cholinergic neurotransmission via both preganglionic and postganglionic mechanisms (Tanaka and Grunstein, 1984; Sekizawa *et al.*, 1987; Hall *et al.*, 1989). Their ability to modulate responses of inflammatory cells suggests that the tachykinins have a role in inflammation and may exert an immunomodulatory influence (Payan *et al.*, 1984b; Payan, 1989). The tachykinins degranulate mast cells, although this is not thought to occur via a receptor-mediated mechanism (Foreman and Jordan, 1983; Repke and Bienert, 1988). Sub P increases human T lymphocyte proliferation (Payan *et al.*, 1984b), stimu-lates guinea-pig alveolar macrophages (Brunelleschi *et al.*, 1990), and stimulates polymorphonuclear neutrophil (PMN) leukocyte or monocyte chemotaxis (Marasco *et al.*, 1981; Ruff *et al.*, 1985) and cytokine release from human monocytes (Lotz *et al.*, 1988).

The tachykinins may be involved in the aetiology of a variety of chronic inflammatory disorders including asthma, arthritis and inflammatory bowel disease, diseases involving hypersensitivity responses (Barnes, 1986, 1988; Payan, 1989; Spampinato and Ferri, 1991). In asthma,

Sub P may play a role via the axon reflex, whereby as a result of epithelial damage, afferent nerve endings containing sensory peptides are exposed. Stimulation of these nerve fibres by pro-inflammatory substances such as Bk or leukotrienes released from inflammatory cells would result in the release of the tachykinins to exert their myriad effects including bronchospasm, mucus secretion, microvascular leakage and cough (Barnes, 1986, 1988; Payan, 1989; Rogers *et al.*, 1988).

4.6 ENDOTHELIN

In 1988, Yanagisawa and co-workers described the isolation, cloning, sequencing and preliminary pharmaco-logical characterization of a novel, 21-amino acid vaso-constrictor peptide, designated ET-1 which was present in supernatants of porcine aortic endothelial cells. It was proposed to by synthesized in the cytoplasm from the precursor peptide, prepro-ET-1, which is cleaved by a dibasic-pair-specific endopeptidase to produce an inter-mediate 39 (porcine) or 38 (human) amino acid peptide called big ET-1 (Yanagisawa *et al.*, 1988; Masaki and Yanagisawa, 1992). Big ET-1 is then processed by a putative ECE, to produce the active 21-amino acid peptide. Conversion of big ET-1 is essential for expression of full biological activity. The identity and location of ECE remains uncertain, although evidence has been provided in endothelial cells that there may be different forms of the enzyme, two in the cytosol and one membrane bound (Ikegawa *et al.*, 1990; Okada *et al.*, 1990; Sawamura *et al.*, 1990; Matsumura *et al.*, 1991). It appears that the most likely candidate as the physio-logical ECE is the membrane-bound form which is inhibited by EDTA, EGTA and phosphoramidon, but not by thiorphan (a specific inhibitor of the neutral metalloprotease NEP), suggesting that ECE is a unique metalloprotease (Ikegawa *et al.*, 1990; Okada *et al.*, 1990; Masaki and Yanagisawa, 1992). The three isoforms of ET are encoded by three distinct genes (Inoue *et al.*, 1989) and, although they are structurally very similar, with all containing two characteristic disulphide bonds, they possess different biological activities. This provided the first evidence for multiple ET receptor subtypes (Masaki and Yanagisawa, 1992).

Evidence to date suggests that only ET-1 is produced by the endothelium, whereas all three isoforms are expressed in other tissues, including brain and kidney (Bloch *et al.*, 1989; Ohkubo *et al.*, 1990; Sakurai *et al.*, 1992). The focus of research has been on ET-1. Importantly, the airway epithelium is now known to be a site of ET synthesis and release. In the pulmonary system there is increasing information on the activities of ET, which has been implicated in the pathogenesis of asthma and pulmonary hypertension (Cernacek and Stewart, 1989). ET-like immunoreactivity is detected in air-way epithelium of several species including humans (MacCumber *et al.*, 1989; Rozengurt *et al.*, 1990;

Springall *et al.*, 1991). In the rat and mouse, intense staining is observed in mucous, serous and Clara cells, but not in basal or most ciliated cells (Rozengurt *et al.*, 1990). Furthermore, ET is released under basal conditions from cultured porcine, canine (Black *et al.*, 1989), human and bronchial epithelial cells (Mattoli *et al.*, 1990) and guinea-pig tracheal epithelial cells (Ninomiya *et al.*, 1991). Autoradiographic and binding studies, in addition to molecular biological analysis, have revealed the presence of ET receptors/binding sites and ET expression in a variety of tissues, including those in the cardiovascular system, kidneys, lungs, adrenal glands, intestine and brain, with the lungs along with the kidneys possessing the highest density of binding sites (Henry *et al.*, 1990; Kitamura *et al.*, 1990; Sakamoto *et al.*, 1991).

Binding and autoradiographic analyses have revealed large numbers of high-affinity ET-binding sites in the respiratory tract of several species (Henry *et al.*, 1990; McKay *et al.*, 1991). In human airways, labelling with ^{125}I-labelled ET-1 was localized largely to airway and vascular smooth muscle, with very low levels of diffuse binding associated with cartilage, connective tissues and the submucosal layer. Significant labelling was also associated with nerve trunks. Although ET is primarily synthesized in the epithelium, few specific binding sites were observed in this cellular layer (Tschirhart *et al.*, 1991). The expression of ET synthesized in endocrine cells of human lung changes with lung development, suggesting a role in growth regulation. With regard to asthma, an elevation in the levels of ET in BAL has been observed in stable asthmatics (Mattoli *et al.*, 1991) and also in one patient during a status asthmatic attack (Nomura *et al.*, 1989). Furthermore, markedly increased expression of ET was observed within airway epithelium and also in vascular endothelium of endobronchial biopsy specimens from asthmatics compared with non-asthmatics (Thorsen, 1986).

At present there is only preliminary and, in some cases, conflicting information on the effects of ET on relevant parameters of airway function other than bronchospasm, including airway smooth muscle proliferation (Noveral *et al.*, 1992), airway hyper-reactivity (Lagente *et al.*, 1990), microvascular permeability and oedema (Macquin-Mavier *et al.*, 1989; Filep *et al.*, 1991; Pons *et al.*, 1991) and mucus secretion (Shimura *et al.*, 1992).

5. *Interactions Between Epithelium and Inflammatory Mediators*

5.1 AIRWAY SMOOTH MUSCLE TONE

The first report concerning the potential modulatory influence of the epithelium on the level of tone in the airway smooth muscle was published by Orehek *et al.* in 1975, who demonstrated that scraping the mucosal surface of the guinea-pig trachea stimulated the release of prostaglandins. It was hypothesized that the local release of prostaglandins played an important role in the regulation of airway smooth muscle tone (Orehek *et al.*, 1975). These novel observations were not investigated further for a decade, when experiments in canine bronchus demonstrated the modulatory influence of mechanical removal of the epithelium on contractile responses to several agonists and also on relaxant responses to isoproterenol (Flavahan *et al.*, 1985). These workers also hypothesized that the airway epithelium has the potential to release a substance(s) that exerts an inhibitory influence on adjacent smooth muscle cells. Subsequent to this report, there was confirmation from various different groups of investigators, using airway and non-airway smooth muscle preparations from various species including humans, of the modulatory effects of the epithelium on isolated tissue responsiveness (for reviews, see Farmer, 1987; Goldie *et al.*, 1988, 1990; Vanhoutte, 1988). Accordingly, this is now an accepted and widely studied phenomenon for which there is a burgeoning literature.

The evidence suggests strongly that the epithelium influences airway smooth muscle activity via several mechanisms, with their contributions varying depending on the bronchoactive substance involved. These include: (1) acting as a diffusion barrier *in vivo*; (2) serving as a site of metabolism for some substances; (3) acting as a source of inhibitory and contractile substances. In relation to epithelium-derived inhibitory substances, these are thought to include both prostanoids (e.g. PGE2) and non-prostanoids. The latter material, which has been designated EpDIF or EpDRF, has been the focus of much research, with several laboratories attempting to isolate, identify and pharmacologically and biochemically characterize this substance. Unfortunately, EpDIF has proven to be an elusive target.

This section will focus on the influence and mechanism(s) whereby the epithelium modulates the responsiveness of airway smooth muscle to various inflammatory mediators outlined above. Many studies involved determining the effects of mechanical removal of the epithelium on responsiveness of isolated airway smooth muscle.

5.1.1 Histamine

Several studies indicate that epithelium removal produces two- to eight-fold increases in sensitivity of guinea-pig tracheal smooth muscle to histamine; the maximum contraction has been demonstrated to be either elevated (Hay *et al.*, 1986a; Murlas, 1986; Braunstein *et al.*, 1988; Prié *et al.*, 1990) or unaffected (Goldie *et al.*, 1986; Holroyde, 1986). Prié *et al.* (1990) also reported an increase in responsiveness to histamine in guinea-pig isolated bronchus following epithelium removal. In a study with trachea obtained from guinea-pigs sensitized to ovalbumin, Montano and co-workers (1988) reported that the potentiating effect of epithelium removal on responsiveness to histamine was greater in these preparations than in tissues from non-sensitized controls. In canine bronchi (Flavahan *et al.*, 1985; Manning *et al.*, 1990), epithelium removal produced a significant increase in airway sensitivity and maximum response to histamine. Epithelium removal from human isolated bronchi increases the sensitivity to histamine by approximately two-fold (Aizawa *et al.*, 1988; Fernandes *et al.*, 1990). Braunstein and colleagues (1988) demonstrated that histamine caused a greater release of PGE2 from epithelium-intact compared with epithelium-free guinea-pig trachea, and they concluded that the modulatory influence of the epithelium was in large part attributable to the release of relaxant PGE_2 therefrom. However, another report provided no evidence for histamine-induced release of PGE_2, $PGF_{2\alpha}$ or PGI_2 from guinea-pig tracheal epithelium (Hay *et al.*, 1988). Furthermore, since the epithelium contains histaminase, it may modulate histamine-induced responses in guinea-pig trachea by acting as a site for its metabolism (Lindström *et al.*, 1991). The existence of a non-H_1 or non-H_2 histamine receptor which stimulated the release of a non-prostanoid EpDIF was suggested (Güc *et al.*, 1988c).

5.1.2   Peptidoleukotrienes

There is conflicting information on the influence of the epithelium on responsiveness of airway tissues to the peptidoleukotrienes. In the first report, in guinea-pig trachea, sensitivity to LTC_4 and LTD_4 was increased two- to four-fold after epithelium removal, whereas LTD_4-induced responses were unaltered (Hay *et al.*, 1987a). Indomethacin mimics the effect of epithelium removal on sensitivity to LTC_4, but not LTD_4, suggesting that the contractile effect of LTC_4 is modulated by a smooth muscle inhibitory prostanoid derived from the epithelium, whereas that of LTD_4 is influenced by a non-prostanoid substance (Hay *et al.*, 1987a).

In contrast, another study utilizing guinea-pig trachea showed that the influence of epithelium removal on the LTD_4 concentration–response curve (6.3-fold leftward shift) was inhibited by flurbiprofen, a cyclooxygenase inhibitor (Hisayama *et al.*, 1988).

Prié and colleagues (1990) reported that, in guinea-pig isolated trachea and bronchus, responsiveness to all three peptidoleukotrienes was increased following epithelial denudation. LTD_4 and also histamine stimulated the release of PGE_2 and 6-keto-$PGF_{1\alpha}$, although this was not influenced by removal of the epithelium (Prié *et al.*, 1990). In contrast, no effect of LTD_4 on release of PGE_2, $PGF_{2\alpha}$ and PG_1 from guinea-pig trachea was observed in either the presence or absence of the epithelium (Hay *et al.*, 1988). This would suggest that the modulatory effect of the epithelium involved the release of a non-prostanoid substance(s).

Buckner *et al.* (1990) showed that epithelium removal produced a three-fold increase in sensitivity to LTD_4 in human bronchus, via an indomethacin-sensitive mechanism, but had no influence on the responsiveness to LTC_4 or

LTE_4. This suggested that an epithelium-derived inhibitory cyclooxygenase product, perhaps PGE_2, modulated LTD_4-induced responses in human airways.

In the same study it was noted that epithelium removal potentiated sensitivity of isolated guinea-pig trachea to LTC_4 and LTD_4, but caused an approximately three-fold decrease in sensitivity to LTD_4. This latter effect was abolished by lipoxygenase inhibitors, suggesting that the inhibitory effect of epithelium removal on responses to LTE_4 might be due to removal of an epithelium-derived smooth muscle excitatory product of 5-lipoxygenase, release of which is stimulated by LTE_4 (Buckner *et al.*, 1990).

5.1.3   Platelet-activating Factor

In precontracted guinea-pig trachea, PAF induced an epithelium-dependent relaxation and also stimulated the release of PGE_2 (Brunelleschi *et al.*, 1987). PAF-induced synthesis of PGE_2 and relaxation in this tissue was abolished by epithelium removal and markedly decreased by cyclooxygenase inhibition. These data suggest that PAF causes relaxation of guinea-pig isolated trachea by a mechanism involving the synthesis and release of PGE_2 from epithelial cells (Brunelleschi *et al.*, 1987).

5.1.4   Bradykinin

Although a potent bronchoconstrictor in animals and asthmatic humans, Bk has little direct spasmogenic activity on airway smooth muscle from all species examined, except guinea-pigs and ferrets (Farmer, 1991). In the latter species, Bk-induced tracheal contraction is unaffected by epithelium removal (Dusser *et al.*, 1988). Conversely, in guinea-pig trachea, Bk causes epithelium-dependent relaxation; following epithelium removal, the kinin induces concentration-dependent contraction (Bewley *et al.*, 1988; Farmer *et al.*, 1989; Bramley *et al.*, 1990). In the presence of indomethacin, Bk contracts guinea-pig trachea containing and lacking an intact epithelium (Farmer *et al.*, 1989; Bramley *et al.*, 1990).

Bramley and colleagues (1990) also demonstrated that PGE_2, like Bk, relaxes intact trachea but contracts denuded preparations. Furthermore, the ability of Bk to increase PGE_2 production in this tissue is dramatically reduced following denudation, and it has been proposed that the epithelium-dependent relaxation of guinea-pig trachea, in response to Bk, is mediated by this prostanoid.

Conditioned culture medium from canine tracheal epithelial cells, stimulated with Bk, causes a dramatic inhibition of contractile responses of canine trachea in response to electrical stimulation of intramural cholinergic nerves (Barnett *et al.*, 1988). Bk alone or control culture medium are without effect on neurogenic contractions in this tissue. Similarly, if the epithelial cells are incubated with indomethacin, Bk fails to generate the epithelium-derived smooth muscle inhibitory factor. Moreover, supernatants from Bk-treated epithelial cells do not influence tracheal sensitivity to exogenous ACh (Barnett *et al.*, 1988).

Accordingly, the ability of epithelial culture medium to inhibit neurogenic contractions is probably due to Bk stimulating production of an epithelium-derived product of cyclooxygenase. This EpDIF is likely to be PGE_2, which is generated by canine airway epithelium (Stuart-Smith and Vanhoutte, 1988b; Aizawa et al., 1990; Liu et al., 1990). Barnett and colleagues (1988) also showed that Bk stimulated an almost eight-fold increase in PGE_2 synthesis by their epithelial cells. PGE_2 is known to inhibit neurogenic contractions in the dog airway (Walters et al., 1988). These data are, therefore, consistent with the hypothesis that Bk may indirectly modulate cholinergic neurotransmission in canine airways via production of epithelium-derived PGE_2.

5.1.5   Tachykinins

Overall, the epithelium exerts a profound effect on the responsiveness of guinea-pig trachea to the tachykinins, of which Sub P has been the most widely studied. The modulatory influence occurs via two mechanisms: (1) by acting as a metabolic site, since it contains large amounts of NEP and (2) by releasing an inhibitory substance(s); the former appears to be the predominant mechanism. Thus, removing the epithelium produced 20–150-fold increases in sensitivity to Sub P (Tschirhart and Landry, 1986; Devillier et al., 1988; Fine et al., 1989; Frossard et al., 1989); an elevation of the maximum contraction has also been reported (Fine et al., 1989; Frossard et al., 1989). Inhibitors of NEP such as phorphoramidon or thiorphan decreased and, in some instances, abolished the effects of epithelium removal (Sekizawa et al., 1987; Devillier et al., 1988; Djokic et al., 1989; Fine et al., 1989; Maggi et al., 1990), suggesting that loss of epithelial NEP activity and, thus, enzymatic degradation of Sub P, is the principal mechanism underlying the increased responsiveness.

It has been shown that, in the presence of indomethacin, removal of guinea-pig tracheal epithelium caused a 148-fold increase in sensitivity to Sub P. This was markedly reduced in the presence of phosphoramidon to an 18-fold increase in sensitivity. Thus, although loss of NEP is the main mechanism by which epithelial denudation increases responsiveness to Sub P, there may also be a significant influence of a non-prostanoid inhibitory substances (Fine et al., 1989).

In guinea-pig bronchus, in which neurokinins mediate non-cholinergic neuronal bronchoconstriction (Lundberg et al., 1983b; Lundberg and Saria, 1987), epithelium removal enhanced responsiveness to substance P (Maggi et al., 1990) and to stimulation of intramural non-cholinergic nerve endings (Djokic et al., 1989). Epithelium removal also increases the sensitivity of ferret trachea (Sekizawa et al., 1987) and human bronchus (Naline et al., 1989) to Sub P, although to a lesser extent than in guinea-pig airway smooth muscle.

Phosphoramidon also potentiated responsiveness of human bronchial smooth muscle to substance P (Black

et al., 1988; Naline et al., 1989). Naline and colleagues (1989) found that epithelium removal increased sensitivity to Sub P and, to a lesser degree, to NkA. Inhibition of NEP abolished the effect of epithelium removal on sensitivity to Sub P (in the presence of phosphoramidon). However, epithelium removal still increased the sensitivity to NkA (Naline et al., 1989). Thus, the airway epithelium modulated responsiveness of human airways to neurokinins, in part by enzymatic degradation by NEP and, at least for NkA, by another inhibitory mechanism.

Precontracted rat trachea relaxes in response to Sub P, and this is markedly inhibited after removal of the epithelium, suggesting that in this tissue Sub P induces the release of an epithelium-derived smooth muscle relaxant (Frossard and Muller, 1986).

5.1.6   Endothelin

ET-1 is a potent spasmogenic agent in guinea-pig airway smooth muscle (Hay, 1989; Henry et al., 1990), an effect that is enhanced by removing the epithelium (Hay, 1989). The ability of epithelium removal to increase tracheal responsiveness to ET is evident in the presence of indomethacin, captopril, leupeptin, or bacitracin, indicating a lack of involvement of epithelium-derived prostaglandin, ACE, endopeptidase or aminopeptidase, respectively (Hay, 1989). In the presence of phosphoramidon, however, the effect of epithelium removal is abolished. Thus, the increased responsiveness of denuded guinea-pig trachea to ET appears to be due to removal of metabolism influences of NEP associated with the epithelium (Hay, 1989). The ETs have been demonstrated to be good substrates for NEP (Vijayaraghavan et al., 1990; Fagny et al., 1991).

The release of ET from cultured tracheal and bronchial epithelium from several species has been demonstrated (Black et al., 1989; Mattoli et al., 1990; Ninomiya et al., 1991). It has also been proposed that the epithelium-dependent augmentation of ACh-induced contractions of canine trachea in situ, induced by eosinophil major basic protein, is mediated by an epithelium-derived contractile factor (Brofman et al., 1989). An appropriate candidate for this substance is ET-1.

5.1.7   Antigen

It has been demonstrated that antigen-induced contraction in human and guinea-pig isolated airways is mediated exclusively via the release of histamine and leukotrienes (Adams and Lichtenstein, 1979). Given the interactions between the epithelium and these two mediators, it might be anticipated that the epithelium should exert effects on antigen-induced responses.

Removal of the epithelium produced six- to eight-fold increases in sensitivity of tracheal smooth muscle, isolated from actively sensitized guinea-pigs, to the contractile effects of ovalbumin (Hay et al., 1986b; Undem et al., 1988). This phenomenon may be the result of the effect of epithelium on responses to the released mediators, i.e.

histamine and leukotrienes. Alternatively, it is possible that epithelium-derived substances, e.g. EpDIF, influence antigen-induced mediator release from mast cells. With regard to this, epithelium removal from superfused tracheal preparations caused an approximately 30% increase in ovalbumin-induced release of histamine, as well as a 50% reduction in release of PGE_2 (Undem *et al.*, 1988).

However, one has to consider that ovalbumin is a large protein and, accordingly, loss of the protective barrier influence of the epithelium may enhance responses to antigen challenge. The effects of epithelium removal on antigen-induced responses (six-fold increase in sensitivity) of guinea-pig trachea observed by Undem and co-workers (1988) are unlikely to be due to loss of inhibitory prostanoids since, in the presence of indomethacin, denudation produced a more dramatic 300-fold increase in sensitivity to ovalbumin. In superfusion cascade studies, no direct evidence for the release of an inhibitory factor from the guinea-pig tracheal epithelium in response to antigen or histamine was obtained.

In the same study, epithelium removal had little effect on the rate or magnitude of ovalbumin-induced release of immunoreactive peptidoleukotrienes, PGD_2 (both mast cell-derived mediators), $PGF_{2\alpha}$ or thromboxane and had no effect on total histamine content. However, histamine release was enhanced, whereas PGE_2 and 6-keto-$PGF_{1\alpha}$ release were attenuated following epithelium removal (Undem *et al.*, 1988). The effect of epithelium removal is due in significant part to the loss of an epithelial diffusion barrier that normally restricts access of large antigenic molecules to tissue mast cells (Undem *et al.*, 1988).

5.1.8   Nerve-Induced Responses

Several studies have investigated the effects of the epithelium on nerve-induced contractile responses of isolated airway smooth muscle from various species. The focus has been on the cholinergic nerve-mediated contractions, and conflicting effects of epithelium removal in tissues from different species have been noted. In guinea-pig trachea, removal of the epithelium was reported to have no effect (Holroyde, 1986) or to potentiate, via a non-prostanoid mechanism (Murlas, 1986), responses to EFS. In this tissue, neuronal release of $[^3H]ACh$, in response to EFS, was substantially increased following epithelium removal (Wessler *et al.*, 1990). In bovine trachea, removal of the epithelium was without effect on EFS-induced responses (Barnes *et al.*, 1985), whereas in dog bronchus the profile of the response is altered, with no "fade" observed in denuded preparations, perhaps via a prejunctional and/or postjunctional mechanism (Flavahan *et al.*, 1988). In contrast, in porcine third- and fourth-order bronchial smooth muscle, epithelium removal elicits a rightward shift in the frequency–response curve for neurogenic contractions (Stuart-Smith and Vanhoutte, 1988a). This is in contrast to the leftward

shift in the concentration–response curves for exogenous ACh in porcine bronchi. It was suggested that there exists an epithelium-derived contracting factor that augments contractile responses to endogenous ACh either prejunctionally or postjunctionally (Stuart-Smith and Vanhoutte, 1988b). The release of a cyclooxygenase constrictor substance from a hamster lung epithelial cell line was recently described (Wilkens *et al.*, 1992).

Ullman and colleagues (1988, 1990) demonstrated that, in ferret trachea containing epithelium, phasic contractions exhibited successive decreases in magnitude during repetitive application of EFS or direct stimulation of the vagus nerve, whereas in denuded preparation this was not observed. Furthermore, responses in denuded trachea were inhibited by bathing fluid transferred from neurally stimulated, intact preparations (Ullman *et al.*, 1988). The epithelium-dependent inhibitory influence present in transferred buffer fluid was reduced, but not abolished, after treatment of donor tissues with indomethacin, suggesting that both prostanoid and non-prostanoid inhibitor factors are released from the epithelium in response to vagus nerve stimulation (Ullman *et al.*, 1988).

In guinea-pig bronchus, removal of epithelium results in a small potentiation of responses elicited by the release of tachykinins from non-cholinergic nerve endings (Djokic *et al.*, 1989). This appears to be attributed to loss of epithelial NEP, as phosphoramidon abolishes the effects of epithelium removal (Djokic *et al.*, 1989). In a recent study examining the effects of the epithelium on membrane potential, neurotransmission and the level of tone in canine trachea and bronchus, an inhibitory effect of the epithelium on EFS-induced responses was observed in the presence of indomethacin. Furthermore, addition of isolated and dispersed epithelial cells produced hyperpolarization of the smooth muscle membrane, an effect which was not altered by indomethacin (Xie *et al.*, 1992).

5.2   EPITHELIUM-DERIVED INHIBITORY FACTOR

As indicated above, the major focus of research on the modulatory role of the epithelium on inflammatory mediators has been the examination of its influence on the responsiveness of isolated airway smooth muscle. In particular, there has been increasing interest in the putative, non-prostanoid, EpDIF. This remains a controversial topic with conflicting information and evidence regarding the existence, identity and relevance of this factor. Some of this is discussed below.

5.2.1   Evidence for Epithelium-derived Inhibitory Factor

1. The loss of an epithelial diffusion barrier is unlikely to alter the sensitivity or magnitude of equilibrium responses to agents that are not subject to degradation. However, the rate of the response may be affected. In

studies employing strips of isolated airway, substances added to the bathing medium have free access to the smooth muscle.

2. The effects of epithelium removal are to some extent selective, in that the responses to some, but not all, bioactive agents are affected.

3. The increases in responsiveness to stimulation of intramural nerve endings is unlikely to result from loss of a barrier.

4. In addition to evidence from studies with open airway preparations, some experiments utilizing perfused whole trachea from guinea-pigs and rats provided further evidence for an inhibitory factor released from the epithelium (Güc *et al.*, 1988b).

5. The most direct evidence has been obtained from studies demonstrating the transfer of epithelium-dependent relaxant activity in response to various agonists in superfusion/perfusion bioassay preparations involving vascular and non-vascular smooth muscles (Ilhan and Sahin, 1986; Güc *et al.*, 1988a; Fernandes *et al.*, 1989).

5.2.2 Evidence Against Epithelium-derived Inhibitory Factor

1. The differential effects of epithelium on the responsiveness of airway smooth muscle preparations to agonists has been used as an argument against, rather than for, EpDIF. For example, it was proposed that small inorganic ions such as K^+ would pass through epithelial tight junctions, and that contractions induced by these agents, unlike those produced by larger organic molecules, would not be influenced by the epithelium (Holyoyde, 1986).

2. Overall there has been limited success in experiments with superfusion/perfusion systems to demonstrate directly the release and transfer of an epithelium-derived inhibitory substance using a guinea-pig airway smooth muscle as a bioassay tissue (Holroyde, 1986; Undem *et al.*, 1988). Vanhoutte's group has provided preliminary evidence for the release and transfer of an effective epithelium-derived relaxant of canine airway smooth muscle (Vanhoutte, 1988). However, great difficulty was encountered in reproducing these results (N.A. Flavahan, personal communication). In fact, greater success has been achieved in demonstrating airway epithelium-dependent relaxation of non-airway rather than of airway smooth muscle, perhaps indicating the existence of several EpDIFs which have differences in labilities and tissue specificities.

3. Gunn and Piper (1991) recently presented interesting preliminary evidence that the airway epithelium-dependent nature of relaxation of a strip of rat aorta, mounted within a guinea-pig tracheal cylinder, may be related to hypoxia of the vascular smooth muscle as a result of contraction of the airway tubes and epithelial uptake of oxygen, rather than to the release of an inhibitory factor. However, not all airway smooth muscle spasmogens induce relaxation of rat aorta in coaxial bioassay experiments and relaxations are not seen in coaxial assemblies that use rat trachea as the donor tissue (Fernandes and Goldie, 1990). Furthermore, recent pharmacological and biochemical evidence was provided which is at odds with the postulate of a significant role of hypoxia in mediating epithelium-dependent relaxation of vascular smooth muscle in coaxial bioassay systems (Hay *et al.*, 1992; Spina *et al.*, 1992).

4. There is conflicting evidence from studies employing guinea-pig or rat perfused tracheal tubes for the existence of EpDIF, with some studies indicating much greater effects of epithelium removal when applied luminally (mucosal side) as opposed to serosally (advential side), suggesting a substantial diffusion barrier (Fedan *et al.*, 1990; Small *et al.*, 1990), whereas others found minimal differences (Pavlovic *et al.*, 1989). Interestingly, in guinea-pig tracheal strip preparations in which agents were applied specifically via mucosal or adventitial sides, marked effects of epithelium removal on responses to ACh or histamine occurred when the agonist was administered via the mucosal route whereas no effect occurred when given adventitially (Iriarte *et al.*, 1990). This would again suggest a significant diffusion barrier role for the epithelium.

With regard to studies examining the effects of epithelium removal on agonist-induced responses in isolated airway smooth muscle preparations, a cautionary note was provided by the results of Nadel and colleagues, who observed that, in contrast to enzymatic removal, mechanical removal of the epithelium from isolated airways from several species resulted in disruption of mast cells. It is possible that the release of mediators from mast cells may influence the activity of other cells including smooth muscle (Franconi *et al.*, 1990).

5.3 MUCOUS GLANDS

The influence of the epithelium on mucus secretion, measured as [³H]glucosamine-labelled mucus glyco-conjugate, from feline tracheal submucosal glands has been investigated. The secretory response of isolated glands was decreased by addition of isolated epithelium to the level observed in tracheal explants via a mechanism which did not appear to be mediated by either cyclooxygenase or lipoxygenase products of arachidonic acid metabolism (Sasaki *et al.*, 1989). These intriguing findings warrant further investigation and suggest that an epithelium-derived, non-prostanoid inhibitory factor can modulate a parameter in the pulmonary system other than the level of tone in airway smooth muscle. Whether this substance is the same as the postulated EpDIF remains to be determined.

5.4   INFLAMMATORY CELLS

In addition to exerting a significant modulatory effect on the bronchoconstrictor activity of mediators released from a variety of inflammatory cells, the epithelium may have a profound influence on the recruitment and influx of inflammatory cells into the airways (Rennard et al., 1991). Inflammatory cells may also influence epithelial function. For example, they may alter ion transport and mucus secretion via the release of mediators into the epithelial cell layer (vide supra).

The influence of the airway epithelium on inflammatory cell migration and direction may occur via at least two mechanisms: (Abe et al., 1967) the release of chemotactic substances and (Abraham et al., 1991) the presence of adhesion molecules. There is much more information on the former than the latter mechanism. Thus, various groups of investigators have provided evidence that the airway epithelium releases various substances which are chemotactic for neutrophils, monocytes, lymphocytes, eosinophils, basophils and mast cells (Rennard et al., 1991). These factors include LTB_4 (Holtzman et al., 1983a) and cytokines such as GM-CSF and G-CSF (Jordana et al., 1989; Smith et al., 1989; Cox et al., 1990), in addition to several, as yet, unidentified materials (Rennard et al., 1991).

Neutrophils and eosinophils have been demonstrated to adhere to cultured airway epithelial cells. For example, neutrophil adherence to bovine bronchial epithelium, which was increased by exposure to cigarette smoke, lipopolysaccharide or phorbol ester, was associated with epithelial cell cytotoxicity (Robbins et al., 1990). The markedly increased adherence of neutrophils to human tracheal epithelial cells induced by human parainfluenza virus type 2 appears to involve both ICAM-1-dependent and -independent mechanisms (Tosi et al., 1991a). Furthermore, inflammatory cytokines such as TNF and IL-1 may stimulate ICAM-1 expression on primary cultures of human tracheal epithelial cells (Tosi et al., 1991b). Mast cells selectively adhered to canine epithelial cells, to surface epithelial or submucosal glands, but not to basal membrane or connective tissue, compared with type I and IV collagen or fibronectin (Varsano et al., 1988). ICAM-1 expression was increased in airway epithelium in vitro and in vivo in a monkey model of airway hyper-reactivity following antigen inhalation concomitant with enhanced eosinophil infiltration to the epithelial environment (Wegner et al., 1991).

5.5   EPITHELIAL DAMAGE

For many years it has been recognized that damage to the airway epithelium associated with inflammation is a classic feature of asthma (Cutz et al., 1978; Dunnill, 1960; Laitinen et al., 1985, 1991). Thus, the sputum of asthmatics contains epithelial cells, and their number is elevated during acute exacerbations of the disease. Bronchial biopsies from asthmatic individuals revealed separation of basal and luminal components of the epithelium due to the effects of oedema fluid, with increased intercellular spaces at the base of the epithelium and detachment of basal cells from a thickened basement membrane and evidence of oedema in lamina propria (Laitinen et al., 1985; Laitinen et al., 1991). Indications of epithelial cell regeneration such as squamous metaplasia and ciliogenesis are present. Ciliated cells are those affected most with loss of cilia, intracellular oedema and cell destruction (Laitinen et al., 1985; Laitinen et al., 1991). Oedema of the epithelium may affect nerve profiles near the airway lumen. It has been reported that the number of epithelial mucous cells are increased in asthmatics (Laitinen et al., 1991). There is evidence of epithelial inflammation and the influx of various inflammatory cells, in particular eosinophils, but also including mast cells (Laitinen, 1989).

The precise mechanism(s) which is responsible for the damage to and sloughing of the airway epithelium has not been completely elucidated. However, largely as a consequence of the extensive pioneering work by Gleich and colleagues (1979), the significant contribution of the cytotoxic products of eosinophils, notably MBP, has been proposed. Infiltration of eosinophils into the airways, including the epithelial layer, is another feature of asthma (Cutz et al., 1978; Beasley et al., 1989) and a correlation between eosinophil number and the changes in pulmonary function has been reported (Horn et al., 1975). Eosinophils are thought to play a critical role in the pathogenesis of asthma. MBP produces airway epithelial damage which is histologically similar in nature to that observed in asthmatic airways (Dunnill, 1960; Gleich et al., 1979; Laitinen et al., 1985; Motojima et al., 1989), and which is associated with increased responsiveness of isolated airway tissues (Flavahan et al., 1988; Brofman et al., 1989).

However, there are other possible mechanisms via which epithelial damage may occur in asthmatic airways, including release of toxic substances from other inflammatory cells. Elevated neutrophil levels have been observed, temporally associated with the late-phase asthmatic response following antigen provocation (De Monchy et al., 1985). Furthermore, in one patient an association between epithelial damage and extracellular neutrophil elastase was noted. However, in other patients, neutrophil infiltration and extracellular neutrophil elastase deposition was very limited (Fujisawa et al., 1990). The link between increased epithelial permeability, enhanced microvascular permeability, submucosal oedema and desquamation and damage to the epithelium has not been evaluated fully.

5.6   PATHOPHYSIOLOGICAL SIGNIFICANCE

The pathophysiological significance of epithelial damage in respiratory disorders, most notably asthma, has not been definitively elucidated and remains a subject of considerable speculation and debate. Nevertheless, there is widespread support for the notion that damage or dysfunction to the epithelium contributes significantly to the airway hyper-reactivity that is a feature of asthma (T. Haahtela, A. Laitinen and L.A. Laitinen, submitted).

The increased exposure of nerve fibres in the asthmatic airways, as a result of epithelial damage, will permit enhanced stimulation by pro-inflammatory substances such as Bk, which may result in reflex bronchoconstriction via vagal or local pathways. Alternatively, and probably more importantly, stimulation of sensory nerve C fibres may result in the establishment of an axon reflex involving antidromic conduction and release of pro-inflammatory sensory neuropeptides, including the tachykinins, which may result in numerous deleterious effects including neurogenic inflammation, enhanced mucus secretion and bronchoconstriction. Loss of NEP as a result of damage to epithelial cells will enhance the effects of the tachykinins. Overall, these events have the potential to create a significantly amplified inflammatory response in the affected region of the airways.

Despite the relative uncertainty about the precise mechanisms and their relative importance, whereby the epithelium is involved in the pathogenesis of respiratory disorders, there is unequivocal agreement that, by virtue of its location and biological characteristics, it plays a central role in the complex inflammatory processes which are characteristic of airway diseases, such as asthma. In fact it has been suggested that injury to the epithelial cell is the initial event during airway inflammation (Holzman, 1985).

6. Conclusions

It is now universally recognized that the mammalian airway epithelium does not merely serve as a diffusion barrier, but is a complex, multifunctional and extremely dynamic cellular layer consisting of several different specialized cell types. By virtue of its strategic location, being the first site of interaction with airborne substances, in addition to its diverse biological activities, the epithelium plays a critical role in the homeostatic control of the pulmonary system. In addition, dysfunction of, or an alteration in the integrity of, the epithelium appears to be a key event in the pathogenesis of several respiratory disorders, most notably asthma. Thus, the epithelium appears to be a key player in the initiation — perhaps as a result of epithelial cell injury — progression and regulation of the inflammatory process. Various postulates have been developed to explain the mechanism by which the epithelium exerts its powerful modulatory role. It has to be emphasized that the barrier function of the epithelium appears to be the most important component of its modulatory activity.

Many of the modulatory actions of the epithelium are associated with interactions with inflammatory cells and their constituents. The focus of this chapter has been an examination of the ability of the epithelium to influence the bronchoconstrictor effects of mediators released from inflammatory cells. The evidence, the vast majority of which has come from *in vitro* studies, indicates that the epithelium possess prostanoid and non-prostanoid (e.g. ET) stimulatory and inhibitory (e.g. EpDIF) substances, and the net influence will result from the combined effects of these inputs. In diseased airways, it is possible that the balance is tilted in favour of enhanced bronchospasm. The identity of the postulated EpDIF is unknown, and the importance and pathophysiologically significance of EpDIF and other epithelium-derived relaxant and constrictor substances is unclear. The evidence to date suggests that it is minor. Vascular smooth muscle appears to be more sensitive than airway smooth muscle to the effects of EpDIF and it is conceivable that it is the airway mucosal microcirculation which is the locus for the physiological effects of EpDIF. Thus, EpDIF, by virtue of its vasodilator activity, may be involved in the clearance of endogenous and exogenous substances from the airway mucosa, thereby protecting the airway smooth muscle, nerves, inflammatory cells and airway mucous glands from the effects of these materials.

In addition to perhaps directly influencing the broncho-active activity of products of the inflammatory cells, the epithelium may also play an important role in the recruitment and trafficking of these cells via the release of chemotactic and biologically active substances such as cytokines and LTB_4. Damage to the epithelium may also expose intraepithelial sensory nerve fibres to the effects of inflammatory cell mediators, such as Bk, setting up local axon reflexes which will contribute to amplification of the inflammatory response.

From a therapeutic standpoint, one of the main challenges of future research in this complex area will be the identification of molecular targets and strategies for the design of novel drugs targeted at the epithelium, which may control the inflammation that is characteristic of many diseases of the pulmonary system.

7. References

Abe, K., Watanabe, N., Kumagai, N., Mouri, T., Seki, T. and Yoshinaga, K. (1967). Circulating plasma kinin in patients with bronchial asthma. Experientia 23, 626–627.

Abraham, W.M., Burch, R.M., Farmer, S.G., Ahmed, A. and Cortes, A. (1991). A bradykinin antagonist modifies allergen-induced mediator release and late bronchial responses in sheep. Am. Rev. Respir. Dis. 143, 787–796

Adams, G.K. and Lichtenstein, L. (1979). *In vitro* studies of antigen-induced bronchospasm: effect of antihistamine and SRS-A antagonist on response of sensitized guinea pig and human airways to antigen. J. Immunol. 122, 555–562.

Adelroth, E., Morris, M.M., Hargreave, F.E. and O'Byrne, P.M. (1986). Airways responsiveness to leukotrienes C_4 and D_4

and to methacholine in patients with asthma and normal control. N. Engl. J. Med. 315, 480–484.

Advenier, C., Naline, E., Drapeau, G. and Regoli, D. (1987). Relative potencies of neurokinins in guinea pig trachea and human bronchus. Eur. J. Pharmacol. 139, 133–137.

Advenier, C., Devillier, P., Matran, R. and Naline, E. (1988). Influence of epithelium on the responsiveness of guinea-pig isolated trachea to adenosine. Br. J. Pharmacol. 93, 295–302.

Aizawa, H., Miyazaki, N., Shigematsu, N. and Tomooka, M. (1988). A possible role of airway epithelium in modulating hyperresponsiveness. Br. J. Pharmacol. 93, 139–145.

Aizawa, T., Sekizawa, K., Aikawa, T., Maruyama, N., Itabashi, S., Tamura, G., Sasaki, H. and Takishima, T. (1990). Eosinophil supernatant causes hyperresponsiveness of airway smooth muscle in guinea pig trachea. Am. Rev. Respir. Dis. 142, 133–137.

Anderson, M.P., Sheppard, D.N., Berger, H.A. and Welsh, M.J. (1992). Chloride channels in the apical membrane of normal and cystic fibrosis airway and intestinal epithelia. Am. J. Physiol. 263 (Lung Cell. Mol. Physiol. 7), L1–L14.

Andersson, R.G.G. and Grundstrom, N. (1983). The excitatory non-cholinergic, non-adrenergic nervous system of the guinea-pig airways. Eur. J. Respir. Dis. 64(Suppl. 131), 141–157.

Archer, C.B., Page, C.P., Paul, W., Morley, J. and McDonald, D.M. (1984). Inflammatory characteristics of platelet activating factor (PAF-acether) in human skin. Br. J. Dermatol. 110, 45–50.

Arrang, J.M., Garbarg, M. and Schwartz, J.C. (1983). Auto-inhibition of brain histamine release mediated by a novel class (H3) of histamine receptor. Nature 302 832–837.

Arrang J.M., Garbarg, M., Lancelot, J.C., Lecomte, J.M., Pollard, H., Robba, M., Schunack, W. and Schwartz, J.C. (1987). Highly potent and selective ligands for histamine H3–receptors. Nature, 327, 117–123.

Baraniuk, J.N. (1991). Neural control of human nasal secretion. Pulm. Pharmacol. 4, 20–31.

Barnes, N.C., Piper, P.J. and Costello, J.F. (1984). Comparative effects of inhaled leukotriene C4, leukotriene D4, and histamine in normal human subjects. Thorax 39, 500–504.

Barnes, P.J. (1986). Asthma as an axon reflex. Lancet i, 242–245.

Barnes, P.J. (1988). Neuropeptides and airway smooth muscle. Pharmac. Ther. 36, 119–129.

Barnes, P.J. and Basbaum, C.B. (1983). Mapping of adrenergic receptors in the trachea by autoradiography. Exp. Lung Res. 5, 183–192.

Barnes, P.J. and Henson, P.M. (1991). In "The Lung: Scientific Foundations" (eds R.G. Crystal and J.B. West), pp 49–59. Raven Press, New York.

Barnes, P.J., Cuss, F.M. and Palmer, J.B. (1985). The effect of airway epithelium on smooth muscle contractility in bovine trachea. Br. J. Pharmacol. 86, 685–691.

Barnes, P.J., Dewar, A. and Rogers, D.F. (1986). Human bronchial secretion: effect of substance P, muscarinic and adrenergic stimulation *in vitro*. Br. J. Pharmacol. 89, 767P.

Barnes, P.J., Chung, K.F. and Page, C.P. (1988). Inflammatory mediators and asthma. Pharm. Rev. 40, 49–84.

Barnett, K., Jacoby, D.B., Nadel, J.A. and Lazarus, S.C. (1988). The effects of epithelial cell supernatant on contractions of isolated canine tracheal smooth muscle. Am. Rev. Respir. Dis. 138, 780–783.

Basbaum, C.B. (1986). Regulation of airway secretory cells. Clinics Chest Med. 7, 231–237.

Basbaum, C.B. and Finkbeiner, W.E. (1989). In "Lung Cell Biology" (ed. D. Massaro), pp 37–79. Marcel Dekker, New York.

Basbaum, C.B., Grillo, M.A. and Widdicombe, J.H. (1984). Muscarinic receptors: evidence for a nonuniform distribution in tracheal smooth muscle and exocrine glands. J. Neurosci. 4, 508–520.

Beasley, R., Roche, W.R., Roberts, J.A. and Holgate, S.T. (1989). Cellular events in the bronchi in mild asthma and after bronchial provocation. Am. Rev. Respir. Dis. 139, 806–817.

Benveniste, J., Henson, P.M. and Cochrane, C.G. (1972). Leukocyte-dependent histamine release from rabbit platelets: the role of IgE, basophils, and a platelet-activating factor. J. Exp. Med. 136, 1356–1360.

Bewley, J., Bhoola, K.D., Crothers, D.M. and Cingi, M.I. (1988). Biphasic responses of the isolated guinea-pig tracheal muscle strip to kallidin and bradykinin. Br. J. Pharmacol. 92, 597P.

Bhaskar, K.R., O'Sullivan, D.D., Seltzer, J., Rossing, T.H., Drazen, J.M. and Reid, L.M. (1985). Density gradient study of bronchial mucus aspirates from healthy volunteers (smokers and nonsmokers) and from patients with tracheostomy. Exp. Lung Res. 9, 289–308.

Bhoola, K.D., Figueroa, C.D. and Worthy, K. (1992). Bio-regulation of kinins: Kallikreins, kininogens, and kininases. Pharmacol. Rev. 44, 1–80.

Bisgaard, H., Groth, S. and Madsen, F. (1985). Bronchial hyperreactivity of leukotriene D_4 and histamine in exogenous asthma. Br. Med. J. 290, 1468–1471.

Black, J.L., Johnson, P.R.A. and Armour, C.L. (1988). Potentiation of the contractile effects of neuropeptides in human bronchus by an enkephalinase inhibitor. Pulm. Pharmacol. 1, 21–23.

Black, P.N., Ghatei, M.A., Takahashi, K., Bretherton-Watt, D., Krausz, T., Dollery, C.T. and Bloom, S.R. (1989). Formation of endothelin by cultured airway epithelial cells. FEBS Lett. 255, 129–132.

Bloch, K.D., Eddy, R.L., Shows, T.B. and Quertermous, T. (1989). cDNA cloning and chromosomal assignment of the gene encoding endothelin 3. J. Biol. Chem. 264, 18156–18161.

Boat, T.F. and Matthews, L.W. (1973). In "Sputum: Fundamentals and Clinical Pathology" (ed. M.J. Dulfano), p 243. Charles C. Thomas, Springfield, IL.

Bolin, R.W., Martin, T.R. and Albert, R.K. (1987). Lung endothelial and epithelial permeability after platelet-activating factor. J. Appl. Physiol. 63, 1770–1775.

Borgeat, P. and Samuelsson, B. (1979). Arachidonic acid metabolism in polymorphonuclear leukocytes: effects of ionophore A23187. Proc. Natl Acad. Sci. USA 76, 2148–2170.

Borson, D.B., Corrales, R., Varsano, S., Gold, W., Viro, N., Caughey, G., Ramachandran, J. and Nadel, J.A. (1986). Enkephalinase inhibitors potentiate SP-induced secretion of $^{35}SO_4$-macromolecules from ferret trachea. Exp. Lung Res. 21, 21–36.

Boucher, R.C., Johnson, J., Inoue, S. Hulbert, W. and Hogg, J.C. (1980). The effect of cigarette smoke on the permeability of guinea pig airways. Lab. Invest. 43, 94–100.

Bouthillier, J., DeBlois, D. and Marceau, F. (1987). Studies on the induction of pharmacological responses to des-Arg^9-bradykinin *in vitro* and *in vivo*. Br. J. Pharmacol. 92, 257–264.

Boyd, M.R. (1977). Evidence for the Clara cell as a site of

cytochrome P450-dependent mixed function oxidase activity in lung. Nature 269, 713–715.

Bramley, A.M., Samhoun, M.N. and Piper, P.J. (1990). The role of the epithelium in modulating the responses of guinea-pig trachea induced by bradykinin *in vitro*. Br. J. Pharmacol. 99, 762–766.

Braquet, P., Shen, T.Y., Touqui, L. and Vargaftig, B.B. (1987). Perspectives in platelet-activating factor research. Pharmacol. Rev. 39, 97–145.

Braunstein, G., Labat, C., Brunelleschi, S., Benveniste, J., Marsac, J. and Brink, C. (1988). Evidence that the histamine sensitivity and responsiveness of guinea-pig isolated trachea are modulated by epithelial prostaglandin E_2. Br. J. Pharmacol. 95, 300–308.

Bray, M.A. (1983). The pharmacology and pathophysiology of leukotriene B_4. Br. Med. Bull. 39, 249–254.

Breeze, R.G. and Turk, M. (1984). Celluar structure, function and orgnization in the lower respiratory tract. Environ. Health Perspect. 55, 3–24.

Breeze R.G. and Wheeldon, E.B. (1977). The cells of the pulmonary airways. Am. Rev. Respir. Dis. 116, 705–777.

Brinkman, G.L. (1968). The mast cell in normal human bronchus and lung. J. Ultrastruct. Res. 23, 115–123.

Brofman, J.D., White, S.R., Blake, J.S., Munoz, N.M., Gleich, G.J. and Leff, A.R. (1989) Epithelial augmentation of trachealis contraction caused by major basic protein of eosinophils. J. Appl. Physiol. 66, 1867–1873.

Brokaw, J.J., Hillenbrand, C.M., White, G.W. and McDonald, D.M. (1990). Mechanism of tachyphylaxis associated with neurogenic plasma extravasation in the rat trachea. Am. Rev. Respir. Dis. 141, 1434–1440.

Brunelleschi, S., Haye-Legrand, I., Labat, C., Norel, X. and Benveniste, J. (1987). Platelet-activating factor-acether-induced relaxation of guinea-pig airway muscle: role of prostaglandin E_2 and the epithelium. J. Pharmacol. Exp. Ther. 243, 356–363.

Brunelleschi, S., Vanni, L., Ledda, F., Giotti, A., Maggi, and Fantozzi, R. (1990). Tachykinins activate guinea-pig alveolar macrophages: involvement of NK2 and NK1 receptors. Br. J. Pharmacol. 100, 417–420.

Buck, S.H. and Burks, T.F. (1986). The neuropharmacology of capsaicin: review of some recent observations. Pharmacol. Res. 38, 179–226.

Buckner, C.K., Krell, R.D., Laravuso, R.B., Coursin, D.B., Bernstein, P.R. and Will, J.A. (1986). Pharmacological evidence that human intralobar airways do not contain different receptors that mediate contractions to leukotriene C_4 and leukotriene D_4. J. Pharmacol. Exp. Ther. 237, 558–562.

Buckner, C.K., Fedyna, J.S., Robertson, J.L., Will, J.A., England, D.M., Krell, R.D. and Saban, R. (1990). An examination of the influence of the epithelium on contractile responses to peptidoleukotrienes and blockade by ICI 204,219 in isolated guinea pig trachea and human intralobar airways. J. Pharmacol. Exp. Ther. 252, 77–85.

Bunnett, N.W. (1987). Release and breakdown: postsecretory metabolism of peptides. Am. Rev. Respir. Dis. 136, S27–S34.

Cammussi, G., Pawlowski, I., Tetta, C., Roffirello, C., Alberton, M., Brentjens, J. and Andres, G. (1983). Acute lung inflammation induced in the rabbit by local instillation of 1–*o*-octadecyl-2–acetyl-*sn*-glyceryl-3–phosphorylcholine or of native platelet activating factor. Am. J. Pathol. 112, 78–88.

Carstairs, J.R. and Barnes, P.J. (1986). Autoradiographic mapping of substance P receptors in lung. Eur. J. Pharmacol. 127, 295–296.

Cernacek, P. and Stewart, D.J. (1989). Immunoreactive endo-thelin in human plasma: marked elevations in patients with cardiogenic shock. Biochem. Biophys. Res. Commun. 161, 562–567.

Chang, S.C., Hsu, H.K., Perng, R.P., Shiao, G.M. and Lin, C.Y. (1990). Increased expression of MHC class II antigens in rejecting canine lung allografts. Transplantation 49, 1158–1163.

Christiansen, S.C., Proud, D. and Cochrane, C.G. (1987). Detection of tissue kallikrein in the bronchoalveolar lavage fluid of asthmatic subjects. J. Clin. Invest. 79, 188–197.

Christiansen, S.C., Proud, D., Sarnoff, R.B., Juergens, U., Cochrane, C.G. and Zuraw, B.L. (1992). Elevation of tissue kallikrein and kinin in the airways of asthmatic subjects after endobronchial allergen challenge. Am. Rev. Respir. Dis. 145, 900–905.

Churchill, L., Chilton, F.H., Resau, J.H., Bascom, R., Hubbard, W.C. and Proud, D. (1989). Cyclooxygenase metabolism of endogenous arachidonic acid by cultured human tracheal epithelial cells. Am. Rev. Respir. Dis. 140, 449–459.

Cluzel, M., Undem, B.J. and Chilton, F.H. (1988). Release of platelet activating factor and the metabolism of leukotriene B4 by the human neutrophil when studied in a cell superfusion model. J. Immunol. 143, 3659–3665.

Coles, S.J., Neill, K.H., Reid, L.M., Austen, K.F., Nu, Y., Corey, E.J. and Lewis, R.A. (1983). Effects of leukotrienes C_4 and D_4 on glycoprotein and lysozyme secretion by human bronchial mucosa. Prostaglandins 25, 155–170.

Court, E.N., Goadby, P., Hendrick, D.J., Kelly, C.A., Kingston, W., Stenton, S.C. and Walters, E.H. (1987). Platelet-activating factor in bronchoalveolar lavage fluid from asthmatic patients. Br. J. Clin. Pharmacol. 24, 259P.

Cox, G., Ohtoshi, T., Gauldie, J., Denburg, J. and Jordana, M. (1990). Human bronchial epithelial cells cause differentiation of HL-60 cells and prolong the in vitro survival of eosinophils. Am. Rev. Respir. Dis. 141, A108.

Cuss, F.M., Dixon, C.M. and Barnes, P.J. (1986). Inhaled platelet activating factor in man: effects on pulmonary function and bronchial responsiveness. Lancet ii, 189–191.

Cutz, E., Levison, H. and Cooper, D.M. (1978). Ultrastructure of airways in children with asthma. Histopathology 2, 407–421.

Dahlén, S.E., Björk, J., Hedqvist, P., Arfors, K.E., Hammarström, S., Lindgren, J.A. and Samuelsson, B. (1981). Leukotrienes promote plasma leakage and leukocyte adhesion in post-capillary venules: in vivo effects with relevance to the acute inflammatory response. Proc. Natl Acad. Sci. USA 78, 3887.

Dale, H.H. and Laidlaw, P.P. (1910). The physiological action of b-imidazolylethylamine. J. Physiol. 41, 318–344.

Dale, H.H. and Laidlaw, P.P. (1919). Histamine shock. J. Physiol. 52, 355.

Davidson, A.E., Lee, T.H., Scanlon, P.D., Solway, J., McFadden, E.R., Ingram, R.H., Corey, E.J., Austen, K.F. and Drazen, J.M. (1987). Bronchoconstrictor effects of leukotriene E_4 in normal and asthmatic subjects. Am. Rev. Respir. Dis. 135, 333–337.

DeBlois, D., Bouthillier, J. and Marceau, F. (1988). Effect of glucocorticoids, monokines and growth factors on the spontaneously developing responses of the rabbit isolated aorta to des Arg[9] bradykinin. Br. J. Pharmacol. 93, 969–977.

De Monchy, J.G.R., Kauffman, H.F., Venge, P., Koeter, G.H., Jansen, H.M., Sluiter, H.J. and De Vries, K. (1985). Bronchoalveolar eosinophilia during allergen-induced late asthmatic reactions. Am. Rev. Respir. Dis. 131, 373–377.

Demopoulos, C.A., Pinckard, R.N. and Hanahan, D.J. (1979). Platelet activating factor. Evidence for 1–O–alkyl-2–acetyl-*sn*-glyceryl-3-phosphorylcholine as the active component (a new class of lipid chemical mediators). J. Biol. Chem. 254, 9355–9359.

Devillier, P., Advenier, C., Drapeau, G., Marsac, J., and Regoli, D. (1988). Comparison of the effects of epithelium removal and of an enkaphalinase inhibitor on the neurokinin-induced contractions of guinea-pig trachea. Br. J. Pharmacol. 94, 675–684.

Devereux, T.R., Serabjit-Singh, C.J., Slaughter, S.R., Wolf, C.R., Philpot, R.M. and Fouts, J.R. (1981). Identification of cytochrome P450 Isozymes in nonciliated bronchial epithelial (Clara) and alveolar type 11 cells isolated from rabbit lung. Exp. Lung Res. 2, 221–230.

Dixon, C.M.S. and Barnes, P.J. (1989). Bradykinin-induced bronchoconstriction: inhibition by nedocromil sodium and sodium cromoglycate. Br. J. Clin. Pharmacol. 27, 831–836.

Dixon, C.M.S. and Ind, P.W. (1990). Inhaled sodium metabisulphite induced bronchoconstriction: inhibition by nedocromil sodium and sodium cromoglycate. Br. J. Clin. Pharmacol. 30, 371–376

Dixon, C.M.S., Fuller, R.W. and Barnes, P.J. (1987). The effect of an angiotensin converting enzyme inhibitor, ramipril, on bronchial responses to inhaled histamine and bradykinin in asthmatic subjects. Br. J. Clin. Pharmacol. 23, 91–93.

Dixon, R.A.F., Jones, R.E., Diehl, R.E., Bennett, C.D., Kargman, S. and Rouzer, C.A. (1988). Cloning of the cDNA for human 5–lipoxygenase. Proc. Natl Acad. Sci. USA 85, 416–420.

Djokic, T.D., Nadel, J.A., Dusser, D.J., Sekizawa, K., Graf, P.D. and Borson, D.B. (1989). Inhibitors of neutral endopeptidase potentiate electrically and capsaicin-induced noncholinergic contraction in guinea pig bronchi. J. Pharmacol. Exp. Ther. 247, 7–11.

Dunnill, M.S. (1960). The pathology of asthma with specific reference to changes in the bronchial mucosa. J. Clin. Pathol. 13, 27–33.

Dusser, D.J., Nadel, J.A., Sekizawa, K., Graf, P.D. and Borson, D.B. (1988). Neutral endopeptidase and angiotensin converting enzyme inhibitors potentiate kinin-induced contraction of ferret trachea. J. Pharmacol. Exp. Ther. 244, 531–536.

Eady, R.P. and Jackson, D.M. (1989). Effect of nedocromil sodium on SO_2-induced airway hyperresponsiveness and citric acid-induced cough in dogs. Int. Arch. Allergy Appl. Immunol. 88, 240–243.

Eling, T.E., Danilowicz, R.M., Henke, D.C., Sivarajah, K., Yankaskas, J.R. and Boucher, R.C. (1986). Arachidonic acid metabolism by canine tracheal epithelial cells. J. Biol. Chem. 261, 12841–12849.

Eiser, N.M. (1982). Histamine antagonists and asthma. Pharmac. Ther. 17, 239–250.

Elwood, R.K., Kennedy, S., Belzberg, A., Hogg, J.C. and Pare, P.D. (1983). Respiratory mucosal permeability in asthma. Am. Rev. Respir. Dis. 128, 523–527.

Erdös, E.G. (1990). Some old and new ideas on kinin metabolism. J. Cardiovasc. Pharmacol. 15(Suppl. 6), S20–S24.

Evans, T.W., Dixon, C.M.S., Clarke, B., Conradson, T.-B. and Barnes, P.J. (1988). Comparison of airway and cardiovascular effects of substance P and neurokinin A in man. Br. J. Clin. Pharmacol. 25, 273–275.

Evans, T.W., Rogers, D.F., Aursudkij, B., Chung, K.F. and Barnes, P.J. (1989). Regional and time dependent effects of inflammatory mediators on airway microvascular permeability in the guinea pig. Clin. Sci. 76, 479–485.

Fagny, C., Michel, A., Léonard, I., Berkenboom, G., Fontaine, J. and Deschodt-Lanckman, M. (1991). *In vitro* degradation of endothelin-1 by endopeptidase 24.11 (enkephalinase). Peptides 12, 773–778.

Farmer, S.G. (1987). Airway smooth muscle responsiveness: modulation by the epithelium. Trends Pharmacol. Sci. 8, 8–10.

Farmer, S.G. (1991). Role of kinins in airway diseases. Immunopharmacology 22, 1–20.

Farmer, S.G. and Burch, R.M. (1991). In "Bradykinin antagonists. Basic and Clinical Research" (ed. R.M., Burch), pp 1–31. Marcel Dekker, New York.

Farmer, S.G. and Burch, R.M. (1992). Biochemical and molecular pharmacology of kinin receptors. Annu. Rev. Pharmacol. Toxicol. 32, 511–536.

Farmer, S.G and Togo, J. (1990). Effects of epithelium removal on relaxation of airway smooth muscle induced by vasoactive intestinal peptide and electrical field stimulation. Br. J. Pharmacol. 100, 73–78.

Farmer, S.G., Fedan, J.S., Hay, D.W.P. and Raeburn, D. (1986). The effects of epithelium removal on the sensitivity of guinea-pig isolated trachealis to bronchodilator drugs. Br. J. Pharmacol. 89, 407–414

Farmer, S.G. Burch, R.M., Meeker, S.N. and Wilkins, D.E. (1989). Evidence for a pulmonary bradykinin B_3 receptor. Mol.. Pharmacol. 36, 1–8.

Farmer, S.G., Burch, R.M., Kyle, D.J., Martin, J.A., Meeker, S.N. and Togo, J. (1991). DArg[Hyp^3-Thi^5-DTic^7-Tic^8]-bradykinin, a potent antagonist of smooth muscle BK_2 receptors and BK_3 receptors. Br. J. Pharmacol. 102, 785–787.

Farmer, S.G., Wilkins, D.E., Meeker, S., Seeds, E.A.M. and Page, C.P. (1992). Effects of bradykinin receptor antagonists on antigen-induced respiratory distress, airway hyperresponsiveness and eosinophilia in guinea-pigs. Br. J. Pharmacol. 107, 653–659.

Farquhar M.G. and Palade, G.E. (1963). Junctional complexes in various epithelia. J. Cell Biol. 17, 375–412.

Fedan, J.S., Nutt, M.E. and Frazer, D.G. (1990). Reactivity of guinea-pig isolated trachea to methacholine, histamine and isoproterenol applied serosally versus mucosally. Eur. J. Pharmacol. 190, 337–345.

Feinmark, S.J., Lindgren, J.A., Claesson, H.-E., Malmsten, C. and Samuelsson, B. (1981). Stimulation of human leukocyte degranulation by leukotriene B4 and its omega-oxidized metabolites. FEBS Lett. 136, 141–144.

Fels, A.O.S., Pawlowski, N.A., Cramer, E.B., King, T.K.C., Cohn, Z.A. and Scott, W.A. (1982). Human alveolar macrophages produce leukotriene B4. Proc. Natl Acad. Sci. USA 79, 7866–7870.

Fernandes, L.B. and Goldie, R.G. (1990). Pharmacological evaluation of guinea-pig tracheal epithelium-derived inhibitory factor (EpDIF). Br. J. Pharmacol 100, 614–618.

Fernandes, L.B., Paterson, J.W. and Goldie, R.G. (1989). Co-axial bioassay of a smooth muscle relaxant factor released from guinea-pig tracheal epithelium. Br. J. Pharmacol. 96, 117–124.

Fernandes, L.B., Preuss, J.M.H., Paterson, J.W. and Goldie,

R.G. (1990). epithelium-derived inhibitory factor in human bronchus. Eur. J. Pharmacol. 187, 331–336.

Fernandes, L.B., Preuss, J.M.H. and Goldie, R.G. (1992). Epithelial modulation of the relaxant activity of atriopeptides in rat and guinea-pig tracheal smooth muscle. Eur. J. Pharmacol. 212, 187–194.

Field, J.L., Hall, J.M. and Morton, I.K.M. (1992). Bradykinin receptors in the guinea-pig taenia caeci are similar to proposed BK_3 receptors in the guinea-pig trachea, and are blocked by HOE 140. Br. J. Pharmacol. 105, 293–296.

Filep, J.G., Sirois, M.G., Rousseau, A., Fournier, A. and Sirois, P. (1991). Effects of endothelin-1 on vascular permeability in the concious rat: interactions with platelet activating factor. Br. J. Pharmacol. 104, 797–804.

Fine, J.M., Gordon, T. and Sheppard, D. (1989). Epithelium removal alters responsiveness of guinea pig trachea to substance P. J. Appl. Physiol. 66, 232–237.

Finkbeiner, W.E. and Widdicombe, J.H. (1989). In "Defence Capabilities of the Respiratory Tract" (ed. R.B. Schlesinger), pp 287–362. Raven Press, New York.

Flavahan, N.A., Aarhus, L.L., Rimele, T.J. and Vanhoutte, P.M. (1985). Respiratory epithelium inhibits bronchial smooth muscle tone. J. Appl. Physiol. 58, 834–838.

Flavahan, N.A., Slifman, N.R., Gleich, G.J. and Vanhoutte, P.M. (1988). Human eosinophil major basic protein causes hyperreactivity of respiratory smooth muscle. Am. Rev. Respir. Dis. 138, 685–688.

Ford-Hutchinson, A.W. (1990). Leukotriene B_4 in inflammation. Crit. Rev. Immunol. 10, 1–12.

Ford-Hutchinson, A.W., Bray, M.A., Doig, M.V., Shipley, M.E. and Smith, M.J.H. (1980). Leukotriene B: a potent and aggregating substance released from polymorphonuclear leukocytes. Nature 286, 264–265.

Foreman, J. and Jordan, C. (1983). Histamine release and vascular changes induced by neuropeptides. Agents Actions 13, 105–116.

Franconi, G.M., Rubinstein, I., Levine, E.H., Ideda, S. and Nadel, J.A. (1990). Mechanical removal of airway epithelium disrupts mast cells and releases granules. Am. J. Physiol. 259, L372–L377.

Frossard, N. and Muller, F. (1986). Epithelial modulation of tracheal smooth muscle responses to antigenic stimulation. J. Appl. Physiol. 61(4), 1449–1456.

Frossard, N., Rhoden, K.J. and Barnes, P.J. (1989). Influence of epithelium on guinea pig airway responses to tachykinins: role of endopeptidase and cyclooxygenase. J. Pharmacol. Exp. Ther. 248, 292–299.

Frossard, N., Stretton, C.D. and Barnes, P.J. (1990). Modulation of bradykinin responses in airway smooth muscle epithelial forks. Agents Actions 31, 204–209.

Fujisawa, T., Kephart, G.M., Gray, B.H. and Gleich, G.J. (1990). The neutrophil and chronic allergic inflammation. Am. Rev. Respir. Dis. 141, 689–697.

Fuller, R.W., Conradson, T.B., Dixon, C.M.S., Crossman, D.C. and Barnes, P.J. (1987a). Sensory neuropeptide effects in human skin. Br. J. Pharmacol. 92, 781–788.

Fuller, R.W., Dixon, C.M.S., Cuss, F.M.C. and Barnes, P.J. (1987b). Bradykinin-induced bronchoconstriction in humans: mode of action. Am. Rev. Respir. Dis. 135, 176–180.

Fuller, R.W., Maxwell, D.L., Dixon, C.M.S., McGregor, G.P., Barnes, V.F., Bloom, S.R. and Barnes, P.J. (1987c). The effects of substance P on cardiovascular and respiratory function in human subjects. J. Appl. Physiol. 62, 1473–1479.

Gail, D.G. and Lenfant, C.J.M. (1983). Cells of the lung: biology and clinical implications. Am. Rev. Respir. Dis. 127, 366–387.

Gimbrone, M.A., Jr, Brock, A.F. and Schafe, A.I. (1984). Leukotriene B_4 stimulates polymorphonuclear leukocyte adhesion to cultured vascular endothelial cells. J. Clin. Invest. 74, 1552–1555.

Girard, J.P. and Moret, P. (1963). Action de la sérotonine et de la bradykinine en aérosols chez les asthmatiques. Helv. Med. Acta 30, 520–526.

Glanville, A.R., Tazelaar, H.D., Theodore, J., Imoto, E., Rouse, R.V., Baldwin, J.C. and Robin, E.D. (1989). The distribution of MHC class I and II antigens on bronchial epithelium. Am. Rev. Respir. Dis. 139, 330–334.

Gleich, G.J., Frigas, E., Loegering, D.A., Wassom, D.L. and Steinmuller, D. (1979). Cytotoxic properties of eosinophil major basic protein. J. Immunol. 123, 2925–2927

Goetze, A.M., Fayer, L., Bouska, I., Bornemeier, D. and Carter, G.W. (1985). Puification of a mammalian 5-lipoxygenase from rat basophilic leukemia cells. Prostaglandins 29, 689–701.

Goetzl, E.J. (1980). Mediators of immediate hypersensitivity derived from arachidonic acid. N. Engl. J. Med. 303, 822–825.

Golden, J.A., Nadel, J.A. and Boushey, H.A. (1978). Bronchial hyperirritability in healthy subjects after exposure to ozone. Am. Rev. Respir. Dis. 118, 287–294.

Goldenberg, H., Huttinger, M., Kollner, U., Kramar, R. and Pavelka, M. (1978). Catalase positive particles from pig lung. Biochemical preparations and morphological studies. Histochemistry 56, 253–264.

Goldie, R.G. (1990). Receptors in asthmatic airways. Am. Rev. Respir. Dis. 141, S151–S156.

Goldie, R.G., Papadimitriou, J.M. Paterson, J.W., Rigby, P.J., Self, H.M. and Spina, D. (1986). Influence of the epithelium on responsiveness of guinea-pig isolated trachea to contractile and relaxant agonists. Br. J. Pharmacol. 87, 5–14.

Goldie, R.G. Fernandes, L.B., Rigby, P.J. and Paterson, J.W. (1988). Epithelial dysfunction and airway hypereactivity in asthma. In "Mechanisms in Asthma: Pharmacology, Physiology and Management." (eds C.L. Armour and J.L. Black) Alan Liss, Inc. New York. Prog. Clin. Biol. Res. 263, 317–329.

Goldie, R.G., Fernandes, L.B., Farmer, S.G. and Hay, D.W.P. (1990a). Airway epithelium-derived inhibitory factor. Trends Pharmacol. Sci. 11, 67–70

Goldie, R.G., Pedersen, K.E., Rigby, P.J. and Paterson, J.W. (1990b). PAF receptors in guinea-pig and human lung. Agents Actions 31(suppl.), 243–246.

Grasso, P., Williams, M., Hodgson, R., Wright, M.G. and Gandolli, S.D. (1971). The histochemical distribution of aniline hydroxylase in rat tissues. Histochem. J. 3, 117–126.

Griffin, M., Weiss, J.W. and Leitch, A.G. (1983). Effects of leukotriene D on the airways in asthma. N. Engl. J. Med. 308, 436–439.

Güc, M.O., Ilhan, M. and Kayaalp, S.O. (1988a). The rat anococcygeus muscle is a convenient bioassay organ for the airway epithelium-derived relaxant factor. Eur. J. Pharmacol. 148, 405–409.

Güc, M.O., Ilhan, M. and Kayaalp, S.O. (1988b). Epithelium-dependent relaxation of guinea-pig tracheal smooth muscle by carbachol. Arch. Int. Pharmacodyn. 294, 241–247.

Güc, M.O., Ilhan, M. and Kayaalp, S.O. (1988c). Epithelium-dependent relaxation of guinea-pig tracheal smooth muscle by histamine: evidence for non-H_1- and non-H_2-histamine receptors. Arch. Int. Pharmacodyn. 296, 57–65.

Gumbiner, B. (1987). Structure, biochemistry and assembly of epithelial tight junctions. Am. J. Physiol. 253, C749–C758.

Gunn, L.K and Piper, P.J. (1991). Potential sources of artefact in the co-axial bioassay. Eur. J. Pharmacol. 203, 405–412.

Hall, A.K., Barnes, P.J., Meldrum, L.A. and Maclagan, J. (1989). Facilitation by tachykinins of neurotransmission in guinea-pig pulmonary parasympathetic nerves. Br. J. Pharmacol. 97, 274–280.

Hansbrough, J.R., Atlas, A.B., Turk, J. and Holtzman, M.J. (1989). Arachidonate 12-lipoxygenase and cyclooxygenase: PGE isomerase are predominant pathways for oxygenation in bovine tracheal epithelial cells. Am. J. Respir. Cell. Mol. Biol. 1, 237–244.

Harkema, J.R., Mariassy, A., St. George, J., Hyde, D.M. and Plopper, C.G. (1991). In "The Airway Epithelium" (eds S.G. Farmer and D.W.P. Hay), pp 3–39. Marcel Dekker, New York.

Hay, D.W.P. (1989). Guinea-pig tracheal epithelium and endothelin. J. Pharmacol. 171, 241–245.

Hay, D.W.P., Farmer, S.G., Raeburn, D., Robinson, V.A., Fleming, W.W. and Fedan, J.S. (1986a). Airway epithelium modulates the reactivity of guinea-pig respiratory smooth muscle. Eur. J. Pharmacol. 129, 11–18.

Hay, D.W.P., Raeburn, D., Farmer, S.G., Fleming, W.W. and Fedan, J.S. (1986b). Epithelium modulates the reactivity of ovalbumin-sensitized guinea-pig airway smooth muscle. Life Sci. 38, 2461–2468.

Hay, D.W.P., Farmer, S.G., Raeburn, D., Muccitelli, R.M., Wilson, K.A. and Fedan, J.S. (1987a). Differential effects of epithelium removal on the responsiveness of guinea-pig tracheal smooth muscle to bronchoconstrictors. Br. J. Pharmacol. 92, 381–388.

Hay, D.W.P., Muccitelli, R.M., Tucker, S.S., Vickery-Clark, L.M., Wilson, K.A., Gleason, J.G., Hall, R.F., Wasserman, M.A. and Torphy, T.J. (1987b). Pharmacologic profile of SK&F 104353: a novel, potent and selective peptidoleukotriene receptor antagonist in guinea-pig and human airways. J. Pharmacol. Exp. Ther. 243, 474–481.

Hay, D.W.P., Muccitelli, R.M., Horstemeyer, D.L. and Raeburn, D. (1988). Is the epithelium-derived inhibitory factor in guinea-pig trachea a prostanoid? Prostaglandins 35, 625–638.

Hay, D.W.P., Muccitelli, R.M., Page, C.P. and Spina, D. (1992). Correlation between airway epithelium-induced relaxation of rat aorta in the co-axial bioassay and cyclic nucleotide levels. Br. J. Pharmacol. 105, 954–958.

Heino, M., Monkare, S., Haahtela, T. and Laitinen, L.A. (1982). An electronmicroscopic study of the airways in patients with farmer's lung. Eur. J. Respir. Dis. 36, 52–61.

Henke, D., Danilowicz, R.M., Curtis, J.F., Boucher, R.C. and Eling, T.E. (1988). Metabolism of arachidonic acid by human nasal and bronchial epithelial cells. Arch. Biochem. Biophys. 267, 426–436.

Henry, P.J., Rigby, P.J., Self, G.J., Preuss, J.M.H. and Goldie, R.G. (1990). Relationship between endothelin-1 binding site densities and constrictor activities in human and animal airway smooth muscle. Br. J. Pharmacol. 100, 786–792.

Henson, P.M. (1970). Release of vasoactive amines from rabbit platelets induced by sensitized mononuclear leukocytes and antigen. J. Exp. Med. 131, 287–291.

Hers, J.F.P.H. (1966). Disturbances of the ciliated epithelium due to influenza virus. Am. Rev. Respir. Dis. 93, 162–171.

Herxheimer, H. and Stresemann, E. (1961). The effect of bradykinin aerosol in guinea-pigs and in man. J. Physiol. 158, 38P.

Hisayama, T., Takayanagi, I., Nakazato, F. and Hirano, K. (1988). Epithelium selectively controls hypersensitization of the response of smooth muscle to leukotriene D_4 by endogenous prostanoid(s) in guinea-pig trachea. Naunyn-Schmiedeberg's. Arch. Pharmacol. 337, 296–300.

Holroyde, M.C. (1986). The influence of epithelium on the responsiveness of guinea-pig isolated trachea. Br. J. Pharmacol. 87, 501–507.

Holtzman, M.J. (1985). In: "Progress in Respiratory Research, Asthma and Bronchial Hyperreactivity" (eds H. Herzog and A.P. Perruchoud), pp 165–172. Karger, Basel.

Holtzman, M.J. (1991). In "The Airway Epithelium" (eds S.G. Farmer and D.W.P. Hay), pp 65–115. Marcel Dekker, New York.

Holtzman, M.J., Aizawa, H., Nadel, J.A. and Goetzl, E.J. (1983a). Selective generation of leukotriene B_4 by tracheal epithelial cells from dogs. Biochem. Biophys. Res. Commun. 114, 1071–1076.

Holtzman, M.J., Fabbri, L.M. and O'Byrne, P.M. (1983b). Importance of airway inflammation for hyperresponsiveness induced by ozone. Am. Rev. Respir. Dis. 127, 686, 690.

Holtzman, M.J., Hansbrough, J.R., Rosen, G.D. and Turk, J. (1988). Uptake, release, and novel species-dependent oxygenation of arachidonic acid inhuman and animal airway epithelial cells. Biochim. Biophys. Acta 963, 401–413.

Hoover, R.L., Kamovsky, M.J., Austen, K.F., Corey, E.J. and Lewis, R.A. (1984). Leukotriene B4 action on endothelium mediates augmented neutrophil/endothelial adhesion. Proc. Natl Acad. Sci. USA 81, 2191–2193.

Horn, B.R., Robin, E.D., Theodore, J.A. and Van Kessel, A. (1975). Total eosinophil counts in the management of bronchial asthma. N. Engl. J. Med. 292, 1152–1155.

Hua, X.Y., Dahlen, S.E., Lundberg, J.M., Hammerstrom, S. and Hedqvist, P. (1985). Leukotrienes C_4 and E_4 cause widespread and extensive plasma extravasation in the guinea pig. Naunyn-Schmiedeberg's Arch. Pharmacol. 330, 136–141.

Hulbert, W.C., Walker, D.C., Jackson, A. and Hogg, J.C. (1981). Airway permeability to horseradish peroxidase in guinea pigs: the repair phase after injury by cigarette smoke. Am. Rev. Respir. Dis. 123, 320–326.

Hulbert, W.C., McLean, H. and Hogg, J.C. (1985). The effect of acute airway inflammation on bronchial reactivity in guinea pigs. Am. Rev. Respir. Dis. 132, 7–11.

Hung, K.-S., Hertweck, M.D., Hardy, J.D. and Loosli, C.G. (1973). Ultrastructure of nerves and associated cells in bronchiolar epithelium of the mouse lung. J. Ultrastruct. Res. 43, 426.

Hwang, S.-B.(1987). Specific receptor sites for platelet activating factor on rat liver plasma membranes. Arch. Biochem. Biophys. 257, 339–344.

Hwang, S.-B. (1988). Identification of a second putative receptor of platelet activating factor from human polymorphonuclear leukocytes. J. Biol. Chem. 263, 3225–3233.

Hwang, S.B., Lam, M.H. and Shen, T.Y. (1985). Specific binding sites for platelet activating factor in human lung tissues. Biochem. Biophys. Res. Comm. 128, 972–979.

Ichinose, M. and Barnes, P.J. (1989). Inhibitory histamine H_3-receptors on cholinergic nerves in human airways. Eur. J. Pharmacol. 163, 383–386.

Ichinose, M., Belvisi, M.G. and Barnes, P.J. (1990). Histamine H_3-receptors inhibit neurogenic microvascular leakage in airways. J. Appl. Physiol. 68, 21–25.

Ikegawa, R., Matsumura, Y., Tsukahara, Y., Takaoka, M. and

Morimoto, S. (1990). Phosphoramidon, a metalloproteinase inhibitor, suppresses the secretion of endothelin-1 from cultured endothelial cells by inhibiting a big endothelin-1 converting enzyme. Biochem. Biophys. Res. Commun. 171, 669–675.

Ilhan, M. and Sahin, I. (1986). Tracheal epithelium releases a vascular smooth muscle relaxant factor: demonstration by bioassay. Eur. J. Pharmacol. 131, 293–296.

Ilowite, J.S., Bennett, W.D., Sheetz, M.S., Groth, M.L. and Nierman, D.M. (1989). Permeability of the bronchial mucosa to ^{99m}Tc-DPTA in asthma. Am. Rev. Respir. Dis. 139, 1139–1143.

Inoue, A., Yanagisawa, M., Kimura, S., Kasuya, Y., Miyauchi, T., Goto, K. and Masaki, T. (1989). The human endothelin family: three structurally and pharmacologically distinct isopeptides predicted by three separate genes. Proc. Natl Acad. Sci. USA 86, 2863–2867.

Irani, A.A., Schechter, N.M., Craig, S.S., DeBlois, G. and Schwartz, L.B. (1986). Two types of human mast cells that have distinct neutral protease compositions. Proc. Natl Acad. Sci. USA 83, 4464–4468.

Iriarte, C.F., Pascual, R., Villanueva, M.M., Romn, M., Cortijo, J. and Morcillo, E.J. (1990). Role of epithelium in agonist-induced contractile responses of guinea-pig trachealis: influence of the surface through which drug enters the tissue. Br. J. Pharmacol. 101, 257–262.

Ishikawa, S. and Sperelakis, N. (1987). A novel class (H₃) of histamine receptors on perivascular nerve terminals. Nature 327, 158–160.

Jeffrey, P. and Reid, L. (1973). Intraepithelial nerves in normal rat airways: a quantitative electron microscopic study. J. Anat. 114, 35–45.

Jeffery, P.K. (1983). Morphologic features of airway surface epithelial cells and glands. Am. Rev. Respir. Dis. 128, S14–S20.

Jeffery, P.K., Wardlaw, A.J., Nelson, F.C., Collins, J.V. and Kay, A.B. (1989). Bronchial biopsies in asthma. An ultrastructural, quantitative study and correlation with hyperreactivity. Am. Rev. Respir. Dis. 140, 1745–1753.

Johnson, A.R., Ashton, J., Schulz, W.W. and Erdos, E.G. (1985). Neutral metalloendopeptidases in human lung tissue and cultured cells. Am. Rev. Respir. Dis. 132, 564–568.

Jones, K.G., Holland, J.R., Foureman, G.L., Bend, J.R. and Fouts, J.R. (1983). J. Pharmacol. Exp. Ther. 225, 316–319.

Joos, G., Pauwels, R. and Van der Straeton, M. (1987). Effect of inhaled substance P and neurokinin A on the airways of normal and asthmatic subjects. Thorax 42, 779–783.

Jordana, M., Vancheri, C., Ohtoshi, T., Tsuda, T., Cox, G., Harnish, D., Dolovich, J., Denburg, J. and Gauldie, J. (1989). Hemopoietic function of fibroblasts and epithelial cells derived from normal and inflamed upper respiratory airway tissues. Am. Rev. Respir. Dis. 139, A251.

Kalb, T.H., Mayer, L.F., Teirstein, A.S. and Marom, Z. (1989). Human airway epithelial cells express functional MHC class II molecules. Am. Rev. Respir. Dis. 139, A457.

Kallenberg, C.G., Schilizzi, B.M., Beaumont, F., DeLeij, L. and Poppema, S. (1987). Expression of class II major histocompatibility complex antigens on alveolar epithelium in interstitial lung disease: relevance to pathogenesis of idiopathic pulmonary fibrosis. J. Clin. Pathol. 40, 725–733.

Kerr, M.A. and Kenny, A.J. (1974). The purification and specificity of neutral endopeptidase from rabbit kid-brush border. Biochem. J. 137, 477–488.

Kinnula, V.L., Adler, K.B., Ackley, N.J. and Crapo, J.D. (1992). Release of reactive oxygen species by guinea pig tracheal epithelial cells in vitro. Am. J. Physiol. 262, L708–712.

Kitamura, K., Yukawa, T., Morita, S., Ichiki, Y., Eto, T. and Tanaka, K. (1990). Distribution and molecular form of immunoreactive big endothelin-1 in porcine tissue. Biochem. Biophys. Res. Commun. 170, 497–503.

Kloprogge, E. and Akkerman, J.W.N. (1984). Binding kinetics of PAF-acether (1–O-alkyl-2–acetyl-sn-glycero-3–phosphorylcholine) to intact human platelets. Biochem. J. 223, 901–909.

Knowles, M.R., Stutts, M.J., Yaukaskas, J.R., Gatzg, J.T. and Boucher, R.C. (1986). Abnormal respiratory epithelial ion transport in cystic fibrosis. Clinics Chest Med. 7, 285–297.

Koyama, S., Rennard, S.I., Rickard, K. and Robbins, R.A. (1989). E. coli endotoxin stimulates the release of neutrophil chemotactic activity from and causes cytotoxicity to bovine bronchial epithelial cells in vitro. Am. Rev. Respir. Dis. 139, A371.

Kondo, M., Tamaoki, J. and Takizawa, T. (1990). Neutral endopeptidase inhibitor potentiates the tachykinin-induced increase in ciliary beat frequency in rabbit trachea. Am. Rev. Respir. Dis. 142, 403–406.

Lagente, V., Boichot, E., Mencia-Huerta, J. and Braquet, P. (1990). Failure of aerosolized endothelin (ET-1) to induce bronchial hyperreactivity. Fundam. Clin. Pharmacol. 4, 275–280.

Laitinen, L.A. (1989). In "Glucocorticoids and Mechanisms of Asthma" (eds J.L. Malo, F.E. Hargreave and J.C. Hogg), pp 217–227. Excerpta Medica, New York.

Laitinen, L.A., Laitinen, A., Panula, P.A., Partanen, M., Tervo, K. and Tervo, T. (1983). Immunohistochemical demonstration of substance P in the lower respiratory tract of the rabbit and not of man. Thorax 38, 531–536.

Laitinen, L.A., Heino, M., Laitinen, A., Kava, T. and Haahtela, T. (1985). Damage of the airway epithelium and bronchial reactivity in patients with asthma. Am. Rev. Respir. Dis. 131, 599–606.

Laitinen, L.A., Laitinen, A. and Widdicombe, J.G. (1987). Effects of inflammatory and other mediators on airway vascular beds. Am. Rev. Respir. Dis. 135, S67–S70.

Laitinen, L.A., Heino, M. and Laitinen, A. (1991). In "The Airway Epithelium" (eds S.G. Farmer and D.W.P. Hay), pp 187–204. Marcel Dekker, New York.

Lambrecht, G. and Parnham, M.J. (1986). Kadsurenone distinguishes between different platelet activating factor receptor subtypes on macrophages and polymorphonuclear leucocytes. Br. J. Pharmacol. 87, 287–299.

Lauweryns, J.M., Cokelaere, M. and Theunynck, P. (1972). Neuroepithelial bodies in the respiratory mucosa of various mammals: a light optical, histochemical and ultrastructural investigation. Z. Zellforsch. Mikrosk. Anat. 135, 569.

LeComte, J., Petit, J.M., Mélon, J., Troquet, J. and Marcelle, R. (1962). Propriétés bronchoconstrictrices de la bradykinine chez l'homme asthmatique. Arch. Int. Pharmacodyn. Ther. 137, 232–235.

Lee, C.M., Campbell, N.J., Williams, B.J. and Iversen, L.L. (1986). Multiple tachykinin binding sites in peripheral tissue and brain. Eur. J. Pharmacol. 130, 209–217.

Leikauf, G.D., Driscoll, K.E. and Wey, H.E. (1988). Ozone-induced augmentation of eicosanoid metabolism in epithelial cells from bovine trachea. Am. Rev. Respir. Dis. 137, 435–442.

Lerner, U.H. and Modéer, T. (1991). Bradykinin B1 and B2

receptor agonists synergistically potentiate interleukin-1-induced prostaglandin biosynthesis in human gingival fibroblasts. Inflammation 15, 427–436.

Lewis, T. (1927). "The Blood Vessels of the Human Skin and Their Responses." Shaw, London.

Lindström, E.G., Andersson, R.G.G., Granrus, G. and Grundström, N. (1991). Is the airway epithelium responsible for histamine metabolism in the trachea of guinea pigs? Agents Actions 33, 1–2.

Liu, M.C., Bleecker, E.R., Lichtenstein, L.M., Kagey-Sobotka, A., Niv, Y., McLemore, T.L., Permutt, S., Proud, D. and Hubbard, W.C. (1990). Evidence for elevated levels of histamine, prostaglandin D_2, and other bronchoconstricting prostaglandins in the airways of subjects with mild asthma. Am. Rev. Respir. Dis. 142, 126–132.

Lotz, M., Vaughan, J.H. and Carson, D.A. (1988). Effect of neuropeptides on production of inflammatory cytokines by human monocytes. Science 241, 1218–1221.

Lozewicz, S., Wells, C., Gomez, E., Ferguson, H., Richman, P., Devalia, J. and Danes, R.J. (1990). Morphological integrity of the bronchial epithelium in mild asthma. Thorax 45, 12–15.

Lucas, A. and Douglas, L. (1934). Principles underlying ciliary activity in the respiratory tract. II. A comparison of nasal clearance in man, monkey and other mammals. Arch. Otolaryngol. 20, 518–541.

Luciano, L., Reale, E. and Ruska, H. (1968). Uber eine 'chemorezeptive' sinneszelle in der treachea der ratte. Z. Zellforsch. Mikrosk. Anat. 85, 350.

Lundberg, J.M. and Saria, A. (1987). Polypeptide-containing neurons in airway smooth muscle. Annu. Rev. Physiol. 49, 557–572.

Lundberg, J.M., Martling, C.R. and Saria, A. (1983a). Substance P and capsaicin-induced contraction of human bronchi. Acta Physiol. Scand. 119, 49–53.

Lundberg, J.M., Broden, E. and Saria, A. (1983b) Effects and distribution of vagal capsaicin-sensitive substance P neurons with special reference to the trachea and lungs. Acta Physiol. Scand. 119, 243–252.

Lundberg, J.M., Hokfelt, T., Martling, C.-R., Saria, A. and Cuello, C. (1984). Sensory substance P-immunoreactive nerves in the lower respiratory tract of various mammals including man. Cell Tissue Res. 235, 251–261.

Lynch, J.M. and Henson, P.M. (1986). The intracellular retention of newly-synthesized platelet-activating factor. J. Immunol. 137, 2653–2661.

MacCumber, M.W., Ross, C.A., Glaser, B.M. and Snyder, S.H. (1989). Endothelin visualization of mRNA's by *in situ* hybridization provides evidence for local action. Proc. Natl Acad. Sci. USA 86, 7285–7289.

Macquin-Mavier, I., Levame, M., Istin, N. and Hart, A. (1989). Mechanisms of endothelin-mediated bronchoconstriction in the guinea-pig. J. Pharmacol. Exp. Ther. 250, 740–745.

McDonald, D.M. (1987). Neurogenic inflammation in the respiratory tract: actions of sensory nerve mediators on blood vessels and epithelium of the airway mucosa. Am. Rev. Respir. Dis. 136, S65–S72.

McKay, K.O., Black, J.L. and Armour, C.L. (1991). The mechanism of action of endothelin in human lung. Br. J. Pharmacol. 102, 422–428.

McMenamin, C., Schon-Hegrad, M., Oliver, J., Girn, B. and Holt, P.G. (1991). Regulation of IgE responses to inhaled antigens: cellular mechanisms underlying allergic sensitization versus tolerance induction. Int. Arch. Allergy Appl. Immunol. 94, 78–82.

Maggi, C.A. (1990). Tachykinin receptors in the airways and lung: what should be blocked? Pharmacol. Res. 22, 527–540.

Maggi, C.A., Giuliani, S., Santicioli, P., Regoli, D. and Meli, A. (1987). Peripheral effects of neurokinins: functional evidence for the existence of multiple receptors. J. Autonom. Pharmacol. 7, 11–21.

Maggi, C.A., Patacchini, R., Perretti, F., Meini, S., Manzini, S., Santicioli, P., Del Bianco, E. and Meli, A. (1990). The effect of thiorphan and epithelium removal on contractions and tachykinin release produced by activation of capsaicin-sensitive afferents in the guinea-pig isolated bronchus. Naunyn-Schmiedeberg's Arch. Pharmacol. 341, 74–79.

Manning, P.J., Jones, G.L., Otis, J., Daniel, E.E. and O'Byrne, P.M. (1990). The inhibitory influence of tracheal mucosa mounted in close proximity to canine trachealis. Eur. J. Pharmacol. 178, 85–89.

Mansour, E., Ahmed, A., Cortes, A., Caplan, J., Burch, R.M. and Abraham, W.M. (1992). Mechanisms of metabisulfite-induced bronchoconstriction. Evidence for bradykinin B_2 receptor stimulation. J. Appl. Physiol. 72, 1831–1837.

Mapp, C.E., Polato, R., Maestrelli, P., Hendrick, D.J. and Fabbri, L.M. (1985). Time course of the increase in airway responsiveness associated with late asthmatic reactions to toluene diisocyanate in sensitized subjects. J. Allergy Clin. Immunol. 75, 568–572.

Marasco, W.A., Showell, H.J. and Becker, E.L. (1981). Substance P binds to the formyl peptide chemoaxis receptor on the rabbit neutrophil. Biochem. Biophys. Res. Commun. 99, 1065–1068.

Marceau, F. and Regoli, D. (1991). In "Bradykinin Antagonists. Basic and Clinical Research" (ed. R.M. Burch.), pp 33–49. Marcel Dekker, New York.

Marin, M.G. (1986). Pharmacology of airway secretion. Pharmacol. Rev. 38:273–289.

Marin, M.G. and Culp, D.J. (1986). Isolation and culture of submucosal gland cells. Clinics Chest Med. 7, 239–245.

Marin, M.G., Davis, B. and Nadel, J.A. (1977). Effect of histamine on electrical and ion transport properties of tracheal epithelium. J. Appl. Physiol. 42, 735–738.

Marini, M., Vittori, E., Halemborg, J. and Mattoli, S. (1992). Expression of the potent inflammatory cytokines granulocyte-macrophage-colony-stimulating factor and interleukin-6 and interleukin-8 in bronchial epithelial cells of patients with asthma. J. Allergy Clin. Immunol. 89, 1001–1009.

Marom, Z., Schelhamer, J.H., Bach, M.K., Morton, D.R. and Kaliner, M.A. (1982). Slow reacting substances, leukotrienes C_4 and D_4 increase the release of mucus from human airways *in vitro*. Am. Rev. Respir. Dis. 126, 449–451.

Martling, C.R., Theordorsson-Norheim, E. and Lundberg, J.M. (1987). Occurrence and effects of multiple tachykinins: substance P, neurokinin A, neuropeptide K in human lower airways. Life Sci. 40, 1633–1643.

Masaki, T. and Yanagisawa, M. (1992). Physiology and pharmacology endothelins. Med. Res. Rev. 12(4), 391–421.

Massaro, G.D. (1987). The Clara cell: secretory and developmental studies in the rat. Eur. J. Respir. Dis. 153, 52–55.

Massaro, G.D., Amado, C., Clerch, L. and Massaro, D. (1982). Studies on the regulation of secretion in Clara cells with evidence for chemical nonautonomic mediation of the secretory response to increased ventilation in rat lungs. J. Clin. Invest. 70, 608–613.

Matsumura, Y., Ikegawa, R., Tsukahara, Y., Takaoka, M. and Morimoto, S. (1991). Conversion of big endothelin-1 to endothelin-1 by two types of metalloproteinases of cultured porcine vascular smooth muscle cells. Biochem. Biophys. Res. Commun. 78, 899–905.

Mattoli, S., Mezzetti, M., Riva, G., Allegra, L. and Fasoli, A. (1990). Specific binding of endothelin on human bronchial smooth muscle cells in culture and secretion of endothelin-like material from bronchial epithelial cells. Am. J. Respir. Cell. Mol. Biol. 3, 145–151.

Mattoli, S., Colotta, F., Fincato, G., Mezzetti, M., Mantovani, A., Patalano, F. and Fasoli, A. (1991). Time course of IL1 and IL6 synthesis and release in human bronchial epithelial cell cultures exposed to toluene diisocynate. J. Cell. Physiol. 149, 260–268.

Montano, L.M., Selman, M., Ponce-Monter, H. and Vargas, M.H. (1988). Role of airway epithelium on the reactivity of smooth muscle from guinea pigs sensitized to ovalbumin by inhalatory method. Res. Exp. Med. 188, 167–173.

Motojima, S., Frigas, E., Loegering, D.A. and Gleich, G.J. (1989). Toxicity of eosinophil cationic proteins for guinea-pig tracheal epithelium in vitro. Am. Rev. Respir. Dis. 139, 801–805.

Murlas, C., (1986). Effects of mucosal removal on guinea-pig airway smooth muscle responsiveness. Clin. Sci. 70, 571–575.

Nagy, L., Lee, T.H, Goetzl, E.J., Pickett, W. and Kay, A.B. (1982). Complement receptor enhancement and chemotaxis of human neutrophils and eosinophils by leukotrienes and other lipoxygenase products. Clin Exp. Immunol. 47, 541–547.

Nakamura, T., Morita, Y., Kuriyama, M., Ishihara, K., Ito, K. and Miyamoto, T. (1987). Platelet activating factor in late asthmatic responses. Int. Arch. Allergy Appl. Immunol. 82, 57–61.

Nakanishi, S. (1987). Substance P precursor and kininogen: their structures gene organizations and regulation. Physiol. Rev. 67, 1117–1142.

Naline, E., Devillier, P., Drapeau, G., Toty, L.F., Bakdach, H., Regoli, D. and Advenier, C. (1989). Characterization of neurokinin effects and receptor selectivity in human isolated bronchi. Am. Rev. Respir. Dis. 140, 679–686.

Ninomiya H., Uchida, Y., Ishii, Y., Nomura, A., Kameyama, M., Saotome, M., Endo, T. and Hasegawa, S. (1991). Endotoxin stimulates endothelin release from cultured epithelial cells of guinea-pig trachea. Eur. J. Pharmacol. 203, 299–302.

Nomura, A., Uchida, Y., Kameyama, M., Saotome, M., Oki, K. and Hasegawa, S. (1989). Endothelin and bronchial asthma. Lancet ii, 747–748.

Noveral, J.P., Rosenberg, S.M., Anbar, R.A., Pawlowski, N.A. and Grunstein, M.M. (1992). Role of endothelin-1 in regulating proliferation of cultured rabbit airway smooth muscle cells. Am. J. Physiol. 263, L317–L324.

O'Byrne, P.M., Aizawa, H., Bethel, R.A., Chung, K.F., Nadel, J.A. and Holtzman, M.J. (1984). Prostaglandin F_{2a} increases responsiveness of pulmonary airways in dogs. Prostaglandins 28, 537–543.

O'Flaherty, J.T., Wykle, R.L., Miller, C.H., Lewis, J.C., Waite, M., Bass, D.A., McCall, C.E. and DeChatelet, L.R. (1981). 1–O–alkyl-sn-glyceryl-3–phosphorylcholines. A novel class of neutrophil stimulants. Am. J. Pathol. 103, 70–78.

Ohkubo, S., Ogi, K., Hosoya, M., Matsumoto, H., Suzuki, N., Kimura, C., Onda, H. and Fujino, M. (1990). Specific expression of endothelin-2 (ET-2) gene in a renal adenocarcinoma cell line. FEBS Lett. 274, 136–140.

Okada, K., Miyazaki, Y., Takada, J., Matsuyama, K., Yamaki, T. and Yano, M. (1990). Conversion of big endothelin-1 by membrane-bound metalloendopeptidase in cultured bovine endothelial cells. Biochem. Biophys. Res. Commun. 171, 1192–1198.

Orehek, J., Douglas, J.S. and Bouhuys, A., (1975). Contractile responses of the guinea-pig trachea in vitro: modification by the prostaglandin synthesis inhibiting drugs. J. Pharmacol. Exp. Ther. 194, 554–564.

Page, C.P. In "Asthma: Basic Mechanisms and Clinical Management" (eds P.J. Barnes, I.W. Rodger and N.C. Thompson), pp 283–304. Academic Press, London.

Pavlovic, D., Fournier, M., Aubier, M. and Pariente, R. (1989). Epithelial vs. serosal stimulation of tracheal muscle: role of epithelium. J. Appl. Physiol. 67, 2522–2526.

Payan, D.G. (1989). Neuropeptides and inflammation: the role of substance P. Annu. Rev. Med. 40, 341–352.

Payan, D.G., Brewster, D.R., Missirian-Bastian, A. and Goetzl, E.J. (1984b). Substance P recognition by a subset of human T-lymphocytes. J. Clin. Invest. 74, 1532–1539.

Penney, D.P. (1988). "International Review of Cytology," Vol. III, pp 231–269. Academic Press, New York.

Peters, S.P., Schleimer, R.P., Naclerio, R.M., MacGlashan, D.W., Togias, A.G., Proud, D., Freeland, H.S., Fox, C., Adkinson, N.J. and Lichtenstein, L.M. (1987). The pathophysiology of human mast cells: in vitro and in vivo function. Am. Rev. Respir. Dis. 135, 1196–1200.

Pitts, J.D. and Finbow, M.E. (1986). The gap junction. J. Cell Sci. (Suppl.) 4, 239–266.

Plopper, C.G., Hill, L.H. and Mariassy, A.T. (1980). Ultrastructure of the nonciliated bronchiolar epithelial (Clara) cell of mammalian lung. III. A study of man with comparison of 15 mammalian species. Exp. Lung Res. 1, 171–180.

Polosa, R. and Holgate, S.T. (1990). Comparative airway responses to inhaled bradykinin, kallidin, and [des-Arg⁹]-bradykinin in normal and asthmatic subjects. Am. Rev. Respir. Dis. 142, 1367–1371.

Polosa, R., Phillips, G.D., Lai, C.K.W. and Holgate, S.T. (1990). Contribution of histamine and prostanoids to bronchoconstriction provoked by inhaled bradykinin. Allergy 45, 174–182.

Pongracic, J.A., Churchill, L. and Proud, D. (1991). In "Bradykinin Antagonists. Basic and Clinical Research" (ed. R.M. Burch), pp 273–259. Marcel Dekker, New York.

Pons, F., Loquet, I., Touvay, C., Roubert, P., Chabrier, P.-E., Mencia-Huerta, J.M. and Braquet, P. (1991). Comparison of the bronchopulmonary and pressornactivities of endothelin isoforms ET-1, ET-2, and ET-3 and characterization of their binding sites in guinea-pig lung. Am. Rev. Respir. Dis. 143, 294–300.

Poston, R.N., Chanex, P., Lacoste, J.Y., Litchfield, T., Lee, T.H. and Bousquet, J. (1992). Immunohistochemical characterization of the cellular infiltration in asthmatic bronchi. Am. Rev. Respir. Dis. 145, 918–921.

Prescott, S.M., Zimmerman, G.A. and McIntyre, T.M. (1990). Platelet-activating factor. J. Biol. Chem. 265, 17381–17384.

Prié, S., Cadieux, A. and Sirois, P. (1990). Removal of guinea pig bronchial and tracheal epithelium potentiates the contractions to leukotrienes and histamine. Eicosanoids, 3, 29–37.

Proud, D. and Kaplan, A.P. (1988). Kinin formation: mechanisms and role in inflammatory disorders. Annu. Rev. Immunol 6, 49–84.

Rangachari, P.K. and McWade, D. (1985). Effects of tachykinins on the electrical activity of isolated canine tracheal epithelium: an exploratory study. Reg. Peptides 12, 9–19.

Regoli, D., Drapeau, G., Dion, S. and Couture, R. (1988). New selective agonists for neurokinin receptors: pharmacological tools for receptor characterization. Trends Pharmacol. Sci. 9, 290–295.

Rennard, S.I., Beckmann, J.D. and Robbins, R.A. (1991). In "The Lung: Scientific Foundations" (eds R.G. Crystal and J.B. West), pp 157–167. Raven Press, New York.

Repke, H. and Bienert, M. (1988). Structural requirements for mast cell triggering by substance P-like peptides. Agents Actions 23, 207–210.

Revel, J.P., Nicholson, B.J. and Yancey, S.B. (1985). Chemistry of gap junctions. Annu. Rev. Physiol. 47, 263–279.

Rhoden, K.J. and Barnes, P.J. (1990). Epithelial modulation of non-adrenergic, non-cholinergic and vasoactive intestinal peptide-induced responses: role of neutral endopeptidase. Eur. J. Pharmacol. 171, 247–250.

Robbins, R.A., Shoji, S., Linder, J., Gossman, G.L., Allington, L.A., Klassen, L.W. and Rennard, S.I. (1989). Bronchial epithelial cells release chemotactic activity for lymphocytes. Am. J. Physiol. 257, L109–L115.

Robbins, R.A., Koyama, S., Nelson, K., Gossman, G., Rickard, K. and Rennard, S.I. (1990). Neutrophil adherence to bronchial epithelial cells. Am. Rev. Respir. Dis. 141, A910.

Robinson, N.P., Kyle, H., Webber, S.E. and Widdicombe, J.G. (1989). Electrolyte and other chemical concentrations in the tracheal airway and mucus. J. Appl. Physiol. 66, 2129–2135.

Rogers, D.F., Belvisi, M.G., Aursudkij, B., Evans, T.W. and Barnes, P.J. (1988). Effects and interactions of sensory neuropeptides on airway microvascular leakage in guinea-pigs. Br. J. Pharmacol. 95, 1109–1116.

Rossi, G.A., Sacco, O., Balbi, B., Oddera, S., Mattioni, T., Corte, G., Ravazzoni, C. and Allegra, L. (1990). Human ciliated bronchial epithelial cells: expression of the HLA-DR antigens and of the HLA-DR alpha gene, modulation of the HLA-DR antigens by gamma-interferon and antigen-presenting function in the mixed leukocyte reaction. Am. J. Respir. Cell. Mol. Biol. 3, 431–439.

Rouzer, C.A. and Samuelsson, B. (1985). On the nature of 5-lipoxygenase reaction in human leukocytes: enzyme purification and requirement for multiple stimulatory factors. Proc. Natl Acad. Sci. USA 82, 6040–6044.

Rouzer, C.A., Rands, E., Kargman, S., Jones, R.E., Register, R.B. and Dixon, R.A.F. (1988). Characterization of cloned human leukocyte 5–lipoxygenase expressed in mammalian cells. J. Biol. Chem. 263, 10135–10140.

Rozengurt, N., Springall, D.R. and Polak, J.M. (1990). Localization of endothelin-like immunoreactivity in airway epithelium of rats and mice. J. Pathol. 160, 5–8.

Ruff, M.R., Whal, S.M. and Pert, C.B. (1985). Substance P receptor-mediated chemotaxis of human monocytes. Peptides, 6(Suppl. 2), 107–111.

Said, S.I. (1987). Effector Actions: Influence of neuropeptides on airway smooth muscle. Am. Rev. Respir. Dis. 136, S52–S58.

Sakamoto, A., Yanagisawa, M., Sakurai, T., Takuwa, Y., Yanagisawa, H. and Masaki, T. (1991). Cloning and functional expression of human cDNA for the ET_B endothelin receptor. Biochem. Biophys. Res. Commun. 178, 656–663.

Sakurai, T., Yanagisawa, M. and Masaki, T. (1992). Molecular characterization of endothelin receptors. Trends Pharmacol. Sci. 13, 103–108.

Salari, H. and Chan-Yeung, M. (1989). Release of 15-hydroxyeicosatetraenoic acid (15-HETE) and prostaglandin E_2 (PGE_2) by cultured human bronchial epithelial cells. Am. J. Respir. Cell Mol. Biol. 1, 245–250.

Salari, H. and Schellenberg, R.R. (1991). Stimulation of human airway epithelial cells by platelet activating factor (PAF) and arachidonic acid produces 15-hydroxyeicosatetraenoic acid (15-HETE) capable of contracting bronchial smooth muscle. Pulm. Pharmacol. 4, 1–7.

Saria, A., Theodorsson-Norheim, E., Gamse, R. and Lundberg, J.M. (1985). Release of substance P and substance K-like immunoreactivities from the isolated perfused guinea pig lung. Eur. J. Pharmacol. 106, 207–208.

Saria, A., Martling, C.-R., Yan, Z. Theodorsson-Norheim, E., Gamse, R. and Lundberg, J.M. (1988). Release of multiple tachykinins from capsaicin-sensitive sensory nerves in the lung by bradykinin, histamine, dimethylphenyl piperazinium, and vagal nerve stimulation. Am. Rev. Respir. Dis. 137, 1330–1335.

Sasaki, T., Shimura, S., Sasaki, H. and Takishima, T. (1989). Effect of epithelium on mucus secretion from feline tracheal submucosal glands. J. Appl. Physiol. 66, 764–770.

Sawamura, T., Kimura, S., Shinmi, O, Sugita, Y., Yanagisawa, M., Goto, K. and Masaki, T. (1990). Purification and characterization of putative endothelin converting enzyme in bovine adrenal medulla: evidence for cathepsin D-like enzyme. Biochem. Biophys. Res. Commun. 168, 1230–1236.

Schneeberger, E.E. (1992). In "The Lung: Scientific Foundations" (eds R.G. Crystal and J.B. West), pp 229–234. Raven Press, New York.

Schwartz, J.-C., Garbarg, M. and Pollard, H. (1986). In "Handbook of Physiology," Vol. IV (ed. F.E. Bloom), pp 257–316. American Physiological Society, Bethesda, MD.

Sekizawa, K., Tamaoki, J., Graf, P.D., Basbaum, C.B., Borson, D.B. and Nal, J.A. (1987). Enkephalinase inhibitor potentiates mammalian tachykinin-induced contraction in ferret trachea. J. Pharmacol. Exp. Ther. 243, 1211–1217.

Serhan, C.N., Radin, A., Smolen, J.E., Korchak, H., Samuelsson, B. and Weissmann, G. (1982). Leukotriene B4 is a complete secretagogue in human neutrophils: a kinetic analysis. Biochem. Biophys. Res. Commun. 107, 1006–1012.

Sertl, K., Takemura, T., Tschachler, E. Ferrans, V.J., Kaliner, M.A. and Shevach, E.M. (1986). Dendritic cells with antigen-presenting capability reside in airway epithelium, lung parenchyma, and visceral pleura. J. Exp. Med. 163, 436–451.

Shannon, V.R., Hansbrough, J.R., Takahashi, Y., Ueda, N., Yamamoto, S. and Holtzman, M.J. (1979). Prostaglandin D_2, a neuromodulator. Proc. Natl Acad. Sci. USA 76, 6231–6234.

Shelhamer, J.H., Marom, Z. and Kaliner, M. (1980). Immunologic and neuropharmacologic stimulation of mucous glycoprotein release from human airway *in vitro*. J. Clin. Invest. 66, 1400–1408.

Sheridan, J.D. and Atkinson, M.M. (1985). Physiological roles of permeable junctions: some possibilities. Annu. Rev. Physiol. 47, 337–353.

Shimizu, T., Izumi, T., Seyama, Y., Tadokoro, K., Radmark, O. and Samuelsson, B. (1986). Characterization of leukotriene A4 synthase from murine mast cells: evidence for its identity to arachidonate 5–lipoxygenase. Proc. Natl Acad. Sci. USA 83, 4175–4179.

Shimura, S., Sasaki, T., Okayama, H., Sasaki, H. and Takishima, T. (1987). Effect of SP on mucus secretion of isolated submucosal gland from feline trachea. J. Appl. Physiol. 63, 646–653.

Shimura, S., Ishihara, H., Satoh, M., Masuda, T., Nagaki, N., Sasaki, H. and Takishima, T. (1992). Endothelin regulation of mucus glycoprotein secretion from feline tracheal submucosal glands. Am. J. Physiol., 262, L208–L213.

Shoji, S., Rickard, K.A., Ertl, R.F., Robbins, R.A., Linder, J. and Rennard, S.I. (1989). Bronchial epithelial cells produce lung fibroblast chemotactic factor: fibronectin. Am. J. Respir. Cell Mol. Biol. 1, 13–20.

Shoji, S., Ertl, R.F., Linder, J. and Romberger, D.J. (1990). Bronchial epithelial cells produce chemotactic activity for bronchial epithelial cells. Am. Rev. Respir. Dis. 141, 218–225.

Shore, S.A., Stimler-Gerard, N.P., Coates, S.R. and Drazen, J.M. (1988). Substance P induced bronchoconstriction in guinea-pig. Enhancement by inhibitors of neutral metallo-endopeptidase and angiotensin coverting enzyme. Am. Rev. Respir. Dis. 137, 331–336.

Simonsson, B.G., Skoogh, B.-E., Bergh, N.P., Andersson, R. and Svedmyr, N. (1973). In vivo and in vitro effect of bradykinin on bronchial motor tone in normal subjects and patients with airways obstruction. Respiration 30, 378–388.

Siraganian, R.P. and Osler, A.G. (1971). Destruction of rabbit platelets in the allergic response of sensitized leukocytes. I. Demonstration of a fluid phase intermediate. J. Immunol. 106, 1244–1250.

Small, R.C., Good, D.M., Dixon, J.S. and Kennedy, I. (1990). The effects of epithelium on the actions of cholinomimetic drugs in opened segments and perfused tubular preparations of guinea-pig trachea. Br. J. Pharmacol. 100, 516–522.

Smith, S., Lee, D., Lacy, J. and Coleman, D. (1989). Rat tracheal epithelial cells in primary culture release granulocyte-macrophage colony-stimulating factor (GM-SCF). Am. Rev. Respir. Dis. 139, A613.

Solér, M., Sielczak, M. and Abraham, W.M. (1990). A bradykinin antagonist blocks antigen-induced airway hyper-responsiveness and inflammation in sheep. Pulm. Pharmacol. 3, 9–15.

Soloperto, M., Mattoso, V.L., Fasoli, A. and Mattoli, S. (1991). A bronchial epithelial cell-derived factor in asthma that promotes eosinophil activation and survival as GM-CSF. Am. J. Physiol. 260, L530–L538.

Spampinato, S. and Ferri, S. (1991). Pharmacology of spinal peptides affecting sensory and motor functions: dynorphins, somatostatins and tachykinins. Pharmacol. Res. 23, 113–127.

Spina, D., Fernandes, L.B., Preuss, J.M.H., Hay, D.W.P., Muccitelli, R.M., Page, C.P. and Goldie, R.G. (1992). Evidence that epithelium-dependent relaxation of vascular smooth muscle detected by co-axial bioassays is not attributable to hypoxia. Br. J. Pharmacol. 105, 799–804.

Springall, D.R., Howarth, P.H., Counihan, H., Djukanovic, R., Holgate, S.T. and Polak, J.M. (1991). Endothelin immuno-reactivity of airway epithelium in asthmatic patients. Lancet 337, 697–701.

Spurzem, J.R. Sacco, O. and Rennard, S.I. (1990). The expression of fibronectin and vitronectin receptors is regulated on bronchial epithelial cells. Am. Rev. Respir. Dis. 141, A705.

Steranka, L.R., Farmer, S.G., and Burch, R.M. (1989). Antagonists of B$_2$ bradykinin receptors. FASEB J. 3, 2019–2025.

Stewart, A.G., Dubbin, P.N., Harris, T. and Dusting, G.J. (1990). Platelet activating factor may act as a second messenger in the release of icosanoids and superoxide anions from leukocytes and endothelial cells. Proc. Natl Acad. Sci. USA 87, 3215–3219.

Straus, D.S. and Pang, K.J. (1984). Effects of bradykinin on DNA synthesis in resting NIL8 hamster cells and human fibroblasts. Exp. Cell Res. 151, 87–95.

Stuart-Smith, K. and Vanhoutte, P.M. (1988a). Airway epithelium modulates the responsiveness of porcine bronchial smooth muscle. J. Appl. Physiol. 65, 721–727.

Stuart-Smith, K. and Vanhoutte, P.M. (1988b). Arachidonic acid evokes eithelium-dependent relaxations in canine airways. J. Appl. Physiol. 65, 2170–2180.

Tamaoki, J., Sakai, N., Isono, K., Kanemura, T., Yamawaki, I. and Takizawa, T. (1991). Effects of platelet-activating factor on bioelectric properties of cultured tracheal and bronchial epithelia. J. Allergy Clin. Immunol. 87, 1042–1049.

Tanaka, D.T. and Grunstein, M.M. (1984). Mechanisms of substance P-induced contraction of rabbit airway smooth muscle. J. Appl. Physiol. 57, 1551–1557.

Taylor, P.M. and Rose, M.L. (1989). Expression of MHC antigens in normal human lungs and transplanted lungs with obliterative bronchiolitis. Transplantation 48, 506–510.

Thorsen, S. (1986). Leukotriene B4, a mediator of inflammation. Scand. J. Rheumatol. 15, 225–236.

Tosi, M.F., Hamedani A. and Infeld, M. (1991a). Effects of inflammatory cytokines on human airway epithelial cells, expression of surface icam-1 and adhesion by neutrophils. Clin. Res. 39, 2.

Tosi, M.F., Stark, M.J., Hamedani, A., Smith, C.W., Gruenert, D.C. and Huang, Y.T. (1991b). Increased adhesion by neutrophils to human tracheal epithelial cells infected with parainfluenza virus type 2: role of epithelial ICAM-1 and neutrophil CS11/CD18 adhesions. Pediat. Res. 29, A186.

Tschirhart, E. and Landry, Y. (1986). Airway epithelium releases a relaxant factor: demonstration with substance P. Eur. J. Pharmacol. 132, 103–104.

Tschirhart, E.J., Drijfhout, J.W., Pelton, J.T., Miller, R.C. and Jones, C.R. (1991). Endothelins: functional and autoradiographic studies in guinea-pig trachea. J. Pharmacol. Exp. Ther. 258, 381–387.

Uddman, R. and Sundler, F. (1987). Neuropeptides in the airways: a review. Am. Rev. Respir. Dis. 136, S3–S8,

Ullman, A., Lofdahl, C.-G., Svedmyr, N., Bernsten, L. and Skoogh, B.-E. (1988). Mucosal inhibition of cholinergic contractions in ferret trachea can be transferred between organ baths. Eur. Respir. J. 1, 908–912.

Ullman, A., Lofdahl, Svedmyr, N. and Skoogh, B.-E. (1990). Nerve stimulation releases mucosa-derived inhibitory factors, both prostanoids and nonprostanoids, in isolated ferret trachea. Am. Rev. Respir. Dis. 141, 748–751.

Undem, B.J., Raible, D.G., Adkinson, N.F. and Adams, G.K., III (1988). Effect of removal of epithelium on antigen-induced smooth muscle contraction and mediator release from guinea-pig isolated trachea. J. Pharmacol. Exp. Ther. 244, 659–665.

Valone, F.H. and Goetzl, E.J. (1982). Specific binding by human polymorphonuclear leucocytes of the immunological mediator 1-O-hexadecyl/octayl-2-acetyl-sn- glyceryl-3-phosphorylcholine. Immunology, 48, 141–149.

Valone, F.H., Coles, E., Reinhold, V.R. and Goetzl, E.J. (1982). Specific binding of phospholipid platelet activating factor by human platelets. J. Immunol. 129, 1637–1641.

Vanhoutte, P.M. (1988). Epithelium-derived relaxing factor(s)

and bronchial reactivity. Am. Rev. Respir. Dis. 138, S24–S30.

Van Scott, M.R., Cheng, P.W., Henke, D.C. and Yankaskas, J.R. (1991). In "The Airway Epithelium" (eds S. G. Farmer and D.W.P. Hay), pp 135–167. Marcel Dekker, New York.

Varsano, S., Lazarus, S.C., Gold, W.M. and Nadel, J.A. (1988). Selective adhesion of mast cells to tracheal epithelial cells *in vitro*. J. Immunol. 140, 2184–2192.

Voelker, D.R. and Mason, R.J. (1989). In "Lung Cell Biology" (ed. D. Massaro), pp 487–538. Marcel Dekker, New York.

Von Essen, S., Ertl, R., Koyama, S., Robbins, R. and Rennard, S.I. (1989). Grain sorghum dust extract causes production of neutrophil chemotactic activity by bronchial epithelial cells. Am. Rev. Respir. Dis. 139, A391.

Vijayaraghavan, J., Scicli, A.G., Carretero, O.A., Slaughter, C., Moomaw, C. and Hersh, L.B. (1990). The hydrolysis of endothelins by neutral endopeptidase 24.11 (enkephalinase). J. Biol. Chem. 265, 14150–14155.

Walters, E.H., O'Byrne, P.M., Fabbri, L.M., Graf, P.D., Holtzman, M.J. and Nadel, J.A. (1988). Control of neurotransmission by prostaglandins in canine trachealis smooth muscle. J. Appl. Physiol. 57, 129–134.

Wanner, A. (1986). Mucociliary clearance in the trachea. Clinics Chest Med. 7, 247–258.

Ward, P.E. (1991). In "Bradykinin Antagonists. Basic and Clinical Research" (ed. R.M. Burch), pp 147–170. Marcel Dekker, New York.

Wardlaw, A.J., Moqbel, R., Cromwell, O. and Kay, A.B. (1986). Platelet-activating factor. A potent chemotactic and chemokinetic factor for human eiosinophils. J. Clin. Invest. 78, 1701–1706.

Wegner, C.D., Rothlein, R. and Gundel, R.H. (1991). Adhesion molecules in the pathogenesis of asthma. Agents Actions (Suppl.) 34, 529–544.

Wessler, I., Hellwig, H. and Racké, K. (1990). Stimulation-induced over-flow of [^{3}H]-phosphorylcholine and [^{3}H]-acetylcholine from the isolated guinea-pig trachea: inhibitory role of the epithelium. Br. J. Pharmacol. 99, 54P.

Wharton, J., Polak, J.M., Bloom, S.R., Will, J.A., Brown, M.R. and Pearse, A.G.E. (1979). Substance P-like immunoreactive nerves in mammalian lung. Invest. Cell. Pathol. 2, 3–10.

White, M.V., Slater, J.E. and Kaliner, M.A. (1987). Histamine and asthma. Am. Rev. Respir. Dis. 135, 1165–1176.

Widdicombe, J.H. (1991). In "The Lung: Scientific Foundations" (eds R.G. Crystal and J.B. West), pp 263–271. Raven Press, New York.

Widdicombe, J.G. and Pack, R.J. (1982). The Clara cell. Eur. J. Respir. Dis. 63, 202–220.

Widdicombe, J.H., Ueki, I.F., Emery, D., Margolskee, D., Yergey, J. and Nadel, J.A. (1989). Release of cyclo-oxygenase products from primary cultures of tracheal epithelia of dog and human. Am. J. Physiol. 256, L351–L355.

Wilkens, J.A., Becker, A., Wilkens, H., Emura, M., Riebe-Imre, M., Plein, K., Schöber, S., Tsikas, D., Gutzki, F.M. and Frölich, J.C. (1992). Bioassay of a tracheal smooth muscle-constricting factor released by respiratory epithelial cells. Am. J. Physiol. 263, L137–L141.

Wright, W., Zhang, Y.G., Salome, C.M. and Woolcock, A.J. (1990). Effect on inhaled preservatives on asthmatic subjects. I. Sodium metabisulfite. Am. Rev. Respir. Dis. 141, 1400–1404.

Xie, Z., Hakoda, H. and Ito, Y. (1992). Airway epithelial cells regulate membrane potential, neurotransmission and muscle tone of the dog airway smooth muscle. J. Physiol. 449, 619–639.

Yanagisawa, M., Kurihara, H., Kimura, S., Tomobe, Y., Kobayashi, M., Mitsui, Y., Yazaki, Y., Goto, K. and Masaki, T. (1988). A novel potent vasoconstrictor peptide produced by vascular endothelial cells. Nature 332, 411–415.

Yousem, S.A., Curley, J.M., Dauber, J., Paradis, I., Rabinowich, H., Zeevi, A., Duquesnoy, R., Dowling, R., Zenati, M., Hardesty, R. and Griffith, B.P. (1990). HLA-class II antigen expression in human heart-lung allografts. Transplantation 49, 991–995.

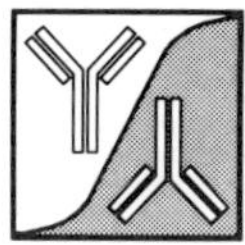

6. Airway Epithelial Inflammation and its Functional Consequences

P. M. O'Byrne *and* Ellinor Ädelroth

1. Introduction

The lining of the airways consists of pseudostratified columnar epithelium that not only serves as a diffusion barrier, but also has several, newly described, metabolic functions. Damage and desquamation of the airway epithelium is a prominent feature of the pathology of asthma and has been described in specimens of airways from patients with severe, fatal asthma, as well as in airway biopsies from patients with very mild disease. The purpose of this chapter will be to describe the characteristic airway inflammation and associated epithelial damage of asthma and discuss the possible functional consequences of this damage.

Asthma is a common disease affecting 10–15% of populations in many countries (Sears, 1990). Despite intense research interest over a number of years, the pathogenesis of asthma remains poorly understood. However, during the last 20–30 years, there has been a growing awareness that the pathogenesis of asthma is linked to the presence of airway inflammation, and that the physiological characteristics of asthma are consequences of events related to the airway inflammatory response. The association between airway inflammation and severe asthma had been recognized by the end of the last century. In 1892, William Osler described, in the first edition of his book *The Principals and Practice of Medicine*, that 'bronchial asthma . . . in many cases is a special form of inflammation of the smaller bronchioles'. It took more than 70 years before a more comprehensive description of airway inflammation in asthma was provided by Dunnill *et al.* (1969), who described findings in patients dying of acute asthma. More recently, the importance of airway inflammation, and associated structural damage, in the pathogenesis of even very mild asthma has been recognized (Laitinen *et al.*, 1985; Beasley *et al.*, 1989; Jeffery *et al.*, 1989; Azzawi *et al.*, 1990; Bousquet *et al.*, 1990; Jeffery *et al.*, 1992).

Immunopharmacology of Epithelial Barriers
ISBN 0–12–288030–7

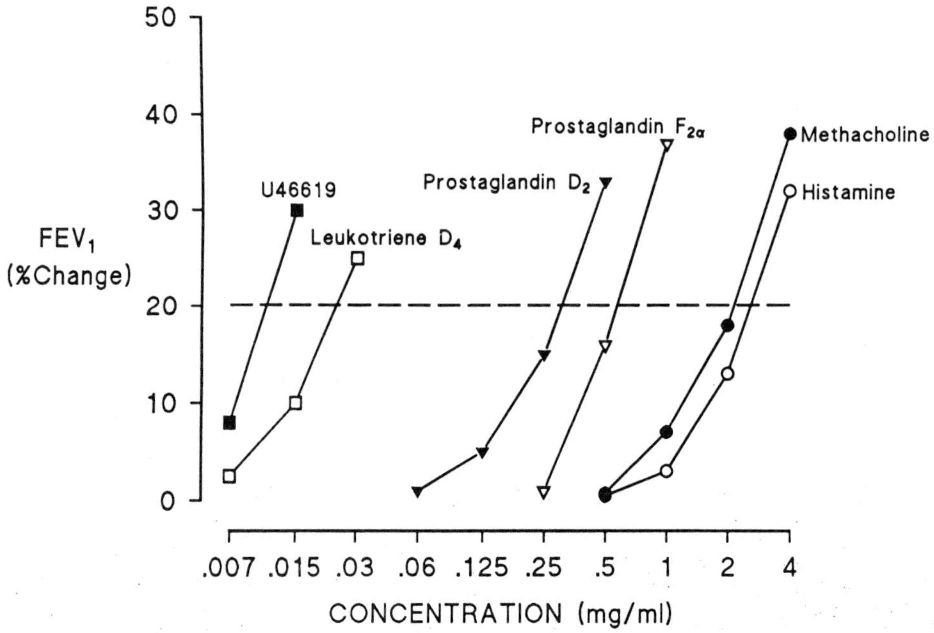

Figure 6.1 Airway responsiveness to LTD₄, PGD₂ and PGF₂α, the thromboxane mimetic U46619, methacholine and histamine in one asthmatic subject. Although the potency of the bronchoconstrictor agonists varies, the subject is hyper-responsive to all of these agonists when compared with normal subjects.

2.   Definition of Asthma

Studies of asthma have been difficult due to the absence of a precise and generally accepted definition of the disease. The most useful current definition of asthma depends on physiological parameters. These have been the presence of variable airflow obstruction which resolves either spontaneously or as a result of treatment, and the presence of airway hyper-responsiveness to a variety of inhaled chemical bronchoconstrictor agonists or naturally occurring stimuli. However, for a number of reasons, including the difficulty in defining variable airflow obstruction in patients with a component of irreversible airflow obstruction, and lack of specificity of the presence of airway hyper-responsiveness, particularly in children (Sears *et al.*, 1986), this definition remains less than ideal.

3.   Airway Hyper-responsiveness and Asthma

The fact that asthmatic subjects develop bronchoconstriction when exposed to cholinergic stimuli has been known since 1921 (Alexander and Paddock, 1921) and to histamine since 1932 (Weiss *et al.*, 1932). In 1945, Tiffeneau and Beauvallet demonstrated that asthmatics bronchoconstrict after lower concentrations of inhaled acetylcholine compared with normal subjects. This was subsequently confirmed with histamine and with the cholinergic agonist methacholine by Curry (1946, 1947).

Asthmatic subjects are now known to have airway hyper-responsiveness to a wide variety of bronchoconstrictor agents (Fig 6.1), such as histamine, acetylcholine, serotonin, sulphidopeptide leukotrienes (Ädelroth *et al.*, 1986) and the stimulatory prostaglandins PGD₂ (Hardy *et al.*, 1984) and PFG₂α (Thomson *et al.*, 1981). Airway hyper-responsiveness is also present to stimuli such as exercise (Anderton *et al.*, 1979) and hyperventilation to cold, dry air (O'Byrne *et al.*, 1982) which are believed to act through bronchoconstrictor mediator release from cells in the airways. Airway hyper-responsiveness not only means that airways require a much smaller concentration of the inhaled bronchoconstrictor agent to induce the same degree of narrowing as in a non-asthmatic airway, but also that asthmatic airways can narrow to a much greater extent than non-asthmatic airways (Woolcock *et al.*, 1984).

Airway hyper-responsiveness is present in almost all subjects with current symptomatic asthma (Hargreave *et al.*, 1985), but it can also be demonstrated in some children who have never had any symptoms of asthma (Sears *et al.*, 1986). The airway hyper-responsiveness in asthmatics can remain stable over many years (Juniper *et al.*, 1982). There are a number of stimuli which can cause increased responsiveness in human subjects. These include allergen (Cartier *et al.*, 1982; Boulet *et al.*, 1983) and ozone (Golden *et al.*, 1978) as well as occupational sensitizing agents such as low molecular weight chemical sensitizers, TDI (Fabbri *et al.*, 1985, 1987) or plicatic acid from the Western red cedar (Chan-Yeung *et al.*,

1982). Exposure to these stimuli and the subsequent airway hyper-responsiveness is associated with the development of, or an increase in, symptoms of asthma (Juniper *et al.*, 1981). Identifying the significance of the different stimuli has been important in the management of subjects with asthma, as removal of the stimulus can cure the disease, especially if an occupational agent has been involved (Chan-Yeung and Lam, 1986). The various stimuli have also proved to be useful in developing experimental models to examine the pathogenesis of airway hyper-responsiveness and its relation to clinical asthma both in humans and animals.

4. Airway Inflammation and Airway Hyper-responsiveness

There have been a number of studies which have associated components of the airway inflammatory response to the pathogenesis of airway hyper-responsiveness. Initial studies, performed in animals, demonstrated a transient association between the development of airway hyper-responsiveness and inflammation as measured by the presence of inflammatory cells in the airways in rabbits after inhaling allergen (Marsh *et al.*, 1985), guinea-pigs after inhaling cigarette smoke (Hulbert *et al.*, 1985) or after ozone (Murlas and Roum, 1985), and dogs after inhaling ozone (Lee *et al.*, 1977) or allergen (Chung *et al.*, 1985). A similar association between the presence of inflammatory cells in the airways and airway hyper-responsiveness was subsequently demonstrated in human subjects after inhaling ozone (Stelzer *et al.*, 1986), TDI (Fabbri *et al.*, 1987) and allergen (De Monchy *et al.*, 1985).

There is also evidence that airway inflammation is responsible for airway hyper-responsiveness in asthmatic subjects. Studies which have provided information on cell populations in bronchoalveolar lavage fluid of mild, stable asthmatics have indicated that an increase in mast cell (Flint *et al.*, 1985; Kirby *et al.*, 1987; Wardlaw *et al.*, 1988; Ädelroth *et al.*, 1990) and/or eosinophils (Kirby *et al.*, 1987; Wardlaw *et al.*, 1988; Ädelroth *et al.*, 1990) exists in asthmatics. Some studies which have related these changes in inflammatory cells to measurements of airway hyper-responsiveness have shown close correlations between numbers of mast cells and eosinophils and the degree of airway hyper-responsiveness in asthmatics (Kirby *et al.*, 1987) although others have not been able to show such relationships (Ädelroth *et al.*, 1990). These studies are, however, consistent with the hypothesis that even in mild asthma, with no clinical evidence of ongoing airway inflammation, airway hyper-responsiveness is maintained as a consequence of some component of the inflammatory response.

5. Epithelium-derived Inflammatory Cell Chemotaxis

Studies in a variety of species, including humans, have demonstrated that inflammatory stimuli, such as inhaled ozone, allergen or TDI cause chemotaxis of eosinophils and/or neutrophils into the airways (Holtzman *et al.*, 1984; Chung *et al.*, 1985; Stelzer *et al.*, 1986; Fabbri *et al.*, 1987). In dogs, the neutrophil infiltration following ozone inhalation is predominantly seen in the epithelial layer in airway biopsies, with much fewer cells per unit area in the submucosa (Holtzman *et al.*, 1984). This suggested that the chemotactic stimulus for neutrophil recruitment may originate either from cells within the airway lumen or, alternatively, from the airway epithelium. Holtzman *et al.* (1983) subsequently demonstrated that canine tracheal epithelial cells can produce the potent chemotactic LT B$_4$. Further studies by these investigators described the release of 15 HETE from human airway epithelial cells (Hunter *et al.*, 1985), which again has been demonstrated to cause neutrophil chemotaxis. The precise cell of origin of these chemotactic mediators has not been identified, but the studies suggest that the airway epithelium may be actively involved in initiating inflammatory cell chemotaxis after some inflammatory stimuli.

6. Epithelium-Derived Inhibitory Mediators

The concept that airway epithelium is a passive barrier across the airway is no longer tenable. It is now clear that the airway epithelium has many important effects in regulating airway function. One of these is to produce and release a variety of mediators which influence the responses of the airways to noxious stimuli. Some of these mediators are stimulatory (see below), while others are inhibitory. The best characterized of the inhibitory mediators are EpDRFs and inhibitory prostaglandins such as PGE$_2$.

The airway epithelium of several species has been reported to release a factor which reduces the contractile responses of airway smooth muscle to a variety of agonists, such as histamine, acetylcholine, methacholine and serotonin (Barnes *et al.*, 1985; Flavahan *et al.*, 1985; Jones *et al.*, 1988). This was initially described from canine bronchial epithelium, and has been termed EpDRF (Flavahan *et al.*, 1985). There is conflicting evidence to suggest that EpDRF is a diffusible mediator. Some investigators have suggested that the reduced contractile responses elicited in the presence of epithelium may be a mechanical effect of the attachment of the epithelium on the smooth muscle rather than due to release of a mediator. For example, the epithelium could alter the passive length–tension characteristics of the smooth muscle strips, resulting in an improper choice of the optimum portion of the length–tension curve. This is unlikely to account for the magnitude of the changes demonstrated in the tissues stripped of epithelium. Also, the epithelium could act as a physical restraint to contraction. In most experiments, however, the muscle is performing an isometric contraction and therefore not altering its length. The epithelium may also be acting as a barrier to diffusion of the agonists to the smooth muscle. Lastly, vascular endothelium is capable of performing many metabolic functions and, in some preparations, is capable of metabolizing acetylcholine and histamine (Shepro and Dunham, 1986). Therefore, the reduced contractile responses in the presence of epithelium may be a result of the epithelium metabolizing these contractile agonists.

The evidence that the reduced contractile responses mediated by the airway epithelium is caused by a transferable mediator is conflicting. For example, Holroyde (1986) could not demonstrate reduced smooth muscle contraction using perfusate taken from baths with both smooth muscle strips and epithelium. However, studies by Hay *et al.* (1986, 1987) suggested allergen-induced EpDRF release, when epithelium was close to, but not attached to, airway smooth muscle. In a recent study, we have demonstrated that the contractile responses of the smooth muscle were reduced both when the airway mucosa was attached to, or was in close proximity to, canine trachealis (Manning *et al.*, 1990). These studies suggest that the reduced contractile responses are caused by a transferable mediator released from the airway epithelium. Also, significant differences existed between the smooth muscle responses from strips with the mucosa attached when compared to strips with the epithelium in close proximity. This suggests that the transferable mediator has a short-lived biological activity. Whether EpDRF is a single chemical entity has not yet been clarified, nor has the biochemical structure of this mediator(s) been elucidated.

Airway epithelial cells have been demonstrated to release PGE_2 in response to stimulation with bradykinin (Leikauf *et al.*, 1985). PGE_2 has potent inhibitory effects in the airways, such as presynaptic modulation and inhibition of acetylcholine release from muscarinic nerves (Walters *et al.*, 1984). In addition, it has been suggested that PGE_2 is able to reduce contractile responses to inhaled histamine, acetylcholine and methacholine (Manning *et al.*, 1989), and reduce bronchoconstrictor responses to exercise in asthmatics (O'Byrne and Jones, 1986). In most asthmatic subjects, if a second exercise challenge of the same intensity is performed within 4 h of the initial challenge, the magnitude of the exercise bronchoconstriction is less. This has been termed exercise refractoriness (Edmunds *et al.*, 1978). Exercise refractoriness is inhibited by pretreatment with indomethacin (O'Byrne and Jones, 1986; Margolskee *et al.*, 1988), suggesting an important role for inhibitory prostaglandins in providing this protective effect in asthmatic airways. Reduced broncho-

constrictor responses (tachyphylaxis) also occur with repeated challenges with inhaled histamine in asthmatic subjects, and this inhibitory effect can also be prevented by indomethacin pretreatment (Manning *et al.*, 1987). Histamine tachyphylaxis can be prevented by pretreatment with H_2 receptor antagonists (Jackson *et al.*, 1988; Manning *et al.*, 1988), as can histamine-stimulated PGE_2 receptors. It appears probable, but is not yet proven, that the source of this inhibitory prostaglandin is the airway epithelium.

The airway epithelium contains enzymes and neutral endopeptidases, capable of metabolizing tachykinins, such as substance P (Dusser *et al.*, 1989), which are bronchoconstrictor and pro-inflammatory mediators. Metabolism of the tachykinins is clearly an important role the epithelium plays in protecting the airways from these effects. Indeed, loss of this protective mechanism has been suggested to be the cause of virus-induced airway hyper-responsiveness in one animal model (Dusser *et al.*, 1989). Therefore, the airway epithelium has the capacity to both release inhibitory mediators and metabolize stimulatory mediators which may alter airway function.

7. Airway Epithelial Damage in Asthma

7.1 MUCOSAL/SUBMUCOSAL PATHOLOGIES IN ASTHMA

The original studies which examined the pathology of asthma of necessity examined airways of patients dying from severe asthma and demonstrated a characteristic pattern of intense airway inflammation of a mixed inflammatory cell infiltrate, often with a predominance of eosinophils, oedema of the submucosa, and extensive, often patchy, epithelial damage and desquamation (Houston *et al.*, 1953; Dunnill, 1960). The airway lumen was often filled with tenacious mucous plugs. Changes that may be more chronic in nature included thickening of the reticular collagen layer of the basement membrane and hypertrophy and/or hyperplasia of the airway smooth muscle.

More recent studies have examined the pathology of the airways of less severe asthmatics, as assessed by the histology of airway biopsies taken during bronchoscopy under stable conditions. Laitinen *et al.* (1985) focused on the airway epithelial damage in stable asthmatics, and demonstrated that extensive damage was present even in mild asthmatic subjects. Subsequent studies (Beasley *et al.*, 1989; Jeffery *et al.*, 1989; Jeffery *et al.*, 1992) have confirmed these initial observations. In two studies, (Beasley *et al.*, 1989; Jeffery *et al.*, 1992) the extent of epithelial damage has been shown to be related to the degree of airway hyper-responsiveness. However, not all studies examining the airway epithelium in mild asthma agree on the fragility of the surface epithelium or have demonstrated abnormalities. Lozewicz *et al.* (1990) have examined airway biopsies from stable asthmatics using both light and electron microscopy, and compared the findings to normal controls. These investigators could not demonstrate any change in structure of the airway epithelium in the asthmatic subjects. Jeffery *et al.* (1992) found loss of surface epithelium to a similar degree of asthmatics and healthy controls, although in this study there was a significant relationship of epithelial loss and airway hyper-responsiveness in the more severe asthmatics. This suggests that the epithelial abnormality is patchy and perhaps more likely to occur in patients with more severe disease. Although not unique to asthma, in addition to the above findings, asthmatic airways have thickening of the subepithelial reticular collagen layer of the basement membrane (Dunnill *et al.*, 1969; Jeffery *et al.*, 1992). As in patients dying of asthma, subjects investigated under stable conditions show an increased inflammatory infiltrate in the airway mucosa compared with healthy controls. The main cell types constituting the inflammatory infiltrate are eosinophils (Beasley *et al.*, 1989; Azzawi *et al.*, 1990; Bousquet *et al.*, 1990; Jeffery *et al.*, 1992), mast cells (Jeffery *et al.*, 1992) or lymphocytes (Jeffery *et al.*, 1989; Azzawi *et al.*, 1990).

The studies involving examination of bronchial biopsies have confirmed and added to findings from studies using BAL in asthmatic subjects. The results indicate that airway inflammation is present in all asthmatics with disease severity from mild to severe, and that epithelial damage is a component of the inflammatory response.

7.2 PATHOGENESIS OF AIRWAY EPITHELIAL DAMAGE

At present, the most likely hypothesis to explain epithelial damage and desquamation in asthmatic airways is that basic proteins released from activated eosinophils damage the epithelium. Levels of MBP have been shown to be increased in sputum from asthmatic patients (Frigas *et al.*, 1981). MBP has also been identified by means of immunofluorescence in tissue sections from airways of patients dying from severe asthma (Filley *et al.*, 1982). Even in sections where identifiable eosinophils could not be seen, evidence showed MBP to be present in epithelium and submucosa giving signs of tissue damage. To further support the hypothesis that the released MBP may be directly toxic to the epithelium, Motojima *et al.* (1989), have demonstrated that MBP in concentrations as low as 10 µg/ml causes ciliostasis when applied to guinea-pig airway epithelium, and higher concentrations cause epithelial damage and desquamation. Similar effects were demonstrated with other eosinophil granule cationic proteins.

The cytotoxic ECP has also been demonstrated by immunohistochemical techniques using the specific monoclonal antibodies EG1 (specific for granular protein) and EG2 (the cleaved form of eosinophil cationic protein) in airway tissue of patients dying of asthma (Azzawi *et al.*, 1989) as well as in subjects with mild atopic asthma (Azzawi *et al.*, 1990) or untreated and more severe disease (Bousquet *et al.*, 1990). In bronchial biopsies investigated by electron microscopy, Jeffery *et al.* (1992) found

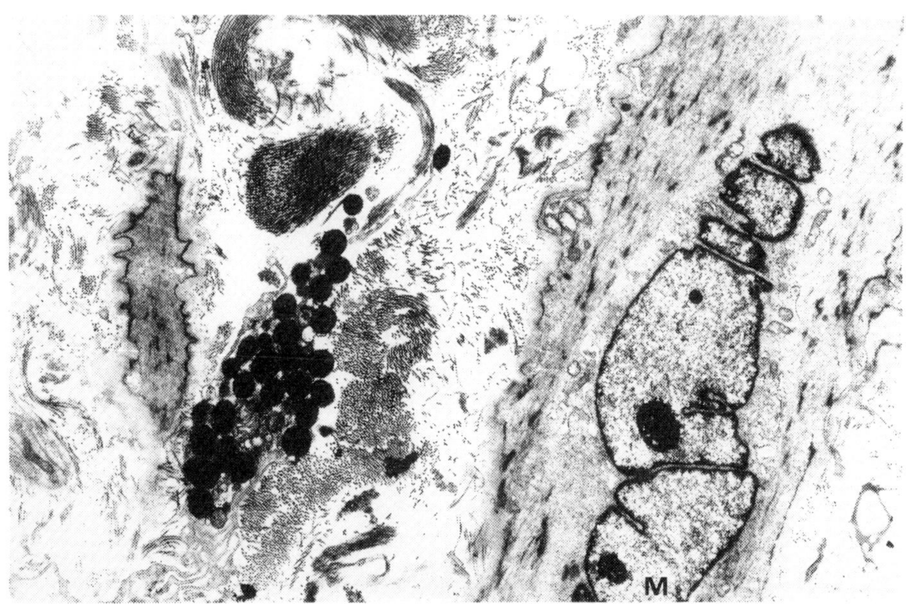

Figure 6.2 Electron micrograph of bronchial mucosa from an atopic asthmatic showing free eosinophil granules close to a smooth muscle fibre (M) (×4850).

eosinophils just below or within the surface epithelium and, in addition, frequent foci of free eosinophil granules interpreted as signs of eosinophil activity and in agreement with results of immunohistological staining with EG1 and EG2 (Azzawi *et al.*, 1990; Bousquet *et al.*, 1990) (Fig.6.2).

Studies using BAL in mild stable asthmatic subjects have given variable results regarding measurable levels of eosinophil proteins; Kirby *et al.* (1987) found an increase in eosinophils in BAL of the asthmatic group, but no increase in MBP levels. However, a subsequent study did show an increase in both eosinophils and MBP in lavage fluid but only in the most hyper-responsive subjects (Wardlaw *et al.*, 1988). Levels of ECP in BAL have been demonstrated to be increased in lavage fluid and correlated to the percentage of eosinophils (Ädelroth *et al.*, 1990; Bousquet *et al.*, 1990).

There may be other explanations to account for the extensive epithelial damage found in severe asthma. For example, Fujisawa *et al.* (1987) have described, in one patient dying from severe asthma, an association between extracellular neutrophil elastase and airway epithelial damage. Thus, other enzymes or mediators released from inflammatory cells could be toxic to the epithelium. Alternatively, submucosal oedema may make the epithelium more likely to desquamate. However, none of these hypotheses has much experimental support.

7.3   SIGNIFICANCE OF AIRWAY EPITHELIAL DAMAGE

There have been several hypotheses proposed to explain how epithelial damage may account for airway hyper-responsiveness in asthma. These have included increased permeability of the airway epithelium, allowing increased access of bronchoconstrictor mediators to receptors on airway smooth muscle; or loss of inhibitory mediators generated by the airway epithelium, thereby increasing the response to inhaled bronchoconstrictor mediators.

Increased permeability of the airway epithelium to high molecular weight substances, such as horseradish peroxidase, has been demonstrated to occur in guinea-pig airways after inhalation of inflammatory stimuli such as cigarette smoke (Boucher *et al.*, 1980). This suggested that in asthmatic subjects, where chronic airway inflammation coexists with epithelial damage, increased permeability of the airway epithelium may account for airway hyper-responsiveness (Hogg, 1981). Several studies have compared the permeability of the respiratory epithelium in normal subjects, asthmatics and asymptomatic smokers. These studies demonstrated that the clearance of inhaled radiolabelled aerosols from the lung into the blood was much faster in the asymptomatic smokers, with normal airway responsiveness, than in either the normal subjects, with normal airway responsiveness, or the

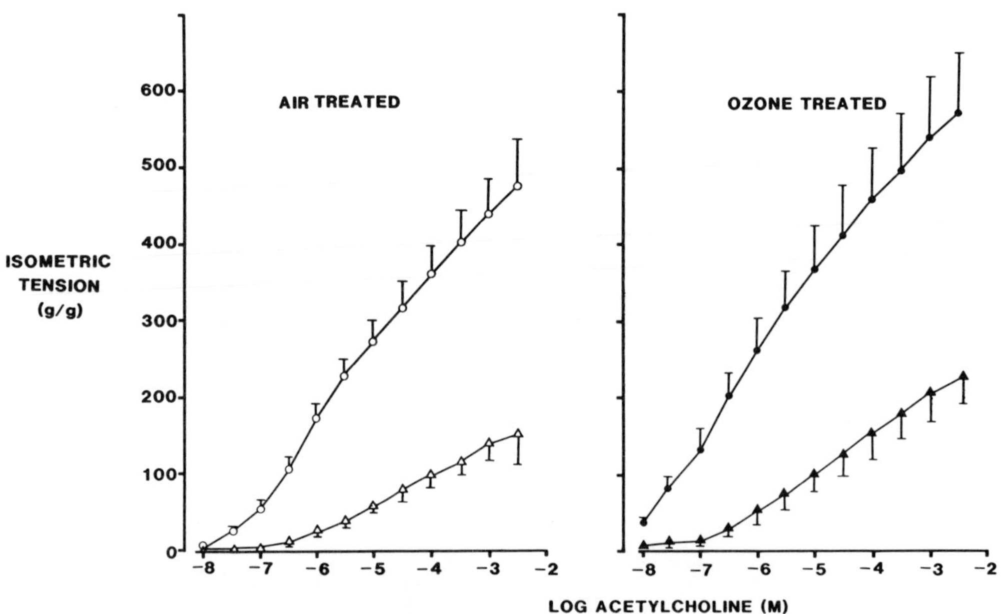

Figure 6.3 *In vitro* trachealis response to acetylcholine from air-treated dogs with (△) and without (○) epithelium attached, and from ozone-treated dogs with (△) and without (○) epithelium attached. Increased constrictive responses are demonstrated from tissues without epithelium from both air-treated ($p<0.0001$) and ozone-treated ($p<0.0001$) dogs. (Reproduced, with permission, from Jones *et al.* (1988).)

asthmatic subjects with airway hyper-responsiveness (Elwood *et al.*, 1983; O'Byrne *et al.*, 1984). These results make it very unlikely that increases in airway epithelial permeability is the cause of airway hyper-responsiveness in asthma.

One attractive hypothesis to explain airway hyper-responsiveness in asthmatic subjects is the loss of EpDRF due to epithelial desquamation in asthmatic subjects resulting in airway hyper-responsiveness to inhaled bronchoconstrictor mediators (Vanhoutte, 1987). The hypothesis that loss of an inhibitory EpDRF may be responsible for airway hyper-responsiveness in asthma has not been possible to test in human subjects. This hypothesis can be tested in animal models of airway hyper-responsiveness (Lee *et al.*, 1977). Also, ozone is a powerful oxidant and has been reported to damage airway epithelium and cause marked influx of inflammatory cells into the epithelium (Holtzman *et al.*, 1984). The ability of inhaled ozone to impair the production or release of EpDRF, leading to the development of airway hyper-responsiveness, has been examined (Jones *et al.*, 1988). This study demonstrated that the function of EpDRF *in vitro*, as determined by the ability to reduce the contractile responses of the trachealis to acetylcholine, histamine or serotonin, was unaltered in dogs with airway hyper-responsiveness *in vivo*, when compared to control dogs (Fig. 6.3). Thus, while epithelial damage occurs following inhalation of ozone, the ability to release EpDRF *in vitro* is maintained. It is unlikely, therefore, that loss of EpDRF is the cause of airway responsiveness after inhaled ozone in dogs.

A final mechanism whereby airway epithelial damage may cause airway hyper-responsiveness has been suggested by a recent study by Dusser *et al.* (1989). These investigators have demonstrated that neutral endopeptid-ase activity is decreased in the airway epithelium from guinea-pigs infected by viruses. This results in an enhanced responsiveness to substance P in these animals compared with non-infected controls. Respiratory viral infections are known to cause airway hyper-responsiveness in human subjects (Empey *et al.*, 1976), and these investigators have suggested that loss of endopeptidases and increased airway responses to tachykinins may be involved.

7.4   EFFECTS OF TREATMENT ON AIRWAY EPITHELIAL DAMAGE

To date there are few reports of the effects of anti-asthmatic treatment on the findings of BAL and/or on the morphology of the airway mucosa as seen by bronchial biopsies.

The first studies involving mucosal biopsies (Thiringer *et al.*, 1975; Lundren, 1977) were performed to investi-gate whether low doses of inhaled glucocorticosteroids would cause bronchial epithelial or mucosal atrophy similar to that reported in the skin in response to topical steroids (Groniowska *et al.*, 1976). Later, Laursen *et al.* (1988) examined mucosal biopsies in asthmatics during treatment with a high dose of inhaled budesonide and compared the findings with those of subjects without asthma undergoing bronchoscopic investigation for malig-nant disease, and Lundgren *et al.* (1988) re-examined most of the patients from their first study (Lundren, 1977) after 10 years of treatment with various doses of inhaled corticosteroids. The results indicated that long-term steroid treatment did not induce any undesired effect but reduced the total cell infiltrate and restored the coverage of the surface by cilia. No details of specific inflammatory cell types were given.

There are a few recent studies where the effects of anti-asthmatic treatment on findings in BAL and bronchial mucosal biopsies have been investigated (Ädelroth *et al.*, 1990; Djukanovic *et al.*, 1991; Burke *et al.*, 1992; Jeffery *et al.*, 1992). In the first study (reported in two parts) (Ädelroth *et al.*, 1990; Jeffery *et al.*, 1992), the effects of treatment were investigated in two aspects. First, in a controlled way, the effects of regular short-term treatment with either an inhaled corticosteroid (budesonide) or an inhaled β₂ agonist (terbutaline) were studied in patients with mild atopic asthma, only requiring occasional use of an inhaled bronchodilator as symptomatic treatment. Secondly, the long-term treatment effects of inhaled steroids were studied in a group of more severe asthmatics who had required regular treatment with inhaled steroids in addition to bronchodilators to achieve adequate symptom control.

In the first group of subjects with mild asthma, bronchoscopy with BAL and bronchial biopsies was performed twice, before and after 4 weeks of randomly allocated treatment with either inhaled budesonide 200 μg daily b.i.d., i.e. a low, yet realistic dose when given without need for symptom control, or terbutaline 500μg q.i.d., i.e. considered a normal average regular dose at the time of the study. In the more severe asthmatics, as in normal control subjects, only one procedure including BAL and mucosal biopsies was undertaken. Cross comparisons were made between the two asthmatic groups and the normal healthy subjects and in the milder asthmatic subjects before and after the controlled treatment.

The BAL component of the study demonstrated a significant increase in the proportion of eosinophils in all asthmatics and of mast cells in the milder asthmatics before commencement of the controlled treatment. In the milder asthmatic subjects, increased levels of ECP in BAL and serum were found compared with the healthy individuals. There were no significant changes in total cell numbers or differential cell counts in the BAL after 4 weeks treatment with either drug. The ECP levels in BAL were, however, decreased by both treatments, although this was only statistically significant for budesonide.

By light microscopic examination, the mucosal biopsies showed similar levels of epithelial loss and thickening of subepithelial reticular collagen before and after the 4 weeks of treatment with either drug. The increased total cell infiltrate of the asthmatics was not affected by either drug, but the increased percentages of eosinophils and mast cells were reduced to levels found in the healthy subjects. The changes were statistically significant for eosinophils and mast cells for the asthmatics treated with budesonide.

By transmission electron microscopy there were significant reductions of mast cells, eosinophils and also of free foci of eosinophil granules. The finding of foci of eosinophil granules lying free in the interstitium was striking, and their frequency was judged as a sign of eosinophilic activity and degranulation, in keeping with other reports of eosinophilic activity in asthma (Azzawi *et al.*, 1990). The frequency with which foci of eosinophil granules were found was decreased by the short-term treatment and thus interpreted as a sign of decreased eosinophil activity. The effects were statistically significant for budesonide, although seen to some extent by both treatments.

The subjects with mild atopic asthma showed extensive pathological changes in the airway mucosa which reversed after regular short-term treatment to a state similar to that found in healthy subjects with no history of asthma. This indicates that regular prophylactic therapy in mild asthma may be beneficial in preventing progression of the disease, at least as judged by its pathology.

Those subjects with more severe asthma, in need of regular inhaled corticosteroids (doses of 100–1000 µg of inhaled steroid for a mean of 3.7 years), showed ECP levels in BAL similar to those of the healthy controls despite an increased proportion of eosinophils in BAL. The total mucosal cell infiltrate resembled that of the controls although the proportion of eosinophils was increased similar to that found in the BAL. There were significantly fewer free foci of eosinophil granules and mast cells than in the milder asthmatic group before their treatment. The thickness of the subepithelial reticular collagenous layer was not altered by long-term inhaled steroid treatment and showed a significantly increased thickness compared with the healthy controls.

In the study by Burke *et al.* (1992) a small group of six asthmatics was investigated before and after 3 months of treatment with inhaled budesonide. Endobronchial biopsies were obtained before and after the treatment period for immunopathological analysis. Before the steroid therapy a T cell-dominated inflammation was seen in all subjects. After the treatment period, a marked reduction was seen in the number of T lymphocytes (CD2, CD5, CD8), CD45RO[+] T cells and macrophages with the phenotype of antigen-presenting cells.

There is as yet little additional information in the literature regarding the effects of treatment on BAL and/or biopsy findings.

Diaz *et al.* (1984) investigated the effects of disodium cromoglycate on BAL findings and reported decreased proportions of eosinophils post-treatment. A recent report by Duddridge *et al.* (1990) on the effects of high-dose inhaled beclomethasone dipropionate, 2000 µg daily for a mean of 2.5 months, on BAL cells suggested decreased proportions of eosinophils and mast cells and increased proportions of lymphocytes. Djukanovic *et al.* (1991) investigated the effects of 6 weeks of treatment with a high dose of inhaled beclomethasone dipropionate on BAL and bronchial mucosa specimens obtained before and after the treatment period. The BAL analysis showed a significant reduction of eosinophil numbers after treatment. No changes in T lymphocyte numbers were noted, but there were significant reductions of lymphocytes expressing the activation marker HLA-DR and IL-2R. In the biopsies, immunostaining with specific antibodies showed a reduction in epithelial and mucosal mast cells, eosinophils and submucosal T lymphocytes after 6 weeks of beclomethasone diproprionate treatment paralleling a significant improvement in the clinical expression of asthma.

Recently, Dahl *et al.* (1991) published an abstract regarding the influence of inhaled salmeterol on signs of bronchial inflammation in a BAL study in asthmatics. Inhaled salmeterol 50 µg b.i.d. or placebo was given to 12 asthmatics in a controlled, double-blind, cross-over study with BAL performed at entry and after each treatment period. The macroscopic appearance of the bronchial mucosa evaluated according to an 'inflammatory index' improved significantly after salmeterol. The total cell number in BAL fluid increased after salmeterol but cell differential counts were similar after both treatment periods. The concentration of ECP in BAL fluid decreased significantly after salmeterol. The authors interpreted the actions of salmeterol as dampening on the inflammatory mechanisms in asthma.

8.   *Conclusions*

Asthma is a chronic inflammatory disease of the airways. The airway inflammation results in airway hyperresponsiveness, resulting in airway narrowing and symptoms of asthma. Airway inflammation can also result in the occlusion of airways by airway oedema and secretions, which is the probable cause of death in most cases of fatal asthma. The airway epithelium is damaged and can desquamate in a patchy way as part of this process in many asthmatics, even those with a mild condition. The currently available evidence suggests that the damage to the epithelium occurs because of the release of toxic products from activated airway inflammatory cells, particularly eosinophils. However, the precise role that the airway epithelium plays in the pathophysiology of airway hyper-responsiveness and asthma remains unclear.

9. Acknowledgement

Dr P.M. O'Byrne is a recipient of a Medical Research Council of Canada Scientist Award.

10. References

Ädelroth, E., Morris, M.M., Hargreave, F.E. and O'Byrne, P.M. (1986). Airway responsiveness to leukotrienes C4 and D4 and to methacholine in patients with asthma and normal controls. N. Engl. J. Med. 315, 436–451.

Ädelroth, E., Rosenhall, L., Johansson, S.., Linden, M. and Venge, P. (1990). Inflammatory cells and eosinophilic activity in asthmatics investigated by bronchoalveolar lavage. The effects of treatment with budesonide or terbutaline. Am. Rev. Respir. Dis. 142, 91–99.

Alexander, H.L. and Paddock, R. (1921). Bronchial asthma: response to pilocarpine and epinephrine. Arch. Intern. Med. 27, 184–191.

Anderton, R.C., Cuff, M.T., Frith, P.A., Cockcroft, D.W., Morse, J.L.C., Jones, N.L. and Hargreave, F.E. (1979). Bronchial responsiveness to inhaled histamine and exercise. J. Allergy Clin. Immunol. 63, 315–320.

Azzawi, M., Jeffery, P.K., Frew, A.J., Johnston, P. and Kay, A.B. (1989). Activated eosinophils in bronchi obtained at post-mortem from asthma deaths. Clin. Exp. Allergy 19, 118.

Azzawi, M., Bradley, B., Jeffery, P.K., Frew, A.J., Wardlaw, A.J., Assoufi, B., Collins, J.V., Durham, S. and Kay, A.B. (1990). Identification of activated T-lymphocytes and eosinophils in bronchial biopsies in stable atopic asthma. Am. Rev. Respir. Dis. 142, 1407–1413.

Barnes, P.J., Cuss, F.M. and Palmer, J.B. (1985). The effect of airway epithelium on smooth muscle contractility in bovine trachea. Br. J. Pharmacol. 86, 685–691.

Beasley, R., Roche, W.R., Roberts, J.A. and Holgate, S.T. (1989). Cellular events in the bronchi in mild asthma and after bronchial provocation. Am. Rev. Respir. Dis., 139, 806–817.

Boucher, R.C., Johnson, J., Inoue, S., Hulbert, W. and Hogg, J.C. (1980). The effect of cigarette smoke on the permeability of guinea pig airways. Lab. Invest. 43, 94–100.

Boulet, L.P., Cartier, A., Thomson, N.C., Roberts, R.S., Dolovich, J. and Hargreave, F.E. (1983). Asthma and increases in nonallergenic bronchial responsiveness from seasonal pollen exposure. J. Allergy Clin. Immunol. 71, 399–406.

Bousquet, J., Chanez, P., Lacoste, J.Y., Barneon, G., Ghavanian, N., Enander, I., Venge, P., Ahlstedt, S., Simony-Lafontaine, J., Godard, P. and Michel, F.B. (1990). Eosinophilic inflammation in asthma. N. Engl. J. Med. 323, 1033–1039.

Burke, C., Power, C.K., Norris, A., Condez, A., Schmekel, B. and Poulter, L.W. (1992). Lung function and immuno-pathological changes after inhaled corticosteroid therapy in asthma. Eur. Respir. J. 5, 73–79.

Cartier, A., Thomson, N.C., Frith, P.A., Roberts, R. and Hargreave, F.E. (1982). Allergen-induced increase in bronchial responsiveness to histamine: relationship to the late asthmatic response and change in airway caliber. J. Allergy Clin. Immunol. 70, 170–177.

Chan-Yeung, M. and Lam, S. (1986). Occupational asthma: state of the art. Am. Rev. Respir. Dis. 133, 686–703.

Chan-Yeung, M., Lam, S. and Koener, S. (1982). Clinical features and natural history of occupational asthma due to Western red cedar (Thuja plicata). Am. J. Med. 72, 411–415.

Chung, K.F., Becker, A.B., Lazarus, S.C., Frick, O.L., Nadel, J.A. and Gold, W.M. (1985). Antigen-induced airway hyper-responsiveness and pulmonary inflammation in allergic dogs. J. Appl. Physiol. 58, 1347–1353.

Curry, J.J. (1946). The action of histamine on the respiratory tract in normal and asthmatic subjects. J. Clin. Invest. 25, 785–791.

Curry, J.J. (1947). Comparative action of acetyl-beta-methyl choline and histamine on the respiratory tract in normals, patients with hay fever and subjects with bronchial asthma. J. Clin. Invest. 26, 430–438.

Dahl, R., Pedersen, B. and Venge, P. (1991). The influence of inhaled salmeterol on bronchial inflammation. A broncho-alveolar study in patients with bronchial asthma. Am. Rev. Respir. Dis. 143, 649A.

De Monchy, J.G.R., Kauffman, H.F., Venge, P., Koeter, G.H., Jansen, H.M., Sluiter, H.J. and De Vries, K. (1985). Bronchoalveolar eosinophilia during allergen-induced later asthmatic reaction. Am. Rev. Respir. Dis. 131, 373–376.

Diaz, P., Galleguillos, F.R., Gonzales, M.C., Pantin, C.F.A. and Kay, A.B. (1984). Bronchoalveolar lavage in asthma. The effect of disodium cromoglycate (cromolyn) on leucocyte counts, immunoglobulins and complement. J. Allergy Clin. Immunol. 74, 41–48.

Duddridge, M., Ward, C.V., Hendrick, D.J. and Walters, E.H. (1990). Effect of treatment with high doses inhaled corti-costeroids on bronchoalveolar lavage (BAL) fluid inflammatory cells in asthma. Thorax 45, 329P.

Djukanovic, R., Wilson, J.W., Britten, K.M. Wilson, S.J., Roche, W.R., Howarth, P.H. and Holgate, S.T. (1991). The effects of inhaled beclomethasone dipropionate (BDP) on inflammatory cells in the asthmatic airways. J. Allergy Clin. Immunol. 87, 135.

Dunnill, M.S. (1960). The pathology of asthma with special reference to changes in the bronchial mucosa. J. Clin. Pathal. 13, 27–33.

Dunnill, M.S., Massarelli, G.R. and Anderson, J.A. (1969). A comparison of the quantitive anatomy of the bronchi in normal subjects, in status asthmaticus, in chronic bronchitis and in emphysema. Thorax, 24 176–179.

Dusser, D.J., Jacoby, D.B., Djokic, T.D., Rubinstein, I., Borson, D.B. and Nadel, J.A. (1989). Virus induces airway hyper-responsiveness to tachykinins: role of neutral endopeptidase. J. Appl. Physiol. 67, 1504–1511.

Edmunds, A.T., Tooley, M. and Godfrey, S. (1978). The refractory period after exercise-induced asthma: its duration and relation to the severity of exercise. Am. Rev. Respir. Dis. 117, 247–254.

Elwood, R.K., Kennedy, S., Belzberg, A., Hogg, J.C. and Pare, P.D. (1983). Respiratory mucosal permeability in asthma. Am. Rev. Respir. Dis. 128, 523–527.

Empey, D.W., Laitinen, L.A., Jacobs, L., Gold, W.M. and Nadel, J.A. (1976). Mechanisms of bronchial hyperreactivity in normal subjects after upper respiratory tract infections. Am. Rev. Respir. Dis. 113, 131–139.

Fabbri, L.M., Di Giacomo, R., Del Vecchio, L., Zocca, E., De Marzo, N., Maestrelli, P. and Mapp, C.E. (1985). Prednisolone, indomethacin and airway responsiveness in toluene diisocyanate sensitized subjects. Clin. Resp. Physiol. 21, 421–426.

Fabbri, L.M., Boschetto, P., Zocca, E., Milani, G., Pivirotto, F., Plebani, M., Burlina, A., Licata, B. and Mapp, C.E. (1987). Bronchoalveolar neutrophilia during late asthmatic reactions induced by toluene diisocyanate. Am. Rev. Respir. Dis. 136, 36–42.

Filley, W.V., Kephart, G.M., Holley, K.E. and Gleich, C.J. (1982). Identification by immunofluorescence of eosinophil granulae major basic protein in lung tissues of patients with bronchial asthma. Lancet ii, 11–15.

Frigas, E., Loegering, D.A., Solley, G.O., Farrow, G.M. and Gleich, G.J. (1981). Elevated levels of eosinophil granule major basic protein in sputum of patients with bronchial asthma. Mayo Clin. Proc. 56, 345–353.

Flavahan, N.A., Aarhuus, L.L., Rimele, T.J. and Vanhoutte, P.M. (1985). Respiratory epithelium inhibits bronchial smooth muscle tone. J. Appl. Physiol. 58, 834–838.

Flint, K.C., Leung, K.B.P., Hudspith, B.N., Brostoff, J., Pearce, F.L. and Johnson, N.M. (1985). Bronchoalveolar mast cells in extrinsic asthma: a mechanism for the initiation of antigen specific bronchoconstriction. Br. Med. J. 291, 923–926.

Fujisawa, T., Kephart, G.M. and Gleich, G.J. (1989). In "Bronchitis IV" (eds H.J. Sluiter and R. Van der Lende), pp 210–221. Van Gorcum, Assen.

Golden, J.A., Nadel, J.A. and Boushey, H.A. (1978). Bronchial hyperirritability in healthy subjects after exposure to ozone. Am. Rev. Respir. Dis. 118, 287–294.

Groniowska, M., Dabrowski, J., Maciejewske, W. and Walski, M. (1976). Electron-microscopic evaluation of collagen fibrils after topical corticosteroid therapy. Dermatologica 152, 147–153.

Hardy, C.C., Robinson, C., Tattersfield, A.E. and Holgate, S.T. (1984). The bronchoconstrictor effect of inhaled prostaglandin D2 in normal and asthmatic men. N. Engl. J. Med. 311, 209–213.

Hargreave, F.E., Ramsdale, E.H. and Dolovich, J. (1985). In "Airway Responsiveness Measurement and Interpretation" (eds F.E. Hargreave and A.J. Woolcock), pp 122–126. Astra Pharmaceuticals, Mississauga.

Hay, D.W.P., Farmer, S.G., Raeburn, D., Robinson, V.A., Fleming, W.W. and Fedan, J.S. (1986). Airway epithelium modulates the reactivity of guinea-pig respiratory smooth muscle. Eur. J. Pharm. 129, 11–18.

Hay, D.W.P., Muccitelli, R.M., Horstemeyer, D.L., Wilson, K.A. and Raeburn, D. (1987). Demonstration of the release of an epithelium-derived inhibitory factor from a novel preparation of guinea-pig trachea. Eur. J. Pharm. 136, 247–250.

Hogg, J.C. (1981). Bronchial mucosal permeability and its relationship to airways hyperreactivity. J. Allergy Clin. Immunol. 67, 421–425.

Holroyde, M.C. (1986). The influence of epithelium on the responsiveness of guinea-pig isolated trachea. Br. J. Pharmacol. 87, 501–507.

Holtzman, M.J., Aizawa, H., Nadel, J.A. and Goetzl, E.J. (1983). Selective generation of leukotriene B4 by tracheal epithelial cells from dogs. Biochem. Biophys. Res. Commun. 114, 1071–1076.

Holtzman, M.J., Fabbri, L.M., O'Byrne, P.M., Gold, B.D., Aizawa, H., Walters, E.H., Alpert, S.E. and Nadel, J.A. (1984). Importance of airway inflammation for hyper-responsiveness induced by ozone. Am. Rev. Respir. Dis. 127, 686–690.

Houston, J.C., De Navasquez, S. and Trounce, J.R. (1953). A clinical and pathological study of fatal causes of status asthmaticus. Thorax 8, 207–213.

Hunter, J.A., Finkbeiner, W.E., Nadel, J.A., Goetzl, E.J., Holtzman, M.J. (1985). Predominant generation of 15–lipoxygenase metabolites of arachidonic acid by epithelial cells from human trachea. Proc. Natl. Acad. Sci. USA 82, 4633–4637.

Hulbert, W.M., McLean, T. and Hogg, J.C. (1985). The effect of acute airway inflammation on bronchial reactivity in guinea pigs. Am. Rev. Respir. Dis. 132, 7–11.

Jackson, P.A., Manning, P.J. and O'Byrne, P.M. (1988). A new role for histamine H2-receptors in asthmatic airways. Am. Rev. Respir. Dis. 138, 784–788.

Jeffery, P.K., Wardlaw, A.J., Nelson, F.C., Coolins, J.V. and Kay, A.B. (1989). Bronchial biopsies in asthma. An ultra-structural, quantitative study and correlation with hyper-reactivity. Am. Rev. Respir. Dis. 140, 1745–1753.

Jeffery, P.K., Godfrey, R.W., Ådelroth, E., Nelson, F., Rogers, A. and Johansson, S.. (1992). Effects of treatment on airway inflammation and thickening of the basement membrane reticular collagen in asthma: a quantitative light and electron microscopic study. Am. Rev. Respir. Dis. 145, 890–899.

Jones, G.L., Lane, C.G. and O'Byrne, P.M. (1988). Release of epithelium-derived relaxation factor (EpDRF) after ozone inhalation in dogs. J. Appl. Physiol. 65, 1238–1243.

Juniper, E.F., Frith, P.A. and Hargreave, F.E. (1981). Airway responsiveness to histamine and methacholine: relationship to minimum treatment to control symptoms of asthma. Thorax 36, 575–579.

Juniper, E.F., Frith, P.A. and Hargreave, F.E. (1982). Long term stability of bronchial responsiveness to histamine. Thorax 37, 288–291.

Kirby, J.G., Hargreave, F.E., Gleich, G.J. and O'Byrne, P.M. (1987). Bronchoalveolar cell profiles of asthmatic and non-asthmatic subjects. Am. Rev. Respir. Dis. 136, 379–383.

Laitinen, L.A., Heino, M., Laitinen, A., Kava, T. and Haahtela, T. (1985). Damage of the airway epithelium and bronchial reactivity in patients with asthma. Am. Rev. Respir. Dis. 131, 599–606.

Laursen, L.C., Taudorf, E., Borgeskov, S., Kobayashi, Y., Jensen, H. and Weeke, B. (1988). Fibreoptic bronchoscopy and bronchial mucosal biopsies in asthmatics undergoing long-term high-dose budesonide aerosol treatment. Allergy 43, 284–288.

Lee, L.Y., Bleeker, E.R. and Nadel, J.A. (1977). Efffect of ozone on bronchomotor response to inhaled histamine in dogs. J. Appl. Physiol. 43, 626–631.

Leikauf, G.D., Ueki, I.F., Nadel, J.A. and Widdicombe, J.H. (1985). Bradykinin stimulates Cl secretion and prostaglandin E_2 release by canine tracheal epithelium. Am. J. Physiol. 248, F48–55.

Lozewicz, S., Wells, C., Gomez, E., Ferguson, H., Richman, P., Devalia, J. and Davies, R.J. (1990). Morphological integrity of the bronchial epithelium in mild asthma. Thorax 45, 12–15.

Lundgren, R. (1977). Scanning electron microscopic studies of bronchial mucosa before and after beclomethasone dipropionate inhalations. Scand. J. Respir. Dis. (Suppl.) 101, 179–187.

Lundgren, R., Söderbderbg, M, Hörstedt, P. and Stenling, R. (1988). Morphological studies of bronchial mucosal biopsies from asthmatics before and after ten years of treatment with inhaled steroids. Eur. Respir. J. 1, 883–889.

Manning, P.J., Jones, G.L. and O'Byrne, P.M. (1987).

Tachyphylaxis to inhaled histamine in asthmatic subjects. J. Appl. Physiol. 63, 1572–1577.

Manning, P.M., Jones, G.L., Lane, C.G. and O'Byrne, P.M. (1988). Histamine-induced prostaglandin E_2 release from canine tracheal smooth muscle is inhibited by H_2 receptor blockade. Am. Rev. Respir. Dis. 137, 373A.

Manning, P.J., Lane, C.G. and O'Byrne, P.M. (1989). The effect of oral prostaglandin E_1 on airway responsiveness in asthmatic subjects. Pulm. Pharm. 2, 121–124.

Manning, P.J., Jones, G.L., Otis, J., Daniel, E.E. and O'Byrne, P.M. (1990). The inhibitory influence of tracheal mucosa mounted in close proximity to canine trachealis. Eur. J. Pharm. 178, 85–89.

Margolskee, D.J., Bigby, B.G. and Boushey, H.A. (1988). Indomethacin blocks airway tolerance to repetitive exercise but not to eucapnic hyperpnea in asthmatic subjects. Am. Rev. Respir. Dis. 137, 842–846.

Marsh, W.R., Irvin, C.G., Murphy, K.R., Behrens, B.L. and Larsen, G.L. (1985). Increases in airway reactivity to histamine and inflammatory cells in bronchoalveolar lavage after the late asthmatic response in an animal model. Am. Rev. Respir. Dis. 131, 875–879.

Motojima, S., Frigas, E., Loegering, D.A. and Gleich, G.J. (1989). Toxicity of eosinophil cationic proteins for guinea pig tracheal epithelium *in vitro*. Am. Rev. Respir. Dis. 139, 801–805.

Murlas, C.G. and Roum, J.H. (1985). Sequence of pathological changes in the airway mucosaa of guinea pigs during ozone-induced bronchial hyperreactivity. Am. Rev. Respir. Dis. 131, 314–320.

O'Byrne, P.M. and Jones, G.L. (1986). The effect of indomethacin on exercise-induced bronchoconstriction and refractoriness after exercise. Am. Rev. Respir. Dis. 134, 69–72.

O'Byrne, P.M., Ryan, G., Morris, M.M., MCormack, D., Jones, N.L., Morse, J.L.C. and Hargreave, F.E. (1982). Asthma induced by cold air and its relation to nonspecific bronchial responsiveness to methacholine. Am. Rev. Respir. Dis. 125, 281–285.

O'Byrne, P.M., Dolovich, M., Dirks, R., Roberts, R.S. and Newhouse, M.T. (1984). Lung epithelial permeability: relation to nonspecific airway responsiveness. J. Appl. Physiol. 57, 77–84.

Osler, W. (1892). "The Principals and Practice of Medicine", pp 497. Appleton, New York.

Sears, M.R. (1990). "Asthma as an Inflammatory Disease", pp 1–34. Marcel Dekker, New York.

Sears, M.R., Jones, D.T., Holdaway, M.D., Hewitt, C.J., Flannery, E.M., Herbison, G.P. and Silva, P.A. (1986). Prevalence of bronchial reactivity to inhaled methacholine in New Zealand children. Thorax 41, 283–289.

Shepro, D. and Dunham, B. (1986). Endothelial cell metabolism of biogenic amines. Ann. Rev. Physiol. 48, 335–345.

Stelzer, J., Bigby, B.G., Stulbarg, M., Holtzman, M.J., Nadel, J.A., Ueki, I.F., Leifauf, G.D., Goetzl, E.J. and Boushey, H.A. (1986). Ozone-induced changes in bronchial reactivity to methacholine and airway inflammation in humans. J. Appl. Physiol. 60, 1321–1326.

Thiringer, G., Malmberg, R., Eriksson, N., Svedmyr, N. and Zettergren, L. (1975). Bronchoscopic biopsies of bronchial mucosa before and after beclomethasone dipropionate therapy. Postgrad. Med. J. 51, 30–31.

Thomson, N.C., Roberts, R., Bandouvakis, J., Newball, H. and Hargreave, F.E. (1981). Comparison of bronchial responses to prostaglandin $F_{2\alpha}$ and methacholine. J. Allergy Clin. Immunol. 68, 392–398.

Tiffeneau, R. and Beauvallet, B. (1945). Epreuve de bronchoconstriction et de bronchodilatation par aerosols. Bull. Acad. Med. 129, 165–168.

Vanhoutte, P.M. (1987). Airway epithelium and bronchial reactivity. Can. J. Physiol. Pharmacol. 65, 448–450.

Walters, E.H., O'Byrne, P.M., Fabbri, L.M., Graf, P.D., Holtzman, M.J. and Nadel, J.A. (1984). Control of neurotransmission by prostaglandins in canine trachealis smooth musssscle. J. Appl. Physiol. 57, 129–134.

Wardlaw, A.J., Dunnette, S., Gleich, G.J., Collins, J.V. and Kay, A.B. (1988). Eosinophils and mast cells in bronchoalveolar lavage in subjects with mild asthma. Am. Rev. Respir. Dis. 137, 62–69.

Weiss, S., Robb, G.P. and Ellis, L.B. (1932). The systemic effects of histamine in man. Arch. Intern. Med. 49, 360–396.

Woolcock, A.J., Salome, C.M. and Yan, K. (1984). The shape of the dose-response curve to histamine in asthmatic and normal subjects. Am. Rev. Respir. Dis. 130, 71–75.

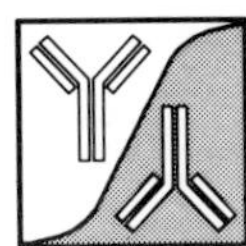

7. Immunopharmacology of Epithelial Cell–Virus Interactions

David B. Jacoby

Immunopharmacology of Epithelial Barriers
ISBN 0–12–288030–7

1. Introduction

Under normal circumstances, the entire body is covered with a layer of epithelial cells. This includes not only the squamous epithelium of the epidermis, but also the columnar epithelial cells lining the respiratory and gastrointestinal tracts.

Because these epithelial cells form the interface between the rest of the body and the external milieu, they are frequently exposed to viruses. Many viruses have developed the ability to target the epithelial cell as their primary site of infection. Conversely, epithelial cells have developed a variety of mechanisms of defence against viral infection. Furthermore, prevention of this initial step of viral infection of the epithelium is the goal of some of the body's other immune responses to viruses.

The host defence against viruses can occur at many levels. In the case of the epithelial cells lining the respiratory tract, the fluid and mucus that overlie the epithelium may form a physical barrier against inhaled viruses. Secretion of ions and water by the epithelial cells themselves is stimulated by a variety of inflammatory mediators (Al-Bazzaz et al., 1981; Leikauf et al., 1985, 1986; Jacoby et al., 1988), and such secretion may therefore be increased during viral infection (Cloutier et al., 1989). This increased secretion of water may allow viral particles to be carried away by mucociliary clearance in the airways.

More specific to the defence against viruses is the production of interferons. Under the influence of interferons, produced either by inflammatory cells, fibroblasts or, perhaps, the epithelial cell itself, the epithelial cell can produce a variety of substances that inhibit replication of viruses (see Section 2.5).

Still more specific to the type of virus, and even to the strain of virus, are the antibody and cytotoxic T lymphocyte responses. Virus-specific antibodies may serve to prevent infection or to limit its spread, depending on the part of the virus recognized. T lymphocytes appear to be important, if not vital, in the ability to clear viral infections once they have become established (see Section 3.5).

This discussion will focus first on interferons, which are produced in response to many viral infections and are effective, to varying degrees, in the defence against many viruses. A discussion of influenza virus and rhinovirus, two of the most commonly encountered viruses affecting epithelial cells, will follow. These will be used to illustrate the interactions of viruses with epithelial cells as well as the immunopharmacology of the response to these infections. Strategies for vaccine design will be discussed as relevant to these viruses, as well as others.

2. Interferon

Interferon was first identified as a factor secreted by fragments of chicken egg chorioallantoic membrane upon exposure to influenza virus. This factor prevented influenza infection of other chorioallantoic membrane fragments (Isaacs and Lindenmann, 1957). While interferons affect a wide variety of antiviral, growth inhibitory and immunomodulatory functions (Pestka et al., 1987; Samuel, 1991; Sen and Lengyel, 1992), the present discussion will focus on their antiviral activities.

Interferons are classified into type I, which map to chromosome 9, and type II, which map to chromosome 12. Type I interferons include interferon α (produced by leucocytes) and interferon β (produced by fibroblasts), while type II interferon (interferon γ) is produced by T lymphocytes and NK cells (Samuel, 1991).

2.1 Production of Interferon

Production of interferon in virus-infected cells is controlled at the transcriptional level by a cytoplasmic protein called NF-κB. This protein exists in the cytoplasm of uninfected cells in a complex with an inhibitor protein called IκB. A variety of stimuli can cause phosphorylation of IκB, causing it to dissociate from NF-κB (Ghosh and Baltimore, 1990). When NF-κB is released from this inhibitor, it can bind to several sites in the DNA of the cell, inducing gene expression for interferon as well as for other cytokines (Lenardo and Baltimore, 1989).

Phosphorylation of IκB can be achieved through several pathways. Perhaps the most relevant with regard to infection with RNA viruses is IκB phosphorylation in response to double-stranded RNA, leading to NF-κB activation. This has been shown quite clearly using synthetic double-stranded RNA (poly(rI)-poly(rC)) (Visvanathan and Goodbourn, 1989). When cells become infected with RNA viruses (particularly those such as influenza and rhinovirus that do not contain reverse transcriptase) an important step in viral replication is production of RNA using the viral RNA as a template. This process, catalysed by viral RNA-dependent RNA polymerase, leads to production of RNA that is capable of hybridizing with the viral RNA and producing at least short segments of double-stranded RNA. Although it has not been demonstrated directly, this double-stranded RNA might be responsible for phosphorylation of IκB, activation of NF-κB, and expression of interferon genes.

The DNA sequence to which NF-κB binds in the interferon β gene is called PRD-II. PRD-II has the sequence: GGGAAATTCC (Visvanathan and Goodbourn, 1989). Interestingly, this is similar to two sequences in the human immunodeficiency virus I and to four sequences in the cytomegalovirus genome (Lenardo and Baltimore, 1989). Binding of activated NF-κB to these sites is an important step in replication of these viruses, markedly increasing the rate of transcription of viral genes.

2.2 Interferon Receptors

The cellular actions of interferons are mediated via binding of interferon to cell surface receptors. These receptors bind interferons with high affinity (K_D = 10^{-10}–10^{-9} M) (Merlin *et al.*, 1985). Competitive binding studies show that interferons α and β bind to the same receptor with comparable affinities, while interferon γ recognizes a different receptor (Fig. 7.1) (Merlin *et al.*, 1985).

2.2.1   The Interferon α/β Receptor

The receptor for human interferons α and β has been cloned and expressed. Mouse 10B H7 cells that were transfected with human interferon α receptor bound interferon α with a K_D of 2×10^{-10} M (Uze *et al.*, 1990). This ability to bind interferon α was associated with a 64 000-fold increase in antiviral activity as compared with non-transfected cells. Interestingly, although it is clear from competitive binding studies that interferons α and β bind to the same receptor, the antiviral activity of interferon β in these transfected cells remained low, suggesting that other, species-specific factors are necessary for expression of interferon β antiviral activity. Credence is lent to this hypothesis by a cell line that is insensitive to interferon α, yet remains sensitive to interferon β (Pellegrini *et al.*, 1989). The factors responsible for the distinct responses to interferon α and interferon β remain to be investigated.

2.2.2   The Interferon γ Receptor

The receptor for human interferon γ which, as stated above, is distinct from the receptor for interferon α and interferon β, has also been cloned and expressed (Aguet *et al.*, 1988). When this receptor was expressed in Chinese hamster ovary cells, the cells bound interferon γ but failed to respond as measured by cellular expression of HLA-B7 antigen. However, when the human interferon γ gene was inserted into the Chinese hamster ovary cell line 153-E7Bx, which is a hamster–human somatic cell hybrid that contains the human chromosome 21q, full responsiveness to human interferon γ was obtained (Jung *et al.*, 1990). Thus, it appears that a product of the long arm of chromosome 21 is required for species-specific responsiveness to interferon γ.

To further elucidate the role of this chromosome 21-derived cofactor, Hemmi *et al.* (1992) constructed a hybrid interferon γ receptor in which the extracellular portion was human interferon γ receptor while the transmembrane and cytoplasmic portions were mouse interferon γ. When this hybrid interferon γ receptor was expressed in mouse L929 cells, the transfected cells bound interferon γ normally, but failed to respond to human interferon γ in terms of expression of major histocompatibility class I antigens, induction of the transcription factor IRF-1, or by becoming resistant to viral infection. However, when the receptor was expressed in a mouse-derived cell line that contained human chromosome 21 (SCC 16-5 cells) the cells exhibited all these responses to human interferon γ. Thus, differences in the intracellular portion of the receptor from different species cannot explain the species-specific functions of interferon γ. It appears from these

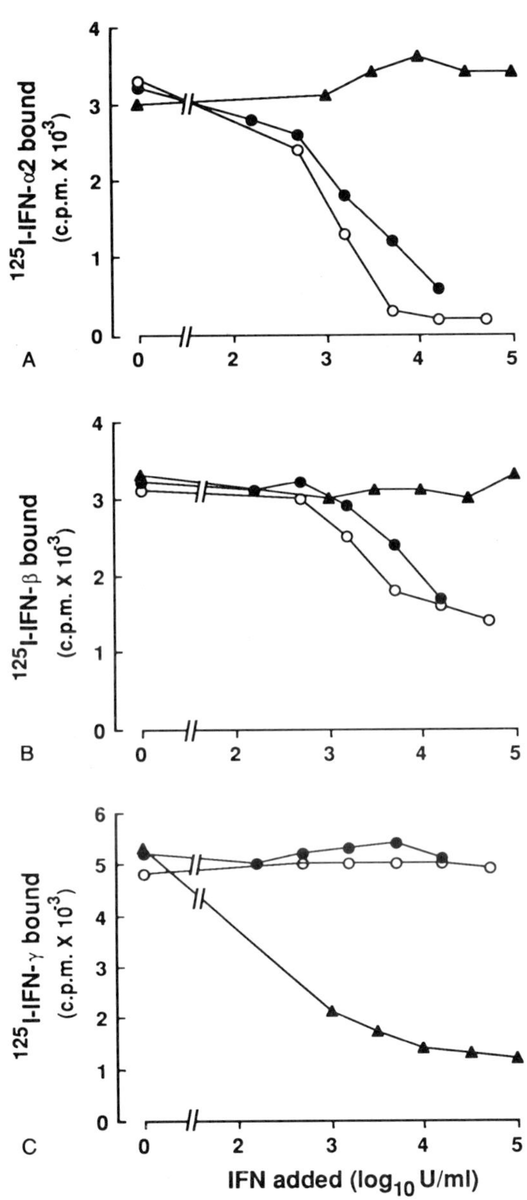

Figure 7.1 Displacement of (A) ^{125}I-labelled interferon α_2, (B) ^{125}I-labelled interferon β, and (C) ^{125}I-labelled interferon γ from Daudi cells by unlabelled interferon α_2 (○), interferon β (●) and interferon γ (▲). Both interferon α and interferon β displace ^{125}I-labelled interferon α_2 and ^{125}I-labelled β from cells, demonstrating that these two interferons bind to the same receptor. In contrast, only interferon γ displaces ^{125}I-labelled interferon γ from cells, demonstrating that interferon γ binds to a separate receptor. (Redrawn, with permission, from Merlin, G., Falcoff, E. and Aguet, M. (1985) ^{125}I-labelled human interferons alpha, beta and gamma: comparative receptor-binding data. J. Gen. Virol. 66, 1149–1152.)

studies that the extracellular portion of the interferon γ receptor must interact with a chromosome 21-derived cofactor in order for the effects of interferon γ to be seen.

2.3   SIGNAL TRANSDUCTION

The transductional mechanisms responsible for the actions of interferons on cells are also unclear (Pfeffer and Tan, 1991). In the past, it was suggested that activation of protein kinase C is a key step in interferon-induced cell signalling. Several investigators showed that staurosporin and other protein kinase inhibitors block interferon-induced gene expression (Mattila *et al.*, 1989; Griffiths *et al.*, 1990; Gumina *et al.*, 1991; Towata *et al.*, 1991). However, the specificity of the inhibitors used for protein kinase C inhibition is not absolute, and the ability of down-regulation of protein kinase C (as, for example, after chronic treatment with 12-O-tetradecaoyl-phorbol 13-acetate) to block the effects of interferon has been inconsistent (Kessler and Levy, 1991). Furthermore, a more selective protein kinase C inhibitor, calphostin C, does not affect the response to interferon α (Kessler and Levy, 1991). This led to the suggestion that other, as yet unidentified, protein kinases may be important in the transduction of the interferon receptors.

It has also been shown that interferon α activates phospholipase A_2, hydrolysing arachidonic acid from cellular phospholipids. Supporting a role for phospholipase A_2 activation is the observation that the phospholipase inhibitor bromphenacyl bromide blocks the interferon-induced binding of a transcriptional factor, ISGF3, to DNA-binding sites (see Section 2.4). Inhibiting arachidonic acid oxidation using eicosatetraenoic acid had the same effect, while inhibitors of cyclooxygenase or lipoxygenase increased the effect of interferon (Hannigan and Williams, 1991). These data suggest that arachidonic acid may be an important second messenger in the response to interferon.

Several recent studies have helped to clarify the interferon receptor transductional mechanism. Using a cell-free system, it has been possible to show that the cell membrane contains the components required for interferon α-stimulated activation, via phosphorylation, of ISGF3α (David and Larner, 1992). The membrane-associated kinase responsible for this phosphorylation is a protein tyrosine kinase called Tyk2, originally described by Firmbach-Kraft *et al.* (1990), but only recently recognized as the kinase responsible for transduction of the interferon α receptor (Fu, 1992; Schindler *et al.*, 1992; Velazquez *et al.*, 1992).

2.4   INTERFERON-INDUCED GENE EXPRESSION

As noted above, binding of interferons to their receptors leads to synthesis of a variety of proteins within the target cell. The 5′ flanking regions of the genes for proteins synthesized in response to interferons α and β all contain a consensus DNA sequence (GGAAANNGAAACT) known as the interferon-stimulated response element (Stark and Kerr, 1992). Such a sequence has not yet been identified for interferon γ-responsive genes.

The interferon-stimulated response elements form binding sites for the attachment of transcriptional factors, which then lead to gene transcription and protein synthesis. As noted above, one such factor is known as ISGF3. ISGF3 exists as two inactive subunits (ISGF3α

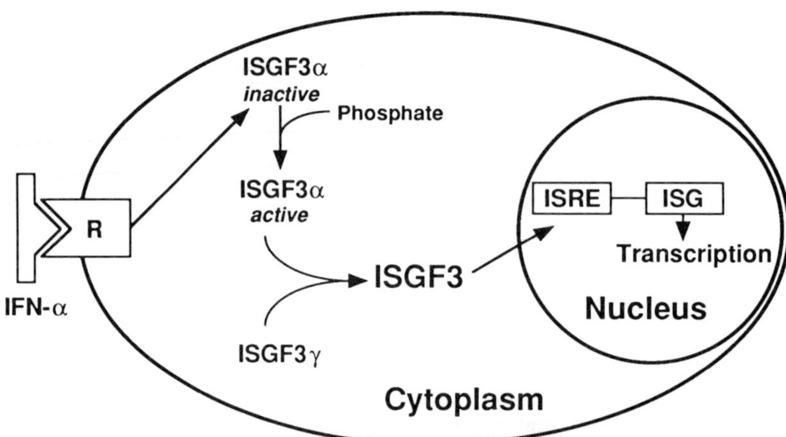

Figure 7.2 Induction of interferon-responsive genes by interferon α. Interferon α binds to its receptor, and the signal is transduced via the tyrosine kinase Tyk2, leading ultimately to phosphorylation of the inactive form of ISGF3α. Phosphorylation activates the subunit, allowing it to combine with the other subunit, ISGF3γ, thus forming active ISGF3. This complex moves from the cytoplasm to the nucleus, where it binds to an interferon-sensitive reponse element (ISRE) in the 5′ flanking region of interferon-sensitive genes, leading to their transcription. (Redrawn, with permission, from Sen and Lengyel (1992).)

and ISGF3γ) in the cytoplasm of unstimulated cells. It appears that phosphorylation of ISGF3α is the staurosporine-sensitive step leading to its activation and association with ISGF3γ. After assembly of the two subunits, ISGF3 translocates to the nucleus, where it binds to interferon-stimulated response elements and initiates transcription of interferon-responsive genes (Fig. 7.2) (Sen and Lengyel, 1992). Mutant cell lines lacking either ISGF subunit are unresponsive to the actions of interferon α (John *et al.*, 1991; McKendry *et al.*, 1991). Furthermore, the formation of the ISGF3 complex can be interfered with by adenovirus E1A proteins (Ackrill *et al.*, 1991; Gutch and Reich, 1991; Kalvakolanu *et al.*, 1991), an example of a virus defending itself against cellular antiviral mechanisms.

2.5 INTERFERON-INDUCIBLE GENES

The antiviral effects of interferons are mediated via induction of the expression of host cell genes. Among these are genes encoding Mx proteins (Staeheli *et al.*, 1984; Staeheli and Haller, 1987), 2′,5′-oligoadenylate synthetase (Kerr and Brown, 1978), p68 protein kinase, and MHC antigens.

2.5.1 Mx Proteins

Influenza infections carry a high mortality in mice. The mortality rate is related to the dose of virus used to induce infection. Lindenmann (1962) observed that the A2G mouse strain was relatively resistant to doses of influenza virus that killed other strains of mice. Inherited resistance to influenza, termed the Mx1 trait, was inherited as a single dominant allele. This trait was ultimately shown to result from the expression of the interferon-inducible Mx1 gene.

The Mx1 protein is a 72 kD nuclear protein. Production of this protein is strongly induced by interferon α and interferon β, and is less strongly induced by interferon γ (Staeheli *et al.*, 1984). It has been shown that mice with mutant Mx genes are susceptible to influenza virus. Conversely, the over-expression of Mx1 in a transgenic mouse confers relative resistance to the lethal effects of intracerebral injection of influenza virus (Fig. 7.3) (Arnheiter *et al.*, 1990).

In the same study, a second line of transgenic mice expressed Mx1 only weakly. These animals had only partial protection against influenza. However, pretreatment with interferon α decreased the mortality rate among these mice from 100 to 25% following exposure to a 10^{-5} dilution of virus. Interferon α pretreatment did not affect mortality in non-transgenic control mice.

Using cultured cells, it has been shown that the antiviral effect of Mx1 is selective for the influenza virus, and does not affect replication of vesicular stomatitis virus or Sindbis virus (Staeheli *et al.*, 1986). The basis of this **selectivity** is unknown, but it is worth noting that the

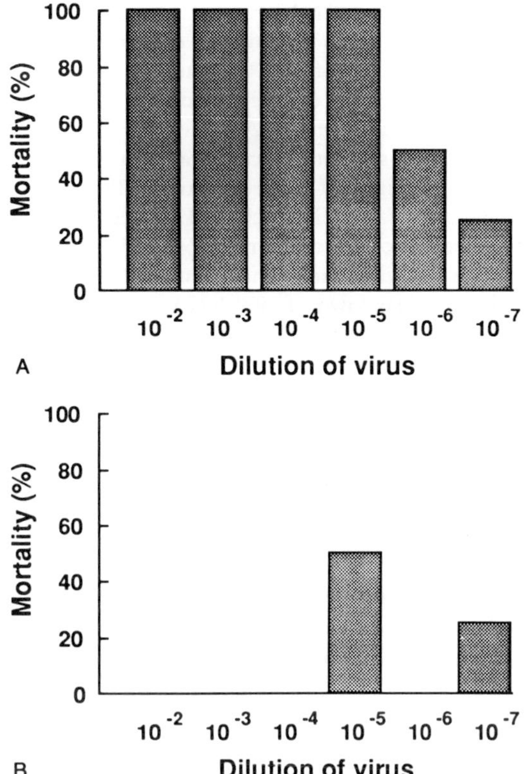

Figure 7.3 Survival of (A) control mice and (B) transgenic mice expressing Mx protein after intracerebral injection with various dilutions of a stock of influenza A virus. (Drawn, with permission, from data in Arnheiter *et al.* (1990).

corresponding protein in humans, the MxA protein, confers resistance to infection with both influenza and vesicular stomatitis viruses (Pavlovic *et al.*, 1990).

A variety of cellular functions for Mx1 and Mx-related proteins have been identified. Mx proteins may function as histocompatibility antigens (Speiser *et al.*, 1990), and may be involved in microtubule function and cell protein sorting (Obar *et al.*, 1990; Rothman *et al.*, 1990). Mx proteins in a variety of species show a high degree of conservation in the amino-terminal region, including a tripartite GTP-binding domain. Therefore, the recently described GTPase activity of Mx proteins may also be important in their anti viral actions (Nakayama *et al.*, 1991).

2.5.2 2′,5′-Oligoadenylate Synthetase

Interferon induces expression of the gene for 2′,5′-oligoadenylate synthetase (Kerr *et al.*, 1974; Kerr and Brown, 1978). This enzyme catalyses the synthesis of oligonucleotides via the following reaction (Kerr and Brown, 1978):

$$(n + 1)\ \text{ATP}\ \phi(2',5')\text{pppA(pA)}n + n\ \text{pyrophosphate}$$

2′,5′-Oligoadenylate synthetase is activated by double-stranded RNA, which may be present in host cells during replication of certain RNA viruses. The $(2',5')pppA(pA)n$ formed by this reaction activates endoribonuclease L, which cleaves RNA of both cellular and viral origin. It is possible that the double-stranded RNA produced during replication of RNA viruses leads to $2',5'$-oligoadenylate synthetase activation in the immediate intracellular vicinity of viral replication. This would lead to activation of endoribonuclease L adjacent to the site of viral replication and might therefore lead to preferential breakdown of viral, rather that cellular, RNA.

It is this breakdown of viral RNA that is responsible for the antiviral activity of $2',5'$-oligoadenylate synthetase. Furthermore, although endoribonuclease L is constitutively expressed in most cells, production of this enzyme can be markedly increased in some cells by interferon (Jacobsen *et al.*, 1983). Thus, both the cellular content of endoribonuclease L and the activity of this enzyme may be increased by the products of $2',5'$-oligoadenylate synthetase.

2.5.3   p68 Protein Kinase

Another protein expressed after exposure to interferon is the p68 kinase (Samuel, 1979). This 68 kD protein kinase does not function until it is activated by double-stranded RNA (it may also have other pathways of activation). It is therefore also referred to as double-stranded RNA-dependent protein kinase. Although dependence on double-stranded RNA is a characteristic shared by p68 kinase and $2',5'$-oligoadenylate synthetase, the structure of the double-stranded RNA-binding sites on the two enzymes are not similar (Patel and Sen, 1992). Once activated, it phosphorylates first itself, and then a subunit of protein synthesis initiation factor eIF-2. This leads to a series of steps culminating in inhibition of protein synthesis. Whether this inhibition has to include both cellular and viral protein synthesis, or whether it can be specific for viral proteins, is unknown. Once again, it is possible that a high concentration of double-stranded RNA in the vicinity of replication of RNA viruses leads to a high concentration of activated p68 kinase (deBenedetti and Baglioni, 1984).

It has been shown that a wide variety of viruses, including influenza (Katze *et al.*, 1988) and adenovirus (Kitajewski *et al.*, 1986), can inhibit the function of the p68 kinase. In the case of the adenovirus, it has been studies of the VAI and VAII genes, which are responsible for inhibiting p68 kinase function, that most directly implicate p68 kinase in the antiviral activity of interferon. The mutant adenovirus *dl*720, which lacks both VAI and VAII genes, grows poorly in human 293 cells. In contrast, growth of this mutant virus is relatively normal in human 293 cells that contain a phosphorylation-resistant mutant eIF-2a, thus demonstrating that it is the eIF-2a subunit that is normally phosphorylated by p68 kinase (Davies *et al.*, 1989). Furthermore, although interferon treatment does not inhibit growth of wild-type adenovirus, it does inhibit the growth of the mutant

*dl*331, which contains a mutant VAI gene (Kitajewski *et al.*, 1986).

2.5.4   Major Histocompatibility Complex Molecules

Interferon treatment increases expression of major histocompatibility molecules (Pestka *et al.*, 1987). Class I major histocompatibility molecules are induced by all interferons, while class II major histocompatibility molecules are induced only by interferon γ. The expression of viral antigens in conjunction with class I major histocompatibility molecules may stimulate antigen-specific cell lysis by cytotoxic T lymphocytes (see Section 3.4.2). The significance of such responses will be apparent after discussion of the immune response to influenza virus.

3. *Influenza Virus*

Thomas Sydenham described the influenza epidemic of August 1679, as follows (Major, 1945):

> At the beginning of this month a cough arose, which was more epidemic than any I had hitherto observed; for it seized nearly whole families at once. Some required little medicine, but in others the cough occasioned such violent motion of the lungs, that sometimes a vomiting and vertigo ensued . . . In short, from the smallness of the expectoration, the violence of the cough and the duration of the coughing fits; it seemed greatly to resemble hooping [*sic*] cough of children.

Influenza continues to be a major cause of morbidity among otherwise healthy people, as well as a cause of death primarily in the elderly and in those with pre-existing pulmonary disease. In the USA, during the period 1957–1986, 19 influenza epidemics caused more than 10 000 deaths each (Centers for Disease Control, 1991), with severe epidemics typically causing four times as many deaths (Centers for Disease Control, 1991). The availability of vaccines has helped somewhat, particularly for patients in high-risk groups. However, the antigenic mutability of the influenza virus has limited the effectiveness of vaccines and requires frequent development of new vaccines. The reasons for this, as well as possible new strategies in the treatment and prevention of influenza, will be described below.

3.1   VIRUS STRUCTURE

Influenza virus is an RNA virus (i.e. its genetic material is RNA rather than DNA). The RNA contained within the virus is negative stranded and must be transcribed into viral mRNA which is then translated into viral proteins. Synthesis of viral RNA takes place in the nucleus of the infected host cell. The RNA that makes up the viral genome is divided into eight segments, each of which

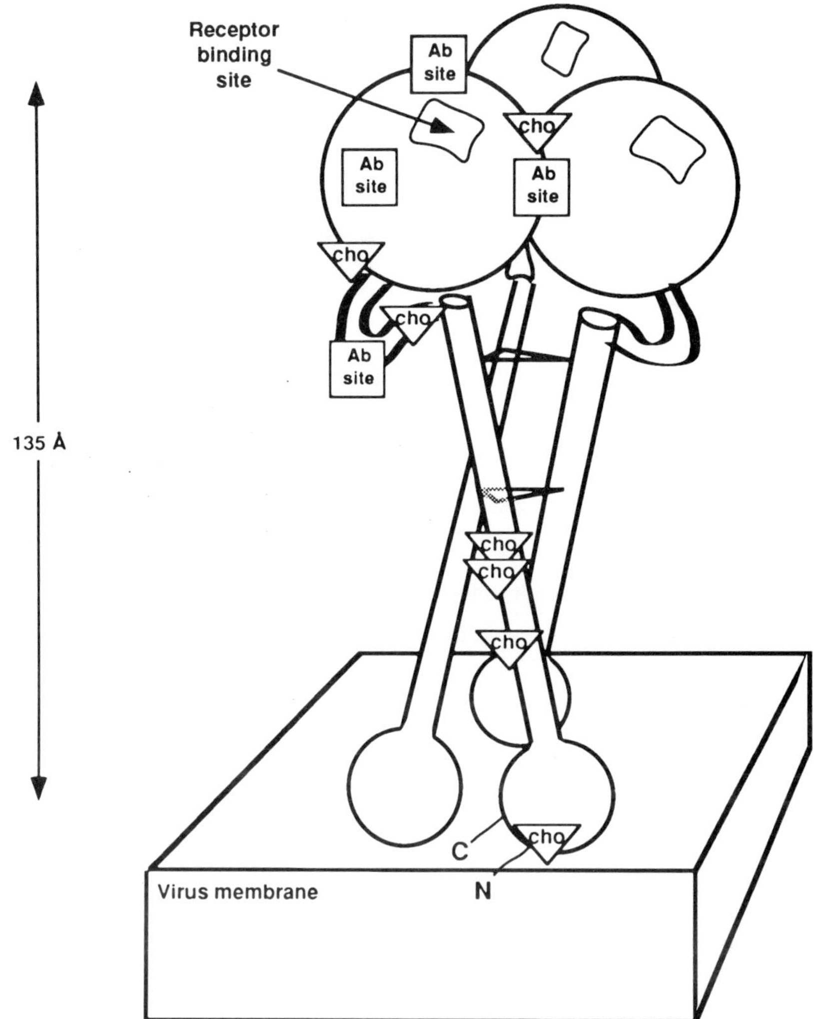

Figure 7.4 Structure of influenza A virus haemagglutinin. Each unit of the trimer consists of (1) a stem containing two antiparallel α helices terminating near the viral membrane in a globular portion consisting of a five-standed β sheet, and (2) a globular head that contains the principal antigenic sites as well as the receptor-binding site. (Redrawn, with permission, from Wilson *et al.* (1981).)

encodes one or two of the viral proteins. Thus, influenza is classified as a segmented, negative-stranded RNA virus.

3.1.1   The Virus Envelope

The viral membrane is derived from the host cell during viral budding, and thus consists of a lipid bilayer. The glycoproteins of the influenza envelope include a haemagglutinin and a neuraminidase, which are arranged radially protruding from the surface of the viral membrane.

3.1.2   Haemagglutinin

As the most plentiful protein on the surface of the influenza virus, the haemagglutinin has two functions. It is responsible for the binding of the virus to glycoprotein or glycolipid receptors on the host cell surface, and it causes fusion of the virus membrane with the endosomal membrane, thereby allowing release of the viral genome

into the host cell. It is antibodies to the haemagglutinin that are responsible for immunity to particular influenza viruses. Conversely, it is the ability of the haemagglutinin to alter its antigenicity that allows the influenza virus to escape such antibodies (see Section 3.3).

The structure of the influenza virus haemagglutinin is shown in Fig. 7.4. The molecule is a trimer of approximately 225 kDa, which is anchored in the virus membrane by a hydrophobic portion of the C terminus. The bulk of the molecule is a large hydrophilic domain outside the virus membrane. It is the globular head region of this external domain that contains not only the binding site for attachment to cellular receptors, but also the principal antigenic determinants of the virus (Wiley *et al.*, 1981; Wilson *et al.*, 1981). Each of the three peptide units of the haemagglutinin is synthesized as a single long chain that is subsequently cleaved by a host cell protease into

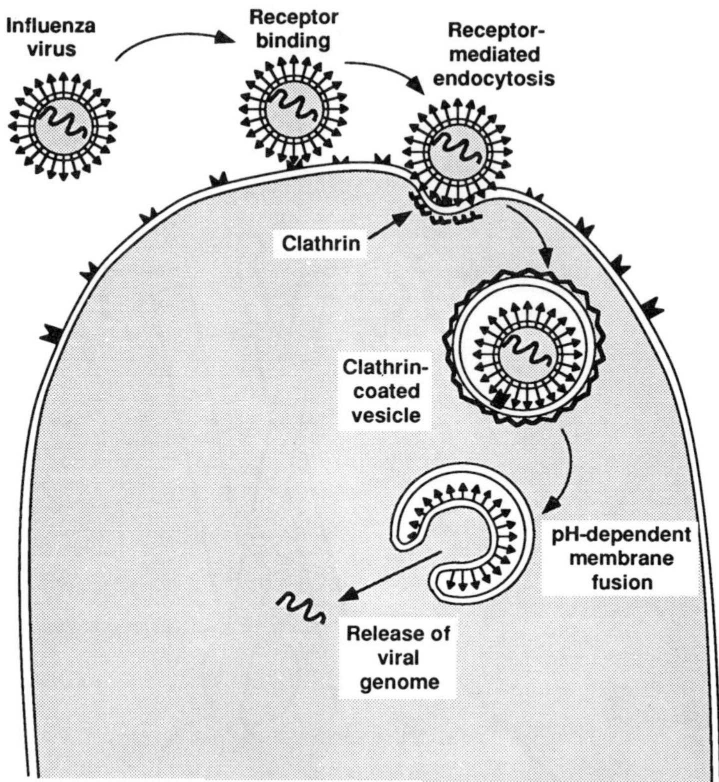

Figure 7.5 Infection of epithelial cell with influenza virus. The virus uses haemagglutinin, arranged as radial spikes on the virus surface, to bind to sialic acid-containing cell receptors. This leads to receptor-mediated endocytosis and formation of a clathrin-coated vesicle. As the interior of the endosome becomes progressively more acidic, the haemagglutinin undergoes a conformational change, exposing the N-terminal residues, which are responsible for fusion of the viral membrane with the endosome membrane. At the same time, acidification of the interior of the virus due to the proton channel properties of the viral M_2 protein causes the nucleoprotein to dissociate from the viral RNA, allowing the viral RNA to be released into the cell.

HA1 and HA2 subunits. This cleavage is required for the haemagglutinin to become active.

3.1.2.1   Receptor Binding

As shown in Fig. 7.5, the first step in infection of a cell is attachment of the virus to cell membrane receptors. In the case of influenza, the receptor is sialic acid, linked to galactose (Weis *et al.*, 1988). While variablility in the haemagglutinin amino acid sequence may affect host cell preference (Yewdell *et al.*, 1986), sialic acid residues appropriate for influenza haemagglutinin binding are present on most cells. The availability of receptors does not, therefore, contribute significantly to species specificity or tissue tropism of influenza virus infection. It is the ability of influenza virus to bind to receptors on the surface of erythrocytes that causes erythrocytes to clump in the presence of the virus, a process known as haemagglutination (see below). Viral attachment is blocked by antibodies to haemagglutinin, accounting for the immunity conferred by such antibodies.

The receptor-binding portion of the haemagglutinin is a depression on the globular head (Weis *et al.*, 1988).

This pocket is made up of portions of multiple discontinuous peptide chains. Its binding to sialic acid is shown in Fig. 7.6.

After the virus has bound to its receptor, it enters the cells via the formation of a clathrin-coated vesicle, as shown in Fig. 7.5. Clathrin-coated vesicles form within 5 minutes of receptor binding (Matlin *et al.*, 1981). This is followed by membrane fusion.

3.1.2.2   Membrane Fusion

Having entered the cell, the virus must release its RNA in order for viral replication to proceed. Given that the virus is a relatively stable structure extracellularly, it is clear that structural changes must take place in the endosome. Such changes appear to be triggered by the low pH in the endosome.

Under conditions of low pH (< 5.6), a conformational change occurs in the haemagglutinin (Skehel *et al.*, 1982). This leads to exposure of the hydrophobic N-terminal peptide of the haemagglutinin, which appears to contain the fusion activity of the molecule. In the presence of the exposed N-terminal peptide, the viral membrane fuses

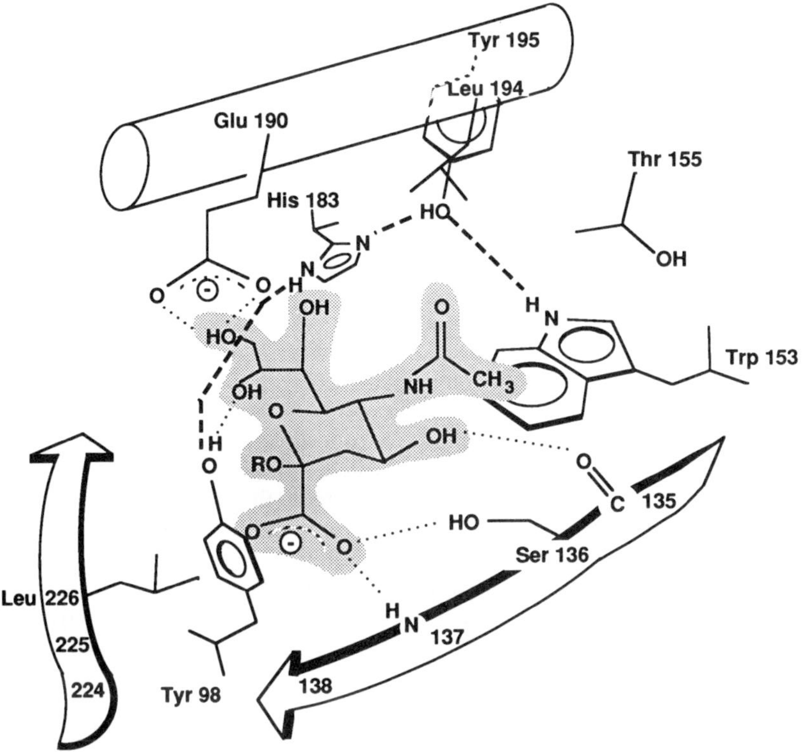

Figure 7.6 The receptor-binding pocket of the influenza A virus haemogglutinin, complexed with the main determinant of its receptor, sialic acid (shaded area). Hydrogen bonds between sialic acid and the receptor are indicated by dotted lines; hydrogen bonds within the haemagglutinin molecule are indicated by bold dashed lines. (Redrawn, with permission, from Weis *et al.* (1988).)

with the vesicle membrane, allowing release of the viral RNA into the cell (see Fig. 7.5; for a more complete review of this process, see Hoekstra, 1990).

3.1.2.3 Haemadsorption and Haemagglutination

As noted above, receptors for the influenza haemagglutinin are present on most cells. The presence of these receptors on erythrocytes makes possible three widely used *in vitro* tests for influenza virus: haemadsorption, haemagglutination, and haemagglutination inhibition.

The haemadsorption test is used to detect viral infection in cells, usually cultured cell monolayers. Because viral haemagglutinin is transported to the membrane of the infected cell in preparation for viral assembly and budding, the presence of haemagglutinin serves as a marker for infection. While this can be detected by immunofluorescence, an easier and more widely used method is to expose the infected monolayer to a dilute suspension of erythrocytes. The presence of haemagglutinin will cause the erythrocytes to bind to the surface of the infected cell's surface. Such tests can also be used to quantify virus by exposing cultures of susceptible cells to serial dilutions of virus-containing medium. After allowing infection to become established, usually for 1 week, the haemadsorption test is used to determine the dilution of virus that causes

infection in 50% of cultures. This is known as the TCID$_{50}$.

A faster, but much less sensitive, test for the presence and quantity of virus is the haemagglutination test. Serial dilutions of virus-containing fluid are mixed with a suspension of erythrocytes, and the mixture is allowed to settle in a round-bottomed tube or well. In the absence of virus, erythrocytes will form a small pellet. In contrast, viruses that have been cross-bridged by viral haemagglutinin settle as a sheet, a difference that is easily detectable.

It should be emphasized that neither the haemadsorption test nor the haemagglutination test identifies the type of virus present. To do so, it is possible to inhibit either haemadsorption or, more commonly, haemagglutination using antisera that recognize a particular viral haemagglutinin. Because antibodies against influenza virus haemagglutinin will not cross-react with haemagglutinins of other viruses, the haemagglutination inhibition test can be used to identify a virus as influenza. Furthermore, because influenza viruses are identified by the antigenicity of their haemagglutinins, strain-specific antisera can identify the strain of influenza present.

3.1.3 Neuraminidase

The other major glycoprotein of the influenza virus envelope is neuraminidase. This enzyme (previously referred to as the "receptor-destroying enzyme") has the capacity to cleave sialic acid residues from these receptors, rendering them incapable of binding the viral haemagglutinin. The function of neuraminidase is still unknown, but it may be necessary to prevent progeny virus particles from rebinding to the host cell surface, thus allowing them to be released to infect other cells. Neuraminidase may also be important in removing sialic acid residues from the carbohydrates of newly formed haemagglutinin and neuraminidase on the progeny virus, thus preventing viral particles from aggregating with one another (Griffin *et al.*, 1983). Furthermore, as the airway epithelial cells are normally covered by a layer of sialic acid-containing mucin, removal of these sialic acid residues may be required to prevent the haemagglutinin from binding to the mucin, thus entrapping the virus (Burnet *et al.*, 1947; Burnet, 1948).

Antibodies directed at antigenic sites on the viral neuraminidase do not prevent infection of cells (Kilbourne *et al.*, 1968), but they do modify the course of the acute disease, apparently by decreasing the amount of progeny virus produced (Schulman, 1975).

3.1.4   Viral M Proteins

The viral membrane contains two other proteins. The M_1 protein, which is the most abundant protein of the influenza virus, bridges the space between the internal ribonucleoprotein structure and the cytoplasmic surface of the virus envelope. The M_2 protein is an integral protein of the virus envelope.

It has recently been shown that the M_2 protein is a pH-sensitive cation channel. Because the M_1 protein is removed from the ribonucleoprotein structure at acidic pH, it has been proposed that the M_2 protein acts as a proton channel, acidifying the inside of the virus and thereby aiding in the release of the viral RNA (Pinto *et al.*, 1992) (Fig. 7.7). This may be particularly important in that the M_2 protein may be the principal target of the anti-influenza drugs amantadine and rimantadine (Hay *et al.*, 1985; Lamb *et al.*, 1985).

3.1.5   Other Proteins

The influenza virus also contains a nucleoprotein that interacts with the viral RNA, three polymerases that are responsible for transcribing negative-stranded viral RNA into positive-stranded mRNA, and two "non-structural proteins" of unknown function. There is some evidence to suggest that the nucleoprotein may be involved in determining the host specificity of influenza viruses (Scholtissek *et al.*, 1985). While the haemagglutinin is the target of neutralizing antibodies, the nucleoprotein appears to be the principal target of influenza-specific cytotoxic T lymphocytes (Gotch *et al.*, 1987) (see Section 3.4).

3.1.6   Classification of Influenza Viruses

The antigenicity of influenza virus proteins is used to classify and name the virus strains. By convention (World Health Organization, 1980), the name of an influenza virus indicates whether the nucleoprotein is type A, B or C. This is followed by the place it was initially isolated, the strain number, and the year of isolation. For type A

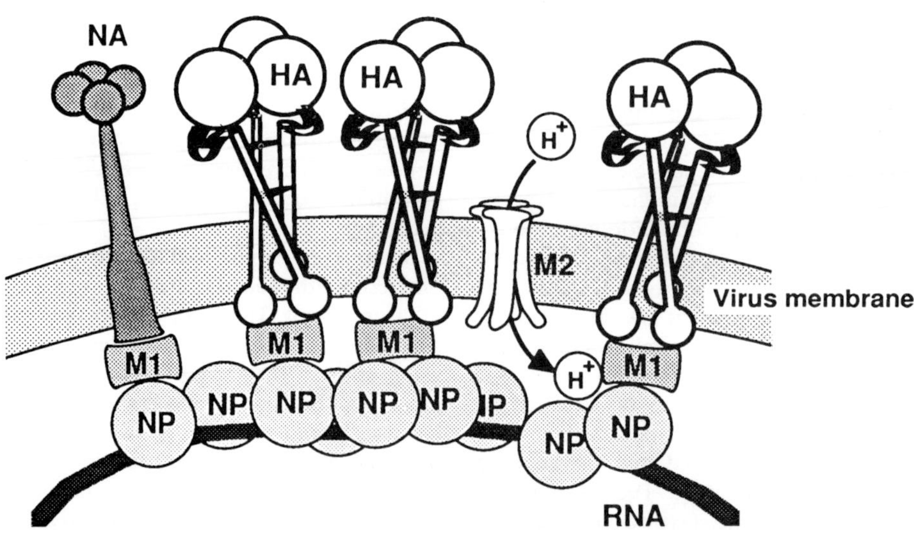

Figure 7.7 Interactions among influenza virus proteins. Two membrane glycoproteins, haemagglutinin (HA) and neuraminidase (NA), are arranged as radial spikes in the viral membrane. The M_1 protein (M1) contacts the internal aspect of the haemagglutinin and the neuraminidase as well as the nucleoprotein (NP). The nucleoprotein exists as a complex with the viral genome (RNA). The M_2 protein (M2) acts as a proton channel, allowing acidification of the interior of the virus, perhaps allowing the M_1 protein to dissociate from the RNA during viral uncoating. (Redrawn, with permission, after Helenius, A. (1992). Unpacking the incoming influenza virus. Cell 69, 577–578.)

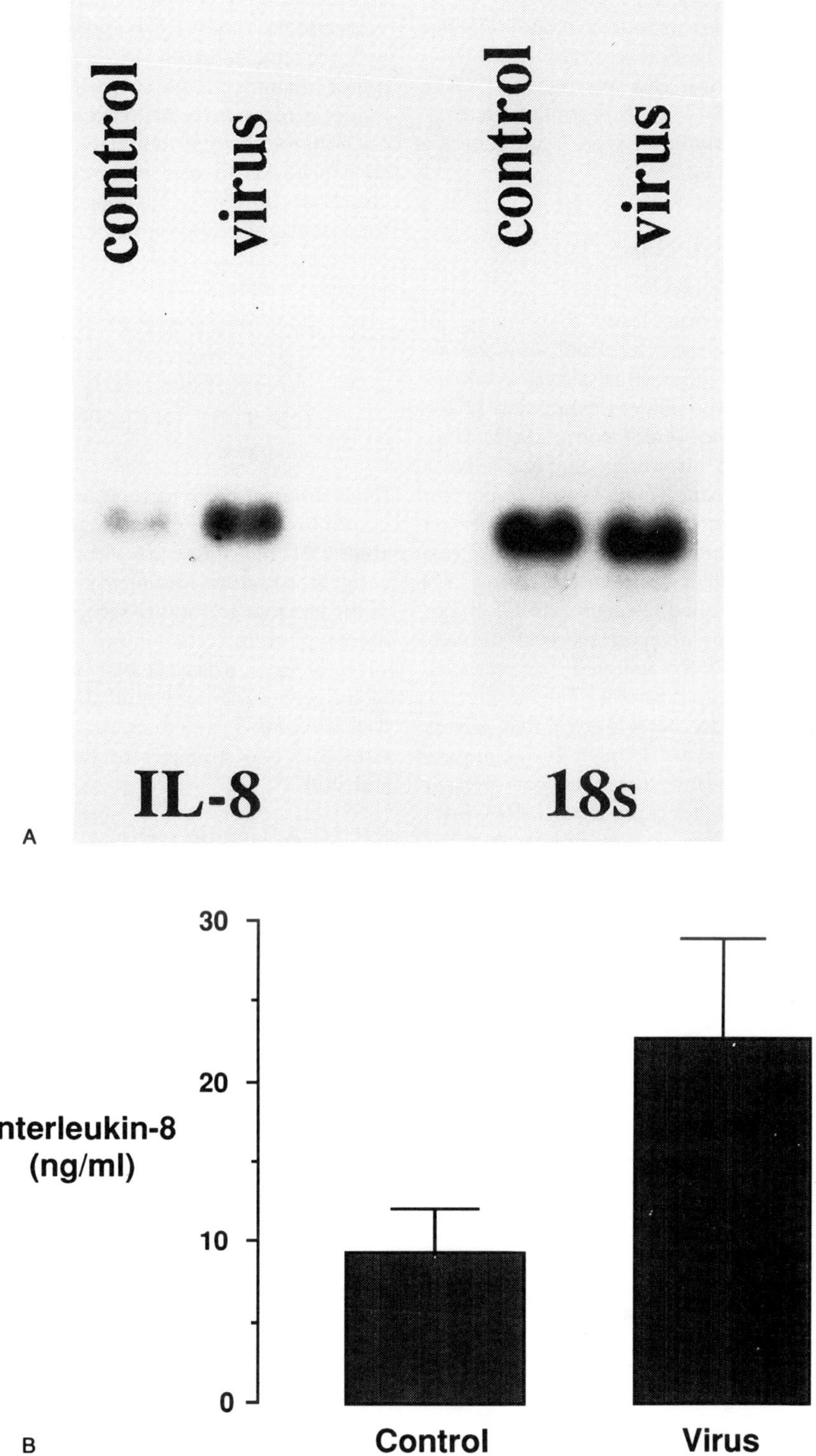

Figure 7.8 Infection of cultured human bronchial epithelial cells with influenza A/Port Chalmers/73 (H3N2) increases expression of the interleukin-8 gene four-fold, and more than doubles release of interleukin-8. (A) Right: Northern blot shows hybridization of an oligonucleotide probe for interleukin-8 with RNA extracted from control (uninfected) or virus-infected epithelial cells. Left: same blot rehybridized with a probe complimentary to 18S RNA shows equivalent RNA loading. (B) Release of interleukin-8 into culture medium, as measured by ELISA. (Reproduced, with permission, from Choi and Jacoby (1992).)

influenza viruses, the antigenic type of haemagglutinin and of neuraminidase is also included. Thus, influenza virus A/Hong Kong/1/68 (H3N2) is a type A influenza (based on the nucleoprotein), and was the first isolate from Hong Kong in 1968. The virus contains haemagglutinin type 3 and neuraminidase type 2, antigenically defined.

3.2   IMMUNE RESPONSES TO INFLUENZA VIRUS

Infection with influenza virus leads to a range of immune responses. On the most localized, non-specific level, we have shown that infection of airway epithelial cells with the influenza virus induces expression of the gene for interleukin-8 (Choi and Jacoby, 1992) (Fig. 7.8A). Increased release of interleukin-8 (Fig. 7.8B), a strongly chemotactic cytokine, may be an important mediator, causing the intense inflammation of the airways characteristic of influenza infection.

The role of epithelial cells as antigen-presenting cells has only recently been studied. Because class I major histocompatibility molecules are expressed constitutively on all somatic cells, it has been assumed that epithelial cells can present antigen to cytotoxic T lymphocytes. Furthermore, it has recently been shown that airway epithelial cells can express class II major histocompatibility complex molecules, and that they can present antigen to helper T lymphocytes (Kalb *et al.*, 1991). Both pathways may be important in viral immunity clearance of viral infections (see Section 3.5).

The production of antibodies directed at antigenic sites on the viral haemagglutinin is the major determinant of immunity to the virus (Stuart-Harris *et al.*, 1985). The presence of an antihaemagglutinin antibody prevents binding of the influenza virus to its receptor on the epithelial cell, and thereby prevents infection. Thus, once an individual has been infected with a particular influenza virus and has mounted an antibody response, reinfection with the same influenza virus is impossible. Unfortunately, minor changes in the structure of the haemagglutinin render it resistant to neutralization by a given antibody. Such minor changes (antigenic drift) are responsible for continual year-to-year repeated epidemics of influenza virus infection. Major changes in the antigenicity of the haemagglutinin (antigenic shift) occur much less frequently; when this occurs, major pandemics of influenza follow.

3.3   ANTIGENIC MUTABILITY OF THE INFLUENZA VIRUS

Elucidation of the structure and function of the haemagglutinin have allowed a fuller understanding of the ability of the influenza virus to repeatedly alter its antigenic structure, leading to repeated epidemics. In view of the fact that antibodies recognizing the influenza virus haemagglutinin confer immunity against that particular strain of virus, it is clear that only a considerable degree of antigenic variation would allow recurrent infections as well as repeated epidemics. As noted above, the influenza virus uses two strategies to vary its antigenicity – drift and shift.

3.3.1   Antigenic Drift

Comparisons of influenza virus A isolates from one year to the next have demonstrated that antibodies capable of neutralizing a particular year's virus neutralize other years' viruses relatively poorly or not at all. Table 7.1 shows such data for isolates in the years 1968–1982 (Stuart-Harris *et al.*, 1985). Ferrets were infected with each of the viruses, and the immune sera obtained from these

Table 7.1 Ability of antisera to inhibit haemagglutination caused by homologous and heterologous influenza virus strains.

	A/Hong Kong/ 1/68	A/England/ 42/72	A/Port Chalmers/ 1/73	A/Scotland/ 840/74	A/Victoria/ 3/75	A/Texas/ 1/77	A/Bangkok/ 1/79	A/Phillippines/ 2/82
A/Hong Kong/1/68	2560	2560	320	80	<20	<20	<20	<20
A/England/42/72	80	1280	320	160	40	<20	<20	<20
A/Port Chalmers/1/7	20	640	640	160	40	40	<20	<20
A/Scotland/840/74	<20	40	40	640	<20	320	40	<20
A/Victoria/3/75	<20	<20	<20	20	1280	320	160	160
A/Texas/1/77	<20	<20	<20	<20	80	1280	640	160
A/Bangkok/1/79	<20	<20	<20	<20	20	320	2560	160
A/Phillippines/2/82	<20	<20	<20	<20	20	160	80	640

Antisera were raised in ferrets infected with each virus strain. Results are expressed as the reciprocal of the greatest dilution of antisera able to inhibit haemagglutination. Each virus strain antiserum was much more potent in inhibiting haemagglutination by that strain than by other strains. Furthermore, it can be seen that there is a progressive decrease over time of the ability to inhibit haemagglutination caused by other strains. This illustrates the phenomenon of antigenic drift.

Reproduced, with permission, from Stuart-Harris, C.H., Schild, G.C. and Oxford, J.S. (1985). "Influenza: The Virus and the Disease". Edward Arnold, Baltimore.

Table 7.2 Amino acid sequences of the HA1 polypeptide of the haemagglutinin of eight variants of Hong Kong (H3N2) influenza virus isolated between 1968 and 1979.

		10		20		30		40
NT68	Q D L P G N D N N T	A T L C L G H H A V	P N G T L V K T I T	D D Q I E V T N A T				
X31	Q D L P G N D N S T	A T L C L G H H A V	P N G T L V K T I T	D D Q I E V T N A T				
ENG69	Q D L P G N D N S T	A T L C L G H H A V	P N G T L V K T I T	N D Q I E V T N A T				
QU70	Q D L P G N D N S T	A T L C L G H H A V	P N G T L V K T I T	N D Q I E V T N A T				
MEM72	Q D F P G N D N S T	A T L C L G H H A V	P N G T L V K T I T	N D Q I E V T N A T				
VIC375	Q D L P G N D N S T	A T L C L G H H A V	P N G T L V K T I T	N D Q I E V T N A T				
TEX77	Q N L P G N D N S T	A T L C L G H H A V	P N G T L V K T I T	N D Q I E V T N A T				
BK79	Q N L P G N D N S T	A T L C L G H H A V	P N G T L V K T I T	N D Q I E V T N A T				

		50		60		70		80
NT68	E L V Q S S S T G K	I C N N P H I R L D	G I D C T L I D A L	L G D P H C D V F Q
X31	E L V Q S S S T G K	I C N N P H I R L D	G I D C T L I D A L	L G D P H C D V F Q
ENG69	E L V Q S S S T G K	I C N N P H I R L D	G I N C T L I D A L	L G D P H C D V F Q
QU70	E L V Q S S S T G K	I C N N P H I R L D	G I D C T L I D A L	L G D P H C D G F Q
MEM72	E L V Q S S S T G K	I C N N P H I R L D	G I N C T L I D A L	L G D P H C D G F Q
VIC375	E L V Q S S S T G K	I C N N P H I R L D	G I N C T L I D A L	L G D P H C D G F Q
TEX77	E L V Q S S S T G R	I C D S P H I R L D	G K N C T L I D A L	L G D P H C D G F Q
BK79	E L V Q S S S T G R	I C D S P H I R L D	G K N C T L I D A L	L G D P H C D G F Q

		90		100		110		120
NT68	N E T W D L F V E R	S K A F S N C Y P Y	D V P D Y A S L R S	L V A S S G T L E F
X31	N E T W D L F V E R	S K A F S N C Y P Y	D V P D Y A S L R S	L V A S S G T L E F
ENG69	D E T W D L F V E R	S K A F S N C Y P Y	D V P D Y A S L R S	L V A S S G T L E F
QU70	N E T W D L F V E R	S K A F S N C Y P Y	D V P D Y A S L R S	L V A S S G T L E F
MEM72	N E T W D L F V E R	S K A F S N C Y P Y	D V P D Y A S L R S	L V A S S G T L E F
VIC375	N E K W D L F V E R	S K A F S N C Y P Y	D V P D Y A S L R S	L V A S S G T L E F
TEX77	N E K W D L F V E R	S K A F S N C Y P Y	D V P D Y A S L R S	L V A S S G T L E F
BK79	N E K W D L F V E R	S K A F S N C Y P Y	D V P D Y A S L R S	L V A S S G T L E F

		130		140		150		160
NT68	I T E G F T W T G V	T Q N G G S N A C K	R G P G S G F F S R	L N W L T K S G S T
X31	I T E G F T W T G V	T Q N G G S N A C K	R G P G S G F F S R	L N W L T K S G S T
ENG69	I T E G F T W T G V	T Q N G G S N A C K	R G P D S G F F S R	L N W L T K S G S T
QU70	I T E G F T W T E V	T Q N G G S N A C K	R G P G S G F F S R	L N W L T K S G S T
MEM72	I N E G F T W T G V	T Q N G G S N A C K	R G P D S G F F S R	L N W L Y K S G S T
VIC375	I N E G F N W T G V	T Q N G G S S A C K	R G P D S G F F S R	L N W L Y K S G S T
TEX77	I N E G F N W T G V	T Q N G G S Y A C K	R G P D N S F F S R	L N W L Y K S E S T
BK79	I N E G F N W T G V	T Q S G G S Y A C K	R G S D N S F F S R	L N W L Y E S E S K

Table 7.2 (continued)

		170		180		190		200
NT68	Y P V L N V T M P N	N D N F D K L Y I W	G V H H P S T N Q E	Q T S L Y V Q A S G
X31	Y P V L N V T M P N	N D N F D K L Y I W	G V H H P S T N Q E	Q T S L Y V Q A S G
ENG69	Y P V L N V T M P N	N D N F D K L Y I W	G V H H P S T N Q E	Q T S L Y V Q A S G
QU70	Y P V L N V T M P N	N D N F D K L Y I W	G V H H P S T N Q E	Q T S L Y V Q A S G
MEM72	Y P V L N V T M P N	N D N F D K L Y I W	G V H H P S T D Q E	Q T S L Y V Q A S G
VIC375	Y P V L N V T M P N	N D N F D K L Y I W	G V H H P S T D K E	Q T N L Y V Q A S G
TEX77	Y P V L N V T M P N	N D N F D K L Y I W	G V H H P S T D K E	Q T N L Y V Q A S G
BK79	Y P V L N V T M P N	N D N F D K L Y I W	G V H H P S T D K E	Q T N L Y V R A S G

		210		220		230		240
NT68	R V T V S T R R S Q	Q T I I P N I G S R	P N V R G L S S R I	S I Y W T I V K P G
X31	R V T V S T R R S Q	Q T I I P N I G S R	P N V R G L S S R I	S I Y W T I V K P G
ENG69	R V T V S T R R S Q	Q T I I P N I G S R	P N V R G L S S R I	S I Y W T I V K P G
QU70	R V T V S T R R S Q	Q T I I P N I G S R	P N V R G L S S R I	S I Y W T I V K P G
MEM72	R V T V S T K R S Q	Q T I I P N I G S R	P N V R G L S S R I	S I Y W T I V K P G
VIC375	K V T V S T K R S Q	Q T I I P N V G S R	P N V R G L S S R I	S I Y W T I V K P G
TEX77	R V T V S T K R S Q	Q T I I P N V G S R	P N V R G L S S G I	S I Y W T I V K P G
BK79	R V T V S T K R S Q	Q T I I P N I G S R	P N V R G L S S R I	S I Y W T I V K P G

		250		260		270		280
NT68	D V L V I N S N G N	L I A P R G Y F K M	R T G K S S I M R S	D A P I D T C I S E
X31	D V L V I N S N G N	L I A P R G Y F K M	R T G K S S I M R S	D A P I D T C I S E
ENG69	D V L V I N S N G N	L I A P R G Y F K M	R T G K S S I M R S	D A P I D T C I S E
QU70	D V L V I N S N G N	L I A P R G Y F K M	R T G K S S I M R S	D A P I D T C I S E
MEM72	D V L V I N S N G N	L I A P R G Y F K M	R T G K S S I M R S	D A P I G T C I S E
VIC375	D V L V I N S N G N	L I A P R G Y F K M	R T G K S S I M R S	D A P I G T C S S E
TEX77	D V L V I N S N G N	L I A P R G Y F K M	R T G K S S I M R S	D A P I G T C S S E
BK79	D V L V I N S N G N	L I A P R G Y F K M	R T G K S S I M R S	D A P I G T C S S E

		290		300		310		320
NT68	C I T P N G S I P N	D K P F Q N V N K I	T Y G A C P K Y V K	Q N T L K L A T G M
X31	C I T P N G S I P N	D K P F Q N V N K I	T Y G A C P K Y V K	Q N T L K L A T G M
ENG69	C I T P N G S I P N	D K P F Q N V N K I	T Y G A C P K Y V K	Q N T L K L A T G M
QU70	C I T P N G S I P N	D K P F Q N V N K I	T Y G A C P K Y V K	Q N T L K L A T G M
MEM72	C I T P N G S I P N	D K P F Q N V N K I	T Y G A C P K Y V K	Q N T L K L A T G M
VIC375	C I T P N G S I P N	D K P F Q N V N K I	T Y G A C P K Y V K	Q N T L K L A T G M
TEX77	C I T P N G S I P N	D K P F Q N V N K I	T Y G A C P K Y V K	Q N T L K L A T G M
BK79	C I T P N G S I P N	D K P F Q N V N K I	T Y G A C P K Y V K	Q N T L K L A T G M

NT68	N V P E K Q T
X31	N V P E K Q T
ENG69	N V P E K Q T
QU70	N V P E K Q T
MEM72	N V P E K Q T
VIC375	N V P E K Q T
TEX77	N V P E K Q T
BK79	N V P E K Q T

Boxed residues are those that have changed from the NT68 isolate. This illustrates the phenomenon of antigenic drift. Note that these minor changes in amino acid sequence allow the viruses to escape neutralization by antibodies directed at the other isolates

Reproduced, with permission, from Webster *et al.* (1982).

animals were tested for their ability to inhibit haemagglutination *in vitro*, as described above. The dilution of sera from animals infected with viruses across the top of the table that will inhibit haemagglutination by each of the virus strains listed at the left is indicated. As can be seen, immune sera efficiently inhibited haemagglutination caused by the corresponding virus strain. In contrast, the sera were much less efficient at neutralizing viruses of other strains. It should also be noted that the cross-reactivity of immune sera for other strains decreases progressively over time between the isolates. This suggests a gradual change in the antigenic structure of the haemagglutinin, a phenomenon known as antigenic drift.

Examining the amino acid sequences of the haemagglutinins of different strains of Hong Kong influenza virus (1968–1979) reveals the structural correlate of this antigenic drift.

As can be seen in Table 7.2, a few changes in the amino acids accounts for the inability of antisera to cross-react from year to year (Webster *et al.*, 1982). While the unstable areas seem to be spread along the peptide chain, folding of the peptide brings them into contact with one another in four "antibody-binding" areas (see Fig. 7.4). The large conserved region 279–328 is also conserved in avian H3 strains; as part of the "stalk" of the haemagglutinin, this conservation of sequence may be necessary to preserve structure.

Interestingly, the same group showed that when antigenic variants were selected by growing virus in the presence of monoclonal antibodies against the haemagglutinins, single amino acid changes were capable of rendering the haemagglutinin incapable of binding the antibody. The changes observed were: 53 AsnϕLys; 133 AsnϕLys; 143 ProϕSer, Thr, Leu, His; 144 Gly ϕAsp; 145 SerϕLys; 205 Ser Tyr (Webster *et al.*, 1982).

Glycosylation of the haemagglutinin molecule, which is a function of the host cell, also affects antigenicity. It has been shown, for example, that when viruses grown in the presence of a monoclonal antibody escape detection using the change 63 LysϕAsn, glycosylation of the Asn residue blocks binding of the antibody. When glycosylation of this virus was prevented by treating cells with tunicamycin, the virus again became recognizable to the monoclonal antibody (Skehel *et al.*, 1984).

The natural selection pressure to change the structure of the influenza virus haemagglutinin has led to positive Darwinian evolution. This is illustrated by comparing the rate of change in nucleotide sequence of the H3 haemagglutinin gene with that of the non-structural gene, which encodes a protein not involved in the immune response to influenza (Fig. 7.9). The rate of substitution in the haemagglutinin gene is much higher (6.7×10^{-3} substitutions per site per year) than that of the non-structural gene (2.3×10^{-3} substitutions per site per year) (Fitch *et al.*, 1991).

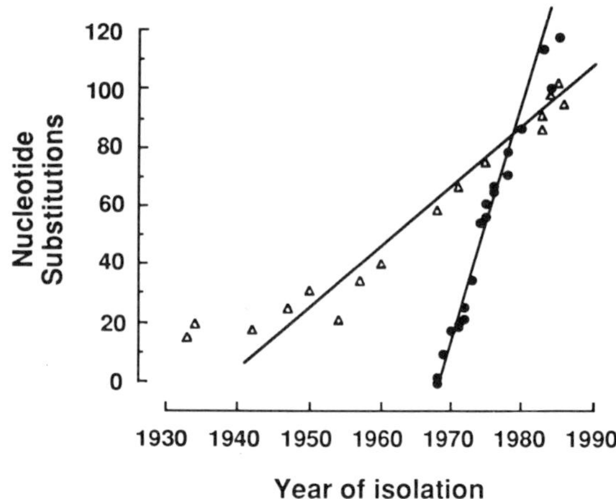

Figure 7.9 Darwinian evolution of the influenza A virus haemagglutinin gene (●). Because antibodies against the haemagglutinin confer immunity against viruses containing that particular haemagglutinin, natural selection pressures favour change in the haemagglutinin gene. When compared with the sequence of the gene for the non-structural protein (△), which is not involved in immunity against the virus, the much more rapid change in the haemagglutinin is evident. (Redrawn, with permission, from Fitch *et al.* (1991).)

3.3.2 Antigenic Shift

In 1918, an influenza virus appeared in which the haemagglutinin structure was markedly different from that of previously circulating influenza viruses. This new haemagglutinin, H1, is similar to that found in influenza virus in pigs, hence H1 influenza viruses are also called swine viruses. History reflects that the emergence of this virus in human hosts caused a severe pandemic in which more than 20 million people died worldwide.

The appearance of a virus containing a haemagglutinin unrelated to previously existing haemagglutinins is known as antigenic shift. This has happened three times in this century, in 1918 (H1, swine strains), in 1957 (H2, Asian strains) and in 1968 (H3, Hong Kong strains). Subsequently, in 1977, viruses of the H1 type, which had formerly been superseded, re-emerged. In each case, the new virus infected a population that had no cross-reacting antibodies, and a severe pandemic ensued. Interestingly, in each case the emergence or re-emergence of the virus took place in China and spread from there around the world. The significance of this observation is unknown.

Comparison of the amino acid sequences of different types of haemagglutinins shows that the changes are much greater than those seen with antigenic drift (Table 7.3) (Winter *et al.*, 1981). The full sequences of these molecules are known, although not presented here. Examination of the full sequences reveals that there are two large conserved regions, the N-terminus of the HA1 subunit and the membrane attachment C-terminal tail.

Table 7.3 Antigenic shift among influenza virus haemagglutinins.

```
H1)                     M K A N L L V L L C A L A A A D A D T I C I G Y H A N N ^ T D T V D T V L E K N
H2)                       M A I I Y L I L L F T A V R G D Q I C I G Y H A N N S T E K V D T N L E R N
H3) M K T I T A L S Y T F C L A L G Q D L P G N D N - S T A T L C L G H H A V P ? G T L V K T I T D D Q
H7)                     M N T Q I L V F A L V A V I P T N A D K I C L G H H A V S N G T K V N T L T E R G
                      <----------- signal peptide ----------->

V T V T H S V N L L E D S H N G K L C R L K G I A P L Q L G K C N I A G W L L G N P E C D P L L P V R
V T V T H A K D I L E K T H N G K L C K L N G I P P L E L G D C S I A G W L L G N P E C D T L L S V P
I E V T N A T E L V Q S S S T G K I C N - N P H R I L D G I D C T L I D A L L G D P H C D - V F Q N E
V E V V N A T E T V E R T N I P K I C S - K G K R T T D L G Q C G L L G T I T G P P Q C D - Q F L E F

S W S Y I V E T P N S E N G I C Y P G D F I D Y E E L R E Q I S S V S S F E R F E I F P K E S S W P N
E W S Y I M E K E N P R D G L C Y P G S F N D Y E E L K H L L S S V K H F E K V K I L P K D - R W T Q
T W D L F V E R S K A F - S N C Y P Y D V P D Y A S L R S L V A S S G T L E F I T E G F - - - T W T G
S A D L I I E R R E G N - D V C Y P G K F V N E E A L R Q I L R G S G G I D K E T M G F - - - T Y S G

H N T T K G V T A A C S H A G K S S F Y R N L L W L - - T E K E G S Y P K L K N S Y V N K K G K E V L
H T T T G G S R - A C A V S G N P S F F R N M V W L - - T K E G S D Y P V A K G S Y N N T S G E Q M L
V T Q N G G S N - A C K R G P G S G F F S R L N W L - - T K S G S T Y P V L N V T M P N N D N F D K L
I R T N G T T S - A C R R S G S S - F Y A E M E W L L S N T D N A S F P Q M T K S Y K N T R R E S A L

V L W G I H H P S N S K D Q Q N I Y Q N E N A Y V S V V T S N Y N R R F T P E I A E R P K V R D Q A G
I I W G V H H P I D E T E Q R T L Y Q N V G T Y V S V G T S T L N K R S T P G I A T R P K V N G Q G G
Y I W G I H H P S T N Q E Q T S L Y V Q A S G R V T V S T R R S Q Q T I I P N I G S R P W V R G L S S
I V W G I H H S G S T T E Q T K L Y G S G N K L I T V G S S K Y H Q S F V P S P G T R P Q I N G Q S G

R M N Y Y W T L L K P G D T I I F F A N G N L I A P R Y A F A L S R G F G S G ? I T S N A S M H E C N
R M E F S W T L L D M W D T I N F E S T G N L I A P E Y G F K I S K R G S S G T M K T E G T L E N C E
R I S I Y W T I V K P G D V L V I N S N G N L I A P R G Y F K M - R T G K S S I M R S D A P I D T C I
R I D F H W L I L D P N D T V T F S F N G A F I A P N R A S F L - R G K S M G I Q S D V Q V D A N C E

T K C Q T P L G A I N S S L P F Q N I H P V T I G E C P K Y V R S A K L R M V T G L R N I P S I Q S R
T K C Q T P L G A I N T T L P F H N V H P L T I G E C P K Y V K S E K L V L A T G L R N V P Q I E S R
S E C I T P N G S I P N D K P F Q N V N K I T Y G A C P K Y V K Q N T L K L A T G M R N V P E K Q T R
G E C Y H S G G T I T S R L P F Q N I N S R A V G K C P R Y V K Q E S L L L A T G M K N V P E P S K K

- - - - - G L F G A I A G F I E G G W
- - - - - G L F G A I A G F I E G G W
- - - - - G L F G A I A G F I E N G W
R E K R G L F G A I A G F I E N G W
        <----------- amino terminus ----------->
```

The sequences shown are the HA1 peptide of haemagglutinins from A/PR/8/34 (H1 subtype), A/Jap/305/57 (H2 subtype), A/Aichi/2/68 (H3 subtype), and the avian influenza A/FPV/Rostock/34 (the fowl plague virus, H7 subtype). Boxed areas show conserved amino acids. Note the greater variability in sequences between haemagglutinin subtypes (antigenic shift) than among variants of a single subtype (antigenic drift, Table 7.2).
Reproduced, with permission, from Winter *et al.* (1981).

Table 7.4 Antigenic shift.

Amino acid conservation (%)

	H1	H2	H3	H7
H1		58 (79)	35 (53)	33 (51)
H2	61 (72)		36 (50)	35 (53)
H3	45 (58)	45 (57)		36 (65)
H7	44 (58)	46 (59)	46 (66)	

Nucleotide conservation (%)

This table shows the poor conservation of amino acid and nucleotide sequences among the four haemagglutinins in Table 7.3. Reproduced, with permission, from Winter *et al.* (1981).

As shown in Table 7.4, neither the nucleotide sequence nor the resulting amino acid sequence of the different subtypes of haemagglutinin is well conserved (Winter *et al.*, 1981). Such sudden, large changes in structure are not likely to be the result of mutation, and attention has been given to the possibility that antigenic shifts represent the emergence of viruses that had previously existed in another reservoir, either animal or avian. There are multiple possible mechanisms underlying such an emergence and several of these may have contributed to previous antigenic shifts.

3.3.2.1   Re-emergence of a Previously Dormant Human Virus

As noted above, the influenza pandemic of 1977 was caused by the re-emergence of H1 strains in Anshan in northern China. The haemagglutinin and neuraminidase of this virus were antigenically similar to those of the H1 viruses that infected humans in the 1950s, before these viruses were superseded by the H2 viruses in 1957 (Kendal *et al.*, 1978). Furthermore, the viruses of the 1977 outbreak have been shown to be genetically identical to the earlier viruses (Nakajima *et al.*, 1978; Scholtissek *et al.*, 1978). The explanation for the phenomenon of a virus disappearing for 27 years and then suddenly re-emerging is unknown. If the virus had been infecting animals in the meantime, antigenic drift would be expected to have occurred and the virus would no longer be serologically or genetically identical. It has been suggested that perhaps the virus became capable of producing a latent infection in humans or in another reservoir (Smith and Palese, 1989), although this has never been reported. The possibility that this outbreak may represent accidental release of a virus that had been frozen in the laboratory has also been raised (Smith and Palese, 1989).

3.3.2.2   Infection of Humans from Animals

In 1977, a group of 500 military recruits in Fort Lee, New Jersey, USA, developed serological evidence of swine (H1N1) virus, with some developing clinical infection, of which one died (Top and Russell, 1977).

Although there was no direct evidence, the subsequent finding that the virus isolated in these cases was antigenically very similar to viruses isolated from pigs suggested the possibility of transmission from pigs to humans. In the same study, it was established with little doubt that influenza virus could be passed between pigs and humans. Genetically identical H1N1 viruses were isolated from humans and pigs on a Wisconsin farm in the USA (Hinshaw *et al.*, 1978).

The frequency with which such interspecies transmission of influenza virus takes place is unknown. Speculation that such transmission may account for the fact that northern China, with a predominantly agrarian society, has provided the new viral haemagglutinin types for all of the pandemics of this century ignores the lack of emergence of new strains from other similarly agrarian societies.

3.3.2.3   Reassortment of Genes

As noted above, influenza is a segmented RNA virus. The RNA that makes up the viral genome is divided into eight parts, each of which encodes one or two proteins. The result of this segmentation of the genome is that if more than one virus is introduced into a cell, there is ample opportunity for the genes encoding some of the proteins for one virus to combine with the genes encoding the other proteins for the other virus, thus producing a reassortant progeny virus.

The haemagglutinins are generally able to bind to receptors irrespective of species. It is therefore thought that other viral proteins, particularly the nucleoprotein, are responsible for the host species preference of the virus (Scholtissek *et al.*, 1985). It is possible, therefore, to envision the reassortment of genes between a human influenza virus, which would donate the genes for the nucleoprotein and whatever else may be involved in determining the host, and an influenza virus from another host species, which would donate the haemagglutinin. In this case, the progeny virus would be capable of infecting humans, but would boast a haemagglutinin not yet seen in human disease.

Such a mechanism has been demonstrated to be responsible for the production of the Hong Kong (H3N2) viruses (Laver and Webster, 1973; Ward and Dopheide, 1981b). The haemagglutinin of the early Hong Kong isolates has been shown to be antigenically (Ward and Dopheide, 1981a) and genetically (Fang *et al.*, 1981) similar to haemagglutinins of equine and duck influenza viruses. The amino acid sequence of the haemagglutinin of influenza A/duck/Ukraine/63 and that of the human influenza A/Aichi/2/68 (H3N2) (an early Hong Kong isolate) are 96% homologous (Fang *et al.*, 1981; Ward and Dopheide, 1981b), suggesting that a duck influenza may have contributed the haemagglutinin responsible for this antigenic shift.

3.4 T LYMPHOCYTE RESPONSES

The immune response to influenza virus, as well as to many other viruses, also includes a response from both cytolytic and helper T lymphocytes. In fact, infection with influenza virus has frequently been used as a model to study antigen presentation and cell-mediated immunity. But the T lymphocyte response to influenza is also of more practical interest. Because of the ability, described above, of the influenza to escape antibody-mediated neutralization by altering the antigenic structure of its haemagglutinin, it has thus far not been possible to produce a vaccine that elicits an antibody response that will be protective against influenza season after season. In contrast, the T lymphocyte response to influenza has been shown to be directed towards viral proteins, particularly the nucleoprotein, that are antigenically more stable than the haemagglutinin (Gotch *et al.*, 1987). Furthermore, recent animal studies suggest that virus-specific T lymphocyte responses can be elicited using synthetic peptides (see Section 3.7.1), thereby circumventing the dangers associated with both live attenuated virus vaccines and intact killed virus vaccines. For all these reasons, interest in the T lymphocyte response to influenza virus continues to grow.

3.4.1 T Lymphocyte Subtypes

There are two major classes of T lymphocytes: cytotoxic T lymphocytes and helper T lymphocytes. When the distinction between these two subclasses was initially made, it was thought that helper T lymphocytes secreted the lymphokines that controlled B lymphocyte and cytotoxic T lymphocyte growth and differentiation, while cytotoxic T lymphocytes were responsible for lysing target cells. It has subsequently been shown that the distinction in function between helper and cytotoxic T lymphocytes is not this clear. Helper T lymphocytes can lyse target cells, a fact that may be important in viral immunity (Fleischer *et al.*, 1985), while cytotoxic T lymphocytes secrete cytokines, among them interferon γ with its crucial role in the immune response to viruses.

Two characteristics of T lymphocytes are currently believed to distinguish helper from cytotoxic cells. Interleukin-1 is required for proliferation of helper T lymphocytes but not of cytotoxic T lymphocytes. Furthermore, helper T lymphocytes recognize antigen complexed with class I major histocompatibility complex molecules, while cytotoxic T lymphocytes recognize antigen complexed with class II major histocompatibility complex molecules. Although perhaps not absolute, helper T lymphocytes generally express the CD4 antigen, while cytotoxic T lymphocytes generally express the CD8 antigen.

3.4.2 Major Histocompatibility Complex Molecules

In contrast to the humoral immune response, in which antibodies recognize antigen *per se*, in free or bound form, T lymphocytes recognize antigen on the surface of antigen-presenting cells only. Furthermore, it is not sufficient for the antigen to be present on the cell surface. It must also be bound to host cell proteins of the major histocompatibility complex.

The molecules of the major histocompatibility complex are divided into two groups, termed class I and class II. These two classes of major histocompatibility antigens differ in many ways, including structurally. Class I major histocompatibility antigens are found on all somatic cells, where they appear to be primarily involved in presentation of antigen to cytotoxic T lymphocytes. This is probably the more important pathway in the cell-mediated immune response to influenza virus infection. In contrast, class II major histocompatibility antigens are expressed on only a limited range of cells, primarily macrophages and B lymphocytes (although recently described on other cells, including airway epithelial cells (Kalb *et al.*, 1991), where they function primarily to present antigen to helper T lymphocytes. It has recently been demonstrated that, in addition to class I major histocompatibility complex molecules, airway epithelial cells can also express class II major histocompatibility complex molecules and present antigen to helper T lymphocytes (Kalb *et al.*, 1991).

3.4.2.1 *Structure and Function of Class I Molecules*

Class I molecules are made up of a 44 kD integral membrane protein complexed with β_2-microglobulin (Bjorkman *et al.*, 1987b). An important structural feature of the class I molecule is a groove, 2.5 nm long and 1.0 nm wide, in the end opposite the membrane-anchoring domain. This groove, which can be pictured as sticking out from the cell, almost certainly acts as the binding site for the antigen to be presented (Bjorkman *et al.*, 1987b).

When the structure of the class I molecule was solved, X ray diffraction data demonstrated extra electron density in the region of the groove that was not accounted for by the structure of the class I molecule itself. This additional electron density was consistent with the presence of an additional peptide bound to the groove. The size and shape of the groove suggests that the peptides bound in it must be short, less than 10 amino acids long, and must be in an extended conformation. Furthermore, charged regions of the antigen-binding groove appear to be important in determining restrictions on the amino acid sequence of peptides that can be bound (van Bleek and Nathenson, 1992).

3.4.3 Antigen Processing for Presentation in Association with Class I Molecules

The major pathway whereby viral antigens become complexed with class I molecules involves *de novo* synthesis of viral proteins in the host cell. As translation of the viral genome assures synthesis of viral proteins in any infected epithelial cell, this requirement will generally be fulfilled.

Class I major histocompatibility complex-restricted cytotoxic T lymphocytes lyse target cells that have been infected with live influenza virus, but fail to lyse cells that have been exposed to a heat-inactivated virus (Braciale and Yap, 1978). As the heat-inactivated virus retains its receptor-binding and membrane-fusing abilities, viral proteins will be released into the host cell cytoplasm. However, no *de novo* synthesis of viral proteins takes place, illustrating the importance of this process in forming the complex of antigen with the class I molecule (however, see also Section 3.4.3.4).

In contrast to the usual synthesis and processing of viral proteins in an infected cell, the viral peptides destined for expression in conjunction with major histocompatibility class I molecules are not glycosylated, nor are they inserted directly into the cell membrane. Furthermore, the whole length of the viral protein is not used as an antigenic determinant in major histocompatibility complex class I-restricted antigen presentation. Rather, peptide fragments, generally nine amino acids long, are produced, transported, combined with class I molecules and β_2-microglobulin, and expressed on the surface of the cell by a specialized set of processes, described below.

The requirement for processing of antigen for presentation to cytotoxic T lymphocytes is supported by several lines of evidence. When cells were transfected with a mutant influenza haemagglutinin gene that produces haemagglutinin lacking an N-terminal signal sequence (and is therefore incapable of being transported normally into the endoplasmic reticulum and inserted into the cell membrane), the cells were still recognized and lysed by cytotoxic T lymphocytes (Townsend *et al.*, 1986a). Furthermore, it has been shown that although vaccinia infection interferes with antigen presentation of influenza haemagglutinin and nucleoprotein (produced by trans-fected genes), when the genes producing the haemagglutinin and nucleoprotein were altered to encourage degradation of the proteins, the vaccinia-induced block in antigen presentation was overcome (Townsend *et al.*, 1988). It has also been shown that blocking transport from the endoplasmic reticulum, using Brefeldin A (Nuchtern *et al.*, 1989; Yewdell and Bennink, 1989) or adenovirus E19 glycoprotein (Cox *et al.*, 1990), prevents presentation of influenza antigens.

Thus, the normal presentation of viral antigen to class I restricted cytotoxic T lymphocytes involves both endogenous synthesis of the viral protein and processing of the protein into an appropriate form. The steps involved in this processing have been elucidated over the past few years, and will be described below.

3.4.3.1 *Synthesis of Nonameric Peptides*

As noted above, the structure of the class I molecule dictates that only short peptides will be presented, as they must fit into the groove formed by the a helices of the class I molecule. In fact, although longer peptides can be bound, the affinity of nonameric peptides for the class I molecule is greater than that of longer peptides. This was demonstrated by Cerundolo *et al.* (1991), who measured association constants (K_a) of nine-amino-acid peptides derived from influenza nucleoprotein for class I molecules, and compared them with the K_a of 14-amino-acid peptides derived from the same protein. Despite containing the same sequence as the nonamers, the addition of five amino acids decreased the K_a by 50-fold ($K_a = 1.9 \times 10^7/\text{M}$ for the nonamer, $3.8 \times 10^5/\text{M}$ for the 14-amino-acid peptide). Thus, longer peptides are at a disadvantage in competing for class I binding sites in the presence of nonameric peptides, as demonstrated by Schumacher *et al.* (1991).

Although these peptides begin by being synthesized as part of proteins during the normal course of viral replication, these proteins are soon broken down. This degradation into short peptides is accomplished by a cytoplasmic structure known as the LMP complex (Monaco and McDevitt, 1982, 1984, 1986). The LMP complex is a large structure, at least part of which is encoded by two genes that map to the class II region of the major histocompatibility complex.

3.4.3.2 *Transport of Nonameric Peptides*

In the same region of the major histocompatibility complex are two genes known as *Tap-1* and *Tap-2* (for "transporter associated with antigen processing"). This class of genes includes the *Ham1* and *Ham2* genes in the mouse and the *mtp1* and *mtp2* genes in the rat (Deverson *et al.*, 1990; Cho *et al.*, 1991; Attaya *et al.*, 1992)), as well as genes in the human major histocompatibility complex known variously as *RING* genes (Trowsdale *et al.*, 1990) or *psf* genes (Spies *et al.*, 1990, 1992).

The proteins encoded by these genes belong to the ATP-binding cassette superfamily of transporters. The proteins TAP-1 and TAP-2 appear to exist as a hetero-dimer (Spies *et al.*, 1992) that functions to transport the short peptides produced by the LMP complex and deliver them to the site of assembly of the peptide-class I molecule complex in the endoplasmic reticulum.

As noted in the discussion of interferon above, there is an increase in expression of major histocompatibility complex molecules in cells exposed to interferon. Although it has not yet been investigated, an intriguing possibility is that, in the face of viral infection, interferon induces expression of other genes of the major histocompatibility complex, increasing cellular content of both LMP and the TAP proteins, thus augmenting the cell's ability to both process and present viral antigen.

3.4.3.3 Assembly of the Peptide–Class I Molecule Complex

The precise mechanisms by which the short peptides that have been formed by the LMP complex and transported by the TAP proteins are joined to the major histocompatibility complex class I molecules and subsequently transported to the cell membrane are not known. It is known that the class I heavy chain binds to both the antigenic peptide and β_2-microglobulin (a protein not encoded in the major histocompatibility complex) (Townsend *et al.*, 1990). This association is thought to take place in the endoplasmic reticulum. As noted above, subsequent transport of these complexes out of the endoplasmic reticulum is required for antigen presentation (Nuchtern *et al.*, 1989; Yewdell and Bennink, 1989; Cox *et al.*, 1990).

It has recently been suggested that an 88 kDa chaperonin-like molecule known as p88 may be involved in assembly of the peptide–class I molecule–β_2-microglobulin complex (Degen and Williams, 1991). The exact role of such molecules has not been determined.

3.4.3.4 Processing of Exogenous Peptides via the Class I Pathway

Evidence presented above suggests that endogenous synthesis of viral proteins, as occurs in virus-infected cells, is the primary source of antigen for presentation in association with major histocompatibility complex class I molecules. Teleologically, it appears desirable for class I-restricted cytotoxic T lymphocytes to recognize only cells that are actually infected with virus, and are therefore synthesizing viral antigens, rather than cells that have been exposed to exogenous, free viral products.

Despite the apparent necessity of endogenous protein synthesis (Braciale and Yap, 1978), however, it is possible that exogenous viral proteins may also become associated with the class I molecule, allowing recognition by cytotoxic T lymphocytes. There has been considerable variability in the results of experiments attempting to sensitize target cells using inactivated influenza virus. Hosaka *et al.* (1985) exposed mouse L cells to heat-inactivated influenza virus. These cells were shown to be susceptible to lysis by cytotoxic T lymphocytes. Interestingly, cytotoxic T lymphocytes specific for the nucleoprotein or for one of the viral polymerases (PB1) were capable of lysing these cells, while cytotoxic T lymphocytes specific for the haemagglutinin were not (Hosaka *et al.*, 1988).

It has also been possible to sensitize cells for lysis using components of the virus. Koszinowski *et al.* (1980) showed that when influenza glycoproteins were introduced into L cells using liposomes, the cells became susceptible to lysis by cytotoxic T lymphocytes. Furthermore, fragments of viral proteins are capable of rendering target cells susceptible to lysis (Townsend *et al.*, 1985; Townsend *et al.*, 1986b).

These and other studies demonstrate that synthesis of viral proteins is not an absolute requirement for recognition of infected cells by class I major histocompatibility complex-restricted cytotoxic T lymphocytes *in vitro*. It is likely that endogenous viral protein synthesis is the primary means by which antigen becomes associated with the class I molecule *in vivo*, and this is probably an important step in the cell-mediated response to viral infections. Nonetheless, the ability of antigen-presenting cells to process exogenous antigen is important because it may allow development of vaccines consisting of antigenic viral peptides that will stimulate cytotoxic T lymphocyte responses.

3.4.4 Alternate Pathway for Class I Major Histocompatibility Complex Antigen Presentation

Recently, Henderson *et al.* (1992) showed that a cell line (CEMx721.174.T2) that lacks the machinery for breakdown of endogenously produced proteins and transport of these peptide fragments into the endoplasmic reticulum is still capable of class I major histocompatibility antigen presentation. The explanation for this surprising finding is that these cells present the signal peptides of the endogenously produced proteins. The signal peptide, a hydrophobic part of a newly synthesized protein that is responsible for its transport to and insertion into the cell membrane, is normally cleaved off by a signal peptidase, and this cleaved fragment can apparently be presented as a complex with class I major histocompatibility molecules and β_2-microglobulin. Pursuing this finding further, it was found that signal sequences were presented by normal cells as class I major histocompatibility antigens (Henderson *et al.*, 1992). The amount of signal peptide presented by normal cells suggested that the signal peptides, although longer than nine amino acids in length, bound to class I molecules with affinities similar to those of nonameric peptides. The explanation for this is unknown.

In that the signal peptide sequences of influenza virus haemagglutinins are quite variable (Wiley and Skehel, 1991), the significance of this pathway for recognizing various influenza viruses is unclear. However, it is possible that recognition of the signal peptides of other proteins of the influenza virus, which might be more consistent in sequence, might allow this alternative pathway to be exploited in viral immunity and vaccine development.

3.5 CYTOTOXIC T LYMPHOCYTES, HELPER T LYMPHOCYTES AND ANTIBODY RESPONSES IN INFLUENZA VIRUS CLEARANCE

The precise role of cell-mediated immunity in the host response to viral infection remains a topic of some debate. It is clear that antibodies that recognize a specific viral haemagglutinin will prevent infection with a virus

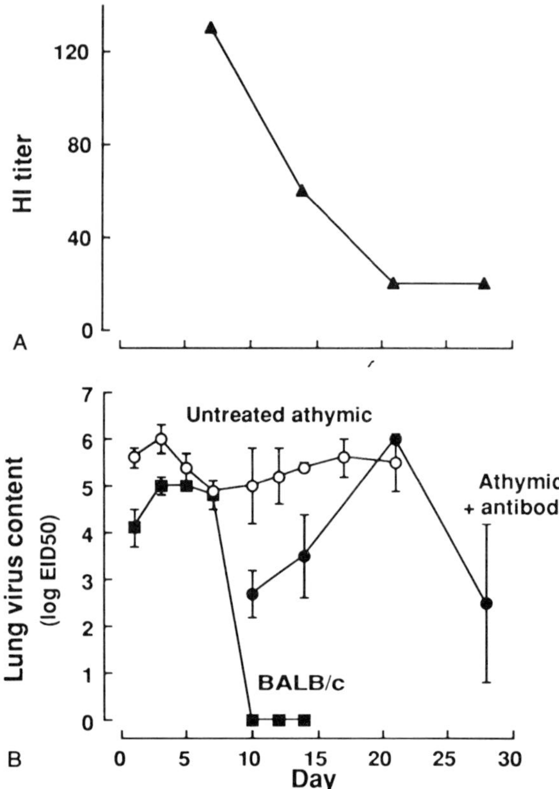

Figure 7.10 Administration of anti-influenza antiserum causes only temporary recovery of from influenza infection in athymic mice. Mice were infected with influenza A/Port Chalmers/73 (H3N2) on day 0. Influenza antiserum was given on day 5 to the athymic plus antibody group only. (A) It is seen that serum antibody, measured as the ability to inhibit haemagglutination by this strain of virus, declines over 2 weeks after intraperitoneal injection of antiserum. (B) Influenza antiserum caused a temporary decrease in virus production in the lung, but virus production again increases as antibody titres decrease. (Redrawn, with permission, after Kris _et al._ (1988).)

containing that haemagglutinin (Stuart-Harris _et al._, 1985). However, once infection is established, the role of antihaemagglutinin antibodies in controlling and eradicating influenza infection is much less clear. It is in the eradication of an existing infection that cell-mediated immunity is likely to be most important.

Kris _et al._ (1988) addressed the question of the role of circulating antihaemagglutinin antibodies in clearing viral infection by administering anti-influenza antiserum to athymic mice 5 days after infection with influenza virus. They found a marked decrease in virus production in the lung for several days after this treatment. However, as the antibodies were cleared from the circulation, the virus production increased in the lungs (Fig. 7.10). Thus, a single treatment with antiserum was incapable of curing a previously established infection.

In contrast, Scherle _et al._ (1992) were able to eradicate an influenza infection from athymic mice using a mixture of monoclonal antibodies against viral haemagglutinin. It is possible that methodological differences account for these divergent results. On the other hand, the authors of the latter study suggested that the presence of polymeric IgA antibodies (which are well transported into the airways) among the monoclonal antibodies they administered may have allowed this treatment to eradicate the infection where the previous group could not.

Another approach to this question was taken by Allan _et al._ (1990), who used a monoclonal antibody to CD4[+] to deplete helper T lymphocytes before infecting mice with influenza virus. This treatment did not affect the clearance of virus from the lungs.

The role of specific cytotoxic T lymphocyte eradication of influenza infection has been widely investigated. Early studies showed that transferring specific cytotoxic T lymphocytes into mice prevented them from dying after otherwise lethal inoculation of influenza virus and decreased the amount of virus produced in the lungs (Yap and Ada, 1977, 1978a,b,c; Yap _et al._, 1978).

The mechanisms by which cytotoxic T lymphocytes might combat viral infections are also under investigation. While lysis of infected cells might perhaps contribute, it is likely that release of cytokines such as interferon γ is an important step. In studies addressing this possibility, pretreatment of mice with cyclosporine blocked the effects of passively transferred cytotoxic T lymphocytes on viral clearance (Schiltknecht and Ada, 1985). The migration of lymphocytes to the lung and their ability to lyse target cells was unaffected by cyclosporine, and the authors suggested that cyclosporine blocked the antiviral effects of cytotoxic T lymphocytes by inhibiting the release of lymphokines.

Interestingly, both specific cytotoxic T lymphocytes and non-specific lymphocyte responses may be important (Mak _et al._, 1983). In this latter category, production of interferon γ by natural killer cells may provide specific protection against influenza by inducing expression of the Mx gene, as well as producing a generalized antiviral state via expression of other interferon-responsive genes (see Section 2.5).

Several recent studies have questioned the concept that cytotoxic T lymphocytes are required for viral clearance. Eichelberger _et al._ (1991), used transgenic mice that were homozygous for disruption of the β_2-microglobulin gene to determine whether class I molecules were required for influenza virus clearance. As noted above, the class I molecule expressed on the surface of the cell is composed of an antigen complexed with a class I heavy chain and β_2-microglobulin. Therefore, these transgenic mice had no functional class I molecules and were therefore incapable of mounting a cytotoxic T lymphocyte response to influenza. In these transgenic mice, as well as mice in whom cytotoxic T lymphocytes were depleted using a monoclonal antibody against CD8, influenza virus was cleared normally from the lungs. The authors suggested

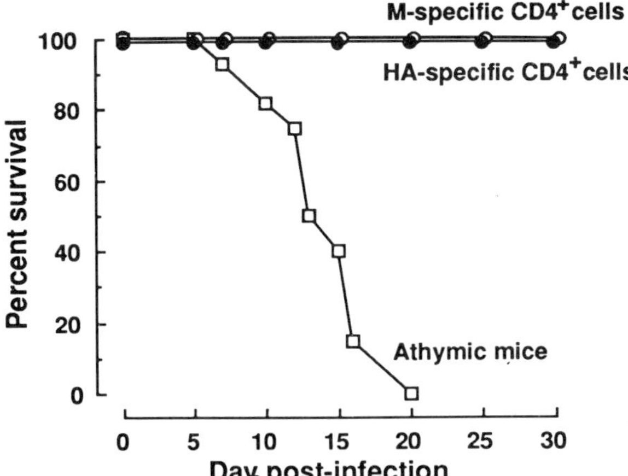

Figure 7.11 Survival of athymic mice infected with influenza A virus (PR/8/34 (H1N1)) in a dose lethal to these mice. Mortality was 100% at 20 days in unreconstituted athymic (□). In constrast, reconstitution with helper T lymphocytes specific for either haemagglutinin (●) or M₁ protein (○) eliminated mortality in athymic mice similarly infected. (Redrawn, with permission, from Scherle *et al.* (1992).)

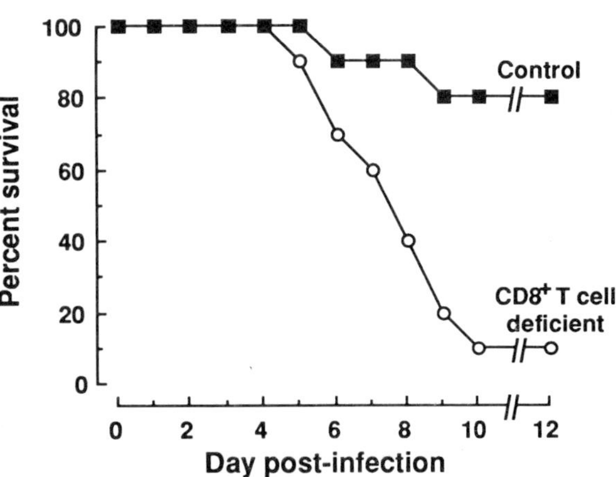

Figure 7.12 Survival of transgenic mice, deficient in β₂-microglubulin (resulting in loss of class I major histocompatibility molecules, and therefore in loss of cytotoxic T lympocyte responses; (○), versus normal controls (■) after infection with influenza A virus (PR/8/34 (H1N1)). (Redrawn, with permission, from Bender *et al.* (1992).)

that redundancy in the immune system allowed recovery from infection in the absence of a cytotoxic T lymphocyte response, and postulated that since these transgenic mice express normal class II major histocompatibility complex molecules, helper T lymphocytes could still recognize antigen and eradicate the viral infection.

Consistent with this possibility are the studies of Scherle *et al.* (1992), who found, not unexpectedly, a high mortality among athymic mice infected with influenza virus. However, rather than using cytotoxic T lymphocytes to reconstitute the immune response in these mice, the investigators used helper T lymphocytes. In this way it was demonstrated that helper T lymphocytes recognizing either the haemagglutinin or the M₁ protein were capable of eliminating influenza virus-induced mortality, even in the absence of cytotoxic T lymphocytes (Fig. 7.11).

The suggestion in this study was that an antibody response to viral antigens is capable of clearing the infection (in contrast to the results of previous studies). This conclusion was further supported by the finding in the same study that transfer of the same helper T lymphocytes into mice with severe combined immuno-deficiency (which are deficient in both T lymphocytes and B lymphocytes) failed to protect against mortality from influenza infections. Furthermore, as noted above, in the hands of these investigators administration of antihaema-gglutinin antibodies that included a polymeric IgA succeeded in clearing the virus from the lungs in both athymic mice and those with severe combined immunodeficiency.

Several studies appear to contradict these findings, and support the concept that it is indeed the cytotoxic T

lymphocyte response that is primarily responsible for viral clearance. Bender *et al.* (1992) used the same β₂-microglobulin-deficient transgenic mice used in the study by Eichelberger *et al.* (1991) above, and showed that, when a more virulent strain of influenza was used to infect the mice, the transgenic mice, which lacked a cytotoxic T lymphocyte response, had increased mortality and delayed viral clearance compared with normal controls (Fig. 7.12).

Based on these data, it may be reasonable to assume that a cytotoxic T lymphocyte response is probably sufficient, but may not be absolutely necessary in all types of infection, to clear the viruses from the lungs. The apparent redundancy of the immune system, with helper T lymphocytes and antibodies also able to clear some infections, is reflected in the apparently contradictory studies above.

Allelic differences in major histocompatibility complex class I molecules have been shown to affect the ability of cytotoxic T lymphocytes to recognize virus-infected cells. In mice containing the K^b allele, cytotoxic T lymphocytes do not demonstrate K^b-restricted lysis of infected target cells (Doherty *et al.*, 1978). Similar findings have been reported for HLA genotypes in humans (McMichael, 1978; Shaw and Biddison, 1979; Shaw *et al.*, 1980), although the phenomenon is not completely understood in either humans or in mice. It is possible that the different cytotoxic T lymphocyte effects reflect differing abilities to recognize specific viral components. Therefore, it may be necessary to include multiple viral peptides in vaccines designed to stimulate a cytotoxic T lympho-cyte response in order to circumvent HLA-linked non-responsiveness.

3.6   OTHER LYMPHOCYTE RESPONSES TO INFLUENZA

Most attention has been paid to the cytotoxic T lymphocyte, as well as to the antibody response, to influenza infection. Nonetheless, a delayed-type hypersensitivity response to influenza can also be demonstrated in mice, particularly upon re-exposure to an influenza virus to which the mouse has already been exposed (Ada *et al.*, 1981). In contrast to the protective roles of cytotoxic and helper T lymphocytes, transfer of the T lymphocytes mediating delayed-type hypersensitivity to influenza virus actually accelerates the rate of death in mice infected with influenza virus (Leung and Ada, 1980; Ada *et al.*, 1981). It has been suggested that this deleterious effect is the result of the pro-inflammatory effects of delayed-type hypersensitivity, leading to increased tissue damage, although direct evidence for this is lacking.

It has also been shown that NK cells can have protective effects in animals infected with influenza (Stein-Streilein *et al.*, 1985) or parainfluenza (Kast *et al.*, 1990) viruses. In parainfluenza-infected mice, stimulation of T lymphocytes using anti-CD3 led to a 60% decrease in mortality, an effect attributed to T lymphocyte activation. However, eliminating NK cells by pretreating with anti-NK1.1 abolished the protective effect of the anti-CD3 treatment, suggesting that a crucial role was played by NK cells.

3.7   CYTOTOXIC T LYMPHOCYTE-STIMULATING VACCINES

Based on currently available data, as reviewed above, it is difficult to be dogmatic concerning the absolute necessity of a cytotoxic T lymphocyte response for clearance of viral and recovery from virus infection. It does seem extremely likely that induction of a cytotoxic T lymphocyte response to a viral antigen will accelerate viral clearance and facilitate recovery from infection, even if other mechanisms may also contribute. This, as well as frustration at the prospect of keeping up with the antigenic drift of the virus with continually updated vaccines aimed at the antibody response, have led to several recent studies suggesting the feasibility of vaccines aimed at stimulating a cytotoxic T lymphocyte response to influenza.

3.7.1   Protein and Oligopeptide Vaccines

Over the past decade, it has become apparent that cytotoxic T lymphocytes can recognize oligopeptides that correspond to parts of the influenza virus antigens (Townsend *et al.*, 1985, 1986a,b). In theory, since the influenza nucleoproteins, which are the principal determinants of the cytotoxic T lymphocyte response (Gotch *et al.*, 1987), are much less variable than the haemagglutinins, either whole nucleoproteins or oligopeptides derived from the nucleoprotein might be expected

to produce immunity to multiple, unrelated influenza viruses. It was in fact found that when mice were immunized with nucleoprotein isolated from the recombinant influenza virus X31 (an H3N2 virus), they recovered more rapidly than non-immunized controls from infection with influenza A/PR/8/34 (H1N1) and were protected against the otherwise lethal effects of such an infection (Wraith *et al.*, 1987). This cross-reactivity is not surprising in that the nucleoprotein of the X31 recombinant is derived from the influenza A/PR/8/34. However, cytotoxic T lymphocytes isolated from mice immunized with X31 nucleoprotein also lysed cells infected with other, unrelated influenza viruses.

A recent study used a peptide derived from influenza nucleoprotein as a vaccine, and examined the effects of various adjuvants on both helper and cytotoxic T lymphocyte responses (Zheng *et al.*, 1992). While injection of the peptide along with the adjuvants alum, saponin and Freund's complete adjuvant produced both helper and cytotoxic T lymphocyte responses, the greatest response was obtained when the peptide was injected in combination with neuraminidase and galactose oxidase.

The theory behind this experiment relates to the finding that binding of T lymphocytes to antigen-presenting cells is dependent upon the temporary formation of covalent Schiff bases between amine groups on one cell and carbonyl groups on another (Rhodes, 1990). Exposure of cells to galactose oxidase produces cell surface aldehydes on terminal D-galactosyl and *N*-acetyl-D-galactosaminyl residues. These substrates are exposed by cleavage of sialic acid residues by the neuraminidase. The aldehydes generated by the actions of these two enzymes on either the antigen-presenting cell or the T lymphocytes are thought to combine with amine groups on the other cell, thereby enhancing antigen presentation and recognition (Zheng *et al.*, 1992).

3.7.2   Vaccinia Virus Vectors

Another approach to inducing a cytotoxic T lymphocyte response to influenza virus has been to use vaccinia virus (a DNA virus that causes cowpox) to deliver the influenza viral antigen (Moss and Flexner, 1987). In this system, a part of the vaccinia virus genome is replaced with a vaccinia promoter and the cDNA encoding the influenza virus protein to be used as an immunogen. When such a vaccinia virus is used to infect an epithelial cell, the inserted gene is transcribed and the protein product is handled normally by the host cell. Infection of epithelial cells with a vaccinia virus vector containing the cDNA for the influenza haemagglutinin leads to production of haemagglutinin that is normally glycosylated and is subsequently transported to the apical surface of the epithelial cell, in the same way haemagglutinin produced during influenza infection would be (Stephens *et al.*, 1986).

A theoretical advantage in such a system is that viral antigen would be produced endogenously in the antigen-presenting cell, and would therefore more naturally feed into the pathway of major histocompatibility complex class I antigen presentation and cytotoxic T lymphocyte stimulation. Indeed, when recombinant vaccinia virus was used to introduce the genes encoding the glycoprotein and nucleoprotein of lymphocytic choriomeningitis virus, mice were protected against subsequently intracerebral injection of the lymphocytic choriomeningitis virus (Hany, *et al.*, 1989).

Unfortunately, the results of such experiments using vaccinia virus vectors expressing influenza proteins have been much less encouraging. Two studies have used recombinant vaccinia viruses expressing influenza nucleo-protein to immunize mice and have shown that, although a cytotoxic T lymphocyte response in these mice could be demonstrated *in vitro*, no protection against infection was conferred by this immunization (Andrew and Couparl, 1988; Stitz *et al.*, 1990). The reason for this dissociation of *in vitro* and *in vivo* results is not known.

3.8   ANTI-IDIOTYPE VACCINES

The usual vaccines used to induce an antibody response to a virus are either inactivated viruses or live attenuated viruses. While such vaccines are generally effective in conferring immunity against the specific virus in question, both types of vaccine have been associated with sig-nificant, if uncommon, untoward effects. Cases of Guillain–Barré syndrome after vaccination with inacti-vated influenza A H1N1 virus vaccine in 1976–1977 (Schonberger *et al.*, 1981), as well as the potential for virulent infection (particularly in immunocompromised patients) that exists with the use of live attenuated virus vaccines, underscore the potential for adverse effects.

An alternative approach that has been used experi-mentally is the production of anti-idiotype antibodies that can serve in the place of viral antigens in the vaccine. While this approach has not been widely used in influenza vaccination, such use has been reported (see Schindler *et al.*, 1992). Anti-idiotype vaccines have been more extensively used experimentally for a variety of other viruses (for reviews, see Rimmelzwaan *et al.*, 1989; Schindler *et al.*, 1992).

3.8.1   Idiotypic Networks

The traditional view of humoral immunity was that a foreign substance (antigen) stimulates production of an antibody that recognizes the structure or part of the structure of the antigen. In the early 1970s, Lindenmann (1973) and Jerne (1974; Jerne *et al.*, 1982) proposed that the immune response involves the production not only of antibodies recognizing antigenic determinants (termed *epitopes* by Jerne), but also of antibodies recogniz-ing unique structures of the epitope-recognizing antibody itself. Thus, the antigen stimulates production of an antibody (termed idiotype, or Ab1) that binds the antigen. The variable regions of these idiotypes are themselves antigenic, and they too stimulate production

of antibodies (termed anti-idiotypes, or Ab2). The antigenic parts of Ab1 include both antigen-binding sites and other parts of the variable region. An anti-idiotype antibody that recognizes part of the variable region outside the antigen-binding site will not interfere with antigen binding to the idiotype, and is referred to as Ab2α. Conversely, if the anti-idiotype antibody recog-nizes the antigen-binding site of the idiotype (termed the *paratope*), it will interfere with antigen binding to the idiotype, and is referred to as Ab2β (Fig. 7.13).

The recognition of antigen by an antibody requires the structure of the paratope to complement the structure of the antigen. Likewise, the paratope of an Ab2β antibody will be complementary to the structure of the paratope of the idiotype, and will thus be similar in structure to the original antigen. Jerne said that such Ab2β antibodies contained an "internal image" of the original antigen.

It was subsequently suggested that such anti-idiotype antibodies might be used as vaccines (Nisonoff and Lamoyi, 1981). The interaction of an antibody with the corresponding antigen is analogous to the interaction of a viral binding protein (e.g. influenza virus haema-gglutinin) with its cellular receptor. Thus, an antibody that recognizes the binding portion of the viral binding protein must be structurally similar to the cellular receptor of that virus. Furthermore, an anti-idiotype antibody that recognizes the paratope of such an antibody must be structurally similar to the original viral binding protein. This allows the anti-idiotype antibody to serve as an antigen that will elicit the formation of an antibody (anti-Ab2β) that recognizes the original viral binding protein. Thus the anti-idiotype antibody could serve as a vaccine to elicit a state of immunity to the virus.

Anti-idiotype vaccines have been used to elicit both antibody and cytotoxic T lymphocyte responses directed towards viral antigens. Poliovirus-neutralizing antibodies were generated via vaccination with an anti-idiotype antibody in the study of UytdeHaag and Osterhaus (1985). They first produced a monoclonal antibody (Ab1) against poliovirus type II. This antipoliovirus antibody was mixed with Freund's complete adjuvant and injected subcutaneously into mice. After multiple booster injections, the spleens from these mice were used to make new hybridomas. Antibodies from these hybridomas (Ab2) were initially screened for ability to bind to Fab fragments of Ab1 using an ELISA. To determine whether the Ab2 was binding to the paratope of Ab1, a similar ELISA was performed in the presence of concentrated poliovirus type II. If the Ab2 were binding to the paratope of Ab1, the addition of poliovirus type II, by binding to the paratope of Ab1, would block binding of Ab2, thus identifying it as an Ab2β. Such an antibody would bear the internal image of the original poliovirus type II antigen. When this anti-idiotype antibody (anti-antipoliovirus) was injected into mice, the mice produced antibodies (Ab3; anti-antiidotype) that neutralized polio-virus type II, but not poliovirus types I or III, *in vitro*.

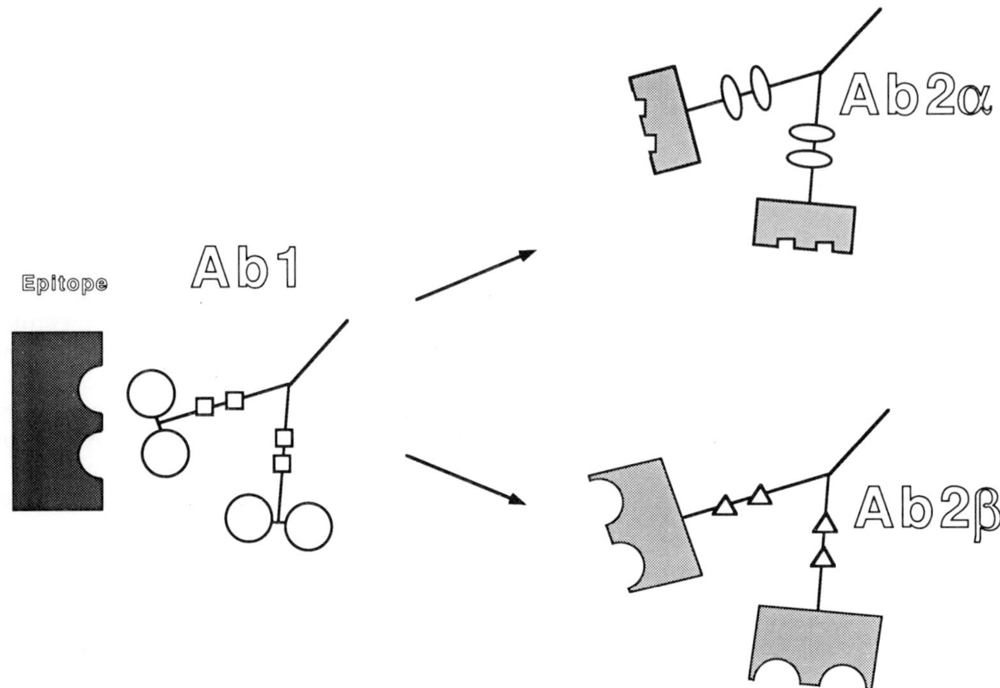

Figure 7.13 An epitope (e.g. a viral antigenic determinant) elicits an antibody response (Ab1). The antigen-binding region (paratope) of the Ab1 molecule has a structure complementary to the structure of the epitope. Ab1 is itself antigenic, and leads to production of antibodies that recognize various parts of the variable region of Ab1. Antibodies that recognize parts of the variable region outside the antigen-binding region (Ab2α) do not interfere with binding of the Ab1 to the epitope. Antibodies that recognize the paratope of Ab1 (Ab2β) will interfere with binding of Ab1 to the epitope. The paratope of Ab2β will have a structure complementary to the structure of the paratope of Ab1, which is complementary to the structure of the epitope. Thus the paratope of Ab2β will be similar to the structure of the epitope (i.e. it will bear the "internal image" of the epitope).

Unfortunately, the titres of this virus-neutralizing Ab3 were too low to protect the mice against the lethal effects of an intracerebral injection of poliovirus type II.

Similar anti-idiotype vaccines have been produced for a variety of other viruses, notably for cytomegalovirus (Keay *et al.*, 1988). None has yet proved useful as an antiviral vaccine. It remains to be seen whether improved adjuvants, such as the neuraminidase–galactose oxidase system described above (Zheng *et al.*, 1992) will allow anti-idiotype antibodies to induce immunity against viral infections (see Section 3.7.1).

It has also been possible to use anti-idiotype vaccines to elicit a cytotoxic T lymphocyte response. Ertl and Finberg (1984) injected mice with helper T lymphocytes that were specific for parainfluenza virus type I (Sendai virus). Splenocytes from these mice were then used to make hybridomas that produced an anti-idiotype antibody. Injection of this antibody into mice led to the production of T-lymphocyte clones that were cytotoxic *in vitro* to cells that were infected with parainfluenza virus. Furthermore, mice that had been so inoculated were protected against an infection with parainfluenza virus that was lethal to uninoculated control mice. Anti-idiotype inoculated mice became infected, but they cleared parainfluenza infections from their lungs more efficiently than uninoculated controls, so that 5 days after intranasal infection with parainfluenza virus their lungs contained only 1/10 000 as much virus as controls.

Similar induction of a cytotoxic T lymphocyte response (in addition to a delayed-type hypersensitivity response) has been reported using an anti-idiotype antibody for reovirus (Sharpe *et al.*, 1984). Whether such vaccines will be effective in causing a cytotoxic T lymphocyte response in humans has yet to be tested.

3.9   Summary

Repeated outbreaks of influenza attest to the ability of the virus to escape from the immune system of the host. This is due primarily to variation in the antigenic structure of the viral haemagglutinin, and has required constant updating of influenza virus vaccines to include new antigens.

The inability to overcome this problem with an antibody-producing vaccine has led to consideration of the possibility of a vaccine that elicits a cytotoxic T lymphocyte response. Such a response may not prevent infection, but will very probably make the course of the disease much milder and shorter. As the primary targets of the cytotoxic T lymphocyte response to influenza virus are fragments of the viral nucleoprotein, which is antigenically much more stable than the haemagglutinin, this approach offers hope for longer-lived protection.

4. Rhinovirus

Like influenza virus, rhinovirus is a major source of morbidity, although not mortality. Colds caused by rhinoviruses account for millions of lost work days annually. Therapy for these colds has been limited to symptomatic treatment with decongestants, analgesics and antipyretics. It has only been recently that any therapy directed at the causative virus has been seriously contemplated.

Rhinovirus is a member of the picornavirus family, which also includes the enteroviruses (polioviruses, coxsackieviruses, hepatitis A and others), apthoviruses (which causes foot-and-mouth disease), cardioviruses and a variety of non-human viruses. The human rhinovirus itself exists in about 100 serotypes, the implications of which will be discussed below.

4.1 RHINOVIRUS STRUCTURE

4.1.1 Rhinovirus Genome

Again like the influenza virus, rhinovirus is an RNA virus. In contrast to the segmented negative-stranded influenza genome, however, the RNA in the rhinovirus is a single positive strand. Thus, the viral genome itself can be translated by the host cell to produce viral proteins. Replication of the viral genome requires production of a negative-stranded intermediate that is then used as a template to produce positive-stranded viral genome.

Where the influenza virus must provide pre-made RNA-dependent RNA polymerases to initiate transcription of negative-stranded RNA into RNA that encodes viral proteins, no such requirement applies to rhinovirus. Because the viral genome is positive stranded, the viral RNA is itself infectious. The principal barrier to infection with free rhinovirus RNA is its susceptibility to degradation by RNases, which abound in nature. In the intact virus, the protein shell provides protection for the genome against RNases (Rueckert, 1991).

Another consequence of the RNA genome is that synthesis of RNA from an RNA template involves the production of short stretches of double-stranded RNA. Furthermore, there is evidence in other picornaviruses that, in infected cells, the viral RNA exists in a double-stranded "replicative form" (Larsen et al., 1980). The role of this double-stranded RNA in viral replication is unknown. However, the presence of double-stranded RNA may be of great importance in activating transcriptional factors such as NF-κB (Lenardo and Baltimore, 1989), which initiates transcription of interferon genes

(see Section 2.1), and in activating 2′,5′-oligoadenylate synthetase and p68 protein kinase, which are important intermediates in the antiviral actions of interferons (see Sections 2.5.2 and 2.5.3).

4.1.2 Rhinovirus Proteins

The RNA core of the rhinovirus is surrounded by an icosahedral protein shell. Three large viral proteins (VP1, VP2 and VP3) make up triangular trimers. Sixty such trimers, arranged in group of five, make up the dodecahedral shell of the rhinovirus (Rossman et al., 1985). A fourth protein, VP4, is formed during cleavage of a precursor protein that also yields VP2; VP4 is located internally.

VP1, VP2 and VP3 all have portions that protrude relative to the otherwise reasonably planar configuration of the surface of the trimer that makes up the outside of the virus shell. These protrusions combine to make up the walls of a cleft or "canyon." The receptor-binding protein of the rhinovirus is thought to be at the bottom of this canyon (Rossman et al., 1985; Rossmann, 1989). The consequences of this arrangement will be discussed below.

4.2 IMMUNITY TO RHINOVIRUSES

Due to the large number of rhinovirus serotypes, most people can look forward to repeated infection with one serotype after another throughout their lives. There appears to be little cross-reactivity of naturally occurring antibodies recognizing one serotype with other rhinovirus serotypes; in fact, the lack of such cross-reactivity has been used to define serotypes (Cooney et al., 1982).

Four antigenic sites have been identified on human rhinovirus 14, the most widely studied serotype. NIm-1A and NIm-1B are located on the external facing surface of the VP1 protein. NIm-II and NIm-III are on VP2 and VP3, respectively (Sherry et al., 1986). It has been shown that changes in a number of single amino acid residues within these sites will render the virus resistant to previously neutralizing antibodies (Sherry et al., 1986). As is the case with influenza virus, the antigenic sites do not include residues involved in the binding of the virus to its receptor. Thus, the rhinoviruses can, like the influenza viruses, present an extremely varied antigenic structure while at the same time maintaining their ability to bind to a single receptor.

The large number of human rhinovirus serotypes has made the development of a vaccine that would confer immunity to all rhinoviruses a daunting challenge. An alternative approach has therefore been tried: using synthetic peptide to immunize against the common, conserved binding protein. This will be discussed below.

Table 7.5 Blocking of attachment of ^{35}S-labelled rhinoviruses by homologous and heterologous serotypes.

Blocking serotype	Blocking of [^{35}S]methionine-labelled challenge									
	1A	2	11	3	5	9	14	15	39	41
1A	+	+/−	−	−	−	−	−	−	−	−
2	+	+	−	−	−	−	−	−	−	−
11	−	−	−	+	+	+	+	+	+	+
3	−	−	−	+	+	+	+	+	+	+
5	+/−	−	−	+	+	+	+	+	+	+/−
9	−	−	−	+	+	+	+	+	+/−	+
14	−	−	−	+/−	+/−	+/−	+	+/−	+/−	+/−
15	−	−	−	+	+	+	+	+	+	+
39	−	−	−	+	+	+/−	+/−	+	+	+/−
41	−	−	−	+	+	+	+	+	+/−	+

HeLa cells were incubated with labelled viruses in the presence of a saturating concentration of unlabelled virus of either the same or different serotype. Serotypes 1A and 2 blocked binding of labelled serotypes 1A and 2 only. The group of viruses of serotypes 3, 5, 9, 14, 15, 39 and 41 blocked binding of all labelled viruses within that group (the inability in these experiments of serotype 11 to be blocked by any virus, including serotype 11, could not be explained). The same study found that serotypes 32, 36, 51, 59, 67 and 89 were also blocked by members of this group of serotypes. Thus, viruses in this group share the same receptor, and are known as "major receptor group rhinoviruses" Reproduced, with permission, from Abraham and Colonno (1984).

4.3   THE RHINOVIRUS RECEPTOR

It was shown in 1984 that the majority of rhinovirus serotypes recognized the same cellular receptor. Abraham and Colonno (1984) examined the binding of ^{35}S-labelled rhinoviruses to HeLa cells and demonstrated saturable binding. They then used unlabelled viruses of various serotypes to compete against labelled viruses of various serotypes for binding to these cells. Some of the results are shown in Table 7.5. These data show that rhinoviruses of different serotypes can compete for the same receptor. The members of this large group of rhinoviruses sharing the same receptor have become known as "major receptor group" rhinoviruses, distinguishing them from the "minor receptor group" rhinoviruses, which bind to a different receptor. Approximately 80% of rhinoviruses are major receptor group rhinoviruses.

It has become apparent during the past few years that the cellular receptor for the major receptor group rhinoviruses is ICAM-1 (Greve *et al.*, 1989; Staunton *et al.*, 1989; Tomassini *et al.*, 1989). This molecule, which is expressed on a variety of cells including epithelial and endothelial cells, binds to receptors on leucocytes, including the LFA-1 receptor (CD11/CD18), allowing the leukocyte to adhere to the cell. The ICAM-1 gene is a member of the immunoglobulin gene superfamily. Structurally, ICAM-1 is a single-chain glycoprotein of 90 kDa containing five immunoglobulin-like domains (D1–D5, with D5 being the end of the molecule that is inserted into the cell membrane).

Two groups reported cloning of the rhinovirus receptor from HeLa cells and showed that it is identical to ICAM-1 (Greve *et al.*, 1989; Tomassini *et al.*, 1989). A third group showed that ^{35}S-labelled rhinovirus bound to COS cells only when they had been transfected with the gene for ICAM-1 (Staunton *et al.*, 1989). It was subsequently

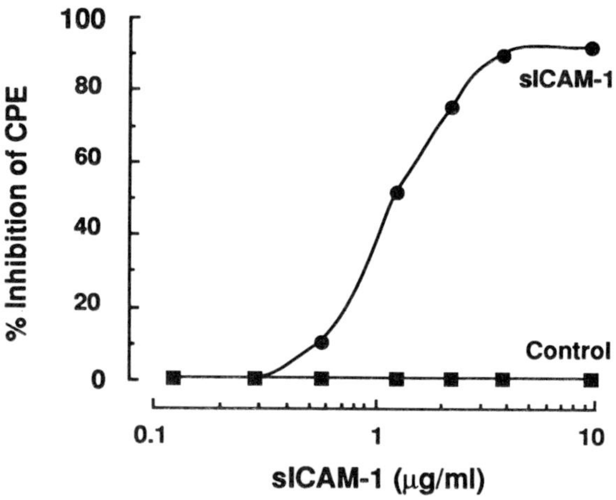

Figure 7.14 Addition of soluble ICAM-1 (sICAM-1) inhibits binding of rhinovirus to cultured epithelial cells, thus inhibiting the cytopathic effect (CPE) of viral infection. This inhibition of viral binding occurs because the rhinovirus enters the cell by binding to ICAM-1 on the cell surface. (Reproduced, with permission, from Marlin *et al.* (1990).)

shown that addition of soluble ICAM-1 to the culture medium blocks infection of HeLa cells by rhinovirus (Fig. 7.14) (Marlin *et al.*, 1990).

The binding site for rhinovirus has also been investigated by these groups, and has been localized to the D1 domain of ICAM-1 (the domain furthest from the cell membrane) (Staunton *et al.*, 1990; McClelland *et al.*, 1991; Register *et al.*, 1991). Some controversy exists concerning the exact residues involved in rhinovirus binding, but they may overlap the area that binds LFA-1.

Binding of rhinovirus to ICAM-1 may have implications for the immune response to rhinovirus infection. Interferon γ, which would be expected to be produced

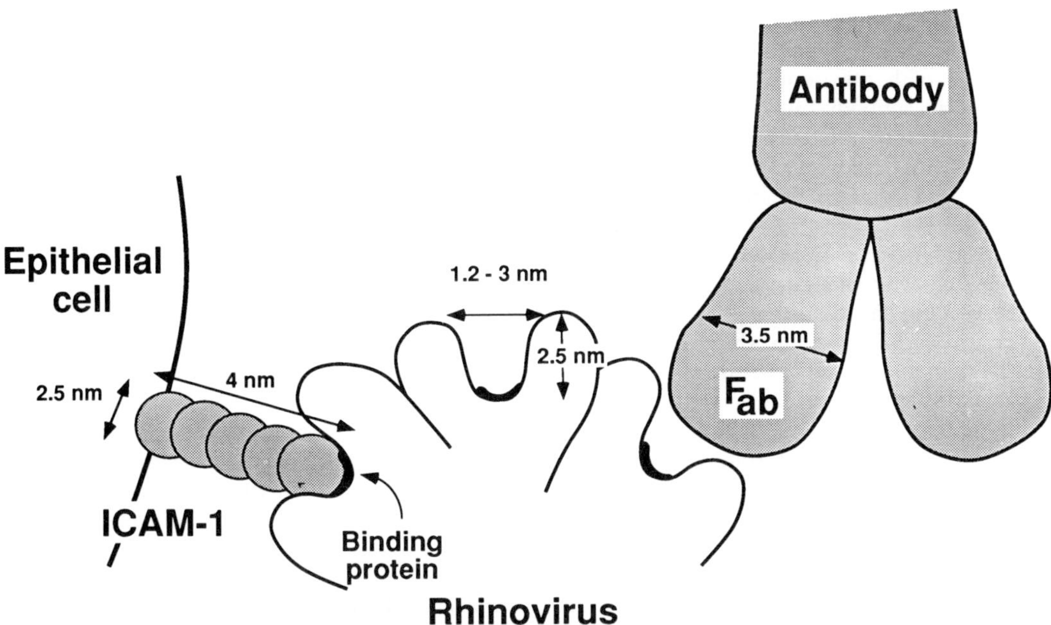

Figure 7.15 Binding of rhinovirus to ICAM-1 on an epithelial cell. The dimensions of the canyon on the surface of the rhinovirus allow access of ICAM-1 to the binding protein of the floor of the canyon. The arm of the antibody molecule, being wider, cannot reach the binding protein. (Redrawn, with permission, after Luo, M., Vriend, G., Kamer, G., Minor, I., Arnold, E., Rossmann, M.G., Boege, U., Scraba, D.G., Duke, G.M., and Palmenberg. (1987) The atomic structure of Mengo virus at 3.0 Å resolution. Science 235, 182–191.)

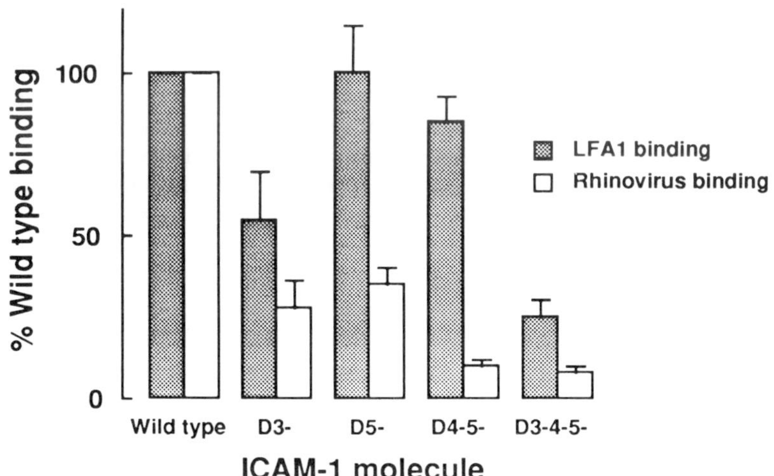

Figure 7.16 Effects of deleting non-binding domains of ICAM-1 (D3, (D3-), D5 (D5-), D4 and D5 (D4-5-), and D3, D4 and D5 (D3-4-5-)) on binding of rhinovirus and LFA-1 to wild-type and mutant ICAM-1. Deletion of D3 decreases expression of the molecule (D3-); this may in part explain the decreased binding with this mutant. Apart from this case, it can be seen that shortening ICAM-1 decreases binding to rhinovirus much more that binding to LFA-1. This may reflect the length of ICAM-1 required to reach the rhinovirus receptor-binding protein at the floor of the canyon. (Redrawn, with permission, after Staunton et al. (1990).)

by lymphocytes during rhinovirus infection, induces expression of ICAM-1, thus possibly enhancing the ability of rhinovirus to infect cells (Piela-Smith *et al.*, 1991). Furthermore, binding of rhinovirus to cells inhibits the subsequent binding of LFA-1-bearing T lymphocytes

(Piela-Smith *et al.*, 1991). This may hamper the local immune response to rhinovirus infection.

In the rhinovirus, the binding protein that recognizes ICAM-1 resides at the bottom of a canyon formed by VP1, VP2 and VP3. The difficulty anticipated in attempting to use the rhinovirus binding protein as a target for immunization is illustrated in Fig. 7.15.

The dimensions of the canyon are such that the epitope-binding arm of an antibody cannot reach the binding protein at the floor of the canyon. In contrast, the dimensions of ICAM-1 allow it to fit between the walls of the canyon, and to reach the floor of the canyon (Rossmann, 1989).

Studies of the binding of ICAM-1 to LFA-1 and to rhinovirus have shown that shortening the ICAM-1 by deleting the non-binding domains (D3, D4 and D5) causes a progressive decrease in rhinovirus binding, to which LFA-1 is relatively insensitive (Fig. 7.16) (Staunton *et al.*, 1990). Thus, an antibody that recognizes the rhinovirus receptor-binding protein might not be effective in protecting against rhinovirus infections because of the inaccessibility of the binding protein to the antibody.

4.4   RHINOVIRUS VACCINES

Attenuation of rhinoviruses has not been observed; thus, this approach to creating a vaccine, which has been successful in other picornaviruses, cannot be used. Therefore, much attention has been focused on the used of synthetic antigenic peptides.

Synthetic peptides with sequences corresponding to known antigenic determinants on the rhinovirus, specifically NIm-II on VP2, have been used to generate antibodies that recognize and neutralize human rhinovirus 2 (Francis *et al.*, 1987; Hastings *et al.*, 1990). Antisera so generated have been shown to bind rhinovirus 2 in a variety of assay systems, and can also block infection of otherwise susceptible cell cultures. It has been suggested that the approach of generating a large number of such immunogenic peptides corresponding to NIm-II in all 100 rhinovirus serotypes might be an approach to generating a rhinovirus vaccine.

An alternative approach, designed to circumvent the need for so large a number of vaccines, has been used by McCray and Werner (1987, 1989). They used synthetic peptides with sequences corresponding to regions of VP1 and VP3 of rhinovirus 14 that are conserved among major receptor group rhinoviruses and were therefore thought to be involved in receptor binding. Such sequences in the virus are also thought to be near to or within the canyon. Antisera obtained by immunizing rabbits with these synthetic peptides were able to neutralize not only rhinovirus 14, but also, to at least some extent, 60% of other major receptor group rhinoviruses tested. As expected, no neutralization of polioviruses, echoviruses, coxsackieviruses or minor receptor group rhinoviruses was found.

While the possibility of using such peptides as vaccines that would immunize against a broad range of rhinoviruses is intriguing, several observations have limited the enthusiasm for this approach. It has been pointed out that the neutralization titres for rhinovirus 14 obtained with these peptides is much lower than that obtained after immunization with intact rhinovirus 14; neutralization titres for other rhinovirus serotypes are even lower. Furthermore, similar experiments using peptides to immunize animals against other picornaviruses, including poliovirus, apthoviruses and hepatitis A virus, have all produced antibodies that were effective in inhibiting virus binding *in vitro*, but conferred poor, if any, protection to inoculated animals *in vivo* (Palmenberg, 1987). The reason for this lack of success may well be the configuration of the proteins making up the canyon, and their relationship to the actual ICAM-1-binding site on the rhinovirus.

Little is known concerning the T lymphocyte response to rhinoviruses. Such responses were investigated recently by injecting mice with rhinovirus 1 (a minor receptor group virus) or rhinovirus 15 (a major receptor group virus) (Hastings *et al.*, 1991). Lymphocytes from lymph nodes removed from mice so inoculated proliferated *in vitro* not only in response to the rhinovirus the animal had been injected with, but also in response to other serotypes of rhinovirus. The proliferating T lymphocytes were identified as CD4$^+$ (helper T lymphocytes). Thus, there may be less variability between rhinovirus serotypes in terms of T lymphocyte recognition than there is in antibody recognition. The viral epitopes responsible for these T lymphocyte responses have yet to be determined.

4.5   ANTI-RHINOVIRUS DRUGS

Although the dimensions of the rhinovirus canyon may ultimately make it impossible to neutralize all major receptor group rhinoviruses with a single anti-binding protein antibody, another avenue exists for interfering with viral binding. A group of structurally related drugs, of which disoxaril is the prototype, interferes with picornavirus binding and uncoating. The principal mode of action in human rhinovirus 14, a major receptor group rhinovirus, is interference with binding of the virus to ICAM-1. This appears to be the result of disoxaril binding to a hydrophobic pocket in VP1 beneath the floor of the rhinovirus canyon, thereby causing a conformational change in the canyon shape. This conformational change interferes with binding of the virus to its cellular receptor, ICAM-1 (Smith *et al.*, 1986; Pevear *et al.*, 1989).

Such drugs have thus far been used mainly for research into the structure and function of picornavirus proteins. Their role as possible therapeutic agents remains to be defined.

5.   *Conclusion*

As the cell most directly exposed to airborne and ingested viruses, the epithelial cell is equipped with a variety of defences, both specific and non-specific, against viral infection. Increasing secretion may help to wash away a potential pathogen, just as the epithelium might dispose of any particle. Somewhat more specific to the defence against viruses is the production of interferons and the expression of interferon-responsive genes such as those encoding the Mx protein, $2',5'$-oligoadenylate synthetase, p69 kinase, and major histocompatibility complex molecules. Still more specific to the type of virus are the antibody response and the T lymphocyte response.

The ability of the virus to evade the immune response occurs at many levels. While interferon is important to the antiviral defence of the cell, several viruses have developed the ability to use activated NF-κB, the key transcriptional activator for interferon production, to promote their own reproduction. Other viruses have developed mechanisms to escape the effects of interferon-inducible proteins, as, for example, in the case of influenza and adenovirus inhibiting the actions of p68 kinase.

The traditional approach to vaccination induces an antibody response, typically against one of the surface proteins of the virus. Such antibodies can be produced in response to an attenuated live virus, to a killed virus vaccine, to synthetic viral peptides, or, most recently, in response to anti-idiotype antibodies that carry the internal image of the viral protein. The antibody response to most epithelial viruses confers immunity to re-infection with the same virus, but this is frequently limited by the ability of the viruses either to change their antigenic structure (as in the case of influenza) or to present an immunologically bewildering array of serotypes (as in the case of rhinovirus). Either of these viral strategies could be overcome by producing an antibody against the binding site on the viral attachment protein. However, in the case of influenza virus, antibody binding is markedly affected by small changes in the amino acid sequence of peptides surrounding the receptor-binding site, while in the case of rhinovirus the structure of the virus does not permit the antibody access of the viral attachment protein.

Such problems have provoked attempts to design vaccines that will induce a cytotoxic T lymphocyte response to viral proteins. In the case of influenza, where the problem has been most extensively studied, such a vaccine has the theoretical advantage that the cytotoxic T lymphocyte response to influenza virus is directed primarily against internal viral proteins, especially against the nucleoprotein. Such proteins are antigenically much less variable over time than are the surface proteins such as the haemagglutinin and the neuraminidase, against which the antibody response is generally directed. Thus, there is the hope that a vaccine that causes a cytotoxic T lymphocyte response directed against the nucleoprotein could confer protection against a wide range of influenza virus strains.

Several approaches to such a vaccine have been used. Proteins isolated from the viruses, or synthetic peptides analogous to portions of such proteins, have been used as immunogens. Recent work suggests that such peptides

may be effective, particularly with new adjuvants such as galactose oxidase. Alternatively, the use of recombinant vaccinia virus vectors to deliver cDNA encoding viral proteins can cause the epithelial cells themselves to produce the appropriate viral proteins, thereby feeding more physiologically into the pathway for presentation of antigen in conjunction with major histocompatibility complex class I molecules. The reasons for the lack of success of this final approach are not yet known, but perhaps the answer will lie in the development of other viral vectors.

Given the many ways viruses have found to circumvent immune surveillance, it may be excusable to feel that the viruses will always be one step ahead of the virologists. However, recent years have witnessed the rapid development both of our knowledge of viral immunology and of technology to make use of such knowledge. Greater understanding of the interactions of viruses with their host cells will lead to new strategies to prevent and treat viral infections.

6. *References*

Abraham, G. and Colonno, R.J. (1984). Many rhinovirus serotypes share the same cellular receptor. J. Virol. 51, 340–345.

Ackrill, A.M., Foster, G.M., Laxton, C.D., Flavell, D., Stark, G.R. and Kerr, I.M. (1991). Inhibition of the cellular response to interferons by products of the adenovirus type 5 E1A oncogene. Nucleic Acids Res. 19, 4387–4393.

Ada, G.L., Leung, K.N. and Ertl, H. (1981). An analysis of effector T cell generation and function in mice exposed to influenza A or Sendai viruses. Immunol. Rev. 58, 5–24.

Aguet, M., Dembic, Z. and Merlin, G. (1988). Molecular cloning and expression of the human interferon-g receptor. Cell 55, 273–280.

Al-Bazzaz, F.J., Yadava, V.P. and Westenfelder, C. (1981). Modification of Na and Cl transport in canine tracheal mucosa by prostaglandins. Am. J. Physiol. 240, F101–F105.

Allan, W., Tabi, Z., Cleary, A. and Doherty, P.C. (1990). Cellular events in the lymph node and lung of mice with influenza. Consequences of depleting CD4$^+$ T cells. J. Immunol. 144, 3980–3986.

Andrew, M.E. and Coupar, B.E.H. (1988). Efficacy of influenza haemagglutinin and nucleoprotein as protective antigens against influenza virus infection in mice. Scand. J. Immunol. 28, 81–85.

Arnheiter, H., Skuntz, S., Noteborn, M., Chang, S. and Meier, E. (1990). Transgenic mice with intracellular immunity to influenza virus. Cell, 62, 51–61.

Attaya, M., Jameson, S., Martinez, C.K., Hermel, E., Aldrich, C., Forman, J., Lindahl, K.F., Bevan, M.J. and Monaco, J.J. (1992). *Ham-2* corrects the class I antigen-processing defect in RMA-S cells. Nature 355, 647–649.

Bender, B.S., Croghan, T., Zhang, L. and Small, P.A. (1992). Transgenic mice lacking class I major histocompatibility complex-restricted T cells have delayed viral clearance and increased mortality after influenza virus challenge. J. Exp. Med. 175, 1143–1145.

Bjorkman, P.J., Saper, M.A., Samraoui, B., Bennett, W.S., Strominger, J.L. and Wiley, D.C. (1987a). The foreign antigen binding site and T cell recognition regions of class I histocompatibility antigens. Nature 329, 312–318.

Bjorkman, P.J., Saper, M.A., Samraoui, B., Bennett, W.S., Strominger, J.L. and Wiley, D.C. (1987b). Structure of the human class I histocompatibility antigen, HLA-A2. Nature 329, 506–512.

Braciale, T.J. and Yap, K.L. (1978). Role of viral infectivity in the induction of influenza virus-specific cytotoxic T cells. J. Exp. Med. 147, 1236–1252.

Burnet, F.M. (1948). Mucins and mucoids in relation to influenza virus action. IV. Inhibition by purified mucoid of infection and hemagglutination with the virus strain WSE. Aust. J. Exp. Biol. Med. Sci. 26, 381–387.

Burnet, F.M., McCrea, J.F. and Anderson, S.G. (1947). Mucin as a substrate of enzyme action by viruses of the mumps influenza group. Nature 160, 404–405.

Centers for Disease Control (1991). "Update on Adult Immunization". Recommendations of the immunization practices advisory committee. MMWR: RR 12: 33–36.

Centers for Disease Control (1992). Update: influenza activity – United States and Worldwide, and composition of the 1992–93 influenza vaccine. Mortality Morbidity Weekly Rep. 41, 315–323.

Cerundolo, V., Elliott, T., Elvin, J., Bastin, J., Rammensee, H.-G. and Townsend, A. (1991). The binding affinity and dissociation rates of peptides for class I major histocompatibility complex molecules. Eur. J. Immunol. 21, 2069–2075.

Cho, S., Attaya, M., Brown, M.G. and Monaco, J.J. (1991). A cluster of transcribed sequences between the *Pb* and *Ob* genes of the murine major histocompatibility complex. Proc. Natl Acad. Sci. USA 88, 5197–5201.

Choi, A.M.K. and Jacoby, D.B. (1992). Influenza virus A infection induces interleukin-8 gene expression in human airway epithelial cells. FEBS Lett. 309, 327–329.

Cloutier, M.M., Wong, D. and Ogra, P.L. (1989). Respiratory syncytial virus alters electrophysiologic properties in cotton rat airway epithelium. Pediatr. Pulmonol. 6, 164–168.

Cooney, M.K., Fox, J.P. and Kenny, G.E. (1982). Antigenic grouping of 90 rhinovirus serotypes. Infect. Immun. 37, 642–647.

Cox, J.H., Yewdell, J.W., Eisenlohr, L.C., Johnson, P.R. and Bennink, J.R. (1990). Antigen presentation requires transport of MHC class I molecules from the endoplasmic reticulum. Science 247, 715–718.

David, M. and Larner, A.C. (1992). Activation of transcription factors by interferon-alpha in a cell-free system. Science, 257 813–815.

Davies, M.V., Furtado, M., Hershey, J.W.B., Thimmappaya, B. and Kaufman, R.J. (1989). Complementation of adenovirus virus-associated RNA I gene deletion by expression of a mutant eukaryotic translation initiation factor. Proc. Natl Acad. Sci. USA 86, 9163–9167.

deBenedetti, A. and Baglioni, C. (1984). Inhibition of mRNA binding to ribosomes by localized activation of dsRNA-dependent protein kinase. Nature 311, 79–81.

Degen, E. and Williams, D.B. (1991). Participation of a novel 88-kDa protein in the biogenesis of murine class I histocompatibility molecules. J. Cell Biol. 112, 1099–1115.

Deverson, E.V., Gow, I.R., Coadwell, W.J., Monaco, J.J., Butcher, G.W. and Howard, J.C. (1990). MHC class II region encoding proteins related to the multidrug resistance family of transmembrane transporters. Nature 348, 738–741.

Doherty, P.C., Biddison, W.E., Bennick, J.R. and Knowles, B.B. (1978). Cytotoxic T cell responses in mice infected with influenza and vaccinia viruses vary in magnitude with H-2 genotype. J. Exp. Med. 148, 534–543.

Eichelberger, M., Allan, W., Zijlstra, M., Jaenisch, R. and Poherty, P.C. (1991). Clearance of influenza virus respiratory infection in mice lacking class I major histocompatibility complex-restricted CD8+ T cells. J. Exp. Med. 174, 875–880.

Ertl, H.C.J. and Finberg, R.W. (1984). Sendai virus-specific T-cell clones: induction of cytolytic T cells by an anti-idiotype antibody directed against a helper T-cell clone. Proc. Natl Acad. Sci. USA 81, 2850–2854.

Fang, R., Min Jou, W., Huylebroeck, D., Devos, R. and Fiers, W. (1981). Complete structure of A/Duck/Ukraine/63 influenza hemagglutinin gene: Animal virus as progenitor of human H3 Hong Kong. Cell 25, 315–323.

Firmbach-Kraft, I., Byers, M., Shows, T., Dalla-Favera, R. and Krolewski, J.J. (1990). Tyk2, prototype of a novel class of non-receptor tyrosine kinase genes. Oncogene 5, 1329–1336.

Fitch, W.M., Leiter, J.M.E., Li, X. and Palese, P. (1991). Positive Darwinian evolution in human influenza A viruses. Proc. Natl Acad. Sci. USA 88, 4270–4274.

Fleischer, B., Becht, H. and Rott, R. (1985). Recognition of viral antigens by human influenza A virus-specific T lymphocyte clones. J. Immunol. 135, 2800–2804.

Francis, M.J., Hastings, G.Z., Sangar, D.V., Clark, R.P., Syred, A. and Clarke, B.E. (1987). A synthetic peptide which elicits neutralizing antibody against human rhinovirus type 2. J. Gen. Virol. 68, 2687–2691.

Fu, X.-Y. (1992). A transcription factor with SH2 and SH3 domains is directly activated by an interferon a-induced cytoplasmic protein tyrosine kinase(s). Cell 70, 323–335.

Ghosh, S. and Baltimore, D. (1990). Activation *in vitro* of NF-kB by phosphorylation of its inhibitor IkB. Nature 344, 678–682.

Gotch, F., McMichael, A., Smith, G. and Moss, B. (1987). Identification of the viral molecules recognized by influenza specific human cytotoxic T lymphocytes. J. Exp. Med. 165, 408–416.

Greve, J.M., Davis, G., Meyer, A.M., Forte, C.P., Yost, S.C., Marlor, C.W., Kamarck, M.E. and McClelland, A. (1989). The major human rhinovirus receptor is ICAM-1. Cell 56, 839–847.

Griffin, J.A., Basak, S. and Compans, R.W. (1983). Effects of hexose starvation and the role of sialic acid in influenzas virus release. Virology 125, 324–334.

Griffiths, C.E., Esmann, J., Fisher, G.J., Voorhees, J.J. and Nickoloff, B.J. (1990). Differential modulation of keratinocyte intercellular adhesion molecule-1 expression by gamma interferon and phorbol ester: evidence for involvement of protein kinase C signal transduction. Br. J. Dermatol. 122, 333–342.

Gumina, R.J., Freire-Moar, J., DeYoung, L., Webb, D.R. and Devens, B.H. (1991). Transduction of the IFN-gamma signal for HLA-DR expression in the promonocytic line THP-1 involves a late-acting PKC activity. Cell. Immunol. 138, 265–279.

Gutch, M.J. and Reich, N.C. (1991). Repression of the interferon signal transduction pathway by the adenovirus E1A oncogene. Proc. Natl Acad. Sci. USA 88, 7913–7917.

Hannigan, G.E. and Williams, B.R. (1991). Signal transduction by interferon-alpha through arachidonic acid metabolism. Science 251, 204–207.

Hany, M., Oehen, S., Schulz, M., Hengartner, H., Mackett, M., Bishop, D.H.L., Overton, H. and Zinkernagle, R.M. (1989). Anti-viral protection and prevention of lymphocytic choriomeningitis or of the local footpad swelling reaction in mice by immunization with vaccinia-recombinant virus expressing LCMV-WE nucleoprotein or glycoprotein. Eur. J. Immunol. 19, 417–424.

Hastings, G.Z., Speller, S.A. and Francis, M.J. (1990). Neutralizing antibodies to human rhinovirus produced in laboratory animals and humans that recognize a linear sequence from VP2. J. Gen. Virol. 71, 3055–3059.

Hastings, G.Z., Rowlands, D.J. and Francis, M.J. (1991). Proliferative responses of T cells primed against human rhinovirus to other rhinovirus serotypes. J. Gen. Virol. 72, 2947–2952.

Hay, A.J., Wolstenholme, A.J., Skehel, J.J. and Smith, M.H. (1985). The molecular basis of the specific action of amantadine. EMBO J. 4, 3021–3024.

Hemmi, S., Merlin, G. and Aguet, M. (1992). Functional characterization of a hybrid human-mouse interferon gamma receptor: evidence for species-specific interaction of the extracellular receptor domain with a putative signal transducer. Proc. Natl Acad. Sci. USA 89, 2737–2741.

Henderson, R.A., Michel, H., Sakaguchi, K., Shabanowitz, J., Appela, E., Hunt, D.F. and Engelhard, V.H. (1992). HLA-A2.1-associated peptides from a mutant cell line: a second pathway of antigen presentation. Science 255, 1264–1266.

Hinshaw, V.S., Jr, B.W.J., Webster, R.G. and Easterday, B.C. (1978). The prevalence of influenza viruses in swine and the antigenic relatedness of influenza viruses from man and swine. Virology 84, 51–62.

Hoekstra, D. (1990). Membrane fusion of enveloped viruses: especially a matter of proteins. J. Bioenerg. Biomembr. 22, 121–155.

Hosaka, Y., Sasao, F. and Ohara, R. (1985). Cell-mediated lysis of heat-inactivated influenza virus-coated murine targets. Vaccine 3, 245–252.

Hosaka, Y., Sasao, F., Yamanaka, K., Bennick, J.R. and Yewdell, J.W. (1988). Recognition of noninfectious influenza virus by class I-restricted murine cytotoxic T lymphocytes. J. Immunol. 140, 606–610.

Isaacs, A. and Lindenmann, J. (1957). Virus interference. I. The interferon. Proc. R. Soc. Lond. B. 147, 258–267.

Jacobsen, H., Czarniecki, C.W., Krause, D., Friedman, R.M. and Silverman, R.H. (1983). Interferon-induced synthesis of 2–5A dependent RNase in mouse JLS-V9R cells. Virology 125, 496–501.

Jacoby, D.B., Ueki, I.F., Loegering, D.A., Gleich, G.J., Widdicombe, J.H. and Nadel, J.A. (1988). Effect of human eosinophil major basic protein on ion transport in canine tracheal epithelium. Am. Rev. Respir. Dis. 137, 13–16.

Jerne, N.K. (1974). Towards a network theory of the immune system. Ann. Immunol. 125C, 373–387.

Jerne, N.K., Roland, J. and Cazenave, P. (1982). Recurrent idiotypes and internal images. EMBO J. 1, 243–247.

John, J., McKendry, R., Pellegrini, S., Flavell, D., Kerr, I.M. and Stark, G.R. (1991). Isolation and characterization of a new mutant human cell line unresponsive to alpha and beta interferons. Mol. Cell. Biol. 11, 4189–4195.

Jung, V., Jones, C., Kumar, C.S., Stefanos, S., O'Connell, S. and Pestka, S. (1990). Expression and reconstitution of a biologically active human interferon-γ receptor in hamster cells. J. Biol. Chem. 265, 1827–1830.

Kalb, T.H., Chuang, M.T., Marom, Z. and Mayer, L. (1991). Evidence for accessory cell function by class II MHC antigen-expressing airway epithelial cells. Am. J. Respir. Cell. Mol. Biol. 4, 320–329.

Kalvakolanu, D.V.R., Bandyopadhyay, S.K., Harter, M.L. and Sen, G.C. (1991). Inhibition of interferon-inducible gene expression by adenovirus E1A proteins: block in transcriptional complex formation. Proc. Natl Acad. Sci. USA 88, 7459–7463.

Kast, W.M., Bluestone, J.A., Heemskerk, M.H.M., Spaargaren, J., Voordouw, A.C., Ellenhorn, J.D.I. and Melief, C.J.M. (1990). Treatment with monoclonal anti-CD3 antibody protects against lethal Sendai virus infection by induction of natural killer cells. J. Immunol. 145, 2254–2259.

Katze, M.G., Tomita, J., Black, T., Krug, R.M., Safer, B. and Hovanessian, A.G. (1988). Influenza virus regulates protein synthesis during infection by repressing autophosphorylation and activity of the cellular 68,000 M_r protein kinase. J. Virol. 62, 3710–3717.

Keay, S., Rasmussen, L. and Merigan, T.C. (1988). Syngenic monoclonal anti-idiotype antibodies that bear the internal image of a human cytomegalovirus neutralization epitope. J. Immunol. 140, 944–948.

Kendal, A.P., Noble, G.R., Skehel, J.J. and Dowdle, W.R. (1978). Antigenic similarity of influenza A (H1N1) viruses from epidemics in 1977–1978 to "Scandinavian" strains isolated in epidemics of 1950–1951. Virology 89, 632–636.

Kerr, I.M. and Brown, R.E. (1978). pppA2′p5′A2′p5′A: an inhibitor of protein synthesis synthesized with an enzyme fraction from interferon-treated cells. Proc. Natl Acad. Sci. USA 75, 256–260.

Kerr, I.M., Brown, R.E. and Ball, L.A. (1974). Increased sensitivity of cell-free protein synthesis to double-stranded RNA after interferon treatment. Nature 250, 57–59.

Kessler, D.S. and Levy, D.E. (1991). Protein kinase activity required for an early step in interferon-alpha signaling. J. Biol. Chem. 266, 23471–23476.

Kilbourne, E.D., Laver, W.G., Schulman, J.L. and Webster, R.G. (1968). Antiviral activity of antiserum specific for an influenza vius neuraminidase. J. Virol. 2, 281–288.

Kitajewski, J., Schneider, R., Safer, S., Munemitsu, S., Samuel, C.E., Thimmapaya, B. and Shenk, T. (1986). Adenovirus VAI RNA antagonizes the antiviral action of interferon by preventing the activation of the interferon-induced eIF-2α kinase. Cell 45, 195–200.

Koszinowski, U.H., Allen, H., Gething, M.-J., Waterfield, M.D. and Klenk, H.-D. (1980). Recognition of viral glycoproteins by influenza A-specific cross-reactive cytolytic T lymphocytes. J. Exp. Med. 151, 945–958.

Kris, R.M., Yetter, R.A., Cogliano, R., Ramphal, R. and Small, P.A. (1988). Passive serum antibody causes temporary recovery from influenza virus infection of the nose trachea and lung of nude mice. Immunology 63, 349–353.

Lamb, R.A., Zebedee, S.L. and Richardson, C.D. (1985). Influenza virus M2 protein is an integral membrane protein expressed on the infected-cell surface. Cell 40, 627–633.

Larsen, G.R., Dorner, A.J., Harris, T.J.R. and Wimmer, E. (1980). The structure of poliovirus replicative form. Nucleic Acids Res. 8, 1217–1229.

Laver, W.G. and Webster, R.G. (1973). Studies on the origin of pandemic influenza. III. Evidence implicating duck and equine influenza viruses as possible progenitors of the Hong Kong strain of human influenza. Virology 51, 383–391.

Leikauf, G.D., Ueki, I.F., Nadel, J.A. and Widdicombe, J.H. (1985). Bradykinin stimulates Cl secretion and prostaglandin E2 release by canine tracheal epithelium. Am. J. Physiol. 248, F48–F55.

Leikauf, G.D., Ueki, I.F., Widdicombe, J.H. and Nadel, J.A. (1986). Alteration of chloride secretion across canine tracheal epithelium by lipoxygenase products of arachidonic acid. Am. J. Physiol. 250, F47–F53.

Lenardo, M.J. and Baltimore, D. (1989). NF-κB: a pleiotropic mediator of inducible and tissue-specific gene control. Cell 58, 227–229.

Leung, K.N. and Ada, G.L. (1980). Production of DTH in the mouse to influenza virus; comparison with conditions for stimulation of cytotoxic T cells. Scand. J. Immunol. 12, 129–139.

Lindenmann, J. (1962). Resistance of mice to mouse adapted influenza A virus. Virology 16, 203–204.

Lindenmann, J. (1973). Speculations on idiotypes and homobodies. Ann. Immunol. 124C, 171–184.

McClelland, A., deBear, J., Yost, S.C., Meyer, A.M., Marlor, C.W. and Greve, J.M. (1991). Identification of monoclonal antibody epitopes and critical residues for rhinovirus binding in domain 1 of intercellular adhesion molecule 1. Proc. Natl Acad. Sci. USA 88, 7993–7997.

McCray, J. and Werner, G. (1987). Different rhinovirus serotypes neutralized by antipeptide antibodies. Nature 329, 736–738.

McCray, J. and Werner, G. (1989). Production and properties of site-specific antibodies to synthetic peptide antigens related to potential cell surface receptor sites for rhinovirus. Methods Enzymol. 178, 676–692.

McKendry, R., John, J., Flavell, D., Muller, M., Kerr, I.M. and Stark, G.R. (1991). High-frequency mutagenesis of human cells and characterization of a mutant unresponsive to both a and g interferons. Proc. Natl Acad. Sci. USA 88, 11455–11459.

McMichael, A. (1978). HLA restriction of human cytotoxic T lymphocytes specific for influenza virus. J. Exp. Med. 148, 1458–1467.

Major, R.H. (1945). "Classic Descriptions of Disease". Charles C. Thomas, Springfield, IL.

Mak, N.K., Schiltknecht, E. and Ada, G.L. (1983). Protection of mice against influenza virus infection: enhancement of nonspecific cellular responses by *Corynebacterium parvum*. Cell. Immunol. 78, 314–325.

Marlin, S.D., Staunton, D.E., Springer, T.A., Stratowa, C., Sommergruber, W. and Merluzzi, V.J. (1990). A soluble form of intercellular adhesion molecule-1 inhibits rhinovirus infection. Nature 344, 70–72.

Matlin, K.S., Reggio, H., Helenius, A. and Simons, K. (1981). Infectious entry pathway of influenza virus in a canine kidney cell line. J. Cell Biol. 91, 601–613.

Mattila, P., Hayry, P. and Renkonen, R. (1989). Protein kinase C is crucial in signal transduction during IFN-gamma induction in endothelial cells. FEBS Lett. 250, 362–366.

Merlin, G., Falcoff, E. and Aguet, M. (1985). 125–I-labeled human interferons alpha, beta and gamma: comparative receptor-binding data. J. Gen. Virol. 66, 1149–1152.

Monaco, J.J. and McDevitt, H.O. (1982). Identification of a fourth class of proteins linked to the murine major histocompatibility complex. Proc. Natl Acad. Sci. USA 79, 3001–3005.

Monaco, J.J. and McDevitt, H.O. (1984). H-2-linked low-molecular weight polypeptide antigens assemble into an unusual macromolecular complex. Nature 309, 797–799.

Monaco, J.J. and McDevitt, H.O. (1986). The LMP antigens: a stable MHC-controlled multisubunit protein complex. Hum. Immunol. 15, 416–426.

Moss, B. and Flexner, C. (1987). Vaccinia virus expression vectors. Ann. Rev. Immunol. 5, 305–324.

Nakajima, K., Desselberger, U. and Palese, P. (1978). Recent human influenza A (H1N1) viruses are closely related genetically to strains isolated in 1950. Nature 274, 334–339.

Nakayama, M., Nagata, K., Kato, A. and Ishihama, A. (1991). Interferon-inducible mouse Mx1 protein that confers resistance to influenza virus is GTPase. J. Biol. Chem. 266, 21404–21408.

Nisonoff, A. and Lamoyi, E. (1981). Implications of the presence of an internal image of the antigen in anti-idiotype antibodies: possible application to vaccine production. Cell. Immunol. Immunopathol. 21, 397–406.

Nuchtern, J.G., Bonifacino, J.S., Biddison, W.E. and Klausner, R.D. (1989). Brefeldin A inplicates egress from endoplasmic reticulum in class I restricted antigen presentation. Nature 339, 223–226.

Obar, R.A., Collins, C.A., Hammarback, J.A., Shpetner, H.S. and Vallee, R.B. (1990). Molecular cloning of the mictrotubule-associated mechanochemical enzyme dynamin reveals homology with a family of GTP-binding proteins. Nature 347, 256–261.

Palmenberg, A. (1987). Antipeptide antibodies: a vaccine for the common cold? Nature 329, 668–669.

Patel, R.C. and Sen, G.C. (1992). Identification of the double-stranded RNA-binding domain of the human interferon-inducible protein kinase. J. Biol. Chem. 267, 7671–7676.

Pavlovic, J., Zurcher, T., Haller, O. and Staeheli, P. (1990). Resistance to influenza virus and vesicular stomatitis virus conferred by expression of human MxA protein. J. Virol. 64, 3370–3375.

Pellegrini, S., John, J., Shearer, M., Kerr, I.M. and Stark, G.M. (1989). Use of a selectable marker regulated by alpha interferon to obtain mutations in the signaling pathway. Mol. Cell. Biol. 9, 4605–4612.

Pestka, S., Langer, J.A., Zoon, K.C. and Samuel, C.E. (1987). Interferons and their actions. Annu. Rev. Biochem. 56, 727–777.

Pevear, D.C., Fancher, M.J., Felock, P.J., Rossmann, M.G., Miller, M.S., Diana, G.D., Treasurywala, A.M., McKinlay, M.A. and Dutko, F.J. (1989). Conformational changes in the floor of the human rhinovirus canyon blocks adsorption to HeLa cell receptors. J. Virol. 63, 2002–2007.

Pfeffer, L.M. and Tan, Y.H. (1991). Do second messengers play a role in interferon signal transduction? Trends Biol. Sci. 16, 321–323.

Piela-Smith, T.H., Aneiro, L. and Korn, J.H. (1991). Binding of human rhinovirus and T cells to intercellular adhesion molecule-1 on human fibroblasts. J. Immunol. 147, 1831–1836.

Pinto, L.H., Holsinger, L.J. and Lamb, R.A. (1992). Influenza virus M$_2$ protein has ion channel activity. Cell 69, 517–528.

Register, R.B., Uncapher, C.R., Naylor, A.M., Lineberger, D.W. and Colonno, R.J. (1991). Human–murine chimeras of ICAM-1 identify amino acid residues critical for rhinovirus and antibody binding. J. Virol. 65, 6589–6596.

Rhodes, J. (1990). Erythrocyte rosettes provide an analogue for Schiff base formation in specific T cell activation. J. Immunol. 145, 463–469.

Rimmelzwaan, G.F., Bunschoten, E.J., UytdeHaag, F.G.C.M. and Osterhaus, A.D.M.E. (1989). Monoclonal anti-idiotypic antibody vaccines against poliovirus, canine parvovirus, and rabies virus. Methods Enzymol. 178, 375–390.

Rossmann, M. (1989). The canyon hypothesis. J. Biol. Chem. 264, 14587–14590.

Rossmann, M.G., Arnold, E., Erickson, J.W., Frankenberger, E.A., Griffith, J.P., Hecht, H.-J., Johnson, J.E., Kamer, G., Luo, M., Mosser, A.G., Ruechert, R.R., Sherry, B. and Vriend, G. (1985). Structure of a human common cold virus and functional relationship to other picornaviruses. Nature 317, 145–153.

Rothman, J.H., Raymond, C.K., Gilbert, T., O'Hara, P.J. and Stevens, T.H. (1990). A putative GTP-binding protein homologous to vertebrate interferon-inducible Mx proteins performs an essential function in yeast vacuolar protein sorting. Cell 61, 1063–1074.

Rueckert, R.R. (1991). In "Fundamental Virology" (eds B.N. Fields and D.M. Knipes), pp 409–451. Raven Press, New York.

Samuel, C.E. (1979). Mechanism of interferon action. Phosphorylation of protein synthesis initiation factor eIF-2 in interferon-treated human cells by a ribosome-associated kinase possessing site specificity similar to hemin-regulated rabbit reticulacyte kinase. Proc. Natl Acad. Sci. USA 76, 600–604.

Samuel, C.E. (1991). Antiviral actions of interferon: interferon-regulated cellular proteins and their surprisingly selective antiviral activities. Virology 183, 1–11.

Scherle, P.A., Palladino, G. and Gerhard, W. (1992). Mice can recover from pulmonary influenza virus infection in the absence of class-I restricted cytotoxic T-cells. J. Immunol. 148, 212–217.

Schick, M.R. and Kennedy, R.C. (1989). Production and characterization of anti-idiotypic antibody reagents. Methods Enzymol. 178, 36–48.

Schiltknecht, E. and Ada, G.L. (1985). Influenza virus-specific T cells fail to reduce lung virus titers in cyclosporin-treated infected mice. Scand. J. Immunol. 22, 99–103.

Schindler, C., Shuai, K., Prezioso, V.R. and Darnell, J.E. (1992). Interferon-dependent tyrosine phosphorylation of a latent cytoplasmic transcription factor. Science 257, 809–813.

Scholtissek, C., Van Hoyningen, V. and Rott, R. (1978). Genetic relatedness between the new 1977 epidemic strains (H1N1) of influenza and human influenza strains isolated between 1947 and 1957 (H1N1). Virology 89, 613–617.

Scholtissek, C., Burger, H., Kistner, O. and Shortridge, K.F. (1985). The nucleoprotein as a possible major factor in determining host specificity of influenza H3N2 viruses. Virology 147, 287–294.

Schonberger, L.B., Hurwitz, E.S., Katona, P., Holman, R.C. and Bregman, D.J. (1981). Guillain–Barre syndrome: its epidemiology and associations with influenza vaccination. Ann. Neurol. 9(Suppl.), 31–38.

Schulman, J.L. (1975). In "The Influenza Viruses and Influenza." (ed. D. Kilbournes), pp 373–393. Academic Press, New York.

Schumacher, T.N.M., DeBruijn, M.L.N., Vernie, L.N., Kast, W.M., Melief, C.J.M., Neefjes, J.J. and Ploegh, H.L. (1991). Peptide selection by MHC class I molecules. Nature 350, 703–706.

Sen, G.C. and Lengyel, P. (1992). The interferon system. J. Biol. Chem. 267, 5017–5020.

Sharpe, A.H., Gaulton, G.N., McDade, K.K., Fields, B.N. and Greene, M.I. (1984). Syngenic monoclonal antiidiotype can induce cellular immunity to reovirus. J. Exp. Med. 160, 1195–1205.

Shaw, S. and Biddison, W.E. (1979). HLA-linked genetic control of the specificity of human cytotoxic T-cell responses to influenza virus. J. Exp. Med. 149, 565–575.

Shaw, S., Shearer, G.M. and Biddison, W.E. (1980). Human cytotoxic T-cell responses to type A and type B influenza viruses can be restricted by different HLA antigens. J. Exp. Med. 151, 235–245.

Sherry, B., Mosser, A.G., Colonno, R.J. and Rueckert, R.R. (1986). Use of monoclonal antibodies to identify four neutralization immunogens on a common cold picornavirus, human rhinovirus 14. J. Virol. 57, 246–257.

Skehel, J.J., Bayley, P.M., Brown, E.M., Martin, S.R., Waterfield, M.D., White, J.M., Wilson, I.A. and Wiley, D.C. (1982). Changes in the conformation of influenza virus haemagglutinin at the pH optimum of virus-mediated membrane fusion. Proc. Natl Acad. Sci. USA 79, 968–972.

Skehel, J.J., Stevens, D.J., Daniels, R.S., Douglas, A.R., Knossow, M., Wilson, I.A. and Wiley, D.C. (1984). A carbohydrate side chain on hemagglutinins of Hong Kong influenza viruses inhibits recognition by a monoclonal antibody. Proc. Natl Acad. Sci. USA 81, 1779–1783.

Smith, F.I. and Pelese, P. (1989). In "The Influenza Viruses" (eds H. Fraenkel-Conrat and R.R. Wagners) Plenum Press, New York.

Smith, T.J., Kremer, M.J., Luo, M., Vriend, G., Arnold, E., Kamer, G., Rossmann, M.G., McKinlay, M.A., Diana, G.D. and Otto, M.J. (1986). The site of attachment in human rhinovirus 14 for antiviral agents that inhibit uncoating. Science 233, 1286–1293.

Speiser, D.E., Zurcher, T., Ramseier, H., Hengartner, H., Staeheli, P., Haller, O. and Zinkernagel, R.M. (1990). Nuclear myxovirus-resistance protein Mx is a minor histocompatibility antigen. Proc. Natl Acad. Sci. USA 87, 2021–2025.

Spies, T., Bresnahan, M., Bahram, S., Arnold, D., Blanck, G., Mellins, E., Pious, D. and DeMars, R. (1990). A gene in the human major histocompatibility complex class II region controlling the class I antigen presentation pathway. Nature 348, 744–746.

Spies, T., Cerundolo, V., Collona, M., Cresswell, P., Townsend, A. and DeMars, R. (1992). Presentation of viral antigen by MCH class I molecules is dependent on a putative petide transporter heterodimer. Nature 355, 644–646.

Staeheli, P. and Haller, O. (1987). Interferon-induced Mx protein: a mediator of cellular resistance to influenza virus. Interferon 8, 1–23.

Staeheli, P., Horisberger, M.A. and Haller, O. (1984). Mx-dependent resistance to influenza viruses is induced by mouse interferons α and β but not γ. Virology 132, 456–461.

Staeheli, P., Haller, O., Boll, W., Lindenmann, J. and Weissman, C. (1986). Mx protein: constitutive expression in 3T3 cells transformed with cloned Mx cDNA confers selective resistance to influenza virus. Cell 44, 147–158.

Stark, G.R. and Kerr, I.M. (1992). Interferon-dependent signaling pathways: DNA elements, transcription factors, mutations, and effects of viral proteins. J. Interferon Res. 12, 147–151.

Staunton, D.E., Merluzzi, V.J., Rothlein, R., Barton, R., Marlin, S.D. and Springer, T.A. (1989). A cell adhesion molecule, ICAM-1, is the major surface receptor for rhinoviruses. Cell 56, 849–853.

Staunton, D.E., Dustin, M.L., Erikson, H.P. and Springer, T.A. (1990). The arrangement of the immunoglobulin-like domains of ICAM-1 and the binding sites for LFA-1 and rhinovirus. Cell 61, 243–254.

Stein-Streilein, J., Witte, P.L., Streilein, J.W. and Guffee, J. (1985). Local cellular defenses in influenza-infected lungs. Cell. Immunol. 95, 234–246.

Stephens, E.B., Compans, R.W., Earl, P. and Moss, B. (1986). Surface expression of viral glycoproteins is polarized in epithelial cells infected with recombinant vaccinia viral vectors. EMBO J. 5, 237–245.

Stitz, L., Schmitz, C., Binder, D., Zinkernagel, R., Paoletti, E. and Becht, H. (1990). Characterization and immunological properties of influenza A virus nucleoprotein (NP): cell-associated NP isolated from infected cells or viral NP expressed by vaccinia recombinant virus do not confer protection. J. Gen. Virol. 71, 1169–1179.

Stuart-Harris, C.H., Schild, G.C. and Oxford, J.S. (1985). "Influenza: The Virus and the Disease". Edward Arnold, Baltimore.

Tomassini, J.E., Graham, D., DeWitt, C.M., Lineberger, D.W., Rodkey, J.A. and Colonno, R.J. (1989). cDNA cloning reveals that the major group rhinovirus receptor on HeLa cells is intercellular adhesion molecule 1. Proc. Natl Acad. Sci. USA 86, 4907–4911.

Top, F.H.J. and Russell, P.K. (1977). Swine influenza A at Fort Dix, New Jersey. J. Infect. Dis. 136, S376–S380.

Towata, T., Hayashi, N., Katayama, K., Takehara, T., Sasaki, Y., Kasahara, A., Fusamoto, H. and Kamada, T. (1991). Signal transduction pathways in the induction of HLA class I antigen expression on Huh 6 cells by interferon-gamma. Biochem. Biophys. Res. Commun. 177, 610–618.

Townsend, A.R.M., Gotch, F.M. and Davey, J. (1985). Cytotoxic T cells recognize fragments of the influenza nucleoprotein. Cell 42, 457–467.

Townsend, A.R.M., Bastin, J., Gould, K. and Brownlee, G.G. (1986a). Cytotoxic T lymphocytes recognize influenza haemagglutinin that lacks a signal sequence. Nature 324, 575–577.

Townsend, A.R.M., Rothbard, J., Gotch, F.M., Bahadur, G., Wraith, D. and McMichael, A. (1986b). The epitopes of influenza nucleoprotein recognized by cytotoxic T lymphocytes can be defined with short synthetic peptides. Cell 44, 959–968.

Townsend, A., Bastin, J., Gould, K., Brownlee, G., Andrew, M., Coupar, B., Boyle, D., Chan, S. and Smith, G. (1988). Defective presentation to class I-restricted cytotoxic T lymphocytes in vaccinia-infected cells is overcome by enhanced degradation of antigen. J. Exp. Med. 168, 1211–1224.

Townsend, A., Elliott, T., Cerundolo, V., Foster, I., Barber, B. and Tse, A. (1990). Assembly of class I molecules analyzed in vitro. Cell 62, 285–295.

Trowsdale, J., Hanson, I., Mockridge, I., Beck, S., Townsend, A. and Kelly, A. (1990). Sequences encoded in the class II region of the MHC related to the 'ABC' superfamily of transporters. Nature 348, 741–744.

UytdeHaag, F.G.C.M. and Osterhaus, A.D.M.E. (1985). Induction of neutralizing antibody in mice against poliovirus type II with monoclonal anti-idiotypic antibody. J. Immunol. 134, 1225–1229.

Uze, G., Lutfalla, G. and Gresser, I. (1990). Genetic transfer of a functional human interferon a receptor into mouse cells: cloning and expression of its cDNA. Cell 60, 225–234.

van Bleek, G.M. and Nathenson, S.G. (1992). Presentation of antigenic peptides by MHC class I molecules. Trends Cell Biol. 2, 202–207.

Velazquez, L., Fellous, M., Stark, G.R. and Pellegrini, S. (1992). A protein tyrosine kinase in the interferon α/β signaling pathway. Cell 70, 313–322.

Visvanathan, K.V. and Goodbourn, S. (1989). Double-stranded RNA activates binding of NF-κB to an inducible element in the human β-interferon promoter. EMBO J. 8, 1129–1138.

Ward, C.W. and Dopheide, T.A. (1981a). Amino acid sequence and oligosaccharide distribution of the hemagglutinin from an early Hong Kong variant A/Aichi/2/68 (X31). Biochem. J. 193, 953–962.

Ward, C.W. and Dopheide, T.A. (1981b). Evolution of the Hong Kong influenza A subtype. Biochem. J. 195, 337–340.

Webster, R.G., Laver, W.G., Air, G.M. and Schild, G.C. (1982). Molecular mechanisms of variation in influenza viruses. Nature 296, 115–121.

Weis, W., Brown, J.H., Cusack, S., Paulson, J.C., Skehel, J.J. and Wiley, D.C. (1988). Structure of the influenza virus haemagglutinin complexed with its receptor, sialic acid. Nature 333, 426–431.

Wiley, D.C. and Skehel, J.J. (1991). In "Fundamental Virology" (eds B.N. Fields and D.M. Knipes), pp 63–85. Raven Press, New York.

Wiley, D.C., Wilson, I.A. and Skehel, J.J. (1981). Structural identification of the antibody-binding sites of Hong Kong influenza haemagglutinin and their involvement in antigenic variation. Nature 289, 373–378.

Wilson, I.A., Skehel, J.J. and Wiley, D.C. (1981). Structure of the haemagglutinin membrane glycoprotein of influenza virus at 3Å resoluation. Nature 289, 366–373.

Winter, G., Fields, S. and Brownlee, G.G. (1981). Nucleotide sequence of the haemagglutinin gene of a human influenza virus H1 subtype. Nature 292, 72–75.

Wraith, D.C., Vessey, A.E. and Askonas, B.A. (1987). Purified influenza virus nucleoprotein protects mice from lethal infection. J. Gen. Virol. 68, 433–440.

World Health Organization (1980). A revision of the system of nomenclature for influenza viruses: a WHO memorandum. Bull. WHO 59, 585–591.

Yap, K.L. and Ada, G.L. (1977). Cytotoxic T cells specific for influenza virus-infected target cells. Immunology 32, 151–159.

Yap, K.L. and Ada, G.L. (1978a). Cytotoxic T cells in the lungs of mice infected with an influenza A virus. Scand. J. Immunol. 7, 73–80.

Yap, K.L. and Ada, G.L. (1978b). The recovery of mice from influenza A virus infection: Adoptive transfer of immunity with influenza virus-specific cytotoxic T lymphocytes recognizing a common virion antigen. Scand. J. Immunol. 8, 413–420.

Yap, K.L. and Ada, G.L. (1978c). The recovery of mice from influenza virus infection: adoptive transfer of immunity with immune T lymphocytes. Scand. J. Immunol. 7, 389–397.

Yap, K.L., Ada, G.L. and McKenzie, I.F.C. (1978). Transfer of specific cytotoxic T lymphocytes protects mice inoculated with influenza virus. Nature 273, 238–239.

Yewdell, J.W. and Bennink, J.R. (1989). Brefeldin A specifically inhibits presentation of protein antigens to cytotoxic T lymphocytes. Science 244, 1072–1075.

Yewdell, J.W., Caton, A.J. and Gerhard, W. (1986). Selection of influenza A virus adsorptive mutants by growth in the presence of a mixture of monoclonal anti-hemagglutinin antibodies. J. Virol. 57, 623–628.

Zheng, B., Brett, S.J., Tite, J.P., Lifely, M.R., Brodie, T.A. and Rhodes, J. (1992). Galactose oxidation in the design of immunogenic vaccines. Science 256, 1561–1563.

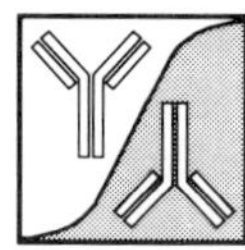

8. Gastroduodenal Defence: Role of Epithelial Factors

Thomas A. Miller, Gregory S. Smith, *and* Jose C. Barreto

1. Introduction

Under normal physiological conditions, the gastroduodenal epithelium is exposed to high intraluminal acid concentrations on a fairly consistent and continual basis. During periods of peak acid output in the human stomach, when a meal is ingested, for example, the luminal hydrogen ion concentration may be more than 150 mmol/l with a pH approaching 1. Similarly, when this acid load is discharged into the proximal duodenum, it is not unusual for the pH to drop as low as 2, despite mixing with duodenal, biliary and pancreatic secretions. When one realizes that such low pHs are capable of denaturing protein, it is remarkable that the gastroduodenal epithelium is generally resistant to such acid exposure without evidence of autodigestion. These findings emphasize the exquisite endogenous processes that are operational within the gastroduodenal mucosa, allowing it to maintain its integrity despite its exposure to this potentially damaging acid milieu. Over a century of research has attempted to identify the mechanisms by which resistance to autodigestion by acid in the stomach and duodenum is effected, but the precise factors involved have continued to remain elusive. Nonetheless, a substantial body of literature has accumulated in the past two decades which has enabled investigators and clinicians alike to gain better insights into those protective processes that appear to be important. It is the purpose of this discussion to review those epithelial factors that seem to be responsible for enabling the gastroduodenal mucosa to resist injury.

2. Protective Processes

Three protective processes have been identified in the gastroduodenal mucosa that contribute greatly to its ability to withstand a potentially damaging insult. These processes can be grouped under the broad headings of mucus and bicarbonate secretion, microcirculatory responses, and epithelial renewal.

2.1 MUCUS AND BICARBONATE SECRETION

It has been appreciated for almost 100 years that an "alkaline mucus" layer lines the gastric mucosa (Flemstrom and Turnberg, 1984). While this layer was thought from its first description to play a role as a protective barrier, it remained for Heatley (1959) to conceptualize how this might occur. He proposed that the mucus secreted by the stomach might result in the development of a pH gradient in which bicarbonate secreted by mucosal cells could effectively neutralize hydrogen ions diffusing from the lumen through it and thereby correspondingly protect the surface epithelium from injury. From the time of this proposal in 1959, over two decades would pass before experimental confirmation of this notion was forthcoming. With the advent of H_2 receptor blockers in the 1970s in which blockage of acid secretion by the stomach clearly demonstrated that the gastric mucosa also secreted substantial amounts of bicarbonate, and the development of sophisticated techniques that enabled the actual measurement of the biochemical components of mucus, it now seems clear that a mucus-bicarbonate barrier does indeed exist under normal physiological conditions and plays a role not only in gastric defence but also in duodenal protection (Allen, 1981; Flemstrom and Garner, 1982; Miller, 1983; Flemstrom and Turnberg, 1984).

Studies with porcine gastric mucus have demonstrated that this substance is water-insoluble and is fundamentally a viscoelastic gel that consists of a polymer of glycoprotein molecules bound together by disulphide bridges (Allen, 1981). Available human studies have demonstrated a similar biochemical composition. It is this glycoprotein matrix that gives rise to the gel-forming properties of mucus. Although, on a weight basis, only 5% of this gel is composed of glycoprotein molecules, with the balance being water, it is this glycoprotein matrix configuration that enables mucus to prevent permeation of pepsin and other large molecules, as well as to impede hydrogen ion diffusion some three- to four-fold when compared with a similar thickness of unstirred water (Allen, 1981; Flemstrom and Garner, 1982; Miller, 1983; Flemstrom and Turnberg, 1984).

This layer of mucus has been shown to be adherent to the mucosa in both the stomach and duodenum and to be of similar molecular configuration in both organs. Various studies have demonstrated the dynamic nature in which the stomach and duodenum secrete mucus, and the means by which this secretion is stimulated and inhibited (Allen, 1981; Flemstrom and Garner, 1982; Flemstrom and Turnberg, 1984; Flemstrom, 1987; Neutra and Forstner, 1987). Thus, luminal acid seems to stimulate mucus release in both epithelia as do PGs. In contrast, cholinergic agents seem to primarily elicit mucus secretion in the gastric mucosa while secretin elicits its stimulatory effects on mucus secretion in the duodenum. Interestingly, agents known to damage the gastroduodenal epithelium have been shown to decrease mucus thickness, inhibit its synthesis, and in some cases actually alter its molecular structure, thereby reducing its gel-like properties. When this latter circumstance occurs, the soluble mucus present in the luminal secretions is largely composed of subunits of the normal mucus glycoprotein polymer. Agents demonstrated to have these adverse effects on mucus include ethanol, bile salts, steroids, aspirin, and various NSAIDS, such as indomethacin. These changes in mucus production in response to various stimulants and damaging agents suggest that mucus plays an important role in mucosal defence (Allen, 1981; Flemstrom and Garner, 1982; Flemstrom and Turnberg, 1984; Flemstrom, 1987; Neutra and Forstner, 1987). This role, however, appears to be one that is linked with bicarbonate secretion, since information concerning the variable thickness of the mucus layer in most species in which it has been studied, and the rate of diffusion of hydrogen ions through it has led to predictions that this layer would need to be renewed approximately 10 times per second to keep up with the rate of acid diffusion (Flemstrom and Turnberg, 1984). Understandably, such a turnover rate is physiologically incomprehensible.

Similar to available information concerning mucus, studies within the past decade or so have clearly demonstrated that both fundic and antral gastric mucosa actively secrete an alkaline fluid in both humans and a wide range of animals (Flemstrom and Garner, 1982; Flemstrom and Turnberg, 1984; Flemstrom, 1987; Neutra and Forstner, 1987). The origin of this secretion appears to be surface epithelial cells. The duodenal mucosa also secretes bicarbonate that is separate from that present in the duodenal lumen secondary to pancreatic bicarbonate secretion. Although Brunner's glands may be the origin of this bicarbonate secretion in some species, the absence of such glands in proximal bullfrog duodenum, in which duodenal bicarbonate secretion has also been observed, indicates that the surface epithelial cells are capable of producing and secreting such a bicarbonate-rich fluid (Flemstrom and Garner, 1982; Flemstrom and Turnberg, 1984; Flemstrom, 1987; Neutra and Forstner, 1987).

A number of agents have been shown to stimulate bicarbonate secretion in both the stomach and duodenum. These include glucagon, PGs, acid, and various nucleotides. As with mucus secretion, certain agents seem to stimulate bicarbonate secretion more commonly in the stomach than in the duodenum, and vice versa. In terms of output, the proximal duodenum appears to secrete nearly twice the amount of bicarbonate as the distal duodenum and several-fold more than that which occurs in the gastric mucosa under similar experimental conditions (Flemstrom and Turnberg, 1984). Various damaging insults have also been shown to inhibit bicarbonate secretion in the stomach and duodenum, similar to effects on mucus. Thus, anoxia, dinitrophenol, and various damaging substances such as ethanol, bile and aspirin commonly inhibit bicarbonate secretion (Flemstrom and Garner, 1982; Flemstrom and Turnberg, 1984; Flemstrom, 1987; Neutra and Forstner, 1987). Interestingly, in both the stomach and duodenum, an increase in the concentration of hydrochloric acid within the luminal bathing solution correspondingly causes an increase in the bicarbonate output from these epithelia (Flemstrom and Garner, 1982; Flemstrom and Turnberg, 1984; Flemstrom, 1987; Neutra and Forstner, 1987). Such findings suggest that alkaline output is regulated by acid load as a potential protective mechanism against mucosal injury.

Stoichiometrically, the magnitude of bicarbonate production in the luminal bulk solution of the stomach approaches only 5–10% of the capabilities of the gastric mucosa to secrete hydrogen ions (Flemstrom and Garner, 1982; Flemstrom and Turnberg, 1984; Flemstrom, 1987; Neutra and Forstner, 1987). Such an amount of bicarbonate would offer little protection from injury if neutralization of the luminal bulk solution was all that occurred. On the other hand, if bicarbonate were secreted into a zone of limited mixing, notably a layer of unstirred mucus gel, the buffering characteristics of bicarbonate at the mucus–mucosal interface could be much more effective. Thus, while mucus and bicarbonate acting independently of each other would have limited effects alone as protective substances in defending the mucosal lining of the stomach and duodenum against injury, acting in concert with each other a very effective defence mechanism could result.

Using pH-sensitive microelectrodes, studies in various animal models as well as humans have clearly shown that a pH gradient does in fact exist across mucus in both the stomach and duodenum (Flemstrom and Garner, 1982; Flemstrom and Turnberg, 1984; Flemstrom, 1987; Neutra and Forstner, 1987). The pH at the cell membrane has been noted to be neutral or slightly alkaline (approximately 7.3) even though the luminal acidity may have a pH of approximately 2. Cyanide, an inhibitor of bicarbonate transport, has been shown to abolish this pH gradient, as have aspirin and bile. This direct demonstration of an alkaline zone against the mucosal surface supports the notion that such a process could protect the mucosa against luminal acid. Since diffusion of hydrogen ions across the mucus gel has been shown experimentally to be markedly slowed, this retardation could enhance the efficiency of neutralization of acid by bicarbonate secreted into the gel at the epithelial surface, and thus help maintain a neutral pH at the luminal cell membrane, even though the pH of the luminal bulk solution may be quite acidic.

Even though the mucus–bicarbonate barrier conceptually has considerable appeal as a mucosal defence mechanism, its true importance in this regard remains to be established. Certainly, it is not the only defence mechanism since, when luminal acidity decreases the pH to 1.5 or less, the ability of mucus to maintain a pH gradient so that a neutral pH exists adjacent to the epithelium is severely compromised (Flemstrom and Garner, 1982; Flemstrom and Turnberg, 1984; Flemstrom, 1987; Neutra and Forstner, 1987). Further, since the mucus–bicarbonate barrier protects only against acid, its

role as a defensive mechanism against other toxic materials, such as ethanol, would be virtually ineffective. Of equal note, the true relationship between mucus–bicarbonate secretion and the presence or absence of morphological epithelial injury remains to be clarified since protection against various injurious insults can be demonstrated under acute conditions even when significant decreases in mucus and bicarbonate secretion can be demonstrated (Henagan et al., 1986; Neutra and Forstner, 1987).

In addition to enabling the establishment of a pH gradient between the gastroduodenal lumen and the surface epithelium, mucus has also been shown to possess hydrophobic properties (at least in the stomach) which allow it to be water-repellent. Recent studies have suggested that this non-wettable property may also contribute to mucosal defence (Slomiany et al., 1985; Goddard et al., 1990). This action of mucus appears to be mediated through its phospholipid components. Gastric injury induced by acid and aspirin have been shown to markedly reduce the hydrophobic properties of mucus coincident with the development of mucosal injury (Lichtenberger et al., 1983; Sarosiek et al., 1984; Lichtenberger et al., 1985; Goddard et al., 1987, 1990). Both intraluminal administration of surface-active phospholipids and PGs have been shown to markedly reduce the injurious effects of these damaging agents at the same time that they restored the hydrophobic properties of the gastric epithelium (Lichtenberger et al., 1983, 1985; Goddard et al., 1990). Thus, these findings suggest that the hydrophobic characteristics of mucus may also contribute to gastric defence independent of the pH gradient known to be established by this substance. Further studies are clearly needed to clarify the exact role, if any, that mucus hydrophobicity plays in helping the stomach, and possibly the duodenum, to resist injury.

2.2 MICROCIRCULATORY RESPONSES

As is true with any tissue, an adequate blood supply is necessary if cellular integrity is to be maintained. The importance of this concept is underscored by the observation that potentially noxious insults to the gastric epithelium may result in little or no damage if mucosal perfusion is maintained, whereas severe epithelial injury may occur when blood flow is reduced (Guth, 1984; Svanes et al., 1984; Leung et al., 1985). Although less studied than the gastric mucosa, similar findings have been obtained in the duodenum, showing again the important relationship between resultant injury and the adequacy of mucosal perfusion (Lillehei et al., 1948; Manabe et al., 1977; Leung et al., 1985).

Interestingly, various microcirculatory responses appear to be operational to insure that damage does not occur in a potentially noxious situation. Several studies have emphasized the important role that active secretion of the gastric epithelium may play in enhancing resistance to gastric damage. In studies with both rabbit and amphibian gastric mucosae, the actively secreting histamine-stimulated stomach was found to be much more resistant to injury induced by exposure to various acid loads than the metiamide-inhibited or resting stomach (Smith *et al.*, 1977; Kivilaakso *et al.*, 1978a). The explanation for these findings was based on the earlier observation that for each hydrogen ion secreted into the lumen of the stomach a corresponding bicarbonate ion is released from the basal side of the epithelium into the bloodstream (Teorell, 1951). This "alkaline tide", as such movement of alkali has been designated, was proposed as being of importance in protecting the mucosa against damage. Studies involving the anatomical arrangement of the vasculature supplying the gastric epithelium have shown that it has a configuration that can readily make available to surface mucous cells the alkaline tide produced upon stimulation of parietal cells (Gannon *et al.*, 1982, 1984). These experimental findings make a strong case for the importance of the alkaline tide physiologically in gastric protection and would explain why the acid loads to which the stomach is exposed during maximal secretion do not routinely damage the surface epithelial cells.

When the permeability of the gastric mucosa is increased secondary to exposing it to various noxious agents, the resultant increase in the back-diffusion of hydrogen ions in the presence of acid also appears to stimulate mucosal blood flow as a protective mechanism against this acid load. Studies in bile acid-treated canine mucosa have shown that the magnitude of the blood flow response is a linear function of the corresponding magnitude of hydrogen ions simultaneously lost from the gastric lumen (Ritchie, 1983). In the absence of acid, no changes in blood flow were noted when bile acids alone were applied to the gastric epithelium in similar concentrations.

It is well known that the mammalian stomach, particularly the submucosa, is densely innervated by a network of afferent neurones which primarily surround submucosal blood vessels (Sharkey *et al.*, 1984; Sternini *et al.*, 1987; Su *et al.*, 1987; Green and Dockray, 1988) and are sensitive to capsaicin, the pungent active ingredient of red peppers (Nagy, 1982; Russell and Burchiel, 1984; Buck and Burks, 1986). Increasing evidence in recent years has suggested that these neurones may participate in gastric mucosal defence. This evidence includes the finding that ablation of capsaicin-sensitive afferent neurones can aggravate lesion formation in various experimental models including gastric damage induced by pylorus ligation (Szolcsanyi and Bartho, 1985), acid distension of the stomach (Szolcsanyi and Bartho, 1985), and in response to various noxious agents such as indomethacin and ethanol (Holzer and Sametz, 1986), and that direct stimulation of these afferent neurones can actually protect the gastric mucosa against damage induced by various injurious insults (Holzer and Lippe, 1988; Holzer *et al.*, 1989; Lippe *et al.*, 1989).

Experimental observations also suggest that stimulation of capsaicin-sensitive fibres in the stomach stimulate gastric mucosal blood flow, and it is through this mechanism that injury is prevented (Limlomwongse *et al.*, 1979; Lippe *et al.*, 1989). When these fibres are ablated, the stimulatory response to enhance mucosal perfusion does not occur, with a resultant inability of the stomach to resist the potentially damaging insult. While further work needs to be done to fully determine the role that capsaicin-sensitive fibres play in gastric defence, these preliminary studies indicate the possibility that exposure to various components of a meal may continually stimulate such neurones to increase mucosal blood flow so that maximal resistance to a potentially damaging situation can be achieved.

In addition to these considerations, various models of gastric damage have contributed greatly to our understanding of events that may occur at the microcirculatory level that ultimately play major roles in allowing injury to occur. Using a haemorrhagic shock model, Menguy (1981) demonstrated a drastic reduction in gastric mucosal ATP levels in rat gastric mucosa following haemorrhage that coincided with the subsequent development of mucosal ulcerations. This energy deficit was observed to be much greater in the gastric fundic mucosa than in the antrum, liver or muscle. If animals were fed prior to inducing haemorrhage, a significant reduction in the magnitude of gastric injury occurred, compared with what was observed in the fasted state, presumably because more energy substrates were available following feeding. Interestingly, it was further observed that such energy deficits secondary to hemorrhagic shock occurred whether or not luminal acid was present (Silen *et al.*, 1981). These studies were interpreted as indicating that cellular necrosis results from a deficit in mucosal energy metabolism secondary to the hemorrhagic shock and that the integrity of the gastric epithelium requires a constant supply of glucose and oxygen to prevent an anaerobic state (Menguy, 1981).

Another means by which ischaemia may cause gastric injury is through disruption of mechanisms which normally enable the stomach to maintain a proper acid–base balance (Silen, 1987). Normally, the pH of the gastric epithelium approaches 7.3–7.4, and is dependent on both the rate of mucosal blood flow and the pH of the arterial blood perfusing the stomach. It is well known that ischaemia secondary to shock or sepsis is closely associated with disturbances in the normal acid–base balance. Thus, it should be no surprise that the pH of the gastric mucosa might be altered by such aberrations and thereby diminish the ability of the gastric lining to protect itself against luminal acid. Of interest, the importance of maintaining an acid–base balance in the stomach was first emphasized as early as 1948 by Cummins and associates. Studies by these investigators demonstrated that production of gastric and duodenal ulcers by systemic acidosis could be induced by constant intragastric infusion of acid. If this systemic acidosis was prevented by intravenous administration of sodium bicarbonate, these experimentally induced ulcers did not occur. Other studies in frog gastric mucosa demonstrated similar findings (Davies and Longmuir, 1948). Additional observations by Cheung and associates (Cheung and Porterfield, 1979; Cheung et al., 1982), using a canine gastric model, showed that acidification of arterial blood significantly enhanced the likelihood of gastric mucosal injury induced by topical bile salts and haemorrhagic shock. Intravenous infusion of sodium bicarbonate prevented such injury. In a haemorrhagic shock model, Kivilaasko and colleagues (1978a), using a microelectrode technique, observed that the pH of the lamina propria, both in rabbit and dog fundic mucosa bathed with acid solution, dropped quickly during shock. In the relatively permeable rabbit gastric epithelium, this pH drop coincided with the development of severe mucosal lesions. While the intraluminal pH decreased more slowly in the less permeable canine fundic mucosa, disruption of the mucosal barrier by 5 mM taurocholate subsequently resulted in a rapid and profound decrease in intramural pH, with concomitant development of extensive mucosal lesion formation. This study indicated that a critical determinant of whether gastric injury will occur depends upon the capacity of the mucosa to remove or buffer the influx of hydrogen ions, which is adversely affected in the ischaemic state. This study also emphasized the important point that the absolute loss of luminal hydrogen ions may be essentially equivalent in the ischaemic stomach as in the well perfused organ but that, in the presence of ischaemia, reductions in intramural pH and the development of subsequent injury are far more profound. Other studies by the same investigators showed that in the isolated amphibian gastric mucosa a reduction in nutrient bicarbonate severely hampers the protective mechanisms by which injury is prevented (Schiessel et al., 1980).

Finally, various damaging substances such as ethanol and aspirin have been shown to have as their targets the vascular endothelium (Robins, 1980; Szabo et al., 1985; Konturek et al., 1988). The precise means by which these agents, and probably many other damaging substances, injure the endothelium remains to be defined. The observation that the endothelium itself releases various substances (some with vasodilatory and/or antiaggregatory properties and others with vasoconstrictive and/or platelet aggregation actions) that modulate blood flow through the microcirculation (Whittle et al., 1981; Whittle and Vane, 1983; Wallace and Whittle, 1986; Wallace et al., 1989) raises the interesting possibility that the presence or absence of damage in a given situation depends on the resultant delicate balance of whether the dilating substance or constricting substance exists in greater concentration. Endothelial-derived substances with vasoconstrictive and/or platelet-aggregatory properties include endothelin-1 (Wallace et al., 1989), PAF (Wallace and Whittle, 1986) and thromboxane (Whittle et al., 1981), while those with vasodilatory and/or anti-aggregatory properties include prostacyclin (Whittle and Vane, 1983) and nitric oxide (MacNaughton et al., 1989; Pique et al., 1989; Whittle et al., 1990; Boeckxstaens et al., 1991). Much further research will be needed to determine the role that these various endothelial factors play in modulating gastroduodenal blood flow under normal conditions, and the extent to which they are perturbed during conditions of gastric and duodenal damage.

2.3   EPITHELIAL RENEWAL

With the exception of the bone marrow, the gastrointestinal epithelium proliferates more rapidly than any other tissue in the body. Surface cells of the stomach, for example, are renewed every 2–3 days while those of the duodenum are replaced every 3–5 days (Wright, 1984). This rapid turnover is probably a reflection of the excessive demands placed on these cells in terms of not only normal digestive and absorptive function but also because of their continual exposure to noxious insults. Thus, it is appropriate to view this rapid renewal as an important protective mechanism in the maintenance of mucosal integrity. The precise mechanisms that control this proliferative activity are only partially understood.

At least one component of this control is hormonally mediated. Studies in the rat stomach have shown that exogenous gastrin, in doses that are submaximal for stimulation of acid secretion, promotes the uptake of radiolabelled leucine, RNA and DNA in gastric oxyntic mucosa, demonstrating direct effects of this hormone on both protein and nucleotide synthesis (Johnson, 1977). Additional studies have shown that this trophic activity is not limited to the stomach; gastrin can also affect proliferative activity in the mucosa of the duodenum as well as other portions of the small and large intestine (Johnson, 1977). The important role that gastrin plays as a trophic hormone was further demonstrated in rats subjected to antrectomy or fed intravenously with various nutrients for extended periods of time, rather than via the oral route. In these experimental settings, the absence of gastrin from antrectomy or its normal daily release during chronic feeding led to subsequent atrophic changes in both the gastric oxyntic mucosa as well as small intestinal epithelium. Interestingly, when gastrin was administered exogenously, these atrophic changes were reversed (Johnson, 1977; Solomon, 1986).

EGF, which exists in both salivary glands and Brunner's glands of duodenal mucosa, has also been shown to exhibit important trophic activity. Similarly to gastrin, this peptide has been shown to stimulate protein synthesis in the stomach and duodenum and DNA synthesis in the gastric oxyntic mucosa of rats (Johnson and Guthrie, 1980). Further, at least two studies have clearly shown that EGF may be important for mucosal defence. In one of these studies, submandibular gland excision of rats was noted to produce more gastric acid production during pentagastrin stimulation, and such animals were found to be more susceptible to ulceration when the mucosa was exposed to acidified bile salts than in corresponding control rats (Skinner and Tepperman, 1981). These findings were interpreted as being due to the absence of EGF in saliva. A second study, utilizing cats and rats, showed that, in both of these species, EGF was capable of protecting the gastric mucosa against the formation of aspirin-induced lesions, and these protective effects correlated with the effects of EGF on DNA synthesis (Konturek *et al.*, 1981). Such observations with gastrin and EGF suggest that these peptides, and perhaps other gastrointestinal hormones, may directly modulate epithelial renewal of both the stomach and duodenum under physiological conditions, and, through such an action, contribute to endogenous mucosal defence.

Other experimental findings suggest that the environment to which the stomach and duodenum is exposed may also influence proliferative processes within their epithelia. Both aspirin and indomethacin, when given in doses that produce no detectable injury or inflammation, directly stimulated epithelial proliferation in the gastric fundus of rats chronically fed with these substances (Eastwood, 1984). Aspirin also appeared to stimulate proliferation of stem cells in the duodenal crypt and to induce acceleration of transit of these cells through the stem cell zone, the anatomical region from which all duodenal epithelial cells arise. These proliferative responses to aspirin and indomethacin may be important in enabling the mucosa to resist injury by these substances and could also explain why long-term administration of a substance like aspirin, in both animals and human beings, results in an apparent decrease in the susceptibility of the gastric epithelium to its damaging effects (Eastwood, 1984). A further finding of interest in these studies is the lack of a proliferative response induced by these substances in the antrum. This finding would explain why gastric injury in the human stomach produced by long-term aspirin administration is often located in this anatomical region rather than elsewhere in the stomach (Eastwood, 1984).

Over the past decade, angiogenesis (i.e. new blood vessel formation) has been shown to be crucially important in the generation of granulation tissue and other wound-healing processes in various models of tissue injury (Folkman, 1985; Folkman and Klagsburn, 1987). Application of these experimental findings to the treatment of some human diseases has already been realized (Folkman, 1986, 1990). The role of angiogenesis in the pathogenesis and healing of gastroduodenal injury has only recently been studied. In a model of chronic cysteamine-induced duodenal ulceration in rats, oral administration of bFGF-CS23, a basic angiogenic polypeptide fibroblast growth factor which was made acid-resistant, was noted to significantly increase the healing rate of such ulcer formation when compared to cimetidine (Folkman *et al.*, 1991). This growth factor was observed to accomplish this goal by stimulating angiogenesis nine-fold in the ulcer bed, independently of decreasing the production of hydrochloric acid or pepsin. In a model of rat gastritis induced by iodoacetamide, bFGF also accelerated healing by enhancing angiogenesis (Stovroff *et al.*, 1991). Comparable studies with cimetidine or PGs were without effect. In studies with sucralfate, an agent with potent gastroprotective and antiulcer properties independent of effects on acid secretion (Hollander and Tarnawski, 1989; Szabo and Hollander, 1989), it was found that this substance was equal to that of bFGF in stimulating angiogenesis and fibroblast proliferation around sterile sponges implanted subcutaneously in the backs of rats (Szabo *et al.*, 1991). This observation, along with the earlier finding *in vitro* that sucralfate binds bFGF with high affinity (Folkman *et al.*, 1990), raised the interesting possibility that one means by which sucralfate mediated its ulcer-healing properties was by enhancing the ability of endogenous growth factors to bind to damaged epithelium. This hypothesis was tested in the rat cysteamine-induced duodenal ulcer model. Sucralfate was found to bind bFGF to the ulcer bed and to protect it from acid degradation (Folkman *et al.*, 1991) Sucralfate was also shown to elevate the level of endogenous bFGF in the ulcer bed. Cimetidine, by reducing gastric acid, was likewise noted to elevate endogenous bFGF in the ulcerated mucosa. It was concluded from these studies that prevention of ulcer formation or accelerated healing of duodenal ulcers by conventional therapy, such as sucralfate and cimetidine, may be bFGF dependent (Folkman *et al.*, 1991). While much additional work needs to be done in the area of angiogenesis to determine its ultimate role in gastroduodenal protection and repair, these preliminary studies suggest that it might be a very important gastroduodenal defence mechanism.

An interesting feature of the surface epithelium of the gastric mucosa is its unique ability to rapidly repair itself when injured, independently of processes that involve protein and nucleotide synthesis as well as proliferation of new cells. Studies utilizing frog and guinea-pig gastric epithelium *in vitro*, in which severe damage to the surface epithelium and gastric pits was induced by exposure to hyperosmolar sodium chloride, demonstrated virtually complete repair within brief periods of time (Silen and Ito, 1985). In guinea-pig stomach, this repair was completed within 30 min; in frog stomach, repair was markedly ongoing by 30 min and was totally compete within 4 h. In studies with rat gastric epithelium under *in vivo* conditions, damage to the surface epithelium was induced by exposure to absolute ethanol for periods of 30–45 s. Marked repair was again noted in a short time span, i.e. 15 min, after exposure to this agent, and was complete in 1 h (Silen and Ito, 1985). The histological features of this "restitution", as this process of rapid repair is called, indicate that epithelial integrity is restored by squamous cells that appear to migrate from the gastric gland region. After complete re-epithelialization occurs, the normal cuboidal and columnar epithelium is reformed.

Further, the adverse effects on the electrophysiological and secretory properties of damaged mucosae studied under *in vitro* conditions were also restored after restitution was complete, indicating re-establishment of functional integrity in addition to morphological integrity (Silen and Ito, 1985). In addition to this rapid repair process that occurs in gastric epithelium, restitution has also been demonstrated to exist in the duodenum as well as other epithelia such as the gallbladder and colon (Silen and Ito, 1985; Silen, 1987).

The mechanism(s) by which restitution occurs in the gastroduodenal epithelium following damage to surface cells remains unknown (Silen and Ito, 1985; Silen, 1987). The fact that it can be demonstrated under *in vitro* conditions suggests that neural, humoral, and blood-borne agents do not play a role. Calcium appears to be important as severe depletion of this substance can prevent restitution, as can a luminal pH of 3.0 or less (Silen and Ito, 1985; Silen, 1987). Complete inhibition of cell migration by cytochalasin-B following injury suggests that the microfilaments of the epithelial cells play a crucial role in enabling movement of cells to cover the denuded epithelial surface (Silen and Ito, 1985; Silen, 1987). Restitution does not appear to be involved with prostaglandins as inhibition of prostaglandin synthesis by indomethacin or administration of prostaglandins has no influence on the speed or completeness of restitution (Silen and Ito, 1985; Silen, 1987).

Since the surface cells of the gastroduodenal epithelium would be the first to undergo damage when confronted with a noxious insult, it is likely that restitution plays a major role in mucosal defence under normal physiological conditions. It is not inconceivable that gastric and duodenal damage sustained by ingestion of various foods or from changes in the intraluminal environment, such as thermal, mechanical or hyperosmolar perturbations, is continually occurring, requiring repair of denuded epithelium. In an effort to prevent greater magnitudes of injury, resurfacing of these epithelial defects by restitution probably occurs physiologically to insure mucosal integrity.

3. Inflammatory Cells and Gastroduodenal Injury

A common feature of injurious processes involving the stomach and duodenum is the accumulation of various inflammatory cells, such as neutrophils, eosinophils and macrophages, at the site of injury. In most forms of gastritis as well as frank ulceration within the stomach and duodenum, infiltration of leucocytes, particularly neutrophils, can be demonstrated histologically as comprising an important cellular component of the damaged epithelium. While such findings have been recognized for decades, the role of these inflammatory cells in gastroduodenal injury and the mechanisms by which they are recruited to the damaged area are only now becoming more fully recognized.

The preponderance of research to date has focused on the role that neutrophils may play in various types of gastric inury. Generally, such experiments can be categorized into models of injury in which gastric damage has been evoked by a topically applied noxious agent, or models of injury in which damage has been induced by some form of ischaemic insult to the stomach followed by restoration of circulating blood flow. With respect to the former type of injury, Wallace and colleagues (1990a) noted that gastric injury in the rat stomach induced by the NSAIDs indomethacin and naproxen could be effectively prevented by making the animal neutropenic prior to treatment with these injurious agents. Neutropenia elicited with a specific antibody directed against rat neutrophils, as well as neutropenia induced by the chemotherapeutic agent methotrexate was equally effective in reducing the susceptibility to gastric damage induced by these anti-inflammatory drugs. In an effort to identify the possible mechanisms by which neutrophils contribute to gastric injury induced by indomethacin and naproxen, further studies using monastral blue (a dye used to determine vascular fragility) demonstrated that vascular endothelial injury preceded the development of the haemorrhagic erosions and that such changes in vascular integrity were not observed in neutropenic rats. Although neutrophils are known to release various lipid mediators, such as leukotrienes and PAF, that can directly affect vascular tone and permeability as well as promote adhesion of neutrophils to the endothelium (Bjork *et al.*, 1983; Wallace and Whittle, 1986; Pihan *et al.*, 1988; Vercellotti *et al.*, 1988; Wallace *et al.*, 1990c), neither of these mediators appeared to play an important role in the indomethacin model of gastric injury used in this study, as shown by the observation that neither a specific leukotriene antagonist nor PAF antagonist significantly altered the susceptibility of the stomach to indomethacin-induced gastric damage. Further, inhibition of PG synthesis by more than 95% failed to reverse the protective effect of neutropenia, eliminating prostanoids as a possible mediator for the protective role of neutropenia.

In addition to these findings with indomethacin and naproxen, neutrophils have been similarly implicated as contributing to gastric damage induced by aspirin and ethanol. Following exposure of the rat gastric mucosa to aspirin, Kitahora and Guth (1987) noted that the application of aspirin to the rat stomach resulted in the appearance of white thrombi in the mucosal microcirculation, subsequent to which there was a slowing of mucosal blood flow. In such areas, haemorrhagic erosions subsequently formed, suggesting that neutrophil margination and activation might have given rise to this lesion formation. That neutrophil aggregates can contribute to vascular engorgement and a resultant hypoperfusion of the gastric mucosa was previously shown by Wallace and Whittle (1986) following the administration of PAF. In other studies by Kvietys and colleagues (1990), perfusion of rat gastric mucosa with progressively increasing concentrations of ethanol (10–30%) resulted in a dose-dependent increase in ^{51}Cr-EDTA clearance from blood to gastric lumen as an index of gastric injury. Rendering the animals neutropenic by the use of an antineutrophil serum greatly attenuated the ethanol-induced injury in a fashion that was directly related to the severity of the neutropenia. To assess further the pro-inflammatory effects of ethanol, an *in vitro* model was employed by these investigators (Kvietys *et al.*, 1990), using cultured bovine microvascular endothelial cells and isolated human neutrophils. These cultured cells were then exposed to varying concentrations of ethanol in which a dose-related increase in neutrophil adherence to endothelial cells was observed that correlated with subsequent endothelial cell injury. It was concluded from both of these studies that ethanol was a pro-inflammatory substance that induces gastric injury through a neutrophil-mediated process.

Although the precise events by which neutrophils initiate gastric injury induced by topical damaging agents remains unknown, a considerable body of evidence suggests that oxygen-derived free radicals may be involved. This is not surprising since phagocytic leucocytes, including neutrophils, eosinophils and macrophages, are known to contain a NADPH oxidase on their cellular membranes that could generate large amounts of super-oxide anions (Grisham and Granger, 1988). While the cellular toxicity attributed to this species of radical is thought to be limited, its dismutation to hydrogen peroxide, and the resultant formation of hydroxyl radicals by the interaction of superoxide anions and hydrogen peroxide in the presence of iron creates a potent oxidizing agent that can induce profound cellular injury (Grisham and Granger, 1988). The fact that gastric mucosae of animals and humans contain substantial amounts of SOD, catalase and glutathione peroxidase (Grisham and Granger, 1988), enzymes that are directly involved in the catabolism of oxygen radicals, implies that the gastric epithelium is exposed frequently enough to the formation of these toxic species that the existence of such enzyme systems is required under normal physiological conditions to protect the gastric mucosa from the oxygen radical-derived damage that may occur under circumstances in which inflammatory injury to the gastric epithelium is generated. In support of this contention, Vaananen *et al.* (1991) examined the role that oxygen radicals might play in acute gastric injury induced by indomethacin by measuring the blood-to-lumen leakage of ^{51}Cr-EDTA, as well as the extent of formation of macroscopically visible haemorrhagic lesions. Exposure of the stomach to indomethacin caused a four-fold increase in leakage of ^{51}Cr-EDTA (an index of damage) when compared with corresponding controls. Subsequent exposure of the stomach to acid resulted in a further increase in such leakage. Treatment with the oxygen radical-scavenging agents SOD (scavenges the superoxide anion), catalase (scavenges hydrogen peroxide) and deferoxamine (an iron chelator) significantly reduced this leakage during both indomethacin exposure and the subsequent exposure to acid. Further studies in which PGE$_2$ was given prior to indomethacin exposure also reduced the ^{51}Cr-EDTA leakage, but only during the recovery period in which acid alone was in contact with the mucosa. In contrast to these findings, allopurinol, an agent known to block the activity of the enzyme xanthine oxidase, which can also generate superoxide production, failed to reduce the ^{51}Cr-EDTA leakage at any time during the experiment. In addition to reducing the leakage of this substance into the gastric lumen, SOD, catalase and PGE$_2$ also significantly reduced the extent of macroscopic lesion formation. These findings, in conjunction with the earlier studies of this group of investigators (Wallace *et al.*, 1990a) in which neutro-penia was found to be protective against indomethacin injury, support the hypothesis that oxygen free radicals, probably derived from neutrophils, contributed to the pathogenesis of indomethacin-induced damage.

The role that oxygen radicals may play in neutrophil-mediated ethanol injury is less certain. Since Kvietys and associates (1990) were unable to demonstrate any protective effects of SOD, catalase or sodium benzoate (a scavenger of the hydroxyl radical) against ethanol-induced injury, even though neutropenia was clearly protective, it was concluded that superoxide anions, hydrogen peroxide, and hydroxyl radicals were not involved in the injurious process. Interestingly, these investigators did observe that DMSO, another known hydroxyl radical scavenger, offered protection against ethanol damage, but, based on the other findings with oxygen radical-scavenging agents, they concluded that the protective effects of DMSO may be mediated through some other mechanism such as inhibition of neutrophil chemotaxis (Antony *et al.*, 1983), adherence (Sekizuka *et al.*, 1989), bacterial killing (Repine *et al.*, 1981), or superoxide production (Beilke *et al.*, 1987), actions which have all been proported to occur by DMSO. Instead, they postulated that proteases released by neutrophils may be responsible for the ethanol injury. Other investigators, however, have reported that oxygen radicals may indeed be involved with ethanol injury. In support of this possibility are the findings that allopurinol (an inhibitor of xanthine oxidase) as well as other oxyradical-scaveng-ing agents including SOD, catalase, sodium benzoate and dimethylthiourea (a scavenger of the hydroxyl radical) can all ameliorate ethanol-induced mucosal injury (Evangelista and Meli, 1985; Pihan *et al.*, 1987; Terano *et al.*, 1989; Smith *et al.*, 1992). It seems clear, therefore, that additional investigation is needed before the true role that oxygen radicals play in gastric injury induced by topical damaging agents, particularly ethanol, can be delineated definitely.

Neutrophils have also been implicated in an ischaemia–reperfusion model of gastric injury (Smith *et al.*, 1987b). Rats subjected to a 30 min ischaemic period (haemorrhage to 27 mmHg mean arterial pressure), and a 60 min reper-fusion period (reinfusion of shed blood), demonstrated a dramatic rise in the leakage of ^{51}Cr-labelled red blood cells into the gastric lumen during the reperfusion period. Pretreatment with a neutrophil antiserum greatly attenuated this red blood cell flux. Additional studies using radioactive microspheres to measure mucosal blood flow demonstrated that in neutrophil-depleted animals higher blood flows were achieved in the ischaemic period than was observed in untreated rats. Bleeding of untreated rats to a mean arterial blood pressure of 40 mmHg resulted in blood flows that were not significantly different from those in antiserum-treated rats in which blood pressure was decreased to 27 mmHg; red blood cell leakage into the gastric lumen was equivalent in these two different shock situations. It was concluded from these observations that neutrophils play an important role in haemorrhagic shock-induced mucosal bleeding and that a major mechanism by which neutrophils may increase damage is by increasing microvascular resistance and thereby exacerbating tissue hypoxia (Smith *et al.*, 1987b).

That oxygen radicals may be the means by which these neutrophil effects are elicited is supported by other studies from this same group of investigators in which ischaemic injury induced in both the rat (Smith *et al.*, 1987a) and cat (Perry *et al.*, 1986) stomachs were shown to be markedly attenuated by various oxygen radical-scavenging agents. In addition, evidence was also presented in these studies that a xanthine oxidase-dependent mechanism of oxygen radical generation was at least in part responsible for generating this injury because a sodium tungstate diet (which inactivates xanthine oxidase), and allopurinol (an agent that blocks xanthine oxidase activity), were also able to prevent gastric damage under these conditions (Perry *et al.*, 1986; Smith *et al.*, 1987a). It was proposed that the sequence of events leading to oxygen radical formation was initiated by ischaemia which leads to hypoxanthine accumulation (due to ATP catabolism) and conversion of the NAD^+-reducing enzyme xanthine dehydrogenase to xanthine oxidase. Upon reperfusion of ischaemic tissue, the newly formed xanthine oxidase then utilizes hypoxanthine and oxygen as a substrate to form the superoxide anion. Other studies by Itoh and Guth (1985), using a similar model of haemorrhagic shock-induced gastric injury, provide additional evidence that a xanthine oxidase-mediated mechanism of oxygen radical formation is probably involved.

Despite the role that oxygen radicals may play in gastric injury, whether mediated via a neutrophil mechanism or by xanthine oxidase, the importance of oxyradicals in the pathogenesis of acid–peptic disorders in the human stomach remains unknown. In a clinical study (Pascu and Dejica, 1987) involving 12 patients with duodenal ulcer disease, in which healing was resistant to conventional therapy, direct injection of oxyradical-scavenging agents into the tissue surrounding the ulcer was undertaken endoscopically. The injectable substance included a mixture of human SOD and catalase. Four patients were dramatically improved, five were partially relieved of symptoms, and three needed surgical intervention to ultimately correct their ulcer diathesis. It was concluded from this study that free radicals may play a role in causing the pain of duodenal ulcers. While these preliminary human results are interesting, it is obvious that much further work is required to determine whether experimental findings in animals, in which oxygen radicals appear to play a role in gastroduodenal injury, can be directly extrapolated to the human situation.

A number of studies in recent years have suggested that chronic gastric inflammation and peptic ulcer disease in humans may be caused by the bacterium *Helicobacter pylori* (formerly *Campylobacter pylori*) (Graham, 1989; Barthel and Everett, 1990; Buck, 1990; Marshall, 1990; Peterson, 1991). This contention is based on the observation that *H. pylori* can be isolated from the mucosa of patients with gastritis and peptic ulcer disease and that such findings disappear when the underlying gastro-duodenal conditions are successfully treated. While such correlations do in fact exist in many patients, a high prevalence rate of *H. pylori* has been noted in asymptomatic patients and a consistent correlation between the presence of infection and evidence of abnormal mucosal pathology has not always been demonstrated. Thus, much controversy presently exists regarding the role that *H. pylori* does in fact play in gastroduodenal disease. Although this bacterium is implicated as a potentially important pathophysiological mediator of gastroduodenal disease, the mechanism by which this agent induces injury to the epithelium remains unknown. Interestingly, a number of studies have shown an intense neutrophilic infiltration in mucosae with *H. pylori*-associated disease (Graham, 1989; Kazol *et al.*, 1991). In view of the aforementioned studies, it is possible that the damage induced by this bacterium may be mediated by toxic oxygen radicals. Obviously, much additional investigation will be required to confirm this speculation.

In contrast to the damaging effects of neutrophil-derived products, IL-1, a family of proteins that are potent mediators of inflammation and are released by macrophages (Dinarello, 1988), appears to be protective against gastroduodenal injury. In rat models of gastric injury induced by indomethacin or ethanol (Wallace *et al.*, 1990b; Robert *et al.*, 1991) and duodenal injury induced by cysteamine (Wallace *et al.*, 1990b), IL-1 was shown to significantly attenuate the damaging effects of these various noxious agents. Because the protective action of IL-1 against the gastric ethanol injury could be reversed by indomethacin pretreatment, a role for endogenous PGs as mediators of protection was suggested. Direct inhibition of gastric acid secretion may also play a role in the protective action of interleukin as IL-1 has been shown to inhibit pentagastrin-stimulated acid secretion (Wallace *et al.*, 1991). The significance of these findings with IL-1 remains uncertain. As is true in other inflammatory processes, these observations are consistent with the notion that the degree of gastric or duodenal injury in a given situation is dependent upon a balance between those mediators released by inflammatory cells that have a primarily destructive action and those that are predominantly protective.

4. *Prostaglandins and Gastroduodenal Defence*

Although a substantial body of literature strongly suggests that the aforementioned protective processes play important roles in defending the epithelium of the stomach and duodenum against injury, the mediator(s) that modulates these protective mechanisms has yet to be defined. It is entirely conceivable that more than one mediator may participate in this role, each functioning under a different set of physiological circumstances. Of all the mediators that have been implicated as playing such a role, PGs have received the most attention (Robert, 1981; Miller, 1983; Robert, 1984; Hawkey and Rampton, 1985). This relates to the large number of studies that have been performed over the last decade or more in which these agents have been shown to prevent or markedly attenuate the damaging effects of almost a limitless number of injurious insults affecting the stomach and duodenum, and at least some evidence, either direct or indirect, that they may be involved endogenously in influencing whether protection or injury occurs in a given experimental situation. This unique protective action of PGs, independent of effects on acid secretion, has been termed "cytoprotection". A review of information relative to the potential role of PGs in gastroduodenal defence is germane to this discussion.

The remarkable ability of PGs to protect the gastroduodenal mucosa against a wide variety of damaging insults, even those induced by 100% ethanol, boiling water and exposure of the epithelium to concentrated acid or alkali, has suggested that they might be mediating their protective effects by preventing common pathogenetic mechanisms of injury (Robert, 1981; Miller, 1983; Robert, 1984; Hawkey and Rampton, 1985). Support for this contention becomes more compelling when it is realized that the major mechanisms thought to be responsible for gastroduodenal defence, namely mucus–bicarbonate secretion, enhancement or maintenance of gastric mucosal perfusion, and enhancement of epithelial renewal (i.e. trophic action (Helander *et al.*, 1985)), have all been shown to be influenced by various PGs, particularly PGE_2 and PGI_2. Further, the finding that PGs of the E, F and I types exist in substantial concentrations in both the gastric and duodenal mucosa of all species studied, including humans, suggest that they have an important function endogenously (Robert, 1981; Miller, 1983; Robert, 1984; Hawkey and Rampton, 1985).

Evidence supporting a role for PGs in gastroduodenal defence has generally centred around three considerations. First, agents known to inhibit cyclo-oxygenase, the primary enzyme system responsible for PG synthesis, also have the capability of eliciting mucosal damage, especially in the stomach. These agents can generally be grouped under the broad heading of NSAIDs, aspirin and indomethacin being particularly noteworthy examples. Interestingly, pretreatment with exogenous PGs concomitant with, or before the administration of, these damaging agents has been shown experimentally to prevent injury by these substances which would normally occur otherwise (Robert, 1981; Miller, 1983; Robert, 1984; Hawkey and Rampton, 1985). The relationship, however, between the damaging effects of these drugs and their consequent effects on PG synthesis is far from tightly linked. Ligumsky and associates (1983) demonstrated that gastric PG synthesis in the rat stomach could be inhibited by up to 95% by aspirin without the development of frank lesion formation. Findings of this nature suggest that aspirin-induced injury, when it occurs, is not solely a process mediated by inhibition of PG synthesis. Conversely, gastric erosions induced experimentally by various forms of stress such as cold restraint and water immersion seem to be associated with reduced mucosal synthesis of PGs (Robert, 1981). Further, several human studies have implicated alterations in endogenous PGs as playing an important role in acid–peptic disease, particularly that involving the duodenum (Robert, 1981; Miller, 1983; Robert, 1984; Hawkey and Rampton, 1985).

The phenomenon of adaptive cytoprotection has also been put forward as evidence that endogenous PGs play an important role in mucosal defence (Robert, 1981; Miller, 1983; Robert, 1984; Hawkey and Rampton, 1985). This process refers to the ability of low doses of topical damaging agents, referred to as mild irritants, that normally do not injure the gastric epithelium, to protect the stomach against subsequent exposure to a higher concentration of the same agent or even differing agents. Studies in rats, orally treated with low concentrations of hydrochloric acid, ethanol or taurocholate, were shown to protect the gastric mucosa against the damaging effects of orally administered, acidified 80 mM taurocholate, which by itself usually damages the gastric epithelium. Because indomethacin, an inhibitor of PG synthesis, abolished this protective effect, and various mild irritants themselves were shown to stimulate PG synthesis under certain circumstances, it was concluded that the protective capabilities of mild irritants were elicited by endogenous PGs. Observations of this nature were taken to indicate that various irritants within a person's normal diet might stimulate PG synthesis and thereby set into motion responsible protective processes (e.g. mucus–bicarbonate secretion, enhanced mucosal blood flow, etc.) to prevent gastric injury. Despite the appealing nature of this hypothesis, several recent studies have failed to confirm that indomethacin does in fact block adaptive cytoprotection and that mild irritants known to prevent gastric injury correspondingly always stimulate PG synthesis (Hawkey *et al.*, 1988; Smith *et al.*, 1991). A recent review has summarized the experimental support for other mediators of adaptive cytoprotection besides PGs (Kauffman, 1991).

A third line of evidence is the observation that specific receptor sites have been identified in the oxyntic mucosa of porcine epithelium to which PGs bind. Such receptor–ligand interaction was shown to influence various cellular processes within that tissue, lending credence to the hypothesis that PGs may in fact play an important role in the gastric epithelium under physiological conditions (Tepperman and Soper, 1981, 1983).

Several models of gastric injury, especially those induced by aspirin and ethanol, appear to have as their target the vascular endothelium (Robins, 1980; Guth *et al.*, 1984; Szabo *et al.*, 1985; Kitahora and Guth, 1987; Konturek *et al.*, 1988). The morphological expression of such injury is haemorrhage in the lamina propria and gastric gland region which macroscopically manifests itself as red streaks. Numerous studies have indicated that PGs can consistently prevent such haemorrhage, suggesting that their protective mode of action is at the level of the microcirculation (Lacy and Ito, 1982; Wallace *et al.*, 1982; Guth *et al.*, 1984; Robert, 1984; Schmidt *et al.*, 1985, 1986). Of equal note, PGs have also been shown to inhibit activation of neutrophils (Wong and Freund, 1980) as well as to prevent the release of superoxide anions from leucocytes (Simpkins *et al.*, 1986; Gryglewski *et al.*, 1987). Since oxyradical formation from neutrophil infiltration is thought to primarily have as its target the vasculature, these effects of PGs provide further evidence that their mode of action may be to specifically prevent vascular injury. The observation that NSAID damage is associated with neutrophil infiltration, and that such substances also inhibit PG synthesis, raises the interesting possibility that the mechanism underlying NSAID damage is merely a tip in the balance of the protective action of PGs on the one hand and the adverse effect resulting from the accumulation of neutrophils at the site of injury on the other hand. Further studies are obviously required to confirm these speculations.

Unfortunately, much of the published information regarding the protective action of PGs against gastroduo-denal injury has assessed protection in terms of the absence or reduction in macroscopically visible necrotic lesions (Miller, 1983). Accordingly, if a given damaging agent induced formation of lesions macroscopically and pretreatment with a given PG prevented such formation, the PG being tested was said to be protective. Recent studies have clearly demonstrated that what might be assumed to be protection macroscopically may not at all correspond to what is actually observed at the microscopic level (Lacy and Ito, 1982; Wallace *et al.*, 1982; Guth *et al.*, 1984; Robert, 1984; Schmidt *et al.*, 1985, 1986). Carefully conducted studies have actually indicated that,

under most circumstances, PG protection is primarily limited to the deeper layers of the mucosa rather than to the surface epithelium. Further, some studies have shown that restitution of damaged surface epithelium does not appear to be influenced by PGs (Silen, 1987). Such findings have been interpreted by some investigators as minimizing the role that PGs may in fact play in gastroduodenal defence (Silen, 1987, 1988). A contrary view, however, is that these observations are consistent with available information, implicating a protective role of PGs primarily against vascular injury. This may in fact be their most important function. Thus, further studies investigating the protective action of PGs should focus more on how PGs protect the microcirculation from injury rather than being preoccupied with the issue of whether they consistently protect all layers of the gastric epithelium from damage. Even though full-thickness protection of the epithelium does not routinely occur with PG treatment (Schmidt and Miller, 1991), this deficiency by no means mitigates against their remarkable ability to reduce the extent of damage by a limitless number of insults that would otherwise evoke profound injury.

5. *Summary*

The gastroduodenal epithelium is regularly exposed to high concentrations of acid but, remarkably, resists autodigestion. The means by which such mucosal integrity is maintained in this acid milieu remain to be defined, but current knowledge would indicate that mucus, bicarbonate, continuous epithelial renewal, and micro-circulatory responses play important roles in defending the epithelium of the stomach and duodenum against injury. The observation that neutrophils are a major component of injury involving the epithelium of these two organs suggests that oxygen-derived free radicals may be an important feature of gastroduodenal injury, since these leucocytes are known to commonly generate the superoxide anion. The precise means by which known protective processes mediate mucosal defence, and whether each is of equal importance, remains uncertain but available data would strongly suggest that perturbations in the microcirculation are an essentially significant antecedent event for most forms of gastroduodenal injury. Since oxygen radicals similarly have as their target the vasculature, understanding mechanisms of injury and protection would seem to be directly linked to understanding those factors that insure adequate perfusion of the gastroduodenal epithelium. The precise role that PGs play in gastroduodenal defence remains uncertain. The presence of these substances in substantial quantities in the mucosa of the stomach and duodenum endogenously, and their ability to stimulate the processes known to be protective against injury, suggest that they may be of extreme importance in maintaining mucosal integrity.

Further study is clearly needed, however, to fully delineate the role that these fatty acids do in fact play in enabling the gastroduodenal mucosa to resist injury.

6. *Acknowledgements*

Supported in part by grant DK 25838 from the National Institutes of Health. The authors gratefully acknowledge the superb secretarial assistance of Ms Inci Akkaya in the preparation of this manuscript.

7. *References*

Allen, A. (1981). In "Physiology of the Gastrointestinal Tract" (ed. L. R. Johnson), pp 617–639, Raven Press, New York.

Antony, V.B., Sahn, S.A. and Repine, J.E. (1983). Dimethylsulfoxide inhibits phagocyte influx into infected pleural spaces and phagocyte locomotion *in vitro*. Inflammation 7, 377–385.

Barthel, J.S. and Everett, E.D. (1990). *Campylobacter pylori* infections: the "gold standard" and the alternatives. Rev. Infect. Dis. 12(Suppl. 1), S107–S114.

Beilke, M.A., Collins-Leck, C. and Sohnle, P.G. (1987). Effects of dimethylsulfoxide on the oxidative function of human neutrophils. J. Lab. Clin. Med. 110, 91–96.

Bjork, J., Lindbom, L., Gerdin, B., Smedegard, G., Arfors, K. and Benveniste, J. (1983). Pafacether (platelet-activating factor) increases microvascular permeability and affects epithelium–granulocyte interaction in microvascular beds. Acta Physiol. Scand. 119, 305–308.

Boeckxstaens, G.E., Pelckmans, P.A., Bogers, J.J., Bult, H., De Man, J.G., Oosterbosch, L., Herman, A.G. and Maercke, Y.M. (1991). Release of nitric oxide upon stimulation of nonadrenergic noncholinergic nerves in the rat gastric fundus. J. Pharmacol. Exp. Ther. 2, 441–447.

Buck, G.E. (1990). *Campylobacter pylori* and gastrointestinal disease. Clin. Microbiol. Rev. 3, 1–12.

Buck, S.H. and Burks, T.F. (1986). The neuropharmacology of capsaicin: Review of some recent observations. Pharmacol. Rev. 38, 179–226.

Cheung, L.Y. and Porterfield, G. (1979). Protection of gastric mucosa against acute ulceration by intravenous infusion of sodium bicarbonate. Am. J. Surg. 7, 106–110.

Cheung, L.Y., Toenjes, A.A. and Sonnenschien, L.A. (1982). Acidification of arterial blood enhances gastric mucosal injury induced by bile salts in dogs. Am. J. Surg. 143, 74–78.

Cummins, G.M., Grossman, M.I. and Ivy, A.C. (1948). An experimental study of the acid factor in ulceration of the gastrointestinal tract in dogs. Gastroenterology 10, 714–726.

Davies, R.E. and Longmuir, N.M. (1948). Production of ulcers in isolated frog gastric mucosa. Biochem. J. 42, 621–627.

Dinarello, C.A. (1988). Interleukin-1. Dig. Dis. Sci. 33(Suppl.), 25S–35S.

Eastwood, G.L. (1984). In "Mechanisms of Mucosal Protection in the Upper Gastrointestinal Tract" (eds A. Allen, G. Flemstrom, A. Garner *et al.*), pp 27–32. Raven Press, New York.

Evangelista, S. and Meli, A. (1985). Influence of antioxidants and radical scavengers on ethanol-induced gastric ulcers in the rat. Gen. Pharm. 16, 285–286.

Flemstrom, G. (1987). In "Physiology of the Gastrointestinal Tract", 2nd Edn (ed. L.R. Johnson), pp 1011–1029. Raven Press, New York.

Flemstrom, G. and Garner, A. (1982). Gastroduodenal HCO_3^- transport: characteristics and proposed role in acidity regulation and mucosal protection. Am. J. Physiol. 242, G183–G193.

Flemstrom, G. and Turnberg, L.A. (1984). Gastroduodenal defense mechanisms. Clin. Gastroenterol. 13, 327–354.

Folkman, J. (1985). Toward an understanding of "angiogenesis": search and discovery. Perspect. Biol. Med. 29, 10–36.

Folkman, J. (1986). How is blood vessel growth regulated in normal and neoplastic tissue? Cancer Res. 46, 467–473.

Folkman, J. (1990). What is the evidence that tumors are 'angiogenesis' dependent? J. Natl Cancer Inst. 82, 4–6.

Folkman, J. and Klagsburn, M. (1987). Angiogenic factors. Science 5, 442–447.

Folkman, J., Szabo, S. and Shing, Y. (1990). Sucralfate affinity for fibroblast growth factor. J. Cell Biol. 111, 223A.

Folkman, J., Szabo, S., Stovroff, M., McNeil, P., Li, W. and Shing, U. (1991). Duodenal ulcer: discovery of a new mechanism and development of angiogenic therapy that accelerates healing. Ann. Surg. 214, 414–425.

Gannon, G., Browning, J. and O'Brien, P. (1982). The microvascular architecture of the glandular mucosa of rat stomach. J. Anat. 135, 667–683

Gannon, G., Browning, J., O'Brien, P. and Rogers, P. (1984). Mucosal microvascular architecture of the fundus and body of human stomach. Gastroenterology 86, 866–875.

Goddard, P.J., Hills, B.A., Lichtenberger, L.M. (1987). Does aspirin damage canine gastric mucosa by reducing its surface hydrophobicity? Am J. Physiol. 15, G421–G430.

Goddard, P.J., Kao, Y.-C.J. and Lichtenberger, L.M. (1990). Luminal surface hydrophobicity of canine gastric mucosa is dependent on a surface mucous gel. Gastroenterology 98, 361–370.

Graham, D.Y. (1989). *Campylobacter pylori* and peptic ulcer disease. Gastroenterology, 96, 615–625.

Green, T. and Dockray, G.J. (1988). Characterization of the peptidergic afferent innervation of the stomach in the rat, mouse, and guinea-pig. Neuroscience 25, 181–193.

Grisham, M.B. and Granger, D.N. (1988). Neutrophil-mediated mucosal injury: role of reactive oxygen metabolites. Dig. Dis. Sci. 33, 6S-15S.

Gryglewski, R.J., Szczeklik, A. and Wandzilak, M. (1987). The effect of six prostaglandins, prostacyclin and iloprost on generation of superoxide anions by human polymorphonuclear leukocytes stimulated by zymozan or formyl-methionylleucyl-phenylalanine. Biochem. Pharmacol. 36, 4209–4212.

Guth, P.H. (1984). In "Mechanisms of Mucosal Protection in the Upper Gastrointestinal Tract" (eds A. Allen, G. Flemstrom, A. Garner *et al.*), pp 253–358. Raven Press, New York.

Guth, P.H., Paulsen, G. and Nagata, H. (1984). Histologic and microcirculatory changes in alcohol-induced gastric lesions in the rat: Effect of prostaglandin cytoprotection. Gastroenterology 87, 1083–1090.

Hawkey, C.J. and Rampton, D.S. (1985). Prostaglandins and the gastrointestinal mucosa: are they important in its function, disease, or treatment? Gastroenterology 89, 1162–1188.

Hawkey, C.J., Kemp, R.T., Walt, R.P., Bhaskar, N.K., Davies, J., Filipowicz, B. (1988). Evidence that adaptive cytoprotection in rats is not mediated by prostaglandins. Gastroenterology 94, 948–954.

Heatley, N.G. (1959). Mucosubstance as a barrier to diffusion. Gastroenterology 37, 313–317.

Helander, H.F., Johansson, C., Blom, H. and Uribe, A. (1985). Trophic actions of E_2 prostaglandins in the rat gastrointestinal mucosa. Gastroenterology, 89, 1393–1399.

Henagan, J.M., Smith, G.S., Schmidt, K.L. and Miller, T.A. (1986). N-Acetyl-cysteine and prostaglandin: comparable protection against experimental ethanol injury in the stomach independent of mucus thickness. Ann. Surg. 204, 698–704.

Hollander, D. and Tarnawski, A. (1989). In "New Pharmacology of Ulcer Disease: Experimental and New Therapeutic Approaches" (eds S. Szabo and G. Mozsik), pp 404–412. Elsevier, New York.

Holzer, P. and Lippe, I.Th. (1988). Stimulation of afferent nerve endings by intragastric capsaicin protects against ethanol-induced damage of gastric mucosa. Neuroscience 27, 981–987.

Holzer, P. and Sametz, W. (1986). Gastric mucosal protection against ulcerogenic factors in the rat mediated by capsaicin-sensitive afferent neurons. Gastroenterology 91, 975–981.

Holzer, P., Pabst, M.A. and Lippe, I.Th. (1989). Intragastric capsaicin protects against aspirin-induced lesion formation and bleeding in the rat gastric mucosa. Gastroenterology 96, 1425–1433

Itoh, M. and Guth, P.H. (1985). Role of oxygen-derived free radicals in hemorrhagic shock-induced gastric lesions in the rat. Gastroenterology 88, 1162–1167.

Johnson, L.R. (1977). New aspects of the trophic action of gastrointestinal hormones. Gastroenterology 72, 788–792.

Johnson, L.R. and Guthrie, P.D. (1980). Stimulation of rat oxyntic gland mucosal growth by epidermal growth factor. Am. J. Physiol. 238, G45–G49.

Kauffman, G.L. (1991). Putative mediator(s) of adaptive cytoprotection? Prostaglandins 41, 201–205.

Kitahora, T. and Guth, P.H. (1987). Effect of aspirin plus hydrochloric acid on the gastric mucosal microcirculation. Gastroenterology 93, 810–817.

Kivilaakso, E., Fromm, D. and Silen, W. (1978a). Effect of the acid secretory state on intramural pH of rabbit gastric mucosa. Gastroenterology 75, 641–648.

Kivilaakso, E., Fromm, D. and Silen, W. (1978b). Relationship between ulceration and intramural pH of gastric mucosa during hemorrhagic shock. Surgery 84, 70–77.

Konturek, S.J., Radecki, T., Brzozowski, T., Piastucki, I., Dembinski, A., Dembinskià-Kiec, A., Zmuda, A., Gryglewski, R. and Gregory, H. (1981). Gastric cytoprotection by epidermal growth factor. Gastroenterology 81, 438–443.

Konturek, S.J., Brzozowski, T., Drozdowicz, D. and Beck, G. (1988). Role of leukotrienes in acute gastric lesions induced by ethanol, taurocholate, aspirin, platelet-activating factor and stress in rats. Dig. Dis. Sci. 33, 806–813.

Kozol, R., Domanowski, A., Jaszewski, R., Czanko, R., McCurdy, B., Prasad, M., Fromm, B. and Calzada, R. (1991). Neutrophil chemotaxis in gastric mucosa: a signal-to-response comparison. Dig. Dis. Sci. 36, 1277–1280.

Kvietys, P.R., Twohig, B., Danzell, J. and Specian, R.D. (1990). Ethanol-induced injury to the rat gastric mucosa: role of neutrophils and xanthine oxidase-derived radicals. Gastroenterology 98, 909–920.

Lacy, E.R. and Ito, S. (1982). Microscopic analysis of ethanol damage to rat gastric mucosa after treatment with a prostaglandin. Gastroenterology 83, 619–625.

Leung, F.W., Itoh, M., Hirabayashi, K. and Guth, P.H. (1985). Role of blood flow in gastric and duodenal mucosal injury in the rat. Gastroenterology 88, 281–289.

Lichtenberger, L.M., Graziani, L.A., Dial, E.J., Butler, B.D. and Hills, B.A. (1983). Role of surface-active phospholipids in gastric cytoprotection. Science 219, 1327–1329.

Lichtenberger, L.M., Richards, J.E. and Hills, B.A. (1985). Effect of 16,16–dimethyl prostaglandin E2 on the surface hydrophobicity of aspirin-treated canine gastric mucosa. Gastroenterology 88, 308–314.

Ligumsky, M., Golanska, E.M., Hansen, D.G. and Kauffman, G.L. (1983). Aspirin can inhibit gastric mucosal cyclo-oxygenase without causing lesions in the rat. Gastroenterology 84, 756–761.

Lillehei, C.W., Dixon, J.L. and Wangensteen, O.H. (1948). Relation of anemia and hemorrhagic shock to experimental ulcer production. Proc. Soc. Exp. Biol. Med. 68, 125–128.

Limlomwongse, L., Chaitauchawong, C. and Tongyai, S. (1979). Effect of capsaicin on gastric acid secretion and mucosal blood flow in the rat. J. Nutr. 109, 773–777.

Lippe, I.Th., Pabst, M.A. and Holzer, P. (1989). Intragastric capsaicin enhances rat gastric acid elimination and mucosal blood flow by afferent nerve stimulation. Br. J. Pharmacol. 96, 91–100.

MacNaughton, W.K., Cirino, G. and Wallace, J.L. (1989). Endothelium-derived relaxing factor (nitric oxide) has protective actions in the stomach. Life Sci. 45, 1869–1876.

Manabe, T., Suzuki, T. and Honjo, I. (1977). Changes of upper gastrointestinal blood flow after hemorrhage in rabbits. Surgery 81, 446–452.

Marshall, B.J. (1990). *Campylobacter pylori*: Its link to gastritis and peptic ulcer disease. Rev. Infect. Dis. 12(Suppl. 1), S87–S97).

Menguy, R. (1981). Role of gastric mucosal energy metabolism in the etiology of stress ulceration. World J. Surg. 5, 175–180.

Miller, T.A. (1983). Protective effects of prostaglandins against gastric mucosal damage: current knowledge and proposed mechanisms. Am. J. Physiol. 245, G601–G623.

Nagy, J.I. (1982). In "Handbook of Psychopharmacology", Vol. 15 (eds L.L. Iversen and S.D. Iversen, and S.H. Snyder), pp 185–235. Plenum Press, New York.

Neutra, M.R. and Forstner, J.F. (1987). In "Physiology of the Gastrointestinal Tract", Vol. 2 (ed. L.R. Johnson), pp 975–1009. Raven Press, New York.

Pascu, O. and Dejica, D. (1987). Oxygen free radicals and duodenal ulcer pain. Preliminary data. Rev. Roum. Med. Int. 25, 81–84.

Perry, M.A., Wadhwa, S., Parks, D.A., Pickard, W. and Granger, D.N. (1986). Role of oxygen radicals in ischemia-induced lesions in the cat stomach. Gastroenterology 90, 362–367.

Peterson, W.L. (1991). Helicobacter pylori and peptic ulcer disease. N. Engl. J. Med. 324, 1043–1048.

Pihan, G., Regillo, C. and Szabo, S. (1987). Free radicals and lipid peroxidation in the ethanol- or aspirin-induced gastric mucosal injury. Dig. Dis. Sci. 32, 1395–1401.

Pihan, G., Rogers, C. and Szabo, S. (1988). Vascular injury in acute gastric mucosal damage: mediatory role of leukotrienes. Dig. Dis. Sci. 33, 625–632.

Pique, J.M., Whittle, B.J.R. and Esplugues, J.V. (1989). The vasodilator role of endogenous nitric oxide in the rat gastric microcirculation. Eur. J. Pharmacol. 174, 293–296.

Repine, J.E., Fox, R.B. and Berger, E.M. (1981). Dimethylsulf-oxide inhibits killing of *Staphylococcus aureus* by polymorphonuclear leukocytes. Infect. Immunol. 31, 510–513.

Ritchie, W.P., Jr (1983). Pathogenesis of acute gastric mucosal injury. View. Dig. Dis. 15, 17–20.

Robert, A. (1981). In "Physiology of the Gastrointestinal Tract" (ed. L.R. Johnson), pp 1407–1434. Raven Press, New York.

Robert, A. (1984). In "Mechanisms of Mucosal Protection in the Upper Gastrointestinal Tract" (eds A. Allen, G. Flemstrom, A. Garner *et al.*), pp 377–382. Raven Press, New York.

Robert, A., Olafsson, A.S., Lancaster, C. and Zhary, W. (1991). Interleukin-1 is cytoprotective, antisecretory, stimulates PGE_2 synthesis by the stomach, and retards gastric emptying. Life Sci. 48, 123–134.

Robins, P.G. (1980). Ultrastructual observations on the pathogenesis of aspirin-induced gastric erosions. Br. J. Exp. Pathol. 61, 497–504.

Russell, L.C. and Burchiel, K.J. (1984). Neurophysiological effects of capsaicin. Brain Res. Rev. 8, 165–176.

Sarosiek, J., Slomiany, B.L., Swierczek, J., Slomiany, A., Jozwiak, Z. and Konturek, S.J. (1984). Effect of acetylsalicylic acid on the constituents of the gastric mucosal barrier. Scand. J. Gastroenterol. 19, 150–153.

Schiessel, R., Merhav, A., Matthews, J.B., Fleischer, L.A., Barzilai, A. and Silen, W. (1980). Role of nutrient HCO_3^- in the protection of amphibian gastric mucosa. Am. J. Physiol. 239, G536–G542.

Schmidt, K.L. and Miller, T.A. (1991). Cytoprotection: fact or fancy? The morphologic basis of gastric cytoprotection by prostaglandins. Exp. Clin. Gastroenterol. 1, 119–131.

Schmidt, K.L., Henagan, J.M., Smith, G.S., Hilburn, P.J. and Miller, T.A. (1985). Prostaglandin cytoprotection against ethanol-induced gastric injury in the rat: a histologic and cytologic study. Gastroenterology 88, 649–659.

Schmidt, K.L., Bellard, R.L., Smith, G.S., Henagan, J.M. and Miller, T.A. (1986). Influence of prostaglandin on repair of rat stomach damaged by absolute ethanol. J. Surg. Res. 41, 367–377.

Sekizuka, E., Benoit, J.N., Grisham, M.B. and Granger, D.N. (1989). Dimethysulfoxide prevents chemoattractant-induced leukocyte adherence. Am. J. Physiol. 256, H594–H597.

Sharkey, K.A., Williams, R.G. and Dockray, G.J. (1984). Sensory substance P innervation of the stomach and pancreas: demonstration of capsaicin-sensitive sensory neurons in the rat by combined immunohistochemistry and retrograde tracing. Gastroenterology 87, 914–921.

Silen, W. (1987). In "Physiology of the Gastrointestinal Tract", 2nd edn (ed. L.R. Johnson), pp 1055–1069. Raven Press, New York.

Silen, W. (1988). What is cytoprotection of the gastric mucosa? Gastroenterology 94, 232–235.

Silen, W. and Ito, S. (1985). Mechanisms for rapid re-epithelialization of the gastric mucosal surface. Ann. Rev. Physiol. 47, 217–229.

Silen, W., Merhav, A. and Simson, J.N. (1981). The pathophysiology of stress ulcer disease. World J. Surg. 5, 165–174.

Simpkins, C.O., Alailima, S.T., Tate, E.A. and Johnson, M. (1986). The effect of enkephalins and prostaglandins on O_2^- release by neutrophils. J. Surg. Res. 41, 645–652.

Skinner, K.A. and Tepperman, B.L. (1981). Influence of desalivation on acid secretory output and gastric mucosal integrity in the rat. Gastroenterology 81, 35–39.

Slomiany, B.L., Piasek, A., Sarosiek, J. and Slomiany, A. (1985). The role of surface and intracellular mucus in gastric mucosal protection against hydrogen ion: compositional differences. Scand. J. Gastroenterol. 20, 1191–1196.

Smith, G.S., Myers, S.I., Bartula, L.L. and Miller, T.A. (1991). Adaptive cytoprotection against alcohol injury in the rat stomach is not due to increased prostanoid synthesis. Prostaglandins 42, 207–223.

Smith, G.S., Barreto, J.C., Schmidt, K.L., Tornwall, M.S. and Miller, T.A. (1992). Protective effect of dimethylthiourea against mucosal injury in rat stomach: implications for hydroxyl radical mechanism. Dig. Dis. Sci. 37, 1345–1355.

Smith, P., O'Brien, P., Fromm, D., Silen, W. (1977). Secretory state of gastric mucosa and resistance to injury by exogenous acid. Am. J. Surg. 133, 81–85.

Smith, S.M., Grisham, M.B., Manci, E.A., Granger, D.N. and Kvietys, P.R. (1987a). Gastric mucosal injury in the rat: role of iron and xanthine oxidase. Gastroenterology 92, 950–956.

Smith, S.M., Holm-Rutili, L., Perry, M.A., Grisham, M.B., Arfors, K.E., Granger, D.M. and Kvietys, P.R. (1987b). Role of neutrophils in hemorrhagic shock-induced gastric mucosal injury in the rat. Gastroenterology 93, 466–471.

Solomon, T.E. (1986). Trophic effects of pentagastrin on gastrointestinal tract in fed and fasted rats. Gastroenterology 91, 108–116.

Sternini, C., Reeve, J.R. and Brecha, N. (1987). Distribution and characterization of calcitonin gene-related peptide immunoreactivity in the digestive system of normal and capsaicin-treated rabbits. Gastroenterology 93, 852–862.

Stovroff, M., Vattay, P., Marino, B., Szabo, S. and Folkman, J. (1991). Healing of experimental gastritis by oral fibroblast growth factor. Surg. Forum XLII, 174–175.

Su, H.C., Bishop, A.E., Poer, F.R., Hamada, Y. and Polak, J.M. (1987). Dual intrinsic and extrinsic origins of CGRPP and NPY-immunoreactive nerves of rat gut and pancreas. J. Neurosci. 7, 2674–2687.

Svanes, K., Varhaug, J.E., Dzienis, H. and Gronbech, J.E. (1984). Gastric mucosal blood flow related to acute mucosal damage. Scand. J. Gastroenterol. 19(Suppl. 105), 62–66.

Szabo, S. and Hollander, D. (1989). Pathways of gastrointestinal protection and repair: mechanisms of action of sucralfate. Am. J. Med. 8(Suppl. 6A), 23–31.

Szabo, S., Trier, J.S., Brown, A. and Schnoor, J. (1985). Early vascular injury and increased vascular permeability in gastric mucosal injury caused by ethanol in the rat. Gastroenterology 88, 228–236.

Szabo, S., Vattay, P., Scarborough, E. and Folkman, J. (1991). Role of vascular factors, including angiogenesis, in the mechanisms of action of sucralfate. Am. J. Med. 91(Suppl. 2A), 158S–160S.

Szolcsanyi, J. and Bartho, L. (1981). In "Advances in Physiological Sciences: Gastrointestinal Defense Mechanisms" (eds G. Mozsik, O. Hanninen and T. Javor), pp 39–51. Pergamon Press and Akademiai Kiado, Oxford and Budapest.

Teorell, T. (1951). The acid–base balance of the secreting isolated gastric mucosa. J. Physiol. (Lond.) 114, 267–276.

Tepperman, B.L. and Soper, B.D. (1981). Prostaglandin E2 binding sites and cAMP production in porcine fundic mucosa. Am. J. Physiol. 241, G313–G320.

Tepperman, B.L. and Soper, B.D. (1983). Subcellular distribution of ^{3}H-prostaglandin E_2 binding sites in porcine gastric mucosa. Prostaglandins 25, 425–441.

Terano, A., Hirashi, H., Ota, S., Shiga, J. and Sugimoto, T. (1989). Role of superoxide and hydroxyl radicals in rat gastric mucosal injury induced by ethanol. Gastroenterol. Jpn. 24, 488–493.

Vaananen, P.M., Meddings, J.B. and Wallace, J.L. (1991). Role of oxygen-derived free radicals in indomethacin-induced gastric injury. Am. J. Physiol. 261, G470–G475.

Vercellotti, G.M., Yin, H.Q., Gustafsson, K.D., Nelson, R.D. and Jacob, H.S. (1988). Platelet-activating factor primes neutrophil responses to agonists: role in promoting neutrophil-mediated endothelial damage. Blood 71, 1100–1107.

Wallace, J.L. and Whittle, B.J.R. (1986). Picomole doses of platelet activating factor predispose the gastric mucosa to damage by topical irritants. Prostaglandins 31, 989–998.

Wallace, J.L., Morris, G.P., Krausse, E.J. and Greaves, S.E. (1982). Reduction by cytoprotective agents of ethanol-induced damage to the rat gastric mucosa: a correlated morphologic and physiologic study. Can. J. Physiol. Pharmacol. 60, 1686–1699.

Wallace, J.L., Cirino, G., De Nucci, G., McKnight, W. and MacNaughton, W.K. (1989). Endothelin has potent ulcerogenic and vasoconstrictor actions in the stomach. Am. J. Physiol. 256, G661–G666.

Wallace, J.L., Keenan, C.M. and Granger, D.N. (1990a). Gastric ulceration induced by nonsteroidal anti-inflammatory drugs is a neutrophil-dependent process. Am. J. Physiol. 22, G462–G467.

Wallace, J.L., Keenan, C.M., Mugridge, K.G. and Parente, L. (1990b). Reduction of the severity of experimental gastric and duodenal ulceration by interleukin-1β Eur. J. Pharm. 186, 279–284.

Wallace, J.L., McKnight, G.W., Keenan, C.M., Byles, N.I.A. and MacNaughton, W.K. (1990c). Effects of leukotrienes on susceptibility of the rat stomach to damage and investigation of the mechanism of action. Gastroenterology 98, 1178–1186.

Wallace, J.L., Cucala, Mugridge, K. and Parente, L. (1991). Secretagogue-specific effects of interleukin-1 on gastric acid secretion. Am. J. Physiol. 261, G559–G564.

Whittle, B.J.R. and Vane, J.R. (1983). In "Progress in Gastroenterology", Vol. IV (eds G.B.J. Glass and P. Sherlock), pp 3–30. Grune and Stratton, New York.

Whittle, B.J.R., Kauffman, G.L. and Moncada, S. (1981). Vasoconstriction with thromboxane A_2 induces ulceration of the gastric mucosa. Nature 292, 472–474.

Whittle, B.J.R., Lopez-Belmonte, J. and Moncada, S. (1990). Regulation of gastric mucosal integrity by endogenous nitric oxide: interactions with prostanoids and sensory neuropeptides in the rat. Br. J. Pharm. 99, 607–611.

Wong, K. and Freund, K. (1980). Inhibition of the *n*-formylmethionyl-leucyl-phenylalanine induced respiratory burst in human neutrophils by adrenergic agonists and prostaglandins of the E series. Can. J. Physiol. Pharmacol. 59, 915–920.

Wright, N.A. (1984). In "Mechanisms of Mucosal Protection in the Upper Gastrointestinal Tract" (eds A. Allen *et al.*), pp 15–20. Raven Press, New York.

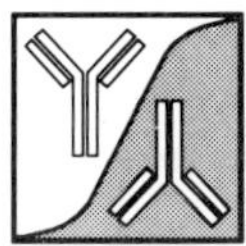

9. Role of the Epithelium in Defence of the Small and Large Intestine

A. G. Cummins *and* I. C. Roberts-Thomson

1. Introduction

The major functions of the intestine are the absorption of fluids, electrolytes and nutrients and the excretion of waste products. These are achieved by the interaction of luminal contents with a large single layer of surface epithelial cells estimated to cover an area of 200–300 m^2 in the human adult. These surface epithelial cells are the site of a mucosal barrier which modulates the passage of molecules not only from the lumen into the lamina propria but also in the reverse direction.

The barrier function of the intestine seems likely to depend on the physical integrity of the epithelial lining, the biochemical integrity of epithelial cells and the function of the intercellular tight junctions. The latter structures, in particular, demonstrate substantial plasticity and can be regulated physiologically by factors such as luminal nutrients (Madara *et al.*, 1990). A degree of plasticity also exists for rates of cell proliferation in intestinal crypts although it is still unclear whether proliferation is largely influenced by cell loss ('pull' phenomenon) or by factors operating at the level of the crypt ('push' phenomenon).

Renewed interest in intestinal permeability has been accompanied by a better understanding of the role of GALT in proliferation and differentiation. As well as the generation of immunological suppression (oral tolerance), GALT has a role in the maturation of fetal small bowel and in proliferative responses to intestinal damage caused by infections or other mechanisms. Possible interactions between intestinal permeability, cell proliferation and immune function may be relevant to a number of intestinal diseases, and will constitute the major focus of this review.

2. Intestinal Permeability

Traditionally, the barrier function of the intestine has been assessed by the ability of molecules to pass from the intestinal lumen into the circulation. In healthy individuals, this is influenced by the molecular weight and shape of the molecule. Indeed, a linear association between log absorption and log molecular weight predicts a large number of small "pores" and a small number of large "pores" with a continuous distribution between the extremes (Kingham and Loehry, 1978). In addition, larger molecules may be taken up by pinocytosis; a process which particularly applies to specialized M cells, which are associated with Peyer's patches (Bockman and Cooper, 1973).

For permeability studies, the ideal probe should be non-toxic, metabolically inert and should be measurable with sensitivity, accuracy and ease. Furthermore, it should not be recognized by the immune system. One such probe is a combination of non-metabolizable sugars; a disaccharide (usually lactulose) and a monosaccharide (usually mannitol or rhamnose). Lactulose is absorbed across tight junctions and at sites of epithelial cell loss while mannitol and rhamnose are largely absorbed through aqueous pores in the cell membrane (Deitch, 1990). In inflammatory disorders, the absorption of lactulose is increased while that of mannitol may decrease. Urinary excretion can be expressed as a ratio to overcome the effects of variables such as changes in intestinal transit or renal function.

A second probe is the use of PEG, which can be given orally and quantified in urine by gas–liquid chromatography (Olaison *et al.*, 1989). Absorption largely occurs in the upper small bowel and, in healthy subjects, is higher for PEG molecules of lower molecular weight. PEG can also be instilled directly into the ileum by a small bowel tube or into the colon at colonoscopy. Abnormal permeability is associated with enhanced absorption and/or a loss of mucosal selectivity against PEG of higher molecular weight. The sites of absorption of PEG probably include areas of cell loss and defects in the function of tight junctions.

Another method for the assessment of permeability involves the use of radioactive isotopes which can be readily quantified in samples of urine. The first of these compounds was ^{51}Cr-EDTA, and this was subsequently followed by ^{99m}Tc-DTPA. Less is known about the sites of absorption of these molecules although enhanced absorption has been demonstrated in inflammatory bowel disease and in other gastrointestinal disorders. Furthermore, simultaneous ingestion of both ^{51}Cr-EDTA and ^{99m}Tc-DTPA showed that results from both tests were positively correlated in a heterogeneous group of subjects (Resnick *et al.*, 1990).

Limitations of the above studies include the presence of variables such as changes in gastric emptying and intestinal transit, changes in renal clearance and inaccurate collection of urine. Furthermore, there is no "gold standard" at present against which the results of various investigations can be evaluated in terms of sensitivity and specificity. One possibility, however, is that the tests are more sensitive in disorders involving the upper small bowel than in disorders involving the ileum and colon.

Under certain circumstances, bacteria may pass from the intestinal lumen into the lamina propria, a process referred to as bacterial translocation. This appears to be due to cell loss and changes in tight junctions caused by mucosal ischaemia and reperfusion, and might explain infections of enteric origin in patients with multiple trauma and burns (Dietch *et al.*, 1987).

Another aspect of permeability involves the passage of molecules from the lamina propria into the intestinal lumen. Several studies have shown that faecal concentrations of labelled leucocytes and some serum proteins (α_1-antitrypsin) can provide a reliable index of the degree and/or extent of inflammation in ulcerative colitis and Crohn's disease (Bartholomeusz and Shearman, 1989). Furthermore, faecal concentrations of cytokines, such as tumour necrosis factor-α, also increase in the presence of bowel inflammation, presumably because of enhanced synthesis in the lamina propria and lysis of inflammatory cells in the gut lumen (Braegger *et al.*, 1992).

3. *Intestinal Proliferation*

The epithelial layer of the small and large intestine is continuously replaced by proliferating cells in intestinal crypts. In both the small and large bowel, the average life span of individual cells from birth to loss into the intestinal lumen is about 48 h. If the rate of cell loss exceeds the rate of cell production, 'gaps' in the surface epithelium may occur which could markedly impair barrier functions.

Rates of cell proliferation have been assessed by a variety of techniques both *in vivo* and *in vitro*. In experimental animals, techniques *in vivo* have included metaphase arrest and uptake of tritiated thymidine and bromodeoxyuridine. The latter two techniques can also be applied *in vitro*. In humans, proliferation in biopsy samples can be determined by microdissection, by flow cytometry or by incubating biopsies with tritiated thymidine or bromodeoxyuridine. A popular alternative is the incubation of tissue sections with a monoclonal antibody against antigens expressed on dividing nuclei. Two of these antigens have been designated K*i*67 and PCNA. Although the relative merits of the above techniques have been debated, support exists for the use of metaphase arrest in experimental animals *in vivo* (Sharp and Wright, 1984) and microdissection for assessment of proliferation in human biopsy specimens (Goodlad *et al.*, 1991).

The physiological factors which influence proliferation are still poorly understood. One experimental model involves an assessment of changes in proliferation which occur with fasting followed by refeeding (Butler *et al.*, 1992). In rats, suppression of proliferation during fasting and enhanced proliferation during refeeding are more prominent in the colon than in the small intestine. Furthermore, the degree of change is greater in the distal than in the proximal colon. Changes in colonic proliferation are not directly related to concentrations of SCFAs but might be influenced by luminal concentrations of bile acids, hormonal factors, physical distension or activation of the enteric or autonomic nervous systems. Another phenomenon of interest is the demonstration of circadian changes in proliferation in both rodents and humans (Buchi *et al.*, 1991). In rats, proliferation in the small bowel decreases in the presence of malnutrition (Rodrigues *et al.*, 1985), but this can be reversed by the strong immune stimulus of infection with *Nippostrongylus brasiliensis* (Cummins *et al.*, 1987). In humans, inflammatory disorders such as coeliac disease, ulcerative colitis and Crohn's disease are unusually associated with enhanced proliferation (Wright *et al.*, 1973; Lipkin and Higgins, 1988).

4. *Metabolism of Epithelial Cells*

The ability of cells to maintain their barrier functions seems likely to depend, at least in part, on the metabolic integrity of cells and the availability of substrates. In contrast to cells in most of the body, intestinal cells have access to nutrients in the lumen as well as access to nutrients from the circulation. The contribution of luminal nutrients to energy needs is about 30–50% in the small intestine and 50–70% in the large intestine. Luminal nutrients in the small intestine include glutamine and glucose while, in the colon, energy is derived from SCFAs produced from the fermentation of dietary fibre (Roediger, 1986).

In the large intestine, concentrations of SCFAs are higher in the proximal colon than in the distal colon (Mitchell *et al.*, 1985). In contrast, the contribution of SCFAs to total CO_2 production by colonocytes appears to be greater in the distal colon than in the proximal colon (Roediger, 1980). During fasting, concentrations of SCFAs fall to low levels in the distal colon, a phenomenon which might be relevant to the predilection of the distal colon to diseases such as ulcerative colitis (Roediger, 1988).

An increasing body of evidence supports a role for SCFAs, particularly butyrate, in protecting and healing the colonic mucosa. For example, luminal infusions of SCFAs promote healing of colonic anastomoses in the rat (Rolandelli *et al.*, 1986) while, in humans, inflammatory lesions of the rectum after colostomy can be improved by enemas containing SCFAs (Harig *et al.*, 1989). These beneficial effects of SCFAs may be due to the provision of a source of energy, to other factors such as increases in regional blood flow or oxygen uptake or changes in the function of the enteric nervous system.

Table 9.1 Complementary defence mechanisms

Physiological defences

    Human milk
    Saliva
    Gastric acidity
    Intestinal microflora
    Lysozyme
    Lactoferrin
    Mucus and anticolonization factors
    Peristalis

Immunological defences

    Secretory IgM and IgA
    Mucosal mast cells and IgE
    Mucosal T cells

5. *Complementary Protective Mechanisms*

The mucosal barrier created by epithelial cells is supplemented by additional factors as outlined in Table 9.1 Breast-feeding has been shown to decrease intestinal permeability in both guinea-pigs and human infants (Weaver *et al.*, 1987). This effect is probably due to the interaction of several components including bifidus factor, complement components C3 and C4, lactic acid, lactoferrin, lysozyme, IgA and leucocytes. A detailed discussion of the outcome of these interactions is outside the scope of this review, but important effects include limitation of antigen passage across epithelial cells and suppression of proliferation of pathogenic organisms. For example, breast-feeding reduces counts of *Escherichia coli* in faeces and suppresses colonization with *Proteus* spp. and *Pseudomonas aeruginosa* (Bullen and Willis, 1971). Other factors such as bifidus factor may promote the growth of non-pathogenic organisms including *Lactobacillus bifidus* (György, 1971). Furthermore, milk has a general trophic effect on intestinal proliferation and maturation, at least in piglets (Widdowson *et al.*, 1976). Additional protective factors may arise from saliva (IgA and lysozyme), gastric secretions (acid, pepsin, lysozyme and lactoferrin) and secretions from Brunner's glands (lysozyme). Finally, intestinal peristalsis has an important 'cleaning' effect on the bowel and minimizes the time for bacterial adherence to cells.

The integrity of the barrier is also influenced by the function of mucosal cells including mast cells and lymphocytes. Mast cells are found in large numbers throughout the small and large intestine. In the mucosal layer, they are located predominantly in the lamina propria and seem likely to participate in several homeostatic and pathological processes. Mediator release can be influenced by antibodies, complement components, factors derived from lymphocytes and macrophages, and neuropeptides released from adjacent nerves. Immediate responses are usually associated with antigen binding to IgE (Stead *et al.*, 1987; Lee *et al.*, 1988).

Mast cell hyperplasia is a prominent feature of helminth infections of the small bowel and may occur in some patients with ulcerative colitis. Mediators released by degranulation have a variety of cytotoxic and inflammatory effects *in vitro* and may have similar functions *in vivo*. In addition, there are interactions with enteric nerves with effects on the transport of electrolytes such as chloride ions (Befus *et al.*, 1988; Perdue *et al.*, 1991). The role of lymphocytes will be discussed in more detail below.

Epithelial cells also contribute to barrier functions by producing factors such as mucus and secretory component. The glycoproteins of mucus have a protein core and long carbohydrate side-chains which account for 70–80% of the molecular weight of the molecule. Colonic glycoproteins tend to be larger in size than glycoproteins from the small intestine and have higher proportions of sialic acid and sulphate. For colonic glycoproteins, at least six subclasses have been described, and changes in mucin species have been associated with disorders such as ulcerative colitis (Podolsky and Fournier, 1988). The release of mucus may be facilitated by several mechanisms, including mucosal histamine, probably derived from mast cells, and interactions between goblet cells and antigen–antibody complexes (Lake *et al.*, 1979; Lake *et al.*, 1980). The formation of a mucus gel acts as a physical barrier to microbes and toxins, facilitates the function of endogenous flora and facilitates the activity of luminal antibody (Neutra and Forstner, 1987).

The addition of secretory component to IgA and IgM facilitates their passage into the intestinal lumen. IgA is secreted at a flow rate of about 3 g/24 h from 10^{10} plasma cells and is in higher concentration than IgM (Brandtzaeg and Baklien, 1976). Over 90% of intestinal IgA is polymeric and contains a J chain produced by plasma cells. IgM may also be polymeric but is of lesser importance because of lower rates of intestinal secretion and higher rates of degradation. The beneficial effects of luminal IgA include prevention of microbe adhesion, neutralization of toxins and promotion of bacterial aggregation. IgA also reduces the uptake of soluble macromolecules in animals (Walker *et al.*, 1972) and seems likely to have similar effects in humans, since an inherited deficiency of IgA is associated with higher concentrations of serum antibodies and circulating immune complexes to food proteins (Cunningham-Rundles *et al.*, 1979).

6. The Immature Mucosal Barrier and Closure

In the human fetus, the intestine is permeable to macromolecules. Absorption of these molecules from amniotic fluid involves the apical endocytic complex, a membranous system in the supranuclear space of enterocytes. This complex appears to be most prominent in the second trimester and is active in endocytosis (Moxley and Trier, 1979). Intestinal permeability in the fetus may also be enhanced by the greater fluidity of microvillous membranes, a phenomenon associated with higher lipid to protein ratios in membranes in rabbits (Pang *et al.*, 1983).

Although immature in terms of permeability, the fetal small intestine has some specialized functions. One of these which is relevant to rodents and ruminants involves the IgG component of maternal milk, which is specifically transported in the jejunum after interaction with receptors for the Fc portion of the molecule. Rodents and ruminants have no placental transfer of antibody and are agammaglobulinaemic before birth. The cessation of transport of immunoglobulin in rodents is termed "closure" and typically occurs in the rat at day 20 of life (Rodewald, 1973).

In contrast, human infants have the benefit of placental transfer of IgG antibody and derive IgA from breast milk in the neonatal period. When applied to humans, the term "closure" has been given a broader definition to encompass the process of decreased permeability that occurs at around the time of birth. Although some authors have claimed that closure is complete at birth (Roberton *et al.*, 1982), others have evidence that the neonatal small intestine is more permeable to food antigens than adult small intestine (Eastham *et al.*, 1978). Permeability is increased, however, in infants who are premature (Roberton *et al.*, 1982). The possibility of promotion of closure by breast-feeding has not been clarified, although breast-fed infants have lower plasma concentrations of IgG antibodies to cows' milk proteins (Fällström *et al.*, 1984).

7. Factors Affecting Growth and Development of the Intestine

Many factors have been proposed as promoters of growth and differentiation in the intestine, and some of these are listed in Table 9.2. The mesenchyme appears to be necessary for intestinal morphogenesis and differentiation during fetal life, but its role in postnatal life remains unclear. Genetic preprogramming may be important but is not an over-riding factor since transplanted small intestine (under the kidney capsule) shows delayed development (Ferguson *et al.*, 1973). The presence of food in the intestinal lumen appears to promote postnatal maturation, at least in rodents (Lee and Lebenthal, 1983). Luminal bacteria also have a role since germ-free animals have an immature small intestine with tall, slender villi, shallow crypts and suppressed proliferation (Abrams *et al.*, 1963). A variety of systemic and local hormones have also been proposed as promoting growth, but their role has been difficult to clarify. Similarly, it

Table 9.2   Proposed growth factors of the intestine

Mesenchyme during fetal life
Genetic preprogramming
Luminal food and bacteria
Systemic and local hormones
Systemic and enteric nerves
Mucosal immune system

seems likely that there are interactions between the systemic and enteric nervous systems and cell proliferation but the mediators of these effects are unclear (Henning, 1981).

8. Intestinal Morphology and Mucosal Immune Activity at Weaning

A focus of our work involves the concept that activation of the immune system provides a physiological drive which promotes maturation of the developing small bowel. In this model, maturation is seen as a beneficial effect of a "controlled enteropathy". The hypothesis is supported by studies in rats which have shown that the small intestine develops postnatally with hyperplasia of villi and crypts (Cummins *et al.*, 1988a,b). This is associated with an increase in mucosal mast cells and intraepithelial lymphocytes. The activity of mucosal mast cells and of T lymphocytes in mesenteric nodes peaks at days 21–22 of life. The former can be assessed by serum concentrations of RMCPII, as shown in Fig. 9.1. The importance of immune activation in maturation has been supported by two studies which have evaluated changes in morphology after immunosuppression with cyclosporin A. In one, cyclosporin delayed the development of crypt hyperplasia in developing rats (Cummins *et al.*, 1989a) while, in the other, the drug prevented crypt hyperplasia in explant cultures of human fetal small intestine (McDonald and Spencer, 1988).

In more precocious species such as the guinea-pig, we have recently confirmed an approximate doubling of villus area and crypt length in the developing guinea-pig that stabilizes at about days 12–16 after birth (unpublished observations). Morphological changes were associated with a decrease in stainable mucosal mast cells, consistent with degranulation, as previously found in the rat. Finally, we have evidence for mucosal T cell activation in human infants after birth which appears to peak at about 4 months, a common time of weaning. Our current studies attempt to dissect whether immune activation is related to luminal food, colonization by bacteria or engraftment in the small intestine of lymphocytes derived from maternal milk (Sheldrake and Husband, 1985).

The relationship between "closure" and changes in morphology has not been clarified. In the rat, "closure" occurs about day 20 while hyperplasia of villi and crypts

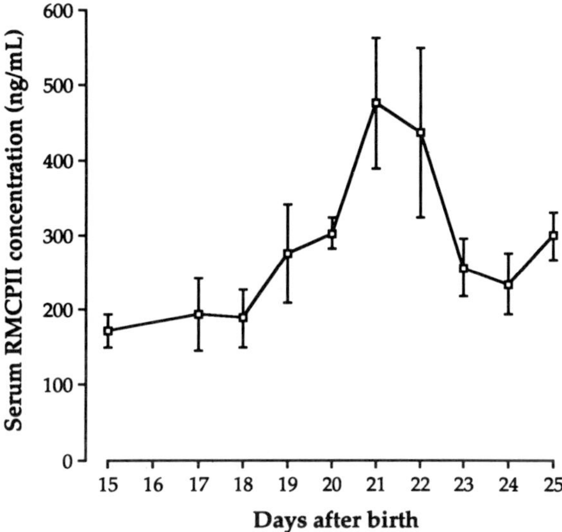

Figure 9.1   Changes in serum concentrations of RMCPII from day 15 to 25 after birth. Each point is the mean ± SD of six to eight animals.

peaks at days 21–22 of life. Thus, "closure" and changes in morphology peak at similar but not identical times. The possibility of transient changes in permeability associated with mast cell activation has not been studied as yet.

9. Immune Activation, Proliferation and Permeability

The induction of graft-versus-host disease may be the best experimental model for evaluating the effect of immune activation on cell proliferation (Ferguson, 1987). Induction of disease is associated with atrophy of villi, hyperplasia of crypts and an increase in cell proliferation in crypts. However, the cytokines or other factors which mediate these effects are unclear. Immune activation also seems likely to be relevant to crypt hyperplasia in coeliac disease and to enhanced proliferation in ulcerative colitis and Crohn's disease. On the other hand, a degree of crypt hypoplasia may occur in individuals who are immunosuppressed, particularly in the presence of HIV (Cummins *et al.*, 1990).

The effect of cytokines on barrier functions has been examined in monolayers of the T84 human intestinal epithelial cell line (Madara and Stafford, 1989). Monolayer resistance was substantially diminished by interferon γ but not by interleukin-1, interleukin-2 or tumour necrosis factor. Various solute flux experiments indicated that changes in resistance were due to an effect of interferon γ on the permeability of tight junctions, perhaps secondary to changes in the cytoskeleton of cells.

10. Gastrointestinal Infections

Some infections are restricted to the gastrointestinal tract, others have gastrointestinal and systemic phases while others gain access to the systemic circulation through the intestine. In gastroenteritis, adherence of organisms to epithelial cells seems likely to be relevant to proliferation and pathogenicity in infections such as enteropathogenic *E. coli* and giardiasis (Roberts-Thomson, 1984; Sherman *et al.*, 1989). Others such as rotavirus attach to and subsequently infect mature cells on the tips of villi (Svensson *et al.*, 1991). Bacteria such as *Yersinia* spp. have a predilection for translocation through M cells (Grutzkau *et al.*, 1990). This results in access to lymphoid tissue where some are phagocytosed by leucocytes and macrophages and begin transcription of genes associated with virulence. Systemic infections such as poliovirus type I and reovirus type I also appear to enter the body by translocation through M cells (Wolf *et al.*, 1987; Sicinski *et al.*, 1990).

Host factors relevant to susceptibility to gastrointestinal infections include genetic factors, antibody responses, leucocyte responses, mast cell responses, availability of nutrients, and changes in the function of the enteric system. For example, humans who carry the leucocyte antigen HLA-B27 are at greater risk for the development of postinfectious arthritis (Laitinen *et al.*, 1977), perhaps because of the slower elimination of organisms such as *Shigella* and *Yersinia* spp. The relevance of antibody responses has been demonstrated in infections such as giardiasis, where prolonged and recurrent infections are frequent in patients with hypogammaglobulinaemia (Roberts-Thomson, 1984). Similarly, patients with impaired lymphocyte responses associated with HIV infection are at a higher risk for a variety of gastrointestinal infections caused by bacteria, viruses and protozoa. Defects in mast cell function have not been well-documented in humans, but congenital mast cell deficiency has been associated with prolonged giardiasis in mice (Erlich *et al.*, 1983). Finally, the availability of nutrients may influence the site of infections such as giardiasis and the outcome of infections such as yersiniosis (Roberts-Thomson, 1984; Stuart *et al.*, 1986).

Gastrointestinal infections seem likely to increase intestinal permeability although this has rarely been studied in detail. Symptomatic HIV infection may also be associated with changes in intestinal function in the absence of enteric pathogens (Ullrich *et al.*, 1989; Guarino *et al.*, 1991). Abnormalities have included an increase in intestinal permeability, increased faecal excretion of α_1–antitrypsin, depressed activities of brush border enzymes and impaired absorption of D-xylose and fat. These effects may be due to bacterial overgrowth in the small bowel leading to activation of monocytes/macrophages and secretion of cytokines.

11. Chronic Inflammatory Bowel Disease

Despite substantial research activity, the pathogenesis of ulcerative colitis and Crohn's disease is still poorly understood. It is also unclear whether the two disorders are separate entities or different manifestations of similar pathogenetic mechanisms. The latter possibility is supported by the aggregation of patients with ulcerative colitis and Crohn's disease within families and by striking similarities in drug therapy.

Hypotheses for the development of chronic inflammatory bowel disease include altered mucosal immunity, altered mucosal permeability, altered bacterial flora, metabolic changes in epithelial cells, immunomodulation by bacterial or dietary products, and infections due to atypical mycobacteria and cell wall-deficient organisms. The possibility that defects in intestinal permeability might predispose to inflammatory bowel disease is supported by the finding of increased intestinal permeability in asymptomatic relatives of patients with Crohn's disease (Hollander *et al.*, 1986). This observation, however, needs to be confirmed in other centres (Bjarnason and Peters, 1987). In dietary studies, patients with Crohn's disease have been shown to have a high intake of sugar, which may increase the permeability of the mucosa to probes such as ^{51}Cr-EDTA and lactulose (Maxton *et al.*, 1986). Furthermore, treatment of Crohn's disease with an elemental diet has been associated with a decrease in permeability as assessed by a reduction in the urinary excretion of ^{51}Cr-EDTA (Teahon *et al.*, 1991). However, patients also showed improvement in symptoms and a reduction in the faecal excretion of indium-labelled leucocytes. In ulcerative colitis, epidemiological studies have shown a lower prevalence in smokers than in non-smokers, perhaps because smoking makes the small bowel and/or colon less permeable to molecules in the bowel lumen (Prytz *et al.*, 1989).

12. Coeliac Disease

Coeliac disease is characterized by hypersensitivity to dietary gluten, which results in intestinal damage in susceptible individuals with particular HLA phenotypes (DR3 or DR7, DQw2). Although the pathogenesis is still debated, it seems likely to involve changes in mucosal

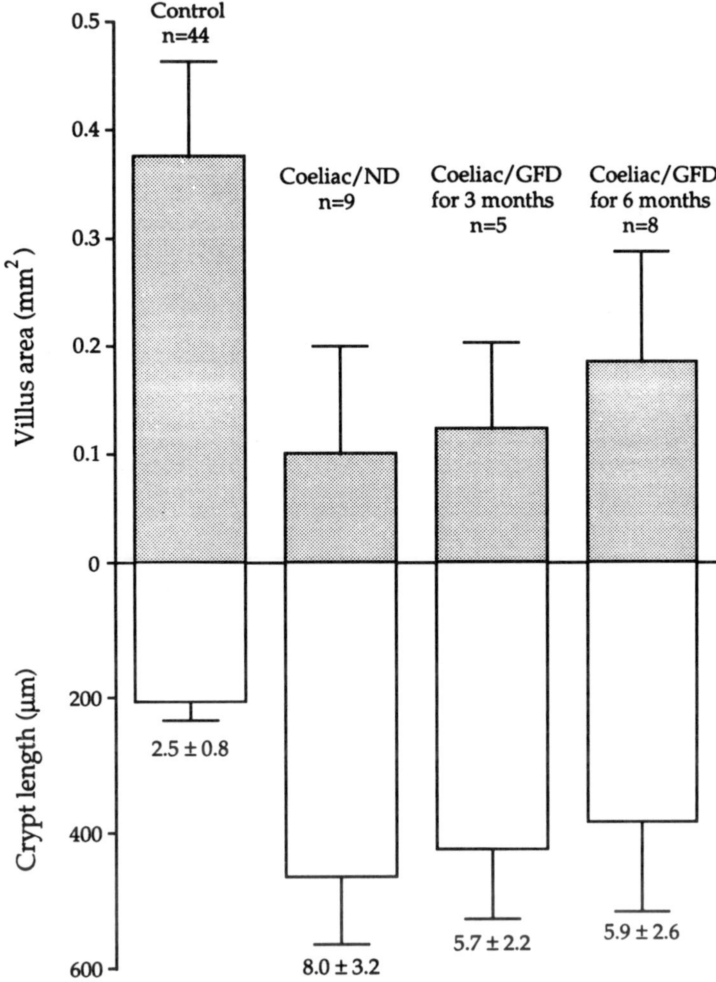

Figure 9.2 Villus area, crypt length and mitotic count per crypt in control subjects, coeliac subjects on a normal diet (ND) and coeliac subjects at 3 and 6 months after the introduction of a gluten-free diet (GFD). Results are expressed as mean ± SD.

immunity with selective failure of oral tolerance to gluten. Inflammatory changes in the small bowel are accompanied by an increase in cell proliferation in crypts (Fig. 9.2) and by an increase in T cell activity as assessed by higher serum concentration of soluble interleukin-2 receptor (Cummins *et al.*, 1991). In addition, intestinal permeability is increased when studied by the absorption of non-metabolizable sugars and by the presence in serum of antibodies to a variety of food proteins other than gluten (Scott *et al.*, 1990; Cummins *et al.*, 1991). The introduction of a gluten-free diet is followed by lower serum concentrations of the interleukin-2 receptor and by an improvement in intestinal permeability. As shown in Fig. 9.3, intestinal permeability improves within 4 weeks, while improvement in small bowel morphology may take 3–6 months. In particular circumstances, the gluten-free diet may need to be supplemented by the addition of immunosuppressive drugs such as corticosteroids,

azathioprine or cyclosporin A. Changes in small bowel morphology in a coeliac patient who developed ulcerative jejunitis and was treated with cyclosporin A are shown in Fig. 9.4 (Cummins and Ferguson, unpublished data). This patient was subsequently diagnosed as having an enteropathy-associated T cell lymphoma.

13. Miscellaneous Disorders

Changes in intestinal permeability accompanied by intolerance to various foods may be relevant to some patients with unexplained gastrointestinal symptoms, often referred to as irritable bowel syndrome. In one study of 14 adults, three had increased serum concentrations of β-lactoglobulin after a standard challenge with cows milk (Paganelli *et al.*, 1990). However, this finding did not correlate with responses to a hypoallergenic diet or

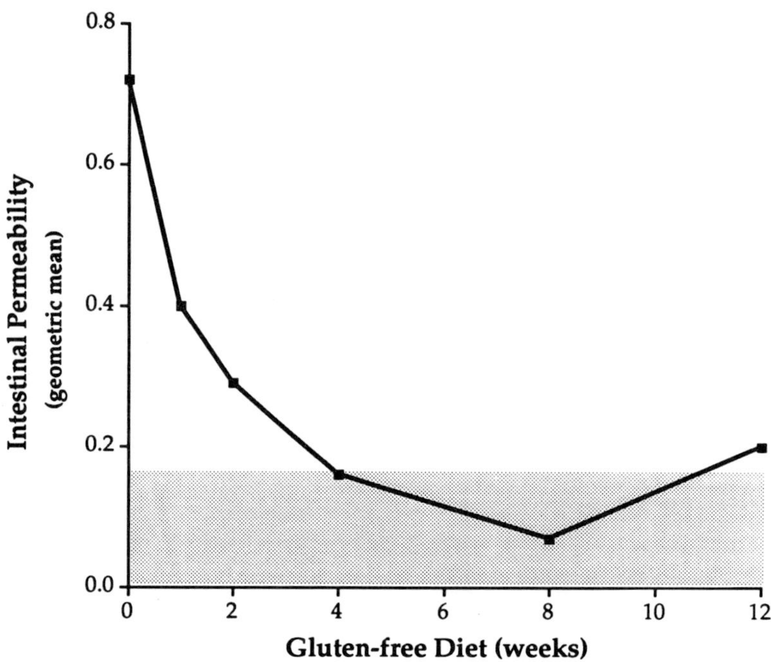

Figure 9.3 Reduction in intestinal permeability in coeliac subjects after the introduction of a gluten-free diet (*n* = 6). Permeability was assessed by the ratio of urinary lactulose to rhamnose. The reference range is shown by the shaded area.

to treatment with disodium cromoglycate. Only two patients had elevated serum levels of IgE consistent with the possibility of food allergy. In another study in children, patients with unexplained recurrent abdominal pain had a higher urinary excretion of ^{51}Cr-EDTA than matched control subjects (Amery and Forget, 1989). Although this area deserves further study, the pathogenesis of functional bowel symptoms seems likely to involve other factors, such as low thresholds for visceral pain, autonomic dysfunction, changes in bowel motility and psychological characteristics including abnormal illness behaviour.

Drugs may also have an effect on intestinal morphology and permeability. This applies particularly to non-steroidal anti-inflammatory drugs which cause small bowel inflammation in 60–70% of patients after 6 or more months of therapy (Bjarnason *et al.*, 1987, 1991). Given as suppositories, they may also cause colitis. At endoscopy, the upper small bowel is often abnormal with superficial ulcers (erosions) and areas of inflammation. This is associated with an increase in intestinal permeability as assessed by the urinary excretion of ^{51}Cr-EDTA (Aabakken *et al.*, 1989).

Permeability may also be affected prior to the development of ulcers, perhaps due to changes in cell membranes or tight junctions. In the majority of patients, the integrity of the small intestine is restored within 1 week of cessation of anti-inflammatory drugs. Rarely, however, the drugs may result in a more severe enteropathy, with prolonged bleeding, stricture formation or perforation. This is

largely restricted to patients with rheumatoid arthritis and may persist for months after cessation of drug therapy.

The integrity of the intestine may also be affected by agents which influence the rate of proliferation of crypt cells. A variety of immunosuppressive drugs including corticosteroids, methotrexate and cyclosporin A cause a decrease in crypt cell proliferation, but possible effects on intestinal permeability have not been studied (Wright *et al.*, 1978; Cummins *et al.*, 1989b; D'Argenio *et al.*, 1989). Irradiation of the bowel causes profound crypt hypoplasia for 1–3 days followed by a prolonged period of crypt hyperplasia, inflammation and damage to villi (Ijiri and Potten, 1983). While the hypoplastic phase represents direct crypt cell damage, the subsequent hyperplastic response is of interest as it can be attenuated by transfer of bone marrow or spleen cells (Cummins *et al.*, 1989b).

14. Summary

The barrier functions of the small and large intestine seem likely to depend on the integrity of the epithelial cell lining and the function of cell membranes and intercellular tight junctions. When the rate of cell loss exceeds the rate of cell production, 'gaps' in the lining may occur which could impair barrier functions. This is best illustrated by changes in morphology and permeability in small bowel which accompany the use of non-steroidal anti-inflammatory drugs. In other settings, the relationship

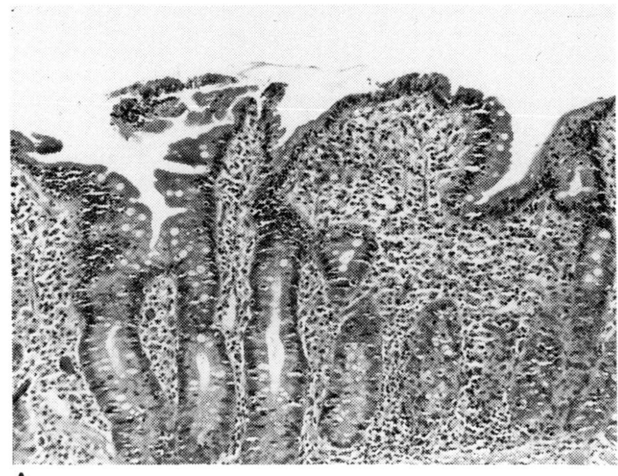

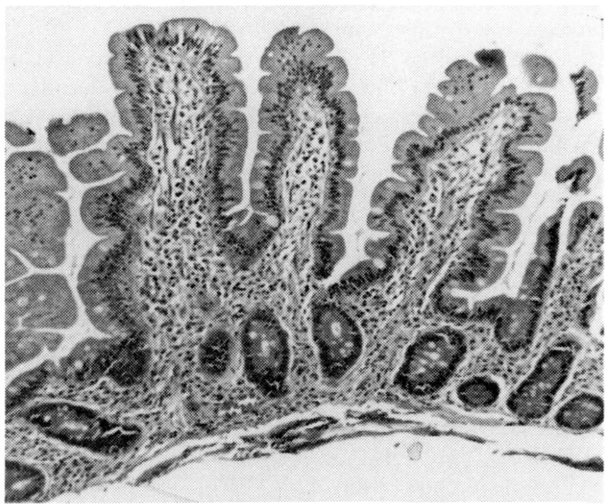

Figure 9.4   Small intestinal biopsies (A) before and (B) after 6 weeks of treatment with cyclosporin A in a coeliac subject who developed ulcerative jejunitis.

between proliferation, inflammation and permeability may be more complex. For example, a degree of activation of the mucosal immune system may be important for maturation of the small intestine and promotion of "closure". In contrast, severe inflammation of the small intestine as occurs in coeliac disease is associated with an increase in intestinal permeability. Apart from cell loss, the factors which influence permeability are poorly understood although evidence exists for an effect of cytokines, especially interferon γ, on permeability through intercellular tight junctions.

An increase in intestinal permeability resulting in epithelial cell damage from bacterial or dietary products in the bowel lumen is an attractive hypothesis for the pathogenesis of chronic inflammatory bowel disease. However, reasons for potential changes in intestinal permeability remain unclear and there is no persuasive explanation for the location and chronicity of inflammation in either ulcerative colitis or Crohn's disease. Useful directions for future research could include a more detailed assessment of the effects of immune activation and mast cell degranulation on colonic permeability and possible interactions with bacterial products and luminal nutrients such as SCFAs.

15. Acknowledgements

Research by the authors has been supported by the National Health and Medical Research Council of Australia, the Anti-Cancer Foundation of the Universities of South Australia and by the Channel 7 children's Medical Research Foundation of South Australia. We are grateful to Mrs Rosemary Goland for secretarial assistance.

16. References

Aabakken, L., Bjornbeth, B.A., Hofstad, B., Olaussen, B., Larsen, S. and Osnes, M. (1989). Comparison of the gastrointestinal side effects of naproxen formulated as plain tablets, enteric-coated tablets, or enteric-coated granules in capsules. Scand. J. Gastroenterol. (Suppl.) 163, 65–73.

Abrams, G.D., Bauer, H. and Sprinz, H. (1963). Influence of the normal flora on mucosal morphology and cellular renewal in the ileum. A comparison of germ-free and conventional mice. Lab. Invest. 12, 355–364.

Amery, W.K. and Forget, P.P. (1989). The role of the gut in migraine: the oral 51-Cr EDTA test in recurrent abdominal pain. Cephalalgia. 9, 227–229.

Bartholomeusz, F.D.L. and Shearman, D.J.C. (1989). Measurement of activity in Crohn's disease. J. Gastroenterol. Hepatol. 4, 81–94.

Befus, D., Fujimaki, H., Lee, T.D.G. and Swieter, M. (1988). Mast cell polymorphisms. Present concepts, future directions. Dig. Dis. Sci. 33, 16S-24S.

Bjarnason, I. and Peters, T.J. (1987). Helping the mucosa make sense of macromolecules. Gut 28, 1057–1061.

Bjarnason, I., Zanelli, G., Smith, T., Prouse, P., Williams, P., Smethurst, P., Delacey, G., Grumped, J. and Levi, A.J. (1987). Nonsteroidal anti-inflammatory drug-induced intestinal inflammation in humans. Gastroenterology 93, 480–489.

Bjarnason, I., Fehilly, B., Smethurst, P., Menzies, I.S. and Levi, A.J. (1991). Importance of local versus systemic effects of non-steroidal anti-inflammatory drugs in increasing small intestinal permeability in man. Gut 32, 275–277.

Bockman, D.E. and Cooper, M.D. (1973). Pinocytosis by epithelium associated with lymphoid follicles in the bursa of Fabricius, appendix and Peyer's patches. Am. J. Anat. 136, 455–478.

Braegger, C.P., Nicholls, S., Murch, S.H., Stephens, S. and MacDonald, T.T. (1992). Tumour necrosis factor alpha in stool as a marker of intestinal inflammation. Lancet 339, 89–91.

Brandtzaeg, P. and Baklien, K. (1976). Immunohistochemical studies of the formation and epithelial transport of immunglobulins in normal and diseased human intestinal mucosa. Scand. J. Gastroenterol. (Suppl.) 36, 1–45.

Buchi, K.N., Moore, J.G., Hrushesky, W.J.M., Sothern, R.B. and Rubin, N.H. (1991). Circadian rhythm of cellular proliferation in the human rectal mucosa. Gastroenterology 101, 410–415.

Bullen, C.L. and Willis, A.T. (1971). Resistance of the breast-fed infant to gastroenteritis. Br. Med. J. 3, 338–343.

Butler, R.N., Bruhn, B., Pascoe, V., Fettman, M.J. and Roberts-Thomson, I.C. (1992). Regional factors affecting proliferation in the large intestine of the rat. Proc. Soc. Exp. Biol. Med. 200, 133–137.

Cummins, A.G., Duncombe, V.M., Bolin, T.D., Davis, A.E. and Yong. J. (1987). The intestinal response of the protein deficient rat to infection with *Nippostrongylus brasiliensis*. Int. J. Parasitol. 17, 1445–1450.

Cummins, A.G., Munro, G.H., Miller, H.R.P. and Ferguson, A. (1988a). Association of maturation of the small intestine at weaning with mucosal mast cell activation in the rat. Immunol. Cell. Biol. 66, 417–422.

Cummins, A.G., Steele, T.W., LaBrooy, J.T. and Shearman, D.J.C. (1988b). Maturation of the rat small intestine at weaning: changes in epithelial cell kinetics, bacterial flora and mucosal immune activity. Gut 29, 1672–1679.

Cummins, A.G., LaBrooy, J.T. and Shearman, D.J.C. (1989a). The effect of cyclosporin A in delaying maturation of the small intestine during weaning in the rat. Clin. Exp. Immunol. 75, 451–456.

Cummins, A.G., Munro, G.H., Huntley, J.F., Miller, H.R.P. and Ferguson, A. (1989b). Separate effects of irradiation and of graft-versus-host reaction on rat mucosal mast cells. Gut 30, 355–360.

Cummins, A.G., LaBrooy, J.T., Stanley, D.P., Rowland, R. and Shearman, D.J.C. (1990). Quantitative histological study of enteropathy associated with HIV infection. Gut 31, 317–321.

Cummins, A.G., Penttila, I.A., LaBrooy, J.T., Shearman, D.J.C., Robb, T.A. and Davidson, G.P. (1991). Recovery of the small intestine in coeliac disease on a gluten-free diet: changes in intestinal permeability, small bowel morphology and T-cell activity. J. Gastroenterol. Hepatol. 6, 53–57.

Cunningham-Rundles, C., Brandeis, W.E., Good, R.A. and Day, N.K. (1979). Bovine antigens and the formation of circulating immune complexes in selective immunoglobulin A deficiency. J. Clin. Invest. 64, 272–279.

D'Argenio, G., Sorrentini, I., Ciacci, C., Spagnuolo, S., Ventriglia, R., De Chiara, A. and Mazzacca, G. (1989). Human serum transglutaminase and coeliac disease: correlation between serum and mucosal activity in an experimental model of rat small bowel enteropathy. Gut 30, 950–954.

Deitch, E.A. (1990). Intestinal permeability is increased in burn patients shortly after injury. Surgery 107, 411–416.

Deitch, E.A., Winterton, J., Li, M. and Berg, R. (1987). The gut as a portal of entry for bacteraemia: role of protein malnutrition. Ann. Surg. 205, 681–692.

Eastham, E.J., Lichuaco, T., Grady, M.I. and Walker, W.A. (1978). Antigenicity of infant formulas: role of immature intestine in protein permeability. J. Pediatr. 93, 561–564.

Erlich, J., Anders, R.F., Roberts-Thomson, I.C., Schrader, J.W. and Mitchell, G.F. (1983). An examination of differences in serum antibody specificities and hypersensitivity reactions as contributing factors to chronic infection with the intestinal protozoan parasite, *Giardia muris*, in mice. Aust. J. Exp. Biol. Med. Sci. 61, 599–615.

Fällström, S., Ahlstedt, S., Carlsson, B., Wettergren, B. and Hanson, L. (1984). Influence of breast feeding on the development of cows' milk protein antibodies and the IgE level. Int. Arch. Allergy Immunol. 75, 87–91.

Ferguson, A. (1987). In "Immunopathology of the Small Intestine" (ed. M. Marsh), pp 225–252. Wiley, Chichester.

Ferguson, A., Gerskowitch, V. and Russell, R.I. (1973). Pre- and post-weaning disaccharidase patterns in isografts of fetal mouse intestine. Gastroenterology 64, 292–297.

Goodlad, R.A., Levi, S., Lee, C.Y., Mandir, N., Hodgson, H. and Wright, N.A. (1991). Morphometry and cell proliferation in endoscopic biopsies: evaluation of a technique. Gastroenterology 101, 1235–1241.

Grutzkau, A., Hanski, C., Hahn, H. and Riecken, E.O. (1990). Involvement of M cells in the bacterial invasion of Peyer's patches: a common mechanism shared by *Yersinia enterocolitca* and other enteroinvasive bacteria. Gut 31, 1011–1015.

Guarino, A., Tarallo, L., Guandalini, S., Troncone, R., Albano, F. and Rubins, A. (1991). Impaired intestinal function in symptomatic HIV infection. J. Pediatr. Gastroenterol. Nutr. 12, 453–458.

György, P. (1971). The uniqueness of human milk. Biochemical aspects. Am. J. Clin. Nutr. 24, 970–975.

Harig, J.M., Soergel, K.H., Komorowski, R.A. and Wood, C.M. (1989). Treatment of diversion colitis with short-chain-fatty acid irrigation. N. Engl. J. Med. 320, 23–28.

Henning, S.J. (1981). Postnatal development: coordination of feeding, digestion and metabolism. Am. J. Physiol. 241, 199–214.

Hollander, D., Vadheim, C.M., Brettholz, E., Peterson, G.M., Delahunty, T. and Rotter, J.I. (1986). Increased intestinal permeability in patients with Crohn's disease and their relatives. A possible aetiologic factor. Ann. Int. Med. 105, 883–885.

Ijiri, K. and Potten C.S. (1983). Response of intestinal cells of differing topographical and hierarchial status to ten cytotoxic drugs and five sources of irradiation. Br. J. Cancer. 47, 175–185.

Kingham, J.G.C. and Loehry, C.A. (1978). Selectivity of small intestinal exudate in celiac disease and Crohn's disease. Am. J. Dig. Dis. 23, 33–38.

Lake, A.M., Block, K.J., Neutra, M.R. and Walker, W.A. (1979). Intestinal goblet cell mucus release II. *In vivo* stimulation by antigen in the immunized rat. J. Immunol. 122, 834–837.

Lake, A.M., Block, K.J., Sinclair, K.J. and Walker, W.A. (1980). Anaphylactic release of intestinal goblet cell mucus. Immunology 39, 173–178.

Laitinen, O., Leirisalo, M. and Skylv, G. (1977). Relation between HLA-B27 and clinical features in patients with *Yersinia* arthritis. Arthritis Rheum. 20, 1121–1124.

Lee, P.C. and Lebenthal, E. (1983). Early weaning and precocious development of small intestine in rats: genetic, dietary or hormonal control. Pediatr. Res. 17, 645–650.

Lee, T.D.G., Swieter, M., and Befus, D. (1988). Mast cells, eosinophils and gastrointestinal hypersensitivity. Immunol. Allergy Clin. North Am. 8, 469–483.

Lipkin, M., and Higgins, P. (1988). Biological markers of cell proliferation and differentiation in human gastrointestinal diseases. Adv. Cancer Res. 50, 1–24.

MacDonald, T.T. and Spencer, J. (1988). Evidence that activated mucosal T cells play a role in the pathogenesis of enteropathy in human small intestine. J. Exp. Med. 167, 1341–1349.

Madara, J.L. and Stafford, J. (1989). Interferon-gamma directly affects barrier function of cultured intestinal epithelial monolayers. J. Clin. Invest. 83, 724–727.

Madara, J.L., Nash, S., Moore, R. and Atiscok, K. (1990).

Structure and function of the intestinal epithelial barrier in health and disease. Monogr. Pathol. 31, 306–324.

Maxton, D.G., Bjarnason, I., Reynolds, A.P., Catt, S.D., Peters, T.J. and Menzies, I.S. (1986). Lactulose, ^{51}Cr-labelled ethylenediaminetetra-acetate, L-rhamnose and polyethyleneglycol 500 as probe markers for assessment *in vivo* of human intestinal permeability. Clin. Sci. 71, 71–80.

Mitchell, B.L., Lawson, M.J., Davies, M., Kerr-Grant, A., Roediger, W.E.W., Illman, R.J. and Topping, D.L. (1985). Volatile fatty acids in the human intestine: studies in surgical patients. Nutr. Res. 5, 1089–1092.

Moxley, P.C. and Trier, J.S. (1979). Development of villus absorptive cells in the human fetal intestine: a morphological and morphometric study. Ana. Rec. 195, 463–482.

Neutra, M.R. and Forstner, J.F. (1987). In "Physiology of the Gastrointestinal Tract", 2nd edn (ed. L.R. Johnson), pp 975–1009. Raven Press, New York.

Olaison, G., Sjodahl, R., Leandersson, P. and Tagesson, C. (1989). Abnormal intestinal permeability pattern in colonic Crohn's disease. Absorption of low molecular weight polyethylene glycols after oral or colonic load. Scand. J. Gastroenterol. 24, 571–576.

Paganelli, R., Fagiolo, U., Cancian, M., Sturniolo, G.C., Scala, E., and D'Offizi, G.P. (1990). Intestinal permeability in irritable bowel syndrome. Effect of diet and sodium cromoglycate administration. Ann. Allergy 64, 377–380.

Pang, K.Y., Bresson, J.L. and Walker, W.A. (1983). Development of the gastrointestinal mucosal barrier. Evidence for structural differences in microvillus membranes from newborn and adult rabbits. Biochim. Biophys. Acta 727, 201–208.

Perdue, M.H., Masson, S., Wershil, B.K and Galli, S.J. (1991). Role of mast cells in ion transport abnormalities with intestinal anaphylaxis. J. Clin. Invest. 87, 687–693.

Podolsky, D.K. and Fournier, D.A. (1988). Emergence of antigenic glycoprotein structures in ulcerative colitis detected through monoclonal antibodies. Gastroenterology 95, 371–378.

Prytz, H., Benoni, C. and Tagesson, C. (1989). Does smoking tighten the gut? Scand. J. Gastroenterol. 24, 1084–1088.

Resnick, R.H., Royal, H., Marshall, W., Barron, R. and Werth, T. (1990), Intestinal permeability in gastrointestinal disorders. Use of oral [99mTc]DTPA. Dig. Dis. Sci. 35, 205–211.

Roberton, D.M., Paganelli, R., Dinwiddie, R. and Levinsky, R.J. (1982). Milk antigen absorption in the preterm and term neonate. Arch. Dis. Child. 57, 369–372.

Roberts-Thomson, I.C. (1984). In "Tropical and Geographic Medicine" (eds A.A.F. Mahmoud and K.S. Warren), pp 319–325. McGraw-Hill, New York.

Rodewald, R. (1973). Intestinal transport of antibodies in the newborn rat. J. Cell Biol. 58, 189–211.

Rodrigues, M.A.M., de Camargo, J.L.V., Coelho, K.I.R., Montenegro, M.R.G., Angeleli, A.Y.O. and Burini, R.C. (1985). Morphometric study of the small intestinal mucosa in young, adult and old rats submitted to protein deficiency and rehabilitation. Gut 26, 816–821.

Roediger, W.E.W. (1980). Role of anaerobic bacteria in the metabolic welfare of the colonic mucosa in man. Gut 21, 793–798.

Roediger, W.E.W. (1986). Metabolic basis of starvation diarrhoea: Implications for treatment. Lancet i, 1082–1083.

Roediger, W.E.W. (1988). What sequence of pathogenic events leads to acute ulcerative colitis? Dis. Colon Rectum 31, 482–487.

Rolandelli, R.H., Koruda, M.J., Settle, R.G. and Rombeau, J.L. (1986). Effects of intraluminal infusion of short-chain fatty acids on the healing of colonic anastomosis in the rat. Surgery 100, 198–204.

Scott, H., Fausa, O., Ek, J., Valnes, K., Bylstad, L. and Brandtzaeg, P. (1990). Measurements of serum IgA and IgA activities to dietary antigens. A prospective study of the diagnostic usefulness in adult coeliac disease. Scand. J. Gastroenterol. 25, 287–292.

Sharp, J.G. and Wright, N.A. (1984). Comparison of tritiated thymidine and metaphase arrest techniques of measuring cell production in rat intestine. Dig. Dis. Sci. 29, 1153–1158.

Sheldrake, R.F. and Husband, A.J. (1985). Intestinal uptake of intact maternal lymphocytes by neonatal rats and lambs. Res. Vet. Sci. 39, 10–15.

Sherman, P., Drumm, B., Karmali, M. and Cutz, E. (1989). Adherence of bacteria to the intestine in sporadic cases of enteropathogenic *Escherichia coli*-associated diarrhoea in infants and young children: a prospective study. Gastroenterology 96, 86–94.

Sicinski, P., Rowinski, J., Warchol, J.B., Jarzabek, Z., Gut, W., Szczygiel, B., Bielecki, K. and Koch, B. (1990). Poliovirus type I enters the human host through intestinal M cells. Gastroenterology 98, 56–58.

Stead, R.H., Bienenstock, J. and Stanisz, A.M. (1987). Neuropeptide regulation of mucosal immunity. Immunol. Rev. 100, 333–359.

Stuart, S.J., Prpic, J.K. and Robins-Browne, R.M. (1986). Production of aerobactin by some species of the genus *Yersinia*. J. Bacteriol. 166, 1131–1133.

Svensson, L., Finlay, B.B., Bass, D., von Bonsdorff, C.H. and Greenberg, H.B. (1991). Symmetric infection of rotavirus on polarised human intestinal epithelial (Caco-2) cells. J. Virol. 65, 4190–4197.

Teahon, K., Smethurst, P., Pearson, M., Levi, A.J. and Bjarnason, I. (1991). The effect of elemental diet on intestinal permeability and inflammation in Crohn's disease. Gastroenterology 101, 84–89.

Ullrich, R., Zeitz, M., Heise, W., L'Age, M., Hoffken, G. and Riecken, E.O. (1989). Small intestinal structure and function in patients infected with human immunodeficiency virus (HIV): evidence for HIV-induced enteropathy. Ann. Intern. Med. 111, 15–21.

Walker, W.A., Isselbacher, K.J. and Bloch, K.J. (1972). Intestinal uptake of macromolecules: effect of oral immunization. Science 177, 608–610.

Weaver, L.T., Laker, M.F., Nelson, R. and Lucas, A. (1987). Milk feeding and changes in intestinal permeability and morphology in the newborn. J. Pediatr. Gastroenterol. Nutr. 6, 351–358.

Widdowson, E.M., Colombo, V.E. and Artavams, C.A. (1976). Changes in the organs of pigs in response to feeding for the first 24 hours after birth. Biol. Neonate 28, 272–281.

Wolf, J.L., Dambrauskas, R., Sharpe, A.H. and Trier, J.S. (1987). Adherence to and penetration of the intestinal epithelium by reovirus type I in neonatal mice. Gastroenterology 92, 82–91.

Wright, N., Watson, A., Morley, A., Appleton, D. and Marks, J. (1973). Cell kinetics in flat (avillous) mucosa of the human small intestine. Gut 14, 701–710.

Wright, N.A., Al-Dewachi, H.S., Appleton, D.R. and Watson, A.J. (1978). The effect of single and multiple doses of prednisolone tertiary butyl acetate on cell population kinetics in the small bowel mucosa of the rat. Virchows Arch. (Cell Pathol.) 28, 339–350.

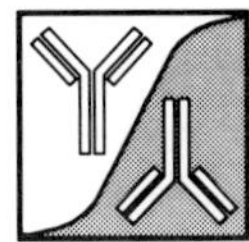

10. Immunoreactivity and Inflammation in Renal Epithelium

D. Gwyn Williams

Immunopharmacology of Epithelial Barriers
ISBN 0–12–288030–7

1. Introduction

The epithelial cells of the kidney fall into two anatomically and physiologically distinct groups, the glomerular and tubular epithelia. They contrast with other epithelial tissues in the body by having no practical relationship to the outside world. Thus, unlike, for example, the epithelia of the skin or the upper respiratory tract, they do not have to provide an effective barrier to the entrance of infectious organisms. Both types of epithelia are, indeed, almost entirely "inward-looking" in that they are mainly concerned with the internal milieu of the body by preventing loss of plasma constituents at the glomerular filtration barrier, and appropriate secretion and reabsorption by the tubular epithelium. A second difference, particularly in the case of the glomerulus, compared to other epithelia, is the special functional relationship of the renal epithelial cells to the underlying basement membrane. This again reflects the importance of the glomerular filtration barrier, to which the glomerular basement membrane makes a crucial contribution.

A third difference is the inability in humans or experimental animals to perform biopsies of the renal epithelia alone, in contrast to the intestinal tract or skin. Of course, glomerular and tubular epithelia are contained in renal biopsy specimens, but, whether these are obtained by the percutaneous or open methods, there is no selective removal of epithelium, as portions of whole kidney are obtained.

Study of the renal epithelia relies on histology, including immunohistological and electron microscopic techniques, and cell culture *in vitro*. The latter allows examination of the production by the cells of immune mediators, and the influence on the cells of mediators or drugs introduced into the culture. As far as the study of renal epithelia in humans is concerned, these approaches have severe limitations. In many instances of renal disease, repeat biopsies of the kidney are not clinically indicated, and are therefore not ethically justified. Sequential study, therefore, is not easily done. The bulk of observations on the role played by epithelial cells themselves in immune-mediated disease comes from animal studies with all the usual difficulties and concerns in making observations in one species and relating them to another. Immunologists and nephrologists have paid more attention, when studying the immune-inflammatory disorders, to the mesangial cells of the glomerulus, the basement membrane, and the infiltrating cells, mainly macrophages, than to the epithelial cells themselves. It is easier to delineate, particularly in cell culture, these other cells types and their function, and much of the genesis of renal inflammation was ascribed, and still is, to these other cells.

Despite these limitations, it is now clear that the renal epithelial cells play an important role in immune-mediated inflammation of the glomeruli and tubules. In this chapter on the structure and function of the GECs and TECs, the changes they undergo in disease, and the roles they themselves play in causing or preventing tissue damage will be described. The effectiveness of immuno-pharmacological agents in modifying these changes will be discussed. Because of the intimate relationship between the epithelia and their basement membranes the latter will be included. The epithelia of the ureters and bladder are not considered here, although they have important properties related to bacterial infection of the lower urinary tract.

2. Structure and Function

2.1 GLOMERULAR EPITHELIAL CELLS AND THE GLOMERULAR BASEMENT MEMBRANE (Fig. 10.1)

The GECs form the outermost layer of the glomerulus. This itself consists of a globular network of capillaries, interposed between the afferent and efferent glomerular arterioles. The capillaries consist of a thin layer of fenestrated endothelial cells supported by the mesangial cells which comprise the mesangium. The endothelial cells are negatively charged due to their content of polyanionic sialoproteins. The GECs surmount the capillary network, with their associated GBM, and comprise the visceral GECs. At the vascular hilum of the glomerulus the GECs reflect onto the inner aspect of Bowman's capsule to become the parietal GECs. The space between the

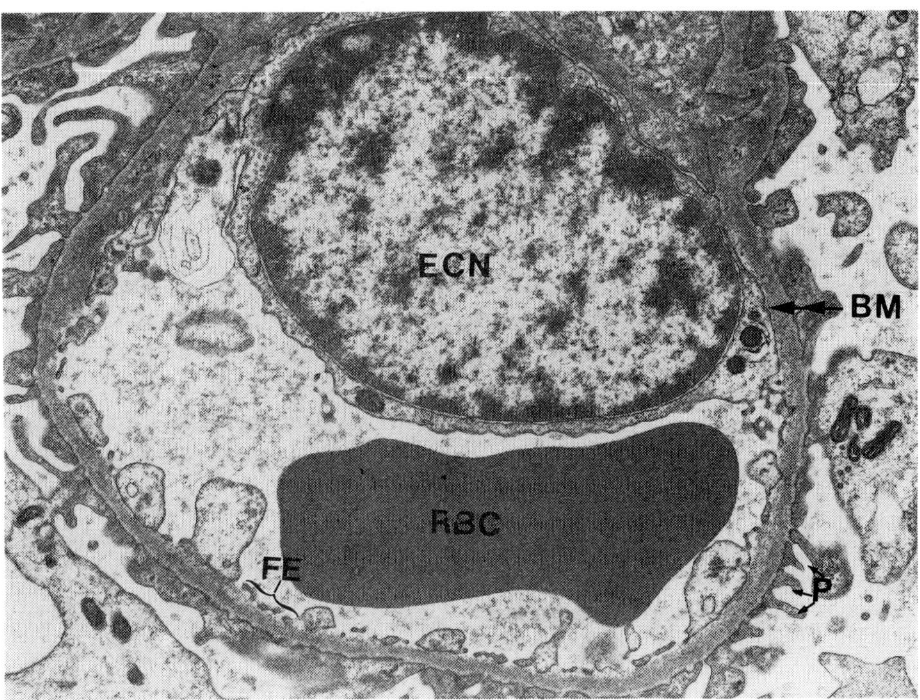

Figure 10.1 Electron micrograph of a capillary loop from a human glomerulus (magnification × 66 000). ECN, endothelial cell nucleus; RBC, red blood cell, within the capillary lumen; FE, fenestrated endothelium; BM, basement membrane (between arrow heads); P, podocytes. (Courtesy of Dr B. Hartley).

glomerulus itself and the surrounding capsule, known as Bowman's space, is therefore lined with the two types of GECs. At the junction of the glomerulus with the tubules the parietal GECs become contiguous with the TECs.

The visceral GECs have been more studied than the parietal cells. They are distinguished by their large size and their so-called foot processes, which give the name "podocytes" to the visceral GECs. The foot processes are formed by long cytoplasmic extensions which subdivide into pedicels. The pedicels are in direct contact with the GBM, and interdigitate with the pedicels of adjacent podocytes. The podocytes contain many lysosymes, actin filaments and myosin-like filaments. Their surface is negatively charged, mainly due to a sialoprotein known as podocalyxin. One of the main functions of the podocytes is to synthesize part of the GBM, the remainder of which is made by the endothelial cells of the capillaries.

Electron microscopy shows the GBM to have a trilaminar structure due to the contribution from the opposing epithelial and endothelial cells. A lamina rara externa lies on the epithelial side, and a lamina rara interna lies on the endothelial side, with the lamina densa in between. The main constituents of the GBM are collagen types IV and V and a number of glycosaminoglycans such as heparan sulphate, fibronectin entactin and laminin. Each layer of the GBM contains anionic sites which are mainly due to the presence of the heparan sulphate

molecules. Neutralization of the negatively charged GBM increases its permeability to proteins.

The podocytes and GBM combine functionally with the endothelium to form the filtration barrier, which filters the plasma in the capillaries of the glomerulus to form the urine in Bowman's space for processing along the renal tubule. Looked at from the point of view of molecules in the capillaries, the first element in the filtration barrier is the pores in the capillary endothelium and its negative charge. The next structure, the GBM, is the most important factor affecting glomerular permeability. It relies on its negative charge to reduce its permeability to negatively charged and neutral molecules such as proteins. The GBM also acts as a size-selective barrier. Thirdly, the visceral GECs provide a charge- and size-selective barrier by their own negative charge and the gaps of 30–60 mm width between the pedicels, which are known as filtration slits.

2.2 TUBULAR EPITHELIAL CELLS AND THE TUBULAR BASEMENT MEMBRANE

The tubular epithelium is a less complex structure than the glomerular epithelium, although its functional requirements are more extensive, i.e. secreting, reabsorbing and metabolizing a very large number of the normal constituents of urine.

The renal tubule is conventionally divided into three parts, which relate to both anatomical and physiological differences, but which are less distinct immunologically. The proximal tubule, further subdivided into the proximal convoluted tubule and the pars recta, is characterized by an increased surface area due to its "brush border", formed by the so-called microvilli. Invaginations of the cell membrane between the bases of the microvilli form the so-called "coated pits", which are coated with clathrin, involved in receptor-mediated endocytosis. The antigen responsible for the experimental form of nephritis known as Heymann nephritis is present in these clathrin-coated pits (see below). Proximal TECs are characterized by endocytic vacuoles and lysosymes, reflecting their function in the reabsorption and degradation of macromolecules from the ultrafiltrate in the tubular lumen. The other main function of the proximal convoluted tubule is to reabsorb electrolytes, water, glucose and amino acids, whereas the pars recta secretes acids and bases into the tubular fluid.

The second portion of the tubule is the thin limb of Henle's loop, which is characterized functionally by its permeability to water. The epithelial cells, particularly in animals, have been classified according to thickness and the presence or absence of microvilli.

The third part of the tubule is the distal tubule, and is itself composed of three parts: the ascending limb of Henle's loop, the macula densa and the distal convoluted tubule. The ascending limb of Henle's loop is much less permeable to water, but very permeable to sodium and chloride. Its epithelial cells seem to be of two populations, with or without small microprojections. The Tamm–Horsfall protein occurs on the luminal membrane. The ascending limb passes through the medulla and cortex to the glomerulus of its parent cortex to form the macula densa, where the epithelial cells of the tubule in contact with the glomerulus become thickened immediately after the distal convoluted tubule begins and continues until it meets its collecting duct.

Although the glomerulus has received more attention as an integral feature of immunological studies, owing to its prominence in human glomerulonephritis and the thrust of experimental work, it is often forgotten that the bulk of the kidney lies in the tubular epithelium, e.g. proximal TECs provide 80% of the renal cortex. As our understanding of the TECs as a source of immuno-modulating factors increases, we shall have to learn to regard the kidney as being proportionately somewhat different from our earlier views.

2.3 IMMUNE-MEDIATED AND INFLAMMATORY DISEASES OF THE GLOMERULAR AND TUBULAR EPITHELIA

The commonest cause of immune-mediated inflammation in the kidney is nephritis, which can be considered in two main histopathological groups, although they may overlap – glomerulonephritis and tubulointerstitial nephritis. The term "glomerulonephritis" covers many different histological appearances, which have been adopted as the main mode of classification. Details of these will not be discussed here, as the factors affecting immune inflammation in the kidney relate to most of these types. Tubulointerstitial nephritis has a much more homogeneous appearance, comprising infiltration of the renal interstitium with lymphocytes and neutrophils, and deposition in the tubular epithelium and on the TBM of immune deposits, consisting of immunoglobulins and complement components. The epithelial cells can be involved in these disorders, the nephritides, in a variety of ways:

1. by providing antigens which generate an inflammatory antibody response, either primarily or as a cross reactive antibody;
2. by acting as a barrier at which circulating immune complexes may be trapped, where the complexes will then cause inflammation;
3. by producing mediators and inhibitors of inflammation;
4. by binding through their receptor sites to mediators or suppressors of the immune response, or to antigens which will initiate the immune response.

As stated previously, much of our knowledge has come from experimental work, and this will inevitably provide most of the content of this section. The problem will be approached not in the conventional histopathological classification of nephritis, but by considering the epithelial cells and their basement membranes as separate structures and examining their roles in nephritis along the above lines.

2.4 GLOMERULAR EPITHELIAL CELLS

2.4.1 Glomerular Epithelial Cells as Antigens

One of the classic types of experimental nephritis is the rat model known as Heymann nephritis, which is produced actively by immunizing with the appropriate antigen derived from rat renal tubules, or passively by injection of antiserum (Heymann *et al.*, 1959; Sugisaki *et al.*, 1973). Extensive studies have now shown the antigen to be contained in a lipoprotein fraction of the brush border of rat tubular epithelium (RTE-a-5). Further purification has identified it as a glycoprotein of 330 kD (gp330), and it is thought to function as a fibronectin receptor. It is present in the cellular endoplasmic reticulum and Golgi apparatus. In GECs the antigen is found in clathrin-coated pits on the base of the podocytes, and are therefore in contact with the GBM, and in TECs in the clathrin-coated pits (apical invaginations) at the base of the microvilli. The disease arises by direct binding of the antibody induced by gp330 via a T cell-dependent mechanism to the antigenic sites on the GECs. Antibody-binding cross-links and insolubilizes the antigens, which are shed and accumulate as detectable immune complexes on the epithelial side of the GBM; it is thought the complexes remain *in situ* and are not cleaved because of binding of gp330 to the GBM. Fractions of rat renal tubule other than RTE-a-5 can induce similar lesions by similar mechanisms, and differences in strain susceptibility suggest that genetic mechanisms govern the expression of the antigens or the function of the autoreactive B cells. Of particular interest is a spontaneous disease in rabbits of similar pathogenesis (Neale *et al.*, 1984).

The microscopic appearance of the glomeruli in Heymann nephritis is that of membranous nephropathy, viz. thickening of the GBM, subepithelial deposits which contain components of complement and immuno-globulins, and a marked absence of inflammatory cells. The epithelial cells themselves remain intact. The tubules show a variable degree of damage. The close similarity between the histological appearances in Heymann nephritis and idiopathic membranous nephropathy led to the rather tenuous assumption that the human disease was caused by similar mechanisms. There is, however, little serological confirmation that this is so. Serum antibodies to tubular antigens are found rarely and, although glomerular eluates contain antibody to renal tubular antigens, technical considerations make these results difficult to interpret. Although gp330 or a gp330–like molecule is found in human tubules, it is not detect-able in human glomeruli, and no immune injury to the tubules is seen in membranous nephropathy, in contrast to Heymann nephritis. The hypothesis that human membranous nephropathy is, however, caused by antibody against GEC antigens remains attractive and valid.

Experimentally, GECs have been shown to be susceptible to injury by a variety of other means. Puromycin and adriamycin produce heavy proteinuria characterized by structural changes in the podocytes and their detachment from the GBM but without inflammatory mediators or cells. (Olson *et al.*, 1981) This last characteristic makes these experimental nephropathies similar to the glomerulopathy in humans known as minimal change nephropathy, so called because it is characterized histo-logically by no inflammatory or destructive changes, and by the absence of any immune mediators, such as immunoglobulins or complement. On electron micro-scopy, the only abnormality is fusion of the foot process of the GECs, but this change is not specific for this disorder, being found in other glomerular disorders associated with heavy proteinuria. The aetiology of minimal change is not known. The combination of its high therapeutic response to steroids, and mounting evidence that patients have various abnormalities of lymphocyte function, such as an impaired response to mitogens, increased suppressor function, and factors in their plasma which cause increased capillary permeability and proteinuria in experimental animals (Clark and

Williams, 1991; Wilkinson *et al.*, 1988), has been inter-preted as evidence that the basic abnormality is secretion by lymphocytes of a lymphokine(s) which increases glomerular permeability and so produces the heavy proteinuria characteristic of this condition. The most recent evidence to support this comes from the cloning of T cell human hybridomas from the T cells of a patient with minimal-change nephropathy, and the secretion by the hybridomas of a factor(s) which caused increased glomerular permeability and proteinuria in rats, and was cytotoxic to epithelial tumour cells in culture (Koyana *et al*, 1991). In rats, proteinuria has been produced by a monoclonal antibody directed against determinants of the filtration slit and podocytes (Orikasa *et al.*, 1988). No human analogues are recognized.

2.4.2   Glomerular Epithelial Cells as Sites of Binding/Trapping of Antigens/Immune Complexes

Extensive studies on various forms of experimental nephritis using defined antigens or preformed complexes have produced numerous models and a vast literature. The severity of the resulting nephritis depends upon many factors ranging from the nature of the antigen, the class of antibody response, the effectiveness of the monocyte phagocyte system in clearing complexes, and local properties of the glomerulus itself. Prominent among these are the charge and filtration function of the GECs. The negative charge on the epithelial cells favours the deposition of cationic antigens or complexes mainly in the subepithelial position. The ability of the glomerular capillary wall to bind antigen alone allows antibody to bind subsequently as a secondary event, the antigen–antibody complexes thus being formed *in situ*. This has now been recognized, at least experimentally, as an important mechanism causing nephritis, not least because it can explain the occurrence of immune complex disease in the absence of circulating immune complexes. One of the important factors influencing the nephritogenicity of complexes is their size, larger molecules in general being more prone to fix in glomeruli and produce inflammation. This is related to the filtration barrier effect of the GECs, small molecules and complexes being more able to pass through the barrier than larger ones. The influence of charge and filtration function on antigen or complex deposition is dynamic, as the very event of charge neutralization and structural damage alters the binding and filtering capacities of the GEC. A histological study has demonstrated a further dynamic aspect – human nephritis GECs can endocytose IgG, thus providing a clearance mechanism for antibody or antigen–antibody complexes (Al-Nawab *et al.*, 1991).

The damage wrought in the glomerular capillary wall is manifest by proteinuria and haematuria. In brief, the deposition of immune aggregates, or antibody to glomerular capillary wall components causes structural changes in the slit pore membranes, and may also neutralize the negative charge on GECs, GBM and endothelium. The effect of puromycin and adriamycin seems to be a combination of GEC detachment and possibly loss of negative charge. In grossly inflammatory glomerulonephritis, the whole structure of the capillary wall may be completely lost, with total disruption of any filtration mechanism.

In humans, the number and proportion of patients with nephritis in infectious nephritis or autoimmune disease. Analysis of the separate influences of the immune response, the haemodynamics of the glomerulus and its particular properties regarding charge and filtration cannot approach that in the experimental animal, but it is reasonable to assume that the human glomerulus behaves similarly.

2.4.3   Glomerular Epithelial Cells as Immune Mediators

The ability to culture glomerular cells, and thus define several of their functions, and techniques such as histochemistry and *in situ* mRNA hybridization have shown that GECs are not passive bystanders in the inflammatory events occurring in glomerulonephritis. They can both augment and diminish the immune response. Inevitably, most of our information comes from experimental work.

Although non-glomerular cells such as PMNs, macrophages and platelets are the main source of proteolytic enzymes in nephritis, the GECs, along with the endothelial and mesangial cells, also produce them. The proteinases, whatever their source, can interact with each other, and can produce damage by degrading the GBM, inactivating proteinase inhibitors or activating other proteinases, increasing the production of immune mediators such as prostaglandins, platelet-activating factor and reactive oxygen metabolites (Verroust *et al.*, 1986). A further role for proteolytic enzymes in glomerulonephritis is their ability to stimulate the proliferation of glomerular cells (Verroust *et al.*, 1986). Proteolytic enzymes produced by the glomerulus include cathepsins, metalloproteinases and gelatinase.

Recently, evidence is accumulating that GECs can secrete cytokines, thus conferring on them the ability to influence inflammatory events, along with the cells of the mononuclear phagocyte system and T cells. TNF-α has been shown to be produced by GECs isolated from normal rats, and rats which also have autologous nephrotoxic serum nephritis (Jevnikar *et al.*, 1991). Although the amounts measured were low, in the microenvironment of the GEC, TNF-α could extend its pro-inflammatory effects, i.e. neutrophil activation, induction of coagulation on endothelium, collagenase production, induction of VCAM, and increasing MHC expression on lymphocytes and endothelial cells. The ability of TNF-α to increase fibroblast growth may be relevant to long-term sequelae of nephritis, such as scarring and sclerosis.

ICAM-1 (CD45) has been detected in the parietal GECs in normal human kidney, conferring the likelihood of increased adhesion of immune-active cells (Muller *et al.*, 1991). The GECs have an involved relationship with the complement system, which has been demonstrated in several different experimental models and in humans to cause glomerular damage. Until recently the cells of the glomerulus were regarded as passive targets of complement-mediated damage, but it is now clear that GECs can produce complement and regulate complement activation, all probably on a very local basis. GECs from normal kidneys synthesize C4 and C3, and have been shown to transcribe the genes for C4 and C3 (Feucht *et al.*, 1989). Production of these two components is increased *in vitro* by IFNγ, indicating the active role GECs can play in inflammation. Confirming the potential of locally produced complement to contribute to inflammation is the increased expression of mRNA for C2, C4, C3 and factor B in the kidneys of lupus mice (Passwell *et al.*, 1988).

There is a considerable body of evidence to show that the terminal complement component complex C5b–9, also known as the membrane attack complex (MAC), is present in the glomeruli of human and experimental nephritic kidneys (Rauterberg, 1987). C5b–9 can act either to cause lysis of the cell membranes, or, in sublytic doses, as a trigger for the cellular production of substances such as, in the case of GECs, prostaglandin E and thromboxane (Petrulis *et al.*, 1981). The ability of C5b–9 to stimulate production of reactive oxygen species and IL-1 from other cells may also apply to GECs. It is of interest that increased urinary excretion of C5b–9 has been shown in membranous nephropathy in humans (Schulze *et al.*, 1991). Complement may also be involved in chronic inflammation and scarring, as C5b–9 has been shown *in vitro* to stimulate synthesis and release of collagen by GECs in humans (Torbohm *et al.*, 1990).

It is not surprising that there are also locally produced factors which can alter complement activation or its effects. The classic pathway is activated by vimentin, present in the GECs. Factors produced by GECs which can diminish complement-mediated injury are DAF (Quigg *et al*, 1989), which increases the decay rate of C5b, and protectin (CD59) (Meri *et al.*, 1991), a recently described glycoprotein which inhibits the effects of C5b–9 on cell membranes. Rat GECs produce a proteoglycan which inhibits complement activation (Quigg *et al.*, 1991) by preventing the formation and accelerating the decay of C3 and C5 alternative pathway convertases. The general property of GECs which reduces complement activation can have an obvious regulating effect on cell damage.

The GECs, in common with neutrophils, erythrocytes, lymphocytes, monocytes and dendritic cells, have receptors for C3b/C4b on their surfaces. The C3b receptor (CR1) is a glycoprotein and is distributed in the normal human kidney on the plasma membrane of the body and pedicels of the podocytes. Although detected in human fetal GBM, it is not found there in normal adult kidneys (Appay *et al.*, 1988). The numbers have been shown to be reduced in a variety of human types of nephritis (Nolasco *et al.*, 1987). CR1 down-regulates complement activation; this receptor also transports complement-bound complexes across cell membranes in other cells, and it is interesting to speculate whether this occurs in GECs. Other complement receptors, CR2 and CR3, are not demonstrable on GECs.

The eicosanoid system has a complex relationship with the glomerulus in glomerulonephritis (Williams and Davies, 1991). GECs exhibit lipoxygenase activity (Jim *et al.*, 1982), and glomerular cells as a whole possess cyclooxygenases (Hassid *et al.*, 1979), and thus have the ability to produce prostaglandins and leukotrienes. Alteration in the charge on GECs, by adding polycations in culture, which mimics what is thought to occur in glomerulonephritis *in vivo*, causes increased production of prostaglandins (Pugliese *et al.*, 1987).

IgG–Fc receptors are present in GECs (Mizoguchi and Horiuchi, 1982) and have an obvious propensity to bind immunoglobulins and antigen–antibody complexes, either when the latter are formed *in situ* or have passed through the filtration barrier. Activation of these receptors may also cause cell proliferation.

The renal epithelia coexist with other cells of the kidney, and the GECs have a close physical relationship with the mesangial cells, with a two-way functional relationship, which in the case of GECs extends to their production of stimulators and inhibitors of mesangial cell growth (Castellot *et al.*, 1985); epidermal cell growth factor has been demonstrated in mouse kidney, and mesangial cells have receptors for this.

2.5   GLOMERULAR BASEMENT MEMBRANE

2.5.1   Glomerular Basement Membrane as an Antigen

It is now clearly proven that the GBM can act as an antigen to cause an antibody-mediated disease in humans and, spontaneously, in some animals. In humans the disease has become widely known as "anti-GBM disease", replacing an earlier term, "Goodpasture's syndrome". Goodpasture noted an association between influenzal pneumonia, pulmonary haemorrhage and acute crescentic nephritis, often causing acute renal failure, in 1918–1919. Forty years later the development of immunological techniques showed that immunoglobulin and complement were deposited on the GBM in a linear fashion, and that the sera of patients contained antibody which fixed to the GBM of normal human kidney and lung sections.

The antigen responsible in humans has been shown to be within the α_3 chain of the non-collagen portion of type IV collagen (Hudson *et al.*, 1989; Turner and Pusey, 1991). In the kidney it is present in the lamina densa. It is also found in the alveolar basement membrane of the lung, although not all patients with anti-GBM disease have pulmonary haemorrhage. This may be due to differences in the way in which the antigen is cross-linked in the lung.

It has recently been recognized that the Goodpasture antigen is absent from the GBM in patients with Alport's syndrome, a hereditary nephritis associated with abnormalities of the lens and cochlear membrane. Alport's syndrome is X linked, and a gene for the α_5 chain of collagen has been localized on the X chromosome in the same region as the Alport gene (Hostikka *et al.*, 1990). A practical consequence of the absence of the Goodpasture antigen in Alport's sydrome is the development in some patients of anti-GBM disease following transplantation.

Animal models of anti-GBM disease have been extensively studied. Among the first were passive immunization of primates with serum from patients, and Steblay nephritis, in which sheep are immunized with homologous or heterologous GBM. More recently, a rat model in which autoimmunity is induced by mercuric chloride has been studied (Pelletier *et al.*, 1987). Autoreactive T cells produce a polyclonal B cell response which results in autoantibody production including antibodies to components of the GBM such as fibronectin, laminin and collagen IV (Pelletier *et al.*, 1988). These produce the first phase of the disease in which there is linear deposition of IgG on the GBM. A second phase follows, in which the histological appearances develop into those of membranous nephropathy, thought to be caused by the deposition of anti-idiotypic antibodies, or of circulating complexes of GBM and anti-GBM antibodies. An interesting aspect of this mercuric chloride-induced model is that there is genetic control of its susceptibility.

In humans, there are well-known drug-associated nephropathies in which the GBM may be acting as an antigen (Fillastre *et al.*, 1988). The best known examples are heavy metals, e.g. gold, and drugs containing sulphydryl groups, e.g. penicillamine. Both can cause a membranous nephropathy, and penicillamine can cause an anti-GBM-like disease, with linear deposition of IgG on the GBM. The membranous nephropathy induced by these drugs is associated with HLA-DR3 and, in the case of gold, also with HLA-B8. The pathogenetic role of the GBM is not clear, but it may bind the drugs as haptens, provide antigens which cross-react with antibodies formed against tubular antigens induced or exposed by the drugs (i.e. analagous to Heymann nephritis) or provide antigens which are released into the circulation, or induce an antibody response, with the formation of circulating immune complexes which then deposit in the glomeruli.

Antibodies to fibronectin have been reported in IgA disease (Cederholm *et al.*, 1986), and antibodies cross-reacting with laminin, heparan, sulphate and type IV collagen in the GBM and streptococcal M protein have been found in post-streptococcal glomerulonephritis (Fillit *et al.*, 1985; Kefalides *et al.*, 1986).

2.5.2 Glomerular Basement Membrane as a Site of Binding/Trapping of Antigens/ Immune Complexes

The charge- and size-related filtration functions of the GBM are, not surprisingly, involved in its ability to act as a focus for antigens or complexes. These will obviously be selected by being cationic or of a certain size. Laminin can bind C3 and C3d. Of particular note is the ability of DNA to bind to the GBM via collagen (Izui *et al.*, 1976), which can relate to nephritis in two different ways. First, DNA, often released following infection, e.g. by the action of lipopolysaccharide, can bind to the GBM and act as a planted antigen, to which anti-DNA antibodies may fix. This is one possible mechanism causing the nephritis of systemic lupus erythematosus. Secondly, a large number of antibodies which cross-react with DNA and substances with similar sequences, e.g. phospholipids and glycoproteins, may be bound to DNA, and the DNA molecule itself binds the DNA–anti-DNA complex to the GBM. Subsequent analysis of the specificity of the cross-reacting antibody may then identify it as binding because of, for example, its affinity for heparan sulphate, which may misrepresent the true cause of its presence on the GBM (Brinkman *et al.*, 1989).

2.5.3 Glomerular Basement Membrane as a Immune Mediator

Being acellular, the GBM alone does not play a large role as a mediator of the immune response. However, when the GBM is degraded by the inflammatory process or is incorporated as an antigen in immune complexes, it stimulates macrophages to produce TNF, IL-1, lysosomal enzymes and proteinases (Vissers *et al.*, 1989), and it can also activate complement (Williams *et al.*, 1987).

2.6 TUBULAR EPITHELIAL CELLS AND TUBULAR BASEMENT MEMBRANE

Immune-mediated and inflammatory diseases of TECs and TBM have several differences compared to diseases of their glomerular counterparts. In humans, they are less important clinically as they occur less frequently and cause many fewer cases of end-stage renal failure. Pathologically, interstitial nephritis, demonstrable as an infiltrate of monocytes in the interstitial tissue of the kidney between the nephrons, is a frequent accompaniment of tubular inflammation, and cell-mediated immunity is therefore presumed a common mediator of injury. In experimental studies the models are not so "clean" as in experimental glomerular disease, as immunization with, for example, TBM will produce anti-

TBM antibody deposition on the TBM, immune complex deposition and interstitial nephritis.

2.6.1 Tubular Epithelial Cells and Tubular Basement Membrane as Antigens

Experimental studies using tubular cells or brush border antigens as immunogens produce various pictures of disease, depending upon the species used. These range through: anti-TBM antibodies, with deposition of IgG on the TBM; interstitial nephritis, with cellular infiltrate; immune complex deposition, with deposits of immuno-globulin and complement on the TBM. This last appearance is found in Heymann nephritis in which the deposits also contain the antigen gp330 (see above); in contrast to the glomerular deposits, however, the tubular deposits do not persist. The mechanism of deposit formation may well be similar, i.e. cross-linking of antigens on tubular cell surfaces by antibodies to form complexes which are then shed from the cells. Of particular interest is a model in which a normally occurring tubular protein secreted into the lumen, the Tamm–Horsfall protein, is used as antigen. It is also known as the glycoprotein uromodulin, the function of which is to regulate the activity of cytokines, particularly IL-1 and TNF. Active or passive immunization produces a nephritis characterized by deposits of the protein, C3 and IgG on the epithelial side of the TBM and cellular infiltrate (Friedman *et al.*, 1982).

This last type of experimental nephritis may be of relevance to humans, in whom antibodies to this protein are found, but as yet not linked to a convincing occurrence of disease. There are sporadic reports in humans of antibodies to TECs in autoimmune disease such as SLE, and in association with chronic fibrosing alveolitis and hepatitis. Anti-TBM antibodies in humans are not common, and are most frequently seen in association with anti-GBM disease. Isolated cases have been reported in association with infection, drugs, and in transplanted patients. Here, two types of anti-TBM occur: anti-TBM which reacts with graft only, and anti-TBM which reacts with both graft and active kidney. The implication is that individuals have immunologically different antigenic structures in the TBM.

The functional importance of changes induced by anti-TBM antibody and anti-TEC antibody varies. Renal failure commonly occurs, and there may or may not be tubular disorders, e.g. renal tubular acidosis. The main consideration is the amount of associated interstitial infiltration with monocytes. Experimentally immuniza-tion with preparations of TBM either isologous or heterologous, induces anti-TBM antibodies with the variable picture described above, and, not surprisingly, anti-GBM antibodies too.

2.6.2 Tubular Epithelial Cells and Tubular Basement Membrane as Sites of Binding/ Trapping of Antigens/Immune Complexes

In experimental models using foreign antigens to cause immune complex disease, the antigen–antibody complexes are deposited at extraglomerular sites which include those

on the TBM between the TECs. A similar abnormality occurs in animals which spontaneously develop SLE. In humans, SLE provides the most common example of immune complex deposition in the tubules, with additional deposit formation in the interstitium in some instances. The deposits contain immunoglobulins, complement and nuclear antigens. Infrequent associations of tubular and interstitial immune deposition have been recorded in idiopathic and drug-induced forms of nephritis.

2.6.3 Tubular Epithelial Cells as Immune Mediators

The TECs have some important differences as a source of immune mediators or agent facilitating immune mediated damage when compared to the GECs. A significant property is the expression of MHC class II in inflammation caused by rejection or nephritis, e.g. in SLE (Fuggle *et al.*, 1986; Muller *et al.*, 1989). In these circumstances the TECs can act as antigen-presenting cells, and so initiate the immune process. When activated in this way, TECs either express TNF on cell membranes or secrete it, and so contribute to inflammation and modulate the immune response by the actions of this cytokine (Wuthrich *et al.*, 1990). TNF is also produced in response to lipopolysaccharide and IL-1. Another cytokine produced by TEC is IL-8, which is chemotactic for both lymphocytes and neutrophils (Schmouder *et al.*, 1992). Human TECs have been shown to produce IL-8, both as expression of mRNA and secretion in response to TNF, IL-1 and lipopolysaccharide, and during allograft rejection. As in the case of GECs, TECs transcribe genes for C4 and C3 (Brooimans *et al.*, 1991) and, in the case of the distal collecting tubules, have been shown to contain CD59 (protectin) (Meri *et al.*, 1991), thus also giving them the ability to produce complement and modulate its activity.

Another way in which TECs can affect the immune response is by the production of the adhesion molecule ICAM-1 (CD45), which promotes interaction with other immunopotent cells, including antigen presentation. ICAM-1 is not expressed in proximal TECs in the normal human kidney, but when nephritis occurs these cells produce ICAM-1, in association with HLA class II molecules as noted above. Murine TECs in culture increase their production of ICAM-1 when exposed to IFNγ, IL-1α, and TNF-α.

The TECs in proximal tubules and the ascending loop of Henle contain cytochrome P-450 mono-oxygenase, and so are capable of producing epoxides from arachidonic acid (Endou, 1983). The direct mediators of tissue damage, reactive oxygen species, are produced by rabbit epithelia *in vitro*, both at basal levels and increasingly so on stimulation by aggregated IgG (Rovin *et al.*, 1990). It is now very clear that the renal epithelia must be viewed as an immunocompetent tissue, capable of antigen presentation, producing phlogistic molecules, regulating inflammation, and responding to extraneously produced molecules which partake in the immune response. The epithelial cells, however, must not be thought of in isolation, as they form a continuum with the endothelial cells, the glomerular capillaries and the mesangial cells, which are also immunocompetent. This area is currently a focus of active research, and there will doubtless be further information rapidly forthcoming which will provide a more detailed and extensive picture of the renal epithelia as an active participant in immune events in the kidney.

3. Mediators of Immune Injury and Inflammation

Several different mediator systems cause inflammation of and damage to the glomerular and tubular epithelia. There is a complex interplay between them, and their effects overlap. As has been noted previously, most of our knowledge comes from animal studies, with the greater part of our knowledge in humans coming from circumstantial evidence obtained from renal biopsies performed during the course of the disease in question. Although the epithelia and basement membranes of the kidney are usually the targets for the immune mediators, they do themselves produce mediators, as noted above, and so contribute to the inflammatory process as a whole.

In this section, a brief description will be given of the various mediator systems, which have a role in renal disease involving the epithelia and basement membranes. Space does not allow a detailed description of their induction, effects and control. The succeeding section, on the immunopharmacology of the diseases affecting renal epithelia, will examine how the drugs available can affect some of these separate components of inflammation. The mediators can be divided into cellular and hormonal.

3.1 CELLULAR MEDIATORS

3.1.1 Polymorphonuclear Neutrophils

Their presence in glomeruli and tubules in human disease, and the reduction or abrogation of injury in depletion experiments, indicate the role these cells have. PMNs produce damage by the release of a host of enzymes and activators, and chemotactic agents for themselves and monocytes. Among the enzymes released are elastase, collagenase, hydrolases, gelatinase, proteinase 3 and myeloperoxidase. In humans, PMNs have been observed to breech the endothelium and thus be in close contact with the GBM, and sometimes to penetrate this structure. In this position the ability of PMN enzymes to destroy collagen is much enhanced. Proteolytic enzymes may act in other ways to cause damage. Some proteins can be activated, e.g. enzymes such as zymogen, which may lead to coagulation, cells can be stimulated to produce reactive oxygen species, and glomerular cells in culture can be stimulated to proliferate. PMNs can activate the complement system and the clotting system, and their chemoattraction of monocytes magnifies the effects which these cells have (see below).

3.1.2   Monocytes

As with PMNs, histology has identified monocytes in human and animal glomerular and tubular–interstitial nephritis. They are particularly important as a component of the crescents which form in the urinary space, combining with the epithelial cells from Bowman's capsule. These cells produce many enzymes, inhibitors of these enzymes, complement components, reactive oxygen species, initiators of coagulation, interleukins, and factors prohibiting and inhibiting growth.

3.1.3   T Lymphocytes

Specific monoclonal antibodies defining T cell surface markers have shown these cells as part of the population of the cells in crescents and proliferating lesions in human nephritis. Most are CD4 T cells. In general, their presence is correlated with that of monocytes. These findings in humans have been mirrored in experimental work. A minority of cells have been found to be cytotoxic T cells and natural killer cells.

3.1.4   Platelets

Platelet antigens have been found in glomeruli, and considerable circumstantial evidence of platelet involvement has been obtained by measuring platelet aggregation, and plasma concentrations of platelet activating factor(s), platelet factor 4 and serotonin, both produced by platelets. Platelets can cause damage to epithelia by direct cellular injury, by increasing adhesion of monocytes, and, of especial interest in the glomerular epithelium, by release of cationic proteins which bind to anionic sites and thus cause loss of epithelial or basement membrane charge leading to proteinuria.

3.2   HORMONAL MEDIATORS

3.2.1   Complement

The complement system is the main effector system for antigen–antibody reactions. Its role in nephritis has been studied extensively. Animal studies have shown its requirement for immune-mediated reactions, either in complement-depleted normal animals or in animals genetically deficient in various components of the complement system. In humans, as also in animals, circumstantial evidence for the involvement of complement in nephritis comes from immunohistological evidence of complement components in glomeruli and tubules. The components most often seen, because they are most frequently sought for, are C3, C4 and C1q, but other species of MAC

are detectable. Much attention has been paid recently to MAC because of its ability to produce lesions in all membranes by altering their charge. Of even greater interest is the recent realization that sublytic doses of MAC, when attached to cell membranes, can regulate intracellular enzymes. Aside from its direct effects on cell membranes, activated complement contributes powerfully to inflammation by producing molecules such as C3a and C5a, which are chemotactic for PMNs and monocytes, anaphylotoxic, vasodilatory, and increase leucocyte production.

There are many different activators of the complement system, and some of these are particularly worthy of mention from the renal point of view. The classic pathway is primarily activated by aggregated immunoglobulins or antigen–antibody complexes, and as these are frequently present in nephritic glomeruli or tubules they could clearly activate complement *in situ*. Immunoglobulin activation of complement is restricted to IgG1, IgG2 and IgG3 (but not IgG4) and IgM. Endotoxins and some viruses can directly activate C1, and activated enzymes such as trypsin can directly activate C3. Direct activation of C4 by proteases, produced by PMNs, and of C5 by reactive oxygen, also generated by PMNs, can increase complement activation when PMNs are present. The alternative pathway exists mainly as an immunoglobulin-independent opsonizing system (although aggregated IgA is effective *in vitro*, and also IgG-activated antibodies where combining with cells infected with virus). It is usually activated by particles, e.g. bacteria or fungi. These organisms, therefore, can cause intrarenal complement activation; perhaps of greater interest is the ability of altered GBM to activate the alternative pathway, and the ability of GECs, via vimentin particles, to activate the classic pathway.

Such a powerful system as complement requires an efficient method of suppressing and checking its activity, and several proteins dedicated to this purpose exist, e.g. C1 inhibitor, factors H and I which inactivate C5b, C3b and C4b, and the S and SP proteins which inhibit MAC. All these have been detected in inflamed renal tissue.

3.2.2   Coagulation

Early observations of fibrin deposition in glomerulonephritis excited much interest because of the possible implications for treatment. Surprisingly, factor VIII is not co-deposited, indicating that the fibrin is not produced via the normal coagulation process causing blood clotting in wounds. Parallel observations of increased fibrin degradation products in the urine and serum confirm the role of fibrin. These observations apply to humans and experimental animals, and, in the latter, the role of fibrin has been demonstrated by using anticoagulants, particularly streptokinase and ancrod, which partly protects against injury.

In the absence of factor VIII, the question arises of how fibrin deposition occurs. A factor, procoagulant activator, is produced by monocytes when stimulated by antigen–antibody complexes and is detectable in TECs, and is presumed responsible (Matsuda *et al.*, 1979).

3.2.3 Reactive Oxygen Species

Hydrogen peroxide, and the hydroxyl, hypochlorous and superoxide anions, are produced by the respiratory burst of PMNs and macrophages. They are metabolites of oxygen with other atoms attached, and are powerful agents designed to destroy bacteria, but can also damage the host's own tissue. There is a extensive system of control of these aggressive molecules, whose half-life is usually therefore very short, e.g. vitamins C and E, glutathione, and superoxide dismutase. Evidence for their involvement in renal disease is based on the protective effect of these substances in experiments in which increasing their availability protected against injury. Differences in susceptibility between the distal TECs and proximal TECs have been noted, dependent upon antioxidant activity (Andreoli, 1991). The TECs respond to reactive oxygen species by diminished production of ATP and cell eicosanoids. The eicosanoid derivatives of arachidonic acid form a complex system of mediators with equally complex effects on renal injury as well as physiology, e.g. by controlling glomerular blood flow and filtration. Their involvement in nephritis is demonstrated by increased production and protective effects, e.g. by indomethacin, in experimental animals. Their effects range from increasing adhesion of PMNs and monocytes to causing basement membrane damage. They are produced in the kidney in mesangial cells, GECs and TECs.

3.2.4 The Kinin System

Hageman factor, bradykinin and kallikrein have all been implicated in glomerular damage in experimental disease, to which they could contribute by causing coagulation, leucocyte chemotaxis and increased vascular permeability.

3.2.5 Antibodies Alone

Although antibodies depend mainly on the complement system and the results of its activation in reuniting other effector systems, they have been shown experimentally to cause mild proteinuria on their own, but without inflammation.

3.2.6 Cytokines

The interleukins, TNF and the interferons all contribute to the inflammatory process in the kidney. Their actions involve cell proliferation, induction of protein synthesis, and producing molecules active in inflammation, e.g. coagulants and eicosanoid derivatives. This involvement provides opportunities for treatment if appropriate blocking agents could be developed.

3.2.7 Growth Factors

The cells of the glomerulus and tubule are capable of responding to growth factors or mitogens such as platelet-derived growth cell factor, epidermal growth factor, fibroblast growth factor and transforming growth factor, the last diminishing rather than increasing proliferation. The growth factors may be relevant to chronic scarring, not only to proliferation during the acute inflammatory response.

4. Immunopharmacology

As stated at the outset of this chapter, the renal epithelia, compared to other epithelia in the body, are relatively inaccessible. The only way in which therapeutic agents can reach the renal epithelia is via the circulation. Drugs can therefore not be applied solely or in the first instance to the renal epithelia; when treated, these cells are inseparable from the other cells of the kidney. In practice, this is not as disadvantageous as it might appear, as most disorders requiring treatment do not involve the renal epithelia alone. Developments in drug targeting may provide drugs which can selectively act on the renal epithelia or basement membranes, e.g. by coupling a drug to a monoclonal antibody directed against a protein peculiar to or highly concentrated in the basement membrane.

The renal disorders requiring immunotherapy can be broadly viewed as nephritis in which therapy may be directed against primary disease in the kidney, e.g. anti-GBM nephritis, or against an underlying disease which has involved the kidney secondarily, e.g. systemic lupus, and rejection of a kidney transplant. Amongst these, only three diseases can be considered as being exclusively of the renal epithelia. These are anti-GBM disease, membranous nephropathy, and minimal-change nephropathy, and even the first two of these can be accompanied by lymphocytic infiltration of the interstitium. All other types of nephritis are characterized by inflammation and/or deposition of immune reactants in the mesangium, and/or vessels, and/or interstitium, as well as in the epithelia and basement membranes.

In general terms, immunotherapy can be regarded as having three approaches, which are complementary. These are the prevention of antigen presentation, of the immune response once antigen presentation has occurred, and of the consequent recruitment of the inflammatory response. The focal point of the immune response is the specific activation of T cells by antigen. Antigens are presented to T cells by macrophages and other antigen-presenting cells as peptide associated with an HLA molecule. A T cell recognizes the peptide/HLA complex through its receptor, which is specific for that complex. This complex of T cell receptor, antigenic peptide and HLA molecule provides a target for immunotherapy because of the polymorphism of the host's molecules and the huge variation in antigenic peptides, allowing, in theory at least, the development of monoclonal antibodies against T cell receptors and HLA molecules, and peptides which will compete with disease-causing peptides for the binding sites. Once the immune response is initiated by the recognition of antigen, the immunotherapeutic targets then become the interleukins, which stimulate the other T cells and B cells to provide cytotoxic and antibody-mediated immune damage, the interleukin receptors on these cells' surfaces, the T and B cells themselves, and antibodies. The final arena for treatment is the many mediators of inflammation invoked by T cells and antibody or antigen–antibody complexes, described above.

In practical terms, the nephrologist's present choice of immunopharmacological agents is limited, and most of them have broad effects, which are not in any way specifically aimed at precise immune events occurring in the kidney. This is disappointing against the background of our increased knowledge of the causative and inflammatory mechanisms of nephritis. The main therapeutic innovations in the last 20 years have been the introduction of plasma exchange and cyclosporin. There have been some successes, such as the improvements in patient survival rate and kidney survival rate in nephritis due to SLE and vasculitis, and in transplantation. However, as a measure of the size of the problem and our still disappointing degree of attainment, it should be noted that nephritis still accounts in the developed countries for 30% of patients requiring dialysis or transplantation for end-stage renal failure, and that the rate of loss of transplanted kidneys is still 10–15% in the first year, with a succeeding loss from the surviving pool of 5% per annum, which equates with an overall 50% survival of transplant kidneys at 5 years. Furthermore, although improvements have occurred these have been due to more rational and careful use of available drugs and better management of their complications, e.g. infections, rather than any innovations, with the exception of the effect of cyclosporin on results in transplantation. Another difficulty in the field of nephritis is the lack of adequate controlled trials containing an appropriate number of subjects to provide data on the therapeutic questions arising in everyday practice. The forms of treatment presently used will first be reviewed, and then those that are developing, or which will realistically be available soon.

4.1   CURRENT TREATMENT

4.1.1   Steroids

One of the most important effects of steroids is on the production of interleukins – the synthesis of TNF, IL-1 and IL-6 is directly reduced, and that of IL-2 indirectly, with consequent effects on T and B cell function. The synthesis of prostaglandins is also reduced. There is, however, little or no reduction in the antibody response to known doses of exogenous antigen, e.g. typhoid vaccine, but there is a reduction in autoantibodies in autoimmune disease. The number of circulating lymphocytes is reduced, but this is due to sequestration in the haemopoietic system, and not to their destruction. The expression of MHC class I antigens on human epithelial cell cultures is reduced, which potentially provides steroids with another immunosuppressive mechanism *in vivo*.

4.1.2   Cyclophosphamide

This drug is cytotoxic through its irreversible alkylation of DNA. All lymphocytes are susceptible to its action, the order of increasing sensitivity being cytotoxic T cells, T helper cells, B cells, and T suppressor cells. The direct effect on B cells means that it can suppress antibody responses to T-dependent and T-independent antigens.

4.1.3   Azathioprine

The active metabolite of azathioprine, which is itself inactive, is thioinosinic acid, which is derived via the intermediate metabolite 6 mercaptopurine. Azathioprine acts by altering purine synthesis, which therefore inhibits formation of DNA and RNA. T cells are the most sensitive, which fail to proliferate, and therefore cannot mount an immune response. There is experimental evidence to suggest that B cells may be directly affected, and *in vivo* in humans there is impairment of specific antibody responses, with greater suppression of IgG production than of IgM.

4.1.4   Chlorambucil and Nitrogen Mustard

These alkylating agents are uncommonly used to treat renal disease, and are used virtually exclusively in systemic lupus nephritis.

4.1.5   Cyclosporin

Cyclosporin mainly acts by inhibiting production of IL-2 in lymphocytes, which have a receptor, cyclophilin, for cyclosporin. Cyclophilin is an enzyme, peptidyl-propyl isomerase, which activates the cell nucleus to transcript the gene for IL-2, and is inhibited by the binding of cyclosporin. The failure of lymphocytes to produce IL-2 abrogates T cell growth and proliferation, production of IL-1 by macrophages and production of immunoglobulins by B cells. Cyclosporin has other actions, i.e. inhibition of gene transcription for IFNγ and the expression of IL-2R. It can affect the immune response after its onset, and its action is reversible.

4.1.6   FK.506

This substance, derived like cyclosporin from fungi, reacts with a binding protein, phosphatidylprolyl isomerase, and inhibits the production of IL-2, IL-3 and IL-4, IFNγ, and the expression of IL-2R.

4.1.7   Rapamycin

This drug warrants particular attention as it acts throughout the T cell cycle, and inhibits B cell activation.

4.1.8 Antilymphocyte Polyclonal and Monoclonal Antibodies

These reagents remove T and B cells, thus diminishing the cytotoxic and humoral effector arms of the immune response. One of the most widely used in transplantation is OKT3, a monoclonal antibody to a portion of the recognition molecule CD3 common to T cells. Administration of OKT3 has two effects – functionally blocking the T lymphocyte receptor, and also an unfortunate side-effect, which is the activation of the targeted cells with release of cytokines, notably IL-1, which can cause fever, myalgia, capillary leak, and sterile meningitis.

4.1.9 Immunoglobulin Therapy

There is increasing interest in the use of intravenous pooled human immunoglobulin for the treatment of several autoimmune diseases. Although the mode of action is not clear, clearance of autoantibodies by anti-idiotypic antibodies or rheumatoid factor antibodies in the pooled immunoglobulins seems likely.

4.1.10 Plasma Exchange

Selective removal of plasma from the circulation, either by plasmapheresis or haemofiltration, is aimed at removing antigen, antibody, antigen–antibody complexes and activated mediators of inflammation.

4.1.11 Anti-inflammatory Drugs: Steroids

As well as affecting the immune response as described above, these have a well described anti-inflammatory action.

4.1.12 Anticoagulants

The fibrinogen deposition in many cases of glomerulonephritis led to the seemingly logical use of anticoagulants. This was before it was realized that the fibrinogen deposition occurred without the co-deposition of factor VIII, and that the conventional anticoagulants would therefore not be expected to influence this process. Despite many contributions to the literature, it was not satisfactorily shown that anticoagulants conferred any benefit.

It is of interest that heparin, used in acute diseases, such as rapidly progressive glomerulonephritis, also has an anticomplement activity and inhibits kallikrein, raising the possibility that any beneficial effects it had may have been due to this, rather than anticoagulation. Another mechanism by which heparin may act is accelerating the removal of glomerular antigen. In experimental chronic serum sickness this effect was demonstrated for heparin independently of any fibrin deposition. It was suggested that because of its structural similarity to heparan sulphate, a component of the GBM, the more anionic heparin disrupts attachments between antigen and the GBM (Furness, 1990). The polyanionic heparin may also provide a source of negatively charged molecules which repairs a decrease in the negative charge of the filtration barrier.

4.1.13 Antiplatelet Drugs

The experimental and human evidence that platelet activation is involved in the pathogenesis of glomerulonephritis led to treatment with drugs such as dipyridamole. As in the case of anticoagulants, there is no firm evidence that they are efficacious.

4.2 USE OF CURRENT DRUGS

The use of current drugs will be discussed briefly and in general terms; the reader is referred to textbooks of nephrology for detailed discussion of the treatment of individual forms of nephritis and of transplanted kidneys.

In the case of nephritis, treatment is usually reserved for patients with:
1. rapidly progressive nephritis, in which there is a deterioration in glomerular filtration over days or weeks, e.g. in vasculitis;
2. a severe nephrotic syndrome, e.g. caused by membranous nephropathy;
3. the nephrotic syndrome, due to minimal-change nephropathy, in which there is a response rate of 85–90% to treatment with steroids which, in many cases, can be given in a single short course, or interrupted short courses.

The approach usually consists of a combination of steroids and immunosuppressive drugs, except in minimal-change nephropathy, in which most cases respond to steroids alone. A less frequently used drug is cyclosporin, which has been used in nephritis which is refractory to steroids and immunosuppressive drugs. There have been recent reports of the use of pooled intravenous human immunoglobulin, but the cases as yet are too few to assess.

In transplantation, the number of drugs used is greater, and so is the number of combinations in which they are used. Cyclosporin and antilymphocyte drugs are used routinely in many units, in contrast to their use in nephritis. The use of FK.506 and rapamycin is being actively explored, although neither is utilized widely yet.

4.3 POTENTIAL TREATMENT

The following brief comments refer to drugs which are being used experimentally, but for which there are reasonable expectations that they will be used in humans, or to drugs which have already been used in a small number of cases. Each is aimed at inhibiting the inception of the immune response at the stage of inception, i.e. by intefering with the complex of antigen plus HLA molecules and T cell receptor.

4.3.1   Anti-Human Leucocyte Antigen Antibody

Monoclonal antibody to HLA in mice has proved effective in conditions such as collagen-induced arthritis (Wooley *et al.*, 1985).

4.3.2   Blocking Peptides

Peptides can be synthesized which bind to the HLA molecule on antigen-preventing cells, but which fail to activate the lymphocytes with the T cell receptor which initiates the disease. These blocking peptides may differ from the pathogenic peptide by only one amino acid. Of interest is a recent observation that in order to function as a blocking agent a peptide does not necessarily have to have a homologous sequence with the disease-causing peptide (Lamont *et al.*, 1990). This not only extends the number of available blocking peptides, but also allows this approach when the antigen-causing disease is unknown.

4.3.3   T Cell Vaccination

Experimental work has shown that clones of T cells derived from the antigen-specific T cell-causing disease can, after modification by chemical or physical changes, act as a vaccine and protect the recipients against development of disease. The mechanism of the induced immunity is thought to be the induction in the hosts of T cells which recognize T cell receptor structures on the vaccinating T cells. An extension of this is to immunize the recipient with peptides from the T cell receptor of the antigen-receiving T cells.

4.3.4   Human Leucocyte Antigen Accessory Molecules

In conjunction with the HLA molecule, which binds to the antigen to present it to the T cell, there are a number of other molecules which either play a role in antigen presentation, or in the process of cell–cell interaction or adhesion. Examples of these molecules are CD4, CD5 and CD7. Monoclonal antibodies to these molecules have been shown experimentally and in some human studies to diminish immune reaction, reduce disease or prevent rejection. Examples of their use in humans are anti-CD4 for treatment of vasculitis and anti-CD7 for preventing rejection of kidney allografts.

4.3.5   Interleukin-2 Receptor

Monoclonal antibodies to IL-2R block activation and proliferation of lymphocytes after the specific antigen-receiving T cell has been activated. IL-2R is expressed on the surface of lymphocytes which are proliferating in response to antigen and is therefore a particular, rather than general, target. A modification of this approach, i.e. attacking the IL-2R, has been the engineering of a chimeric protein consisting of IL-2 and diphtheria toxin, which attaches to and destroys IL-2R-positive cells (Kelley *et al.*, 1988).

4.3.6   Inflammatory and Adhesion Molecules

Monoclonal antibodies to the interleukins, adhesion molecules and other molecules involved in the inflammatory response such as INFγ are increasingly being explored in the modification of immune-mediated disease and rejection. Modification of eicosanoid metabolism by altering the fatty acid content of diet, e.g. supplementation with fish oil, has been tried in experimental and human studies, but with no clear general indication of the outcome.

5. Conclusions

In recent years, nephrologists have come to realize that the GECs and TECs cells do more, in immunological terms, than provide antigens to induce disease, and act as sites for antibody and antigen–antibody complex trapping. They act as modulators of the immune response, by producing inhibitors of mediators, and the mediators themselves. They also, under certain circumstances, can act as antigen-presenting cells and play a central role in initiating the immune response. The development of new ways of modifying the immune response will certainly be applicable to epithelial cells and basement membranes, but mainly as part of the entire approach to treatment of the inflamed glomeruli, tubules and interstitium in nephritis, or the prevention of rejection, which does not single out the epithelial cells alone.

6. References

Al-Nawab, M.D., Jones, N.F., and Davies, D.R. (1991). Glomerular epithelial cell endocytosis of immune deposits in human lupus nephritis. Nephro. Dia. Transplant. 6, 316–323.

Andreoli, S.P. (1991). Reactive oxygen molecules, oxidant injury and renal disease. Pediatr. Nephrol. 5, 733–742.

Appay, M.D., Kazatchkine, M.D., Mounier, F. and Bariety, J. (1988). Presence of the C3b complement receptor (CR1) antigen in the glomerular basement membrane of human fetal kidneys. Nephrol. Dial. Transplant. 2, 162–165.

Brinkman, K., Termaat, R.M., de Jong, J., van den Brink, H.G., Berden, J.H.M. and Smeenk, R.J.T. (1989). Cross-reactive binding patterns of monoclonal antibodies to DNA are often caused by DNA/anti-DNA immune complexes. Res. Immunol. 140, 595–612.

Brooimans, R.A.A., Stegman, P.A., van Dorp, W.T., van der Ark, A.A.J., van der Woude, F.J., van Es, L.A. and Daha, M.R. (1991). Interleukin 2 mediates stimulation of complement C3 biosynthesis in human proximal tubular epithelial cells. J. Clin. Invest. 88, 379–384.

Castellot, J.J. Jr., Hoover, R.L., Harper, P.A. and Karnovsky, M.J. (1985). Heparin and glomerular epithelial cell-secreted heparinlike species inhibit mesangial-cell proliferation. Am. J. Pathol. 120, 427.

Cederholm, B., Wieslander, J., Bygren, P. and Heinegard, D. (1986). Patients with IgA nephropathy have circulating antibodies reacting with structures common to collagen I, II and IV. Proc. Natl Acad. Sci. USA 83, 6151–6155.

Clark, G. and Williams, D.G. (1991). In "Immunology of Renal Diseases" (ed. C.D. Pusey), pp 161–182. Kluwer, New York.

Endou, H. (1983). Cytochrome P450 monoxygenase system in the rabbit kidney; its intranephron localisation and its induction. Jpn. J. Pharmacol. 33, 423–433.

Feucht, H.E., Zwirner, J., Bevec, D., Land, M., Felber, E., Reithmuller, G. and Weiss, E.H. (1989). Biosynthesis of complement C4 messenger RNA in normal human kidney. Nephron 53, 338–342.

Fillastre, J.P., Druet, P. and Mery, J.PH. (1988). In "The Nephrotic Syndrome" (eds J.S. Cameron and R.J. Glassock), p 697. Marcel Dekker, New York.

Fillit, H., Damle, S.P. and Gregory, J.D. (1985). Sera from patients with poststreptococcal glomerulonephritis contain antibodies to glomerular heparan sulfate proteolgycan. J. Exp. Med. 161, 277–289.

Friedman, J., Hoyer, J.R. and Seiler, M.W. (1982). Formation and clearance of tubulointerstitial immune complexes in kidneys of rats immunized with heterologous antisera to Tamm Horsfall protein. Kidney Int. 21, 575–582.

Fuggle, S.V., McWhinnir, D.L., Chapman, J.R., Taylor, H.M. and Morris, P.J. (1986). Sequential analysis of HLA-class II antigen expression in human renal allografts. Transplantation 42, 144–150.

Furness, P.N. (1990). Chronic serum sickness glomerulnephritis: heparin enhances the removal of glomerular antigen. J. Pathol. 161, 233–237.

Harris, R.C., Hoover, R.L., Jacobson, H.R. and Badr, K.F. (1988). Evidence for glomerular actions of epidermal growth factor in the rat. J. Clin. Invest. 82, 1028.

Hassid, A., Konieczkowski, M. and Dunn, M.J. (1979). Prostaglandin synthesis in isolated rat kidney glomeruli. Proc. Natl Acad. Sci. USA 76, 1155–1159.

Heymann, W., Hackel, D.B., Harwood, S., Wilson, S.G.F. and Hunter, J.L.P. (1959). Production of nephrotic syndrome in rats by Freund's adjuvant and rat kidney suspensions. Proc. Soc. Exp. Biol. Med. 100, 660–664.

Hostikka et al. (1990).

Hudson, B.G., Wieslander, J., Wisdom, B.J. and Noelken, M.E. (1989). Goodpasture syndrome: molecular architecture and function of basement membrane antigen. Lab. Invest. 61, 256–269.

Izui, S., Lambert, P.H. and Miescher, P.A. (1976). In vitro demonstration of a particular affinity of glomerular basement membrane and collagen for DNA. A possible basis for a local formation of DNA–anti-DNA complexes in systemic lupus erythematosus. J. Exp. Med. 144, 428–443.

Jevnikar, A.M., Brennan, D.C., Singer, G.G., Heng, J.E., Maslinski, W., Wuthrich, R.P., Glimcher, L.H. and Kelley, V.E.R. (1991). Stimulated kidney tubular epithelial cells express membrane associated and secreted TNFa. Kidney Int. 40, 203–211.

Jim, K., Hassid, A., Sun, F. and Dunn, M.J. (1982). Lipoxygenase activity in rat kidney glomeruli, glomerular epithelial cells and cortical tubules. J. Biol. Chem. 257, 10294–10299.

Kefalides, N.A., Pegg, M.T., Ohno, N., Poon-kina, T., Zabriskie, J. and Fillit, H. (1986). Antibodies to basement membrane collagen and to laminin are present in sera from patients with poststreptococcal glomerulonephritis. J. Exp. Med. 163: 588–602.

Kelley, V.E., Bacha, P., Pankewycz, O., Nichols, J.C., Murphy, J.R. and Strom, T.B. (1988). Interleukin 2–diphtheria toxin fusion protein can abolish cell mediated immunity in vivo. Proc. Natl Acad. Sci. USA 85, 3980–3984.

Koyama, A., Fujisaki, M., Kobayashi, M., Igarashi, M. and Narita, M. (1991). A glomerular permeability factor produced by human T cell hybridomas. Kidney Int. 40, 453–460.

Lamont, A.G., Sette, A., Fujinami, R., Colon, S.M., Miles, C. and Grey, H.M. (1990). Inhibition of experimental autoimmune encephalomyelitis induction in SJL/J mice by using a peptide with high affinity for IA^s molecules. J. Immunol 145, 1687–1693.

Matsuda, M., Aoki, N. and Kawaoi, A. (1979). Localization of urinary procoagulant in the human kidney. Kidney Int. 15, 612–617.

Meri, S., Waldmann, H. and Lachmann, P.J. (1991). Distribution of protectin (CD59), a complement membrane attack inhibitor, in normal human tissues. Lab. Invest. 65, 532.

Mizoguchi, Y. and Horiuchi, Y. (1982). Localization of IgG–Fc receptors in human renal glomeruli. Clin. Immunol. Immunopathol. 24, 320–329.

Muller, C.A., Markovic-Lipkovski, J., Risler, T., Boble, A. and Muller, G.A. (1989). Expression of HLA-DQ, -DR, -DP antigens in normal kidney and glomerulonephritis. Kidney Int. 35, 116–124.

Muller, G.A., Markovic-Lipovski, J. and Muller, C.A. (1991). Intercellular adhesion molecule-1 expression in human kidneys with glomerulonephritis. Clin. Nephrol. 36, 203–208.

Neale, T.J., Woodroffe, A.J. and Wilson, C.B. (1984). Spontaneous glomerulonephritis in rabbits; role of a glomerular capillary antigen. Kidney Int. 26, 701–711.

Nolasco, F.E.B., Cameron, J.S., Hartley, B., Coelho, R.A., Hildredth, G. and Reuben, R. (1987). Abnormal podocyte CR-1 expression in glomerular diseases: association with glomerular cell proliferation and monocyte infiltration. Nephrol. Dial. Transplant. 2, 304–312.

Olson, J.L., Rennke, H.G. and Venkatchalam, M.A. (1981). Alterations in the charge and size selectivity barrier of the glomerular filter in aminonucleoside nephritis in rats. Lab. Invest. 44, 271–279.

Orikasa, M., Matsui, K., Oite, T. and Shimizu, F. (1988). Massive proteinuria induced in rats by a single intravenous injection of a monoclonal antibody. J. Immunol. 141, 807–814.

Passwell, J., Schreiner, G.F., Nonaka, M., Beuscher, H.U. and Colten, H.R. (1988). Local extrahepatic expression of complement genes C3, factor B, C2 and C4 is increased in murine lupus nephritis. J. Clin. Invest. 82, 1676–1684.

Pelletier, L., Hirsch, F., Rossert, J., Druet E. and Druet, P. (1987). Experimental mercury-induced glomerulonepyhritis. Springer Semin. Immunopathol. 9, 359–369.

Pelletier, L., Pasquier, R., Rossert, J., Vial, M.C., Mandet, C. and Druet, P. (1988). Autoreactive T cells in mercury-induced autoimmunity. Ability to induce the autoimmune disease. J. Immunol. 140, 750–754.

Petrulis, A.S., Aikawa, M. and Dunn, M.J. (1981). Prostaglandin and thromboxane synthesis by rat glomerular epithelial cells. Kidney Int. 10, 469–474.

Pugliese, F.,Singh, A.K., Kasinath, B.S., Kreisberg, J.I. and Lewis, E.J. (1987). Glomerular epithelial cell, polyanion neutralization is associated with enhanced prostanoid production. Kidney Int. 32, 57–61.

Quigg, R.J. (1991). Isolation of a novel complement regulatory factor (GCRF) from glomerular epithelial cells. Kidney Int. 40, 668–676.

Quigg, R.J., Nicholson-Weller, A., Cybulsky, A.V., Badalamenti, J. and Salant, D.J. (1989). Decay accelerating factor regulates complement activation on glomerular epithelial cells. J. Immunol. 142, 877–882.

Rauterberg, E.W. (1987). In "The Complement System" (eds K. Rother and G. Till), pp 287–326, Springer-Verlag, New York.

Rovin, B.H., Wurst, E. and Kohan, D.E. (1990). Production of reactive oxygen species by tubular epithelial cells in culture. Kidney Int. 37, 1509–1514.

Schmouder, R.L., Strieter, R.M., Wiggins, R.C., Chensue, S.W. and Kunkel, S.L. (1992). *In vitro* and *in vivo* interleukin-8 production in human renal cortical epithelia. Kidney Int. 41, 191–198.

Schulze, M., Donadio, J.v., Jr, Pruchno, C.J., Baker, P.J., Johnson, R.J., Stahl, R.A.K., Watkins, S., Martin, D.C., Wurzner, R., Gotze, O. and Couser, W.G. (1991). Elevated urinary excretion of the C5b–9 complex in membranous nephropathy. Kidney Int. 40, 533–538.

Sugisaki, T.J., Klassen, J., Andres, G.A., Milgrom, F.J. and McCluskey, R.T. (1973). Passive transfer of Heymann's nephritis with serum. Kidney Int. 3, 66–73.

Torbohm, I., Schonermark, M., Wingen, A.M., Berger, B., Rother, K. and Hansch, G.M. (1990). C5b–8 and C5b–9 modulate the collagen release of human glomerular epithelial cells. Kidney Int. 37, 1098–1104.

Turner, N. and Pusey, C.D. (1991). In "Immunology of Renal Diseases" (ed. C.D. Pusey), pp 229–254. Kluwer, New York.

Verroust, P., Ronco, P.M. and Chatelet, F. (1986). Monoclonal antibodies and identification of glomerular antigens. Kidney Int. 30, 649–655.

Vissers, M.C., Fantone, J.C., Wiggins, R. and Kunkel, S.L. (1989). Glomerular basement membrane-containing immune complexes stimulate tumour necrosis factor and interleukin-1 production by human monocytes. Am. J. Pathol. 134,1–6.

Wilkinson, A.H., Gillespie, C., Hartley, B. and Williams D.G. (1988). Increase in proteinuria and reduction in number of anionic sites on the glomerular basement membrane in rabbits by infusion of human nephrotic plasma *in vivo*. Clin. Sci. 77, 43–48.

Williams, J.D. and Davies, M. (1991). In "Immunology of Renal Diseases" (ed. C.D. Pusey), pp 123–160. Kluwer, New York.

Williams, J.D., Abrahamson, D.R., Davies, M., Harry, T. and Coles, G.A. (1987). In "Renal Basement Membranes in Health and Disease" (eds R.G. Price and B.D. Hudson), pp 375–388. Academic Press, London.

Wooley, P.H., Luthra, W.P., Lafuse, W.P., Huse, A., Stuart, J.M. and David, C.S. (1985). Type II collagen-induced arthritis in mice. III Suppression of arthritis by using monoclonal anti-Ia antisera. J. Immunol. 134, 2366–2376.

Wuthrich, R.P., Glimcher, L.H., Yui, M.A., Jevnikar, A.M., Dumas, S.E. and Kelley, V.E. (1990). MHC class II, antigen presentation and tumour necrosis factor in renal tubular epithelial cells. Kidney Int. 37, 783–792.

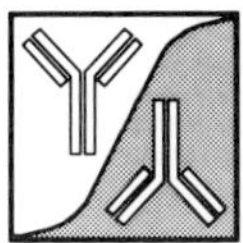

11. Immunological Barriers in the Eye

Jerry Y. Niederkorn

1. Introduction

Although it is only a few centimetres in diameter, the human eye is composed of almost every type of tissue found in the rest of the body as well as additional cellular and noncellular elements found nowhere else (Miller, 1979). This incredibly complex organ is an embryological and anatomical extension of the brain and, like the brain, conducts enormously complex neurological functions. The million ganglion cells of the retina transmit 500 electrical signals along the optic nerve each second; which in computer terms, is equivalent to 1.5×10^9 bits of information per second (Vaughan and Schlimmel, 1971). Like the brain, crucial tissues in the eye possess little or no regenerative capacities and, as such, must be protected from infectious and environmental agents (Fig. 11.1). As we will see later, the eye must also possess unique adaptations to prevent unwitting, self-inflicted immune-mediated damage to its delicate tissues. Thus, the eye employs anatomical, physiological and immunological barriers to shield itself against injury. These barriers are the subject of this brief overview.

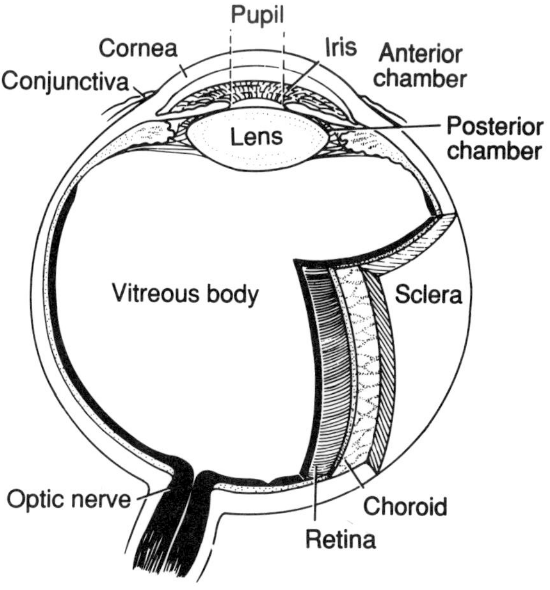

Figure 11.1 Anatomy of the eye.

Immunopharmacology of Epithelial Barriers
ISBN 0–12–288030–7

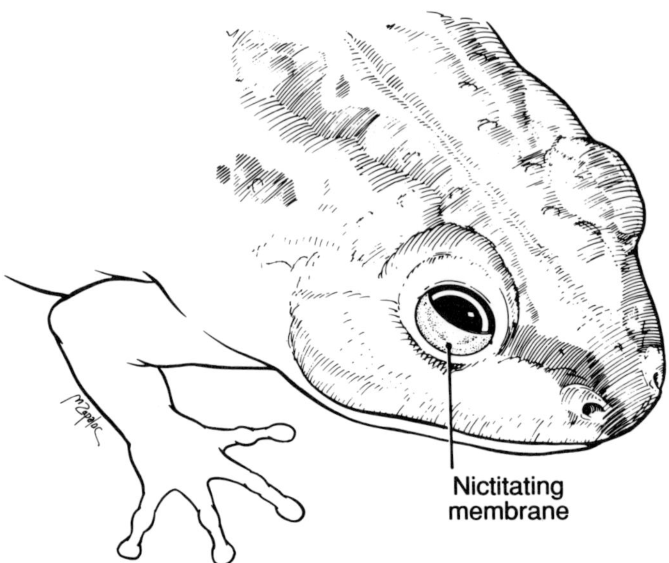

Figure 11.2   The nictitating membrane serves as a anatomical barrier to shield the underlying cornea. In many amphibians and birds it is transparent and highly mobile, allowing clear vision without inhibiting its protective role.

2. *Anatomical Barriers of the Eye*

The most prominent and obvious anatomical barrier that protects the eye from the environment is the eyelid. In addition to the eyelids, lower vertebrates possess a nictitating membrane or 'third eyelid' (Fig. 11.2). The nictitating membrane in amphibians such as the frog is more highly developed and mobile than true eyelids. In the frog and in some birds, the nictitating membrane is transparent centrally to form an avascular window through which vision is possible. In birds the nictitating membrane can be closed without interfering with vision. Moreover, it does not glisten and thus helps animals to avoid detection by predators or prey (Miller, 1979). It has been suggested that, in diving birds, the nictitating membrane not only serves to protect the underlying cornea but also enhances visual acuity by increasing the refraction of the eye when the animal is submerged. In primates and humans, however, the nictitating membrane is vestigial.

3. *Physiological Barriers of the Eye: The Tear Film*

The tear film is a fluid bilayer composed of a thin lipid film (1 µm thick) floating on a large aqueous pool (7 µm thick). The tear film coats the underlying corneal epithelium and serves six important functions: (1) maintains corneal moisture while providing a smooth refracting surface that permits normal, unimpeded light transmission; (2) exerts bactericidal activity; (3) lubricates the lids; (4) transports metabolic products (e.g. CO_2 and O_2) to the corneal epithelium; (5) provides a vehicle for transporting cellular and humoral immune elements to the corneal surface; and (6) dilutes and washes away noxious substances and foreign bodies.

It has been long recognized that tears possess bactericidal properties. Fleming (1922) described the presence of lysozyme in human tears early in this century. Lysozyme (muramidase) destroys bacterial cell membranes by disrupting the *N*-acetylglucosamine-*N*-acetylneuraminic acid in the bacterial cell membrane. The concentration of lysozyme is considerably higher in tears than in serum and suggests that this enzyme is locally synthesized by the lacrimal glands. In addition to lysozyme, the tears contain other antibacterial enzymes including lactoferrin and betalysin.

Recently, it has become apparent that lactoferrin plays an important role in maintaining the protective barrier at the ocular surface. Lactoferrin is a single-chain polypeptide of approximately 80 kD. It has been identified in various secretions such as tears, saliva, pancreatic juice and semen (Masson *et al.*, 1966). Human tears contain approximately 2 mg/ml of lactoferrin, which represents approximately 25% of the total tear proteins (Kijlstra *et al.*, 1983). Tear lactoferrin originates in the lacrimal gland (Fig. 11.3), where it is synthesized by acinar epithelial cells (Franklin and Shepard, 1991). Although levels of tear lactoferrin vary among normal individuals, sharp reductions in the level of tear lactoferrin occur in variety of pathological conditions.

In vitro studies have demonstrated that lactoferrin exercises bacteriostatic activity which can be augmented

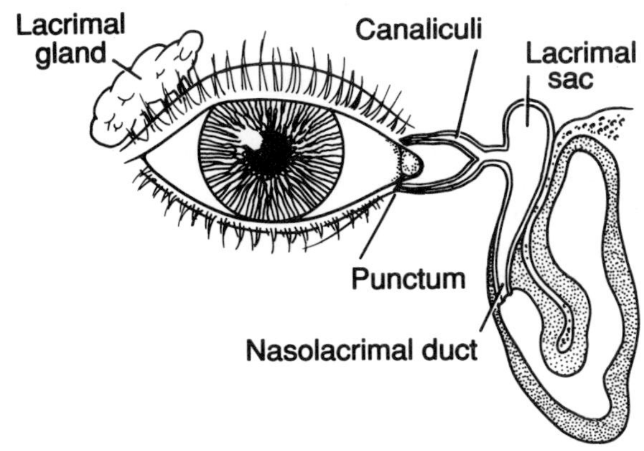

Figure 11.3   The lacrimal gland is the primary site for antibody synthesis of the ocular mucosal immune system and for the secretion of tears.

by IgA antibodies (Bullen *et al.*, 1972). Lactoferrin is also bactericidal for certain bacteria and is believed to act by altering bacterial membrane permeability (Ellison *et al.*, 1988).

Lactoferrin may also protect the ocular surface by preventing immune-mediated damage. Tears contain powerful inhibitors of the complement system, one of which is associated with lactoferrin (Kijlstra *et al.*, 1989). Lactoferrin can inhibit complement activation by blocking the formation of the classic C3 convertase (Kijlstra and Jeurissen, 1982). In addition to inhibiting complement-mediated injury, lactoferrin might also act to prevent damage produced by reactive oxygen species released by activated granulocytes at the ocular surface (Kijlstra, 1991).

4. *The Ocular Mucosal Immune System*

Mucosal epithelial surfaces which are exposed to the external environment are provided with protection from environmental pathogens by the mucosal immune system (Franklin, 1989). Although all of the immunoglobulin isotypes have been detected in normal human tears, only IgA is present in significant quantities (Sullivan and Allansmith, 1984; Franklin, 1989). Secretory IgA found in external secretions is typically produced by plasma cells within the mucosal tissues and differs from serum IgA by the addition of epithelial cell-derived SC (Franklin, 1989). The membrane-bound SC promotes transport of the dimeric IgA across the mucosal surface into the secretion. Secretory IgA has been demonstrated in the tears of a wide variety of mammalian species including humans (Franklin, 1989. In rats, secretory IgA appears to be locally produced by lacrimal glands and is not derived by transudation from the serum (Sullivan and Allansmith, 1984; Peppard and Montgomery, 1987). The presence of tear IgA antibodies specific for oral

microorganisms as well as the demonstration that oral immunization leads to the appearance of tear IgA antibodies (Gregory and Filler, 1987) suggests a linkage between the lacrimal gland and the mucosal immune system of the GI tract.

It has been clearly demonstrated in animal models that topical application of antigen to the ocular surface can stimulate impressive IgA antibody responses in the tears (Montgomery *et al.*, 1984). However, little is known regarding the precise events involved in the processing of topically applied antigens. It is not known, for example, if antigens applied topically are processed by CALT or if such antigens depart from the ocular surface via the nasolacrimal duct and subsequently enter the GI tract where they are processed by GALT. It has been hypothesized that antigen applied to the ocular surface is processed by cells in CALT which subsequently migrate to the lacrimal gland. If topically applied antigens reach the lacrimal gland, they would be encountered by the cellular triumvirate needed to initiate a conventional immune response (i.e. macrophages, T cells and B cells). Moreover, the lacrimal gland possesses a high percentage of IgA-committed plasma cells characteristic of other tissues of the secretory immune system (McGee and Franklin, 1984).

Although there is abundant evidence that IgA-committed B cells migrate to mucosal tissues, much remains to be learned about the events leading to this organ-specific homing. Initially, it was thought than antigen accumulation in mucosal tissues attracted peripheral IgA-committed cells. However, this explanation has been shown to be inadequate (Franklin, 1989). For example, immature glandular tissues containing no plasma cells but which are in constant contact with antigens in the local environment can be transplanted to antigen-free sites such as the anterior chamber of the eye (Parrott and Ferguson, 1975) or subcutaneous sites (Franklin, 1989). Following maturation, these transplanted mucosal tissues become populated with IgA-committed B cells. Other investigations, however, suggest that antigen may be necessary for expansion of IgA B cells that localize in mucosal tissues (Franklin, 1989).

A second explanation to account for the preferential lodging of IgA-committed B cells in mucosal tissues relates to the presence of locally synthesized SC. Dimeric IgA has an affinity to bind with SC. SC is locally produced by mucosal epithelial cells and remains bound to the plasma membrane of these cells. It is feasible that the interaction between IgA expressed in the B cell plasma membrane and SC in the mucosal epithelial cell membrane leads to the retention of IgA B cells in mucosal tissues. However, studies in mice have failed to support this hypothesis.

A third explanation for the selective accumulation of IgA-committed B cells in mucosal glandular tissue involves cellular recognition of postcapillary venules, especially HEVs. It has been repeatedly demonstrated that HEVs play a role in the migration of B lymphocytes into various lymphoid tissues (Butcher, 1986). However, recent attempts to demonstrate a role of postcapillary venules in the homing of lymphoid cells to mucosal glandular tissues have failed. Moreover, adoptive transfer studies in mice have shown that T cells and B cells randomly appear and lodge in lacrimal glands in equal numbers (McGee and Franklin, 1984). Thus, there is no compelling current evidence to suggest that the vascular endothelium of the lacrimal gland expresses addressins or other cell adhesion molecules that preferentially direct IgA-committed B cells to extravasate into this gland.

Recent results from studies in mice (Franklin, 1989) and rats (O'Sullivan and Montgomery, 1990) suggest that the accumulation of IgA-secreting plasma cells in the lacrimal gland occurs by a process distinctly different from that which occurs in organized lymphoid organs. In the latter case, lymphocyte adherence to postcapillary HEVs controls the entry of cells into organized lymphoid organs (Stamper and Woodruff, 1977). By contrast, lymphocyte migration into the lacrimal gland appears to be random (Franklin and Shepard, 1991; O'Sullivan and Montgomery, 1990). However, a growing body of evidence suggests that the retention of lymphocytes in the lacrimal gland is the result of specific interactions between the circulating lymphocytes and the lacrimal gland acinar epithelial cells (Franklin, 1989; O'Sullivan and Montgomery, 1990). A provocative hypothesis offered by Franklin and Shepard (1991) suggests that the lacrimal gland-specific T cells are primarily T helper cells which induce the differentiation of migrating IgA-committed B cells to a non-migratory IgA-secreting plasma cell.

It is generally assumed that the secretory immune system serves a crucial role in protecting the mucosal surfaces against environmental pathogens. However, it has been difficult to directly document the precise effector mechanisms of secretory IgA (Underdown and Schiff, 1986).

Attempts to establish the *modus operandi* of IgA-mediated immunity have fallen into four main categories: (1) *in vitro* experiments examining effector functions such as neutralization or activation of complement-mediated cytolysis; (2) passive transfer of protective immunity to determine if IgA protects naive hosts from challenge infections; (3) correlations between the level of specific IgA and elimination of infectious agents; and (4) studies on the consequences of selective IgA deficiency on disease susceptibility (Underdown and Schiff, 1986).

An obvious mechanism for IgA-mediated protection is by preventing the attachment of pathogenic organisms to mucosal surfaces. Purified secretory IgA antibodies have been shown to prevent the attachment of bacteria (Svanborg-Eden and Svennherholm, 1978) and the internalization of viruses (Taylor and Dimmock, 1985) at mucosal surfaces.

Although IgA antibodies are capable of activating complement through the alternative pathway, the resulting cytolysis is feeble and its role in protective immunity is uncertain (Underdown and Schiff, 1986).

Demonstrating protective immunity via passively transferred secretory IgA has been unsatisfactory due to the technical difficulties in ensuring that blood-borne, transferred IgA gains access to the mucosal surface following passive transfer (Underdown and Schiff, 1986).

Numerous studies have demonstrated a close correlation between the recovery from mucosal infections and the appearance and magnitude of the secretory antibody response (for a review, see Dhar and Ogra, 1985). However, proving a cause and effect relationship between secretory IgA levels and recovery from mucosal infections is tenuous. Both cellular and humoral immune mucosal effector elements may develop in parallel and function either independently or synergistically.

One would predict that a deficiency in the ability to synthesize IgA would result in severe and recurrent infections. Inability to produce IgA is the most common immunoglobulin deficiency in humans, yet the majority of IgA-deficient individuals rarely have major, life-threatening health problems (Underdown and Schiff, 1986).

The role of secretory IgA in protection against ocular infections is unclear. However, the ability of secretory IgA antibody to prevent the binding and internalization of infectious pathogens is an appealing attribute that might be uniquely suited for ocular infections. The two leading causes of preventable infectious blindness, trachoma (Taylor and Bell, 1991) and HSV keratitis (Hyndiuk and Glasser, 1986), occur at the corneal surface and exist as intracellular pathogens. Secretory IgA present in the tears might be an effective method for preventing infection and disease. Interestingly, the pathogenesis of both trachoma (Grayston *et al.*, 1985) and HSV keratitis (Lausch *et al.*, 1985; Newell *et al.*, 1989a,b) is believed to be immune mediated. A growing body of evidence indicates that unbridled DTH responses to *Chlamydia trachomatis* (Grayston *et al.*, 1985) and HSV (Lausch *et al.*, 1985; Newell *et al.*, 1989a,b) produce extensive damage to uninfected, innocent bystander tissues in the infected cornea. Thus, designing an effective vaccine for either ocular disease must take these immunopathological consequences into consideration. A candidate vaccine would ideally promote a prompt and high-titre secretory IgA antibody response that would prevent ocular infections from becoming established and which would eliminate residual HSV or *C. trachomatis* antigens and thereby prevent the arousal of a tissue-damaging DTH response.

5. *The Cornea as an Anatomical Barrier*

The cornea is a transparent avascular tissue that serves as the cellular window for the transmission of light to the

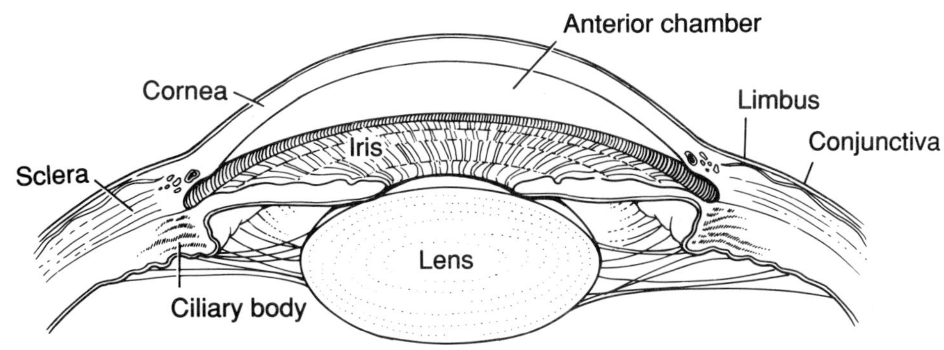

Figure 11.4   The cornea serves as an anatomical barrier to the environment while permitting a clear medium for the unaltered transmission of light to the photoreceptors of the retina.

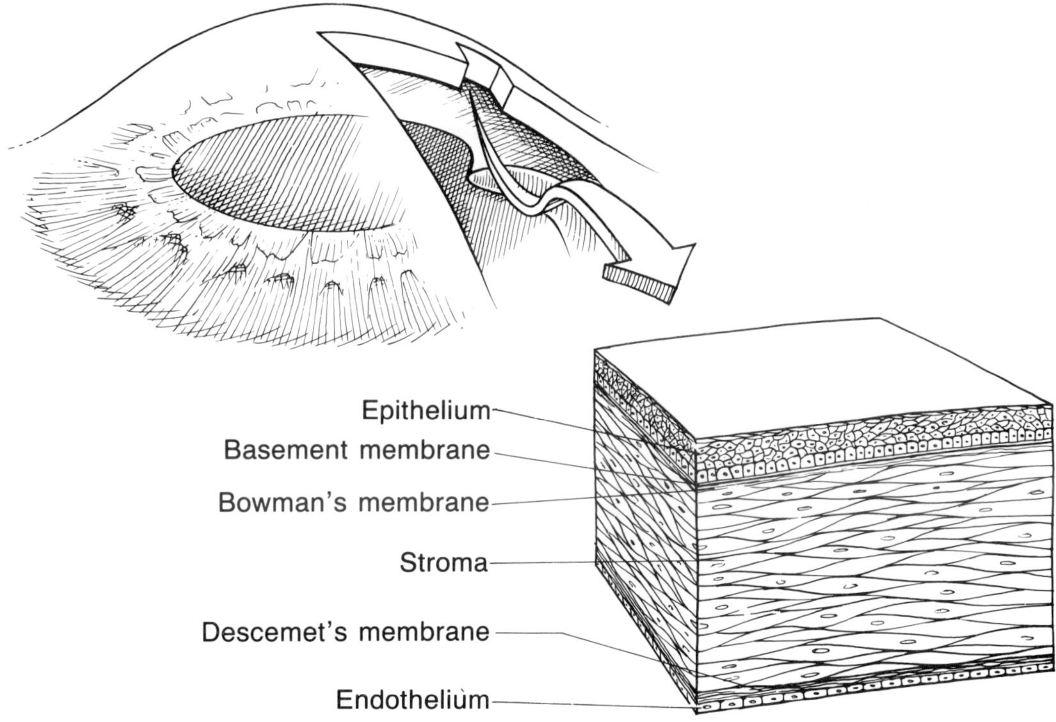

Figure 11.5   The five distinct layers of the cornea. (Reproduced with permission, from Niederkorn and Peeler (1988).)

retina and as a result is crucial for normal vision (Fig. 11.4). The average adult human cornea is 0.65 mm thick at the periphery and 0.54 mm thick in the centre. Five distinct layers make up the cornea: (1) the outer epithelium comprised of five or six layers of stratified squamous epithelial cells; (2) a layer of ground substance and collagen (Bowman's layer); (3) the collagenous stroma; (4) a basement membrane (Descemet's membrane); (5) a single cell layer of endothelium (Fig. 11.5). The corneal stroma accounts for about 90% of the corneal thickness and is comprised of intertwining lamellar collagen fibres approximately 1 μm in diameter.

The transparency of the cornea is a result of its uniform arrangement of stromal collagen lamellae, the avascularity of the entire cornea, and the deturgescence. Deturgescence, or the state of relative dehydration, is maintained by the active "pumping" function of the single-cell layer, the endothelium. A normal cornea is 78% water; however, a mere increase of 9% in the water content results in a doubling in the thickness of the cornea. Maintenance of dehydration is crucial for maintaining normal cornea clarity and thus vision. Destruction of the corneal endothelium is swiftly followed by corneal hydration, oedema, opacity and blindness. Thus, anything that compromises the integrity of the corneal endothelium jeopardizes vision itself. The fragility of corneal function is further emphasized by the inability of human corneal endothelial cells to undergo mitosis. Loss of endothelial cells can only be compensated for by spreading of residual healthy endothelial cells and not by hyperplasia. Thus, the only thing standing between normal vision and blindness is a single layer of cells with virtually no capacity for proliferation.

Maintaining the structural integrity of the cornea requires a dynamic system whereby adherence between various cell types can be required. Considerable interest has been focused on the role of adhesion molecules in inflammation, embryogenesis, metastasis, wound healing and lymphocyte homing (for a review, see Springer, 1990). In particular, the adhesion molecules of the VLA family may be important in the adherence of corneal epithelial and endothelial cells to the underlying stromal matrix. The VLA family is part of the integrin super-family, which also includes the leucocyte integrins (LFA-1, Mac-1, and P150,96) and the GP11b/11a vitronectin receptor family. Ligands for VLA include laminin, collagen, fibronectin, and the tripeptide sequence Arg-Gly-Asp (Santoro *et al.*, 1988). Recently, Lauweryns and co-workers (1991) investigated the distribution of the VLA family of adhesion molecules in the human cornea. Interestingly, it was shown that corneal epithelium, endothelial cells, and stromal keratocytes differentially express various VLA integrims (Lauweryns *et al.*, 1991). Epithelial cells expressed four of the five VLA α subunits and is consistent with their adhesion to the corneal basement membrane which is composed of type IV collagen, laminin, and fibronectin. By contrast, kerato-cytes expressed only the common β_1 chain and no detectable α chains. The impressive expression of all members of the VLA integrin family on the corneal endothelium correlates with the strong adherence of the endothelial monolayer to the underlying Descemet's membrane, which is comprised of collagen, laminin and fibronectin. As mentioned earlier, the corneal endo-thelium serves a critical role in maintaining corneal deturgescence by pumping water from the stroma. Therefore, dense expression of all categories of VLA integrins on corneal endothelial cells promotes firm binding between individual cells and the underlying Descemet's membrane and thus enhances the function of the endothelium.

6. *The Cornea as an Immunological Barrier: Neurogenic Immunology of the Eye*

Once the anatomical barrier of the eye has been breached, noxious agents, either chemical or biological, can produce tissue damage which in turn induces acute inflammation. However, unlike other tissues, the cornea is devoid of mast cells, blood vessels and lymphatic vessels. Therefore, the cornea presents unique anatomical restrictions for the initiation of a normal inflammatory response.

The early steps in corneal inflammation are initiated in the juxtaposed tissues of the limbus and con-junctiva. Following experimental wounding of the central cornea, peripheral blood vessels in the limbus and conjunctiva dilate in response to histamine released by vessel-associated mast cells.

Local exudation of plasma proteins is supplemented with plasma proteins present in the tears. Within 3 h of experimental corneal injury, epithelial cells begin to migrate over the wound surface and undergo proliferation at 24–48 h.

It has been suggested, however, that the majority of inflammatory responses that occur in the injured cornea are directly or indirectly modulated by the peripheral nervous system and accordingly should be termed "neurogenic inflammation" (Waldrep, 1989). The impor-tance of neuropeptides and neurogenic inflammation in the cornea is strongly suggested by the dense innervation of this organ. In fact, the innervation density of some areas of the cornea is 300–600 times that of skin (Rozsa and Beuerman, 1982).

Neurogenic inflammation in the skin results from stimulation of primary, afferent, non-myelinated nerve fibres which contain a variety of neuropeptides (Waldrep, 1989). Of these neuropeptides, Sub P appears to be the major mediator of neurogenic inflammation. Throughout the body there is an intimate association between peripheral nerve endings and mast cells (Kowalski *et al.*, 1988). Intracameral injection of minute quantities of Sub P (i.e. 10 pmol) produces rapid wheal and flare responses through an IgE-independent degranulation of tissue mast cells (Waldrep, 1989). Neurogenic stimulation via Sub P and perhaps other neuropeptides can culminate in increased vascular dilatation and vasopermeability mediated by mast cell products. Other neuropeptides have been implicated in neurogenic inflammation including CGRP and VIP. Evidence suggests that CGRP and Sub P are produced and stored into the same granules and transported to axon terminals, where they are simultaneously released (Waldrep, 1989). VIP can also stimulate the release of Sub P from sensory neurones.

Sub P is the most thoroughly characterized neuro-peptide and is widely distributed throughout the peripheral nervous system. Sub P-containing neurones are especially dense in the trigeminal ganglia, which supply nerve fibres to the eye, mouth, lips, mucosa of the nose, and the dental pulp (Hokfelt *et al.*, 1977). Stimulation of the neurone results in the release of Sub P from the nerve terminal. The released Sub P acts on a variety of cells, directly and indirectly, by inducing the release of additional inflammatory mediators. Sub P further promotes inflam-mation by inducing chemotaxis and activation of poly-morphonuclear neutrophils and monocytes (Stanisz *et al.*, 1987). Lymphocyte activity is also influenced by Sub P. Thus, Sub P serves as a link between the nervous system and the inflammatory cells.

Waldrep (1989) has proposed a model of corneal

neurogenic inflammation based on results from numerous experimental studies. In his model, neurogenic inflammation is triggered by a noxious injury to the corneal surface which in turn stimulates an axonal reflex and the release of Sub P and CGRP at the corneal sensory nerve endings. The released neuropeptides are then believed to initiate a series of events. CGRP stimulates an increased blood flow to the anterior segment of the eye while Sub P triggers the degranulation of limbal and conjunctival mast cells with an ensuing release of vasoactive amines. The vasoactive amines in turn induce vascular dilatation, increased vasopermeability, exudation, and, thus, conjunctival oedema. Neutrophils emigrate from the limbal vessels in response to the chemoattractants released by degranulating mast cells and the chemotactic properties of Sub P. Sub P also activates the infiltrating neutrophils and monocytes which can produce extensive tissue damage by the release of hydrolytic enzymes, oxygen radicals, and arachidonic acid metabolites. This complex choreography involving sensory nerves, neuropeptides, vasoactive products from mast cells, and inflammatory cells could have a major impact on the immunological and anatomical barriers at the corneal surface.

7. Immunological Barriers in the Cornea: A Double-Edged Sword

7.1 IMMUNOPATHOGENESIS OF HERPES SIMPLEX VIRUS KERATITIS

The *raison d'être* for maintaining a robust ocular immune apparatus is to protect the eye from potentially blinding infections. However, an overzealous immune response can culminate in pathological sequelae and possibly irreparable damage to susceptible tissues. Paradoxically, there is a growing body of evidence suggesting that the two leading causes of infectious blindness, trachoma and HSV keratitis, are diseases in which the primary pathological sequelae are immune mediated. Thus, the ocular immune system may unwittingly contribute to the development of the very malady that it was designed to prevent – blindness.

HSV-1 infections remain a leading cause of preventable infectious blindness in the USA and other developed nations (Hyndiuk and Glasser, 1986). Two theories have been offered to account for the pathogenesis of HSV-1 stromal keratitis. The first hypothesis suggests that the pattern of disease is determined by the strain of virus and is not immune mediated (Centifanto-Fitzgerald *et al.*, 1982; Stulting *et al.*, 1985). A second theory based on a large body of experimental evidence in animal models suggests that the genetics of the host (Stulting *et al.*, 1985; Forster *et al.*, 1986; Opremcak *et al.*, 1988) and the immune response of the host to viral antigens are the

determining factors in corneal pathology (for a review, see Pepose, 1991). Since the topic of this review relates to immunological barriers, this discussion will focus on the role of immunological elements in the pathogenesis of HSV keratitis.

Early experimental evidence suggesting that HSV keratitis might be immune mediated came from studies in rabbits. It was shown that preimmunized hosts displayed a higher incidence of HSV keratitis compared to immunologically naive hosts (Williams *et al.*, 1965). Swyers and co-workers (1967) reported that guinea-pigs presensitized by cutaneous infection with HSV developed corneal inflammation when challenged via an intracorneal injection of soluble viral antigen. Along the same lines, Metcalf and co-workers (1979) showed that the absence of an intact immune system prevented the development of corneal lesions in T cell-deficient nude mice. However, reconstitution of nude mice with T cells from HSV-immune animals rendered the hosts susceptible to HSV keratitis (Russel *et al.*, 1984). Ksander and Hendricks (1987) showed that inducing antigen-specific suppression of cell-mediated immunity to HSV antigens prior to corneal challenge infections prevented the development of corneal lesions in A/J mice. Interestingly, HSV stromal keratitis is uncommon in immunodeficient humans, such as AIDS patients or individuals undergoing organ transplantation (Pepose, 1991).

Two fundamental theories have been offered to account for the putative immune-mediated damage produced in HSV keratitis in mice. One hypothesis suggests that corneal lesions are produced by direct cytolysis of virus-infected cells by cytotoxic T lymphocytes (Hendricks *et al.*, 1989; Hendricks and Tumpey, 1990). Hendricks *et al.*, (1989) used novel mutant strains of HSV-1 with altered cell surface glycoproteins which differentially suppressed the development of either DTH responses or CTL responses to HSV-1. The results from this study indicated that corneal lesions were virtually absent in hosts exhibiting suppressed HSV-specific CTL responses. By contrast, suppression of virus-specific DTH responses (while preserving CTL activity) did not prevent HSV keratitis.

A second hypothesis stresses the importance of DTH effector mechanisms in the pathogenesis of HSV-1 keratitis (Lausch *et al.*, 1985; Newell *et al.*, 1989a,b). Using *in vivo* depletion of CD4$^+$ and CD8$^+$ T cell populations, Newell *et al.* (1989b) found that elimination of the CD4$^+$ T helper/DTH subset dramatically reduced the severity of stromal keratitis in HSV-1 corneal infections. Depletion of CD8$^+$ T cells (i.e. CTL/suppressor) had no effect on the development of corneal lesions. By contrast, removal of CD4$^+$ cells abolished DTH and antibody responses to HSV-1 antigens but did not diminish CTL activity. The conclusion from these studies in mice was that corneal lesions were principally caused by DTH responses to HSV antigens and that CTLs played little, if any, role in the pathogenesis of HSV stromal keratitis.

On the surface, these two theories appear to be mutually

exclusive; however, recent results using different strains of HSV-1 offer a logical solution to this apparent paradox (Hendricks and Tumpey, 1990). The explanation lies in the strain of HSV-1 used for these experimental studies. Those investigations which suggested a role for CTL-mediated injury employed the KOS strain of HSV-1 (Ksander and Hendricks, 1987; Hendricks *et al.*, 1989; Hendricks and Tumpey, 1990) while studies implicating DTH-mediated pathogenesis used the RE strain of HSV-1. Further experiments comparing both strains of HSV-1 in CD4$^-$ depleted and CD8$^-$ depleted mice confirmed the suspicion that the strain of virus determines which immunopathological process produces disease. That is, the RE strain of HSV-1 preferentially activates CD4$^+$ T cells and as a result, disease is mediated by DTH effector mechanisms. By contrast, the KOS strain selectively activates CD8$^+$ T cells and, accordingly, keratitis is a result of CTL-mediated damage of virus-infected corneal cells.

7.2   IMMUNOPATHOGENESIS OF TRACHOMA

Trachoma is one of the most common infectious diseases in the world, affecting an estimated 500 million people (Taylor and Bell, 1991). It is also the leading cause of infectious blindness, with 7 million to 10 million people losing their vision as a result of trachoma. In endemic areas, everyone over the age of 1 year has some signs of trachoma and 25% of those over the age of 60 years are blind as a result of this disease (Taylor and Bell, 1991).

Trachoma is a chronic follicular conjunctivitis resulting from infection with the obligate, intracellular bacterial parasite *Chlamydia trachomatis*. Infection is usually acquired in early childhood and produces an acute or subacute, self-limited follicular conjunctivitis. However, repeated reinfection results in chronic inflammation, scarring and blindness, which usually occurs later in life.

Although high levels of specific antibodies against chlamydiae appear in both serum and tears, immunity to persisting infections or reinfection is weak. Repeated infections are common in endemic areas and are thought to play a major role in maintaining chronic follicular conjunctivitis (Grayston *et al.*, 1985). Persistent stimulation of the immune apparatus of the host is believed to promote the development of an exaggerated and misguided immune response. Indeed, many investigators have suggested that the major pathological sequelae of trachoma are immune mediated (Schachter and Dawson, 1978; Grayston *et al.*, 1985). Studies in humans and primates have shown that prior immunization with killed chlamydiae results in more severe trachoma upon reinfection (Bell and Fraser, 1969). Moreover, typical trachoma lesions occur in the absence of cultivable chlamydiae but in the presence of chlamydial antigens (Schachter *et al.*, 1988). Results from studies with guinea-pig and non-human primate models of *C. trachomatis* infections

revealed the importance of repeated infections for the development of ocular lesions characteristic of trachoma (Bell and Fraser, 1969; Monnickendam *et al.*, 1980).

Results from various animal studies support the notion that DTH to chlamydial antigens is intimately involved in the pathogenesis of trachoma. Crude extract of chlamydial antigens elicits severe ocular inflammation which histopathologically resembles human trachoma (Taylor and Dimmock, 1985; Watkins *et al.*, 1986). Moreover, Morrison and co-workers (1989), have isolated a 57 KDa chlamydial protein which evokes intense DTH responses and corneal inflammation similar to trachoma lesions (Watkins *et al.*, 1986; Taylor *et al.*, 1987). Recent findings by Pal and co-workers (1991) demonstrated that chlamydia-specific T cells were 10–100 times higher in the conjunctiva than in peripheral blood of monkeys infected with *C. trachomatis* and, thus, were strategically located to maximize pathogenic responses to ocular infections.

The immunological barriers of the eye may effectively protect the host from reinfection from either *C. trachomatis* or HSV but, in the process, specific unresponsiveness may unwittingly contribute to pathogenesis of both diseases. Ironically, these renegade immunological responses often culminate in blindness.

8. *The Immunobiology of the Cornea*

The embryology, anatomy and function of the cornea and skin are alike in many ways. This is particularly true considering that both of these epithelial surfaces act as barriers to the immediate external environment. One might expect that both the skin and the cornea invoke similar immunological strategies for maintaining the barrier function.

Streilein (1983) has proposed that the epidermis of the skin is endowed with all of the cellular components to initiate and execute a full array of immunological responses. Moreover, the existence of SALT may provide the skin with an immunological apparatus to function independently as a barrier to invasion by microorganisms and insult by other environmental agents (Streilein, 1983). This provocative hypothesis proposes that:

1. a population of T lymphocytes displays a unique affinity for skin and regional lymph nodes;
2. skin contains immunocompetent lymphocytes and APCs and produces immunoregulatory molecules;
3. immune recognition of antigen takes place within the skin;
4. antigen that escapes intracutaneous recognition and processing induces specific unresponsiveness.

8.1   RESIDENT T CELLS IN THE CORNEA?

The prediction that a population of T lymphocytes displays a unique affinity for skin was originally based on the observation that certain T lymphocyte neoplasms demonstrated a propensity to invade and reside within the skin. The term "cutaneous T cell lymphoma" was designated to emphasize the tissue specificity of these neoplasms. Until recently, however, there was scant evidence for the presence of epidermatotrophic T lymphocyte.

It has become clear that a variety of epithelial surfaces, including skin, are endowed with a unique T cell subset bearing the heterodimeric γδ T cell receptor (Janeway *et al.*, 1988; Hein and Mackay, 1991). The γδ T cell receptor-positive T cells do not express heterodimers of the T cell receptor nor do they express the accessory molecules CD8 or CD4 which are associated with class I or class II MHC restriction (Janeway *et al.*, 1988; Hein and Mackay, 1991). The function of so-called double-negative (i.e. CD4⁻, CD8⁻ γδ T cells is uncertain; however, a number of studies have demonstrated that such cells display non-MHC-restricted cytotoxicity (Janeway *et al.*, 1988).

It has been proposed that the primary function of γδ T cells is to function in the immune surveillance of epithelial surfaces (Janeway *et al.*, 1988). This hypothesis is appealing for several reasons. The γδ T cell receptor-positive dendritic epidermal cell is the dominant lymphocyte population isolated from the epithelium, and, in the mouse, accounts for 1% of all of the cells in the epidermis (Kuziel *et al.*, 1987). Teleologically, an MHC non-restricted cytotoxic cell is ideally suited to mediate cytolysis of malignant neoplasms which are notorious for not displaying class I MHC molecules. Epithelial surfaces such as the skin are continuously exposed to environmental carcinogens such as ultraviolet irradiation and would be expected to be at greater risk for neoplasia. This is particularly true for the skin, which is the largest organ in the body in terms of surface area. It is not surprising, therefore, that skin cancer is the most common malignancy in humans. Much additional work will be needed to either confirm or refute this alluring hypothesis.

Although considerable attention has been focused on γδ-bearing cells of the skin, virtually nothing is known about their presence or absence in the cornea. We have been unsuccessful in detecting Thy 1+, CD4⁻, and CD8⁻ cells in the corneal epithelium of rodents (unpublished results). One might predict that the cornea would benefit considerably by the presence of a putative immune surveillance cell similar to the γδ T cell receptor-bearing cells of the murine skin. However, it is possible that the cornea may not require an immune surveillance system like that found in the skin. The cornea is continuously bathed with a tear film that rinses away soluble as well as particulate environmental agents. Moreover, the cornea is rarely exposed to the direct effects of ultraviolet radiation in the same manner as the skin. Finally, it is interesting to note that the cornea resides at the other end of the scale from skin in terms of carcinogenesis. Tumours of the cornea are exceedingly

rare. The central corneal epithelium rarely undergoes neoplastic transformation and tumours occurring in the cornea usually originate in peripheral areas of the limbus and conjunctiva (Miller, 1979; Jakobiec, 1988). Thus, the unique features of the cornea may not require the presence of γδ-bearing cells as sentries of immune surveillance of neoplasms.

8.2 ANTIGEN PRESENTING CELLS IN THE CORNEA

The second functional prerequisite for an organ-specific lymphoid apparatus (e.g. SALT) is the presence of resident antigen-presenting cells and the local production of immunoregulatory molecules. There is an abundant literature on the immunobiology of cutaneous Langerhans cells and their role in antigen presentation in the skin (Streilein, 1983) as well as the ability of keratinocytes to elaborate immunoregulatory molecules such as IL-1 (Ullrich *et al.*, 1990). The cutaneous Langerhans cells are the primary antigen-presenting cells in the skin, where they are uniformly distributed in a reticulum-like configuration. By contrast, Langerhans cells are conspicuously absent from the central regions of the corneal epithelium in adult humans, mice, rabbits and guinea-pigs (Bergstresser *et al.*, 1980; Rodrigues *et al.*, 1981). However, the peripheral region of the cornea is richly endowed with Langerhans cells, which form a circumferential ring at the transitional zone between the central corneal epithelium and limbus (Fig. 11.6).

The distinct absence of Ia+ Langerhans cells in the central cornea is more than a curious anomaly. A growing body of evidence indicates that these cells represent and important immunogenic component of allografts and may explain in part the immunologic privilege of corneal transplants (Niederkorn, 1990). It has been suggested

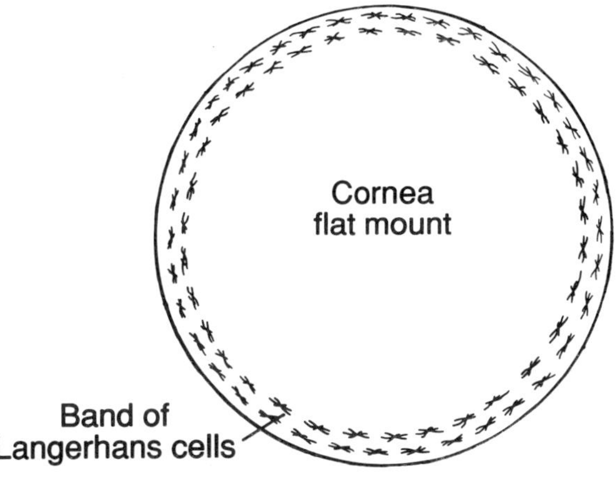

Figure 11.6 Circumferential distribution of dendritic bone marrow-derived Langerhans cells in the limbus. Langerhans cells are conspicuously absent from the central cornea of most mammalian species.

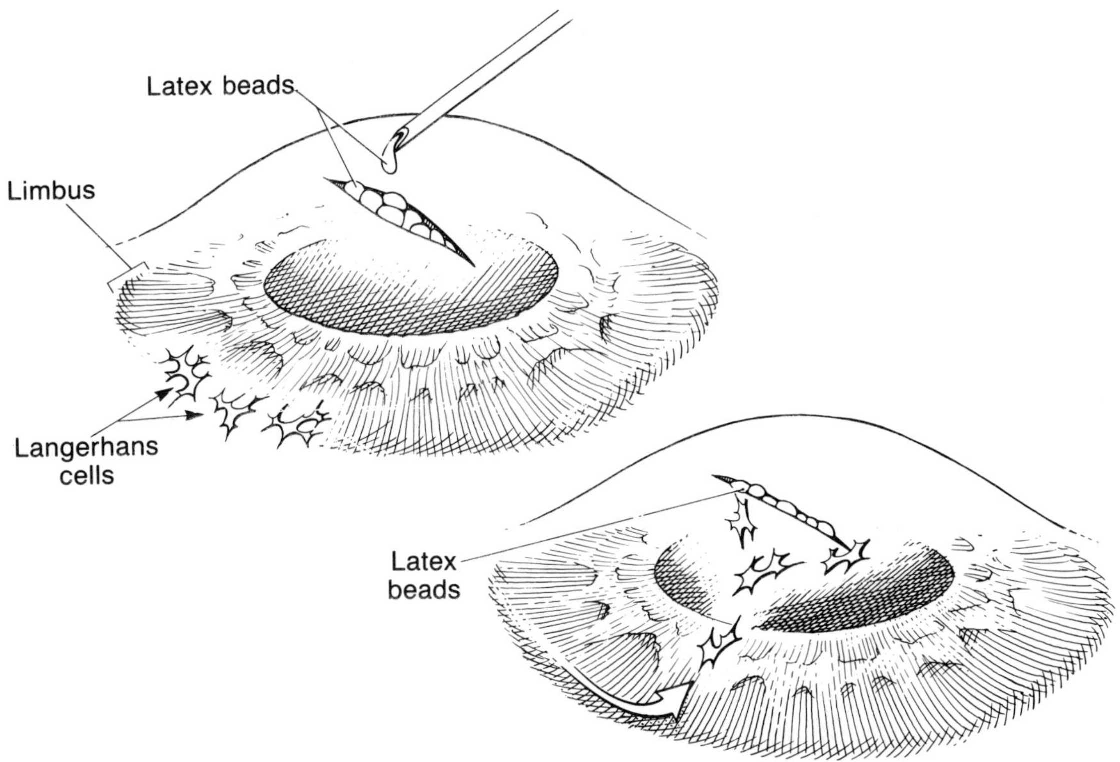

Figure 11.7   Stimulation of centripetal migration of peripheral Langerhans cells into the central cornea by instillation of sterile latex beads into shallow epithelial incisions (Reproduced, with permission, from Niederkorn and Peeler (1988).)

that Ia+ "passenger cells" are the major barrier to successful organ transplantation (Austyn and Steinman, 1988). In recent years there has been a resurgence of interest in the role of passenger cells in the induction of allograft rejection.

The consistent and unique absence of Langerhans cells in the central corneal epithelium provides an opportunity to determine if grafts typically devoid of passenger cells are capable of inducing normal alloimmune responses and if the presence or absence of such cells affects graft survival. Results from the author's laboratory indicate that the absence of resident Langerhans cells has a profound effect on the ability of the graft to induce allospecific DTH responses following heterotropic transplantation to vascularized subdermal graft beds containing lymphatic drainage (Peeler and Niederkorn, 1986). Grafts differing from the host MHC class I and loci II were consistently rejected, but failed to induce detectable DTH responses before, during or after immunological rejection (Peeler and Niederkorn, 1986). Although DTH responses were not induced, Langerhans cell-free grafts could stimulate CTL responses that coincided with graft rejection (Peeler and Niederkorn, 1986). More importantly, it was found that the presence of Langerhans cells in corneal grafts transplanted orthotopically onto an avascular graft bed in the eye had an

enormous effect on graft survival. Studies in a rat model of orthoptic corneal transplantation demonstrated that corneal grafts, presumably devoid of donor Langerhans cells and representing total MHC plus minor histocompatibility antigenic disparities, underwent rejection in 55% of the hosts (Callanan *et al.*, 1989). However, if the grafts were pretreated so as to contain numerous donor-derived Langerhans cells at the time of grafting, rejection increased from 55% to 100% (Callanan *et al.*, 1989).

The well-delineated and symmetrical arrangement of Langerhans cells in the peripheral limbus and the abrupt exclusion of these cells from the central regions of the corneal epithelium suggest that the distribution of these cells is tightly regulated. Moreover, the consistent exclusion of Langerhans cells from the central cornea strongly implies that their absence serves an important biological function other than to benefit ophthalmic surgeons.

Recent studies suggest that the juxtapositioning of corneal Langerhans cells is a dynamic physiological process rather than a static anatomical anomaly. Various stimuli can induce corneal Langerhans cell migration into the central regions of the cornea (Fig. 11.7). These include electrocautery (Williamson *et al.*, 1987), bacterial and viral infections (Hazlett *et al.*, 1986), and phagocytic stimuli (Niederkorn *et al.*, 1989). Since the *raison d'être* of the Langerhan cell is antigen presentation, it is tempting to speculate that the stimuli that induce Langerhans cell migration mimic invasive pathogens and either directly or indirectly beckon Langerhans cells to migrate to the site of "infection". There is some experimental evidence to support this hypothesis. Rabbit corneal cells elaborate an IL-1-like molecule termed CETAF (Grabner *et al.*, 1983). The quantity of this factor is increased 50–70% following phagocytosis of either *Staphylococcus* sp. or latex beads (Niederkorn *et al.*, 1989). Along similar lines, either interferon γ or *Staphylococcus* sp. stimulates the secretion of biologically active IL-1β by human corneal epithelial cells (Shams *et al.*, 1989). The notion that IL-1 or CETAF functioned as a chemoattractant to stimulate the migration of peripheral Langerhans cells to the site of antigen deposition was supported by experiments in which injection of IL-1 into the central corneal epithelium stimulated the undirectional migration of Langerhans cells to the injection site (Niederkorn *et al.*, 1989). The chemotactic response could be inhibited by anti-IL-1 antibody and could not be mimicked by intracorneal injection of other cytokines, such as IL-2, or by irrelevant proteins (e.g. hen egg lysozyme).

The orderly distribution of Langerhans cells in the limbus and the ability of corneal-derived factors to regulate their migration suggests that a dynamic regulatory process governs the positioning and behaviour of limbal Langerhans cells. An appealing although unproven hypothesis is that corneal epithelial cells serve as accessory APCs (Niederkorn, 1990). Since the central cornea is devoid of conventional APCs, such as dendritic cells, some immunological provision must be made to accommodate pathogens and antigens that might penetrate the corneal epithelium. Potential antigens or pathogens (e.g. *Staphylococcus* sp. or HSV) would be phagocytosed by corneal epithelial cells which, in turn, would elaborate increased quantities of IL-1. The soluble IL-1 would then act as a potent chemoattractant for peripheral Langerhans cells residing in the limbus. Langerhans cells would migrate in a unidirectional manner and accumulate at the site of antigen deposition and IL-1 synthesis, where they would encounter the activated corneal epithelial cells. This hypothesis would further suggest that the corneal epithelial cells display the partially processed antigen on their cell membranes for delivery to the infiltrating Langerhans cells. Further antigen processing would occur within the Langerhans cells, which would then migrate to regional lymphoid tissues for the final stage of antigen presentation to T and B lymphocytes.

8.3   CONSEQUENCES OF IMMUNE RECOGNITION OF ANTIGEN IN THE CORNEA

The conspicuous and consistent exclusion of Langerhans cells from the central cornea begs the teleological question: What is the functional significance of this pattern of Langerhans cell juxtapositioning and how does it benefit the cornea and vision in general? One possibility relates to the extraordinary capacity of the Langerhans cells to present antigens to class II-restricted CD4[+] helper cells and induce antigen-specific DTH responses. Sundmacher (1987) proposed that severe HSV stromal keratitis is the result of an intense DTH response to viral antigens and that tissue injury is predominantly a result of an inappropriate and misguided DTH effector response. In a provocative set of experiments, McLeish and co-workers (1989) tested this hypothesis in a mouse model of HSV-1 stromal keratitis. Corneas of experimental mice were subjected to electrocautery as a means of inducing the centripetal migration of peripheral Langerhans cells into the central regions of the cornea. These mice along with normal mice were infected with HSV-1. The results revealed that the presence of Langerhans cells in the central cornea at the time of ocular HSV-1 infection resulted in accelerated appearance of DTH to HSV-1 antigens and a more severe form of stromal keratitis. The authors hypothesized that when HSV-1 enters corneas without Langerhans cells there is a significant lag in the onset of systemic cell-mediated immunity compared to infection through Langerhans cell-containing surfaces, e.g. cauterized corneas or skin. This delay paradoxically protects the eye by allowing the emergence of other immunological effector elements (e.g. interferon γ, antibody) to control infection without inflicting non-specific tissue damage.

The uniform and geometrical arrangement of Langerhans cells in the peripheral limbus suggests that more than mere chance is involved in their distribution. Delaying or denying the Langerhans cells access to antigen would reduce the onset and magnitude of DTH responses to antigens appearing at the ocular surface. One would predict that a faltering of this safeguard would have serious consequences. In this regard, it is interesting to note that the two leading causes of infectious blindness, trachoma and HSV-1 keratitis, are diseases in which the pathogenesis is believed to be mediated by an exaggerated DTH response to pathogenic organisms.

8.4   CONSEQUENCES OF ANTIGEN PENETRATION OF THE CORNEA

The fourth and final prediction of Streilein's SALT hypothesis (Streilein, 1983) suggests that antigen that escapes intracutaneous recognition and processing induces specific unresponsiveness. If one loosely applies this concept to the ocular surface, one would predict that antigens that bypass the cornea and eventually enter the underlying anterior chamber would induce specific unresponsiveness termed anterior chamber-associated immune deviation (ACAID) (Streilin, 1987; Niederkorn, 1990).

9. Summary and Conclusions

The eye is an incredibly complex neurological organ entrusted with perhaps the most precious of the five senses – our vision. The limited capacity to regenerate damaged or diseased tissues requires that the eye possesses an effective system of barriers against both infectious and non-infectious assault. The barriers can be as simple as the highly effective nictitating membrane of lower vertebrates or as complex as the intricate immuno-regulatory circuitry of the anterior chamber. Each of the ocular barriers strives to prevent invasion by pathogens while simultaneously avoiding self-inflicted immune-mediated injury. In some circumstances the teleological wisdom of the ocular immunological barriers appears to go awry and unwittingly contributes to the pathogenesis of the disease it sought to control. Such is the case with HSV keratitis. However, what appears to be a defect in the ocular immune barrier may in fact be a 'fail safe' adaptation to protect against an even greater risk — infection of the central nervous system using the eye as a portal of entry. This latter proposition is admittedly speculative, yet teleologically appealing.

10. References

Asbell, P.A. and Kamenar, T. (1987). The response of Langerhans cells in the cornea to herpetic keratitis. Curr. Eye. Res. 6, 179–188.

Austyn, J.M. and Steinman, R.M. (1988). The passenger leukocyte – a fresh look. Transplant Rev. 2, 139–176.

Bell, S.D. and Fraser, C.E.O. (1969). Experimental trachoma in owl monkeys. Am. J. Trop. Med. Hyg. 18, 568–572.

Bergstresser, P.R., Fletcher, C.R. and Streilein, J.W. (1980). Surface densities of Langerhans cells in relation to rodent epidermal sites with special immunologic properties. J. Invest. Dermatol. 74, 77–80.

Bullen, J.J., Rogers, H.J. and Leigh, L. (1972). Iron-binding proteins in milk and resistance to *Escherichia coli* infection in infants. Br. Med. J. 1, 69–75.

Butcher, E.C. (1986). The regulation of lymphocyte traffic. Cur. Top. Microbiol. Immunol. 128, 85–122.

Callanan, D., Luckenbach, M.W., Fischer, B.J., Peeler, J.S. and Niederkorn, J.Y. (1989). Histopathology of rejected orthotopic corneal grafts in the rat. Invest. Opthalmol. Vis. Sci. 30, 413–424.

Centifanto-Fitzgerald, Y.M., Yamaguchi, T., Kaufman, H.E., Tognon, M. and Roizman, B. (1982). Ocular disease pattern induced by herpes simplex virus is genetically determined by a specific region of viral DNA. J. Exp. Med. 155, 475–489.

Dhar, R. and Ogra, P.L. (1985). Local immune responses. Br. Med. Bull. 41, 28–33.

Ellison, R.T., Giehl, T.J. and La Force, F.M. (1988). Damage of the outer membrane of enteric Gram-negative bacteria by lactoferrin and transferrin. Infect. Immun. 56, 2774–2781.

Fleming, A.A. (1922). On a remarkable bacteriolytic element found in tissues and secretions. Proc. R. Soc. Lond. B 93, 306–317.

Foster, C.S., Tsai, Y., Monroe, J.G., Campbell, R., Cestari, M.,

Wetzig, R., Knipe, D. and Greene, M.I. (1986). Genetic studies on murine susceptibility to herpes simplex keratitis. Clin. Immunol. Immunopath. 40, 313–325.

Franklin, R.M. (1989). The ocular secretory immune system: a review. Curr. Eye Res. 8, 599–606.

Franklin, R.M. and Shepard, K.F. (1991). T-cell adherence to lacrimal gland: the event responsible for IgA plasma cell predominance in lacrimal gland. Region. Immunol. 3, 213–216.

Grabner, G., Luger, T.A., Luger, B.M., Smolin, G. and Oh, J.O. (1983). Biologic properties of the thymocyte activating factor (CETAF) produced by a rabbit corneal cell line (SIRC). Invest. Opthalmol. Vis. Sci. 24, 589–595.

Grayston, J.T., Wang, S-P., Yeh, L.-J. and Kuo, C.-C. (1985). Importance of reinfection in the pathogenesis of trachoma. Rev. Infect. Dis. 7, 717–725.

Gregory, R.L. and Filler, S.J. (1987). Protective secretory immunoglobulin A antibodies in humans following oral immunization with Streptococcus mutans. Infect. Immunol. 55, 2409–2415.

Hazlett, L.D., Moon, M.M., Dawisha, A. and Berk, R.S. (1986). Age alters ADPase positive dendritic (Langerhans) cell response to *P. aeruginosa* ocular challenge. Curr. Eye Res. 5, 343–355.

Hein, W.R. and Mackay, C.R. (1991). Prominence of γδT cells in the ruminant immune system. Immunol. Today 12, 30–34.

Hendricks, R.L. and Tumpey, T.M. (1990). Contribution of virus and immune factors to herpes simplex virus type 1–induced corneal pathology. Invest. Opthalmol. Vis. Sci. 31, 1929–1939.

Hendricks, R.L., Tao, M.S.P. and Glorioso, J.C. (1989). Alterations in the antigenic structure of two major HSV-1 glycoproteins, gC and gB, influence immune regulation and susceptibility to murine herpes keratitis. J. Immunol. 142, 263–269.

Hokfelt, T., Johansson, O., Kellerth, J.-O., Ljungdahl, A., Nilsson, G., Nygards, A. and Pernow, B. (1977). In "Substance P" (eds U.S. von Euler and B. Pernow), pp 117–145. Raven Press, New York.

Hyndiuk, R.A. and Glasser, D.B. (1986). In "Infections of the Eye: Diagnosis and Management" (eds K. Tabburn and R. Hyndiuk), pp 343–368 Little, Brown, Boston.

Jakobiec, R.A. (1988). In "The Cornea" (eds H.E. Kaufman, M.B. McDonald, B.A. Barron and S.R. Waltman), pp 563–598. Churchill Livingstone, New York.

Janeway, C.A., Jr, Jones, B. and Hayday, A. (1988). Specificity and function of T cells bearing γδ receptors. Immunol. Today 9, 73–76.

Kijlstra, A. (1991). The role of lactoferrin in nonspecific immune response on the ocular surface. Regional Immunol. 3, 193–197.

Kijlstra, A. and Jeurissen, S.H.M. (1982). Modulation of classical C3 convertase of complement by tear lactoferrin. Immunology 47, 263–270.

Kijlstra, A., Jeurissen, S.H.M. and Koning, K.M. (1983). Lactoferrin levels in normal human tears. Br. J. Ophthalmol. 67, 199–202.

Kijlstra, A., Kievits, F., Jeurissen, S.H.M. and Veerhuis, R. (1989). In "Modern Trends in Immunology and Immuno-pathology of the Eye" (eds A.G. Secchi and I.A. Fregona), pp 189–193. Masson, Milano.

Kowalski, M.L. and Kaliner. M.A. (1988). Neurogenic inflammation, vascular permeability, and mast cells. J. Immunol. 140, 3905–3911.

Ksander, B.R. and Hendricks, R.L. (1987). Cell-mediated immune tolerance to HSV-1 antigens associated with reduced susceptibility to HSV-1 corneal lesions. Invest. Ophthalmol. Vis. Sci. 28, 1986–1993.

Kuziel, W.A., Takashima, A., Bonyhadi, M., Bergstresser, P.R., Allison, J.P., Tigelaar, R.E. and Tucker, P.W. (1987). Regulation of T-cell receptor γ chain RNA expression in murine Thy-1+ dendritic epidermal cells. Nature 328, 263–265.

Lausch, R.N., Kleinschradt, W.R., Monteiro, C., Kayes, S.B. and Oakes, J.E. (1985). Resolution of HSV corneal infection in the absence of delayed-type hypersensitivity. Invest. Ophthalmol. Vis. Sci. 26, 1509–1515.

Lauweryns, B., van den Oord, J.J., Volpes, R., Foets, B. and Missotten, L. (1991). Distribution of very late activation integrins in the human cornea. Invest. Ophthalmol. Vis. Sci. 32, 2079–2085.

McGee, D.W. and Franklin, R.M. (1984). Lymphocyte migration into the lacrimal gland is random. Cell. Immunol. 86, 75–82.

McLeish, W., Rubsamen, P., Atherton, S.S. and Streilein, J.W. (1989). Immunobiology of Langerhans cells on the ocular surface. II. Role of central corneal Langerhans cells in stromal keratitis following experimental HSV-1 infection in mice. Region. Immunol. 2, 236–243.

Masson, P.L., Heremans, J.F. and Dive, C. (1966). An iron-binding protein common to many external secretions. Clin. Chim. Acta 14, 735–739.

Metcalf, J.F., Hamilton, D.S. and Reichert, R.W. (1979). Herpetic keratitis in athymic (nude) mice. Infect. Immunol. 26, 1164–1171.

Miller, D. (1979). "Ophthalmology. The Essentials", pp 1–25. Houghton Mifflin, Boston.

Monnickendam, M.A., Darougar, S., Treharne, J.D. and Tilbury, A.M. (1980). Development of chronic conjunctivitis with scarring and pannus, resembling trachoma, in guinea-pigs. Br. J. Ophthalmol. 64, 284–290.

Montgomery. P.C., Rockey, J.H., Majumdar, A.S., Lemaitre-Coelho, I.M., Vaerman, J-P. and Ayyildiz, A. (1984). Parameters influencing the expression of IgA antibodies in tears. Invest. Ophthalmol. Vis. Sci. 25, 369–373.

Morrison, R.P., Lyng, K. and Caldwell, H.D. (1989). Chlamydial disease pathogenesis. Ocular hypersensitivity elicited by a genus-specific 57–kD protein. J. Exp. Med. 169, 663–675.

Newell, C.K., Martin, S., Sendele, D., Mercadal, C.M. and Rouse, B.T. (1989a). Herpes simplex virus-induced stromal keratitis: role of T-lymphocyte subsets in immunopathology. J. Virol. 63, 769–775.

Newell, C.K., Sendele, D. and Rouse, B.T. (1989b). Effects of CD4+ and CD8+ T-lymphocyte depletion on the induction and expression of Herpes simplex stromal keratitis. Region. Immunol. 2, 366–369.

Niederkorn, J.Y. (1990). Immune privilege and immune regulation in the eye. Adv. Immunol. 48, 191–226.

Niederkorn, J.Y. and Peeler, J.S. (1988). Regional differences in immune regulation: the immunogenic privilege of corneal allografts. Immunol. Res. 7, 247–255.

Niederkorn, J.Y., Peeler, J.S. and Mellon, J. (1989). Phagocytosis of particulate antigens by corneal epithelial cells stimulates interleukin-1 secretion and migration of Langerhans cells into the central cornea. Region. Immunol. 2, 83–90.

Opremcak, E.M., Wells, P.A., Thompson, P., Daigle, J.A., Rice, B.A., Millin, J.A. and Foster, C.S. (1988). Immunogenetic influence of Igh-1 phenotype on experimental herpes simplex virus type-1 corneal infection. Invest. Ophthalmol. Vis. Sci. 29, 749–754.

O'Sullivan, N.L. and Montgomery, P.C. (1990). Selective interactions of lymphocytes with neonatal and adult lacrimal gland tissues. Invest. Ophthalmol. Vis. Sci. 31, 1615–1622.

Pal, S, Pu, Z., Huneke, R.B., Taylor, H.R. and Whittum-Hudson, J.A. (1991). Chlamydia-specific lymphocytes in conjunctiva during ocular infection: limiting dilution analysis. Region. Immunol. 3, 171–176.

Parrott, D.M.V. and Ferguson, A. (1975). Selective migration of lymphocytes within the mouse small intestine. Immunology, 26, 571–588.

Peeler, J.S. and Niederkorn, J.Y. (1986). Antigen presentation by Langerhans cells in vivo: donor-derived Ia+ Langerhans cells are required for induction of delayed-type hypersensitivity but not for cytotoxic T lymphocyte responses to alloantigens. J. Immunol. 136, 4362–4371.

Pepose, J.S. (1991). Herpes simplex keratitis: role of viral infection versus immune response. Surv. Opthalmol. 35, 345–352.

Peppard, J.V. and Montgomery, P.C. (1987). Studies on the origin and composition of IgA in rat tears. Immunology 62, 193–198.

Rodrigues, M.M., Rowden, G., Hackett, J. and Bakos, I. (1981). Langerhans cells in the normal conjunctiva and peripheral cornea of selected species. Invest. Ophthalmol. Vis. Sci. 21, 759–762.

Rozsa, A.J. and Beuerman, R. (1982). Density and organization of free nerve endings in the corneal epithelium of the rabbit. Pain 14, 105–120.

Russel, R.G., Nasisse, M.P., Larsen, H.S. and Rouse, B.T. (1984). Role of T-lymphocytes in the pathogenesis of herpetic stromal keratitis. Invest. Ophthalmol. Vis. Sci. 25, 938–944.

Santoro, S.A., Rajpara, S.M., Staatz, W.D. and Woods, V.L. (1988). Isolation and characterization of a platelet surface collagen binding complex related to VLA-2. Biochem. Biophys. Res. Commun. 153, 217–221.

Schachter, J. and Dawson, C.R. (1978). "Human Chlamydial Infections", pp 63–96. PSB, Littleton.

Schachter, J., Moncada, J., Dawson, C.R., Sheppard, J., Courtrightt, P., Said, M.E., Zaki, S., Hafez, S.F. and Lorincz, A. (1988). Nonculture methods for diagnosing chlamydial infections in patients with trachoma: a clue to the pathogenesis of the diseases. J. Infect. Dis. 158, 1347–1352.

Shams, N.B.K., Sigel, M.M. and Davis, R.M. (1989). Interferon-gamma, Staphylococcus aureus, and lipopolysaccharide/silica enhance interleukin-1B production by human corneal cells. Region. Immunol. 2, 136–148.

Springer, T.A. (1990). Adhesion receptors of the immune system. Nature 346, 425–434.

Stamper, H.B. and Woodruff, J.J. (1977). An in vitro model of lymphocyte homing. I. Characterization of the interaction between thoracic duct lymphocytes and specialized high-endothial venules of lymph nodes. J. Immunol. 119, 772–780.

Stanisz, A.M., Scicchitano, R., Dazin, P., Bienenstock, J. and Payon, D.G. (1987). Distribution of substance P receptors on murine spleen and Peyer's patch T and B cells. J. Immunol. 1139, 749–754.

Streilein, J.W. (1983). Skin-associated lymphoid tissues

(SALT): origins and functions. J. Invest. Dermatol. 80, 12s–16s.

Streilin, J.W. (1987). Immune regulation and the eye: a dangerous compromise. FASEB J. 1, 199–208.

Stulting, R.D., Kindle, J.C. and Nahmias, A.J. (1985). Patterns of herpes simplex keratitis in inbred mice. Invest. Ophthalmol. Vis. Sci. 26, 1360–1367.

Sullivan, D.A. and Allansmith, M.R. (1984). Source of IgA in tears of rats. Immunology 53, 791–799.

Sundmacher, R. (1987). Clinical aspects of herpetic eye diseases. Curr. Eye Res. 6, 183–188.

Svanborg-Eden, C. and Svennherholm, A.M. (1978). Secretory immunoglobulin A and G antibodies prevent adhesion of *Escherichia coli* to human urinary tract epithelial cells. Infect. Immunity 22, 790–797.

Swyers, J.S., Lausch, R.N. and Kaufman, H.E. (1967). Corneal hypersensitivity to herpes simplex. Br. J. Ophthalmol. 51, 843–846.

Taylor, H. and Dimmock, N.J. (1985). Mechanism of neutralization of influenza virus by secretory IgA is different from that of monomeric IgA or IgG. J. Exp. Med. 161, 198–209.

Taylor, H.R. and Bell, T.A. (1991). In "Hunter's Tropical Medicine" (ed. S.T. Strickland), pp 287–293. W.B. Saunders, Philadelphia.

Taylor, H.R., Johnson, S.L., Schachter, J., Caldwell, H.D. and Prendergast, R.A. (1987). Pathogenesis of trachoma: the stimulus for inflammation. J. Immunol. 138, 3023–3027.

Ullrich, S.E., McIntyre, B.W. and Rivas, J.M. (1990). Suppression of the immune response to alloantigen by factors released from ultraviolet-irradiated keratinocytes. J. Immunol. 145, 489–498.

Underdown, B.J. and Schiff, J.M. (1986). Immunoglobulin A: Strategic defense initiative at the mucosal surface. Ann. Rev. Immunol. 4, 389–417.

Vaughan, H.G. and Schlimmel, H. (1971). In "Visual Prosthesis, The Interdisciplinary Guinea Pigs" (eds R.D. Sterling, E.A. Bering Jr, S.V. Pollack and H.G. Vaughan Jr), pp 65–79. Academic Press, New York.

Waldrep, J.C. (1989). In "Healing Processes in the Cornea" (eds R.W. Beuerman, C.E. Crosson and H.E. Kaufman), pp 27–43. Gulf, Houston.

Watkins, N.G., Hadlow, W.J., Moos, A.B. and Caldwell, H.D. (1986). Ocular delayed hypersensitivity: a pathogenetic mechanism of chlamydial conjunctivitis in guinea pigs. Proc. Natl Acad. Sci. USA 83, 7480–7484.

Williams, L.E., Nesburn, A.B. and Kaufman, H.E. (1965). Experimental induction of disciform keratitis. Arch. Ophthalmol. 73, 112–118.

Williamson, J.S.P., DiMarco, S. and Streilein, J.W. (1987). Immunobiology of Langerhans cells on the ocular surface. Invest. Ophthalmol. Vis. Sci. 28, 1527–1532.

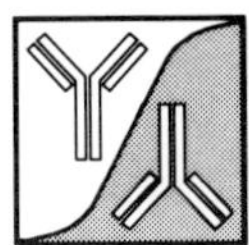

Glossary

Note: This glossary is up to date for the current volume only and will be supplemented with each subsequent volume.

α₁-ACT α₁-Antichymotrypsin
α1–P1 α-1 Proteinase
A Absorbance
Å Angstrom
AA Arachidonic acid
aa Amino acids
AAb Autoantibody
Ab Antibody
Ab1 Idiotype antibody
Ab2 Anti-idiotype antibody
Ab2α Anti-idiotype antibody which binds outside the antigen binding region
Ab2β Anti-idiotype antibody which binds to the antigen binding region
Ab3 Anti-anti-idiotype antibody
Abcc Antibody dependent cytotoxic activity
ABA-L-GAT Arsanilic acid conjugated with the synthetic polypeptide L-GAT
AC Adenylate cyclase
ACAT Acyl-co-enzyme-A acyltransferase
ACAID Anterior chamber-associated immune deviation
ACE Angiotensin-converting enzyme
ACh Acetylcholine
ACTH Adrenocorticotrophin hormone
ADCC Antibody-dependent cell-mediated cytotoxicity
Ado Adenosine
ADP Adenosine diphosphate
AECA Antibodies to endothelial cells
AES Anti-eosinophil serum
Ag Antigen
AGE Advanced glycosylation end-product
AGEPC 1-O-Alkyl-2-acetyl-*sn*-glyceryl-3-phosphocholine
AH Acetylhydrolase
AI Angiotensin I
AII Angiotensin II
AID Autoimmune disease
AIDS Acquired immune deficiency syndrome
A/J A Jackson inbred mouse strain
ALP Anti-leukoprotease
cAMP Cyclic adenosine monophosphate (adenosine 3′, 5′-phosphate)
AM Alveolar macrophage
AML Acute myelogenous leukaemia
AMP Adenosine monophosphate

ANAb Anti-nuclear antibodies
ANCA Anti-neutrophil cytoplasmic auto antibodies
cANCA Cytoplasmic ANCA
pANCA Perinuclear ANCA
AND Anaphylactic degranulation
ANF Atrial natriuretic factor
ANP Atrial natriuretic peptide
Anti-I-A Antibody against class II MHC molecule encoded by I–A locus
Anti-I-E Antibody against class II MHC molecule encoded by I–E locus
anti-Ig Antibody against an immunoglobulin
anti-RTE Anti-tubular epithelium
AP-1 Activator protein-1
APA B-azaprostanoic acid
APAS Antiplatelet antiserum
APC Antigen-presenting cell
APD Action potential duration
ApO-B Apolipoprotein B
APRL Anti-hypertensive polar renal lipid *also known as* PAF
APUD Amine precursor uptake and decarboxylation
AR-CGD Autosomal recessive form of chronic granulomatous disease
ARDS Adult respiratory distress syndrome
AS Ankylosing spondylitis
4–ASA 4-Aminosalicylic acid
5–ASA 5-Aminosalicylic acid
ASA Acetylsalicylic acid (aspirin)
ATHERO-ELAM A monocyte adhesion molecule
ATL Adult T cell leukaemia
ATP Adenosine 3′,5′ triphosphate
AUC Area under curve
AVP Arginine vasopressin

β₂M β-Microglobulin
β-TG β-Thromboglobulin
B Bursa-derived
BAL Bronchoalveolar lavage
B₁receptor Bradykinin receptor subtype
B₂receptor Bradykinin receptor subtype
B₂ (CD18) A leucocyte integrin
BAF Basophil-activating factor
BAL Bronchoalveolar lavage

BALF Bronchoalveolar lavage fluid
BALT Bronchus-associated lymphoid tissue
B cell Bone marrow-derived lymphocyte
BCF Basophil chemotactic factor
BCG Bacillus Calmette-Guérin
bFGF Basic fibroblast growth factor
BG Birbeck granules
BHR Bronchial hyperresponsiveness
BI-CFC Blast colony-forming cells
b.i.d. *Bis in die* (twice a day)
Bk Bradykinin
BM Bone marrow
BMCMC Bone marrow cultured mast cell
BMMC Bone marrow mast cell
BOC-FMLP Butoxycarbonyl-FMLP
bp Base pair
BPB Para-bromophenacyl bromide
BPI Bacterial permeability-increasing protein
BSA Bovine serum albumin

CatG Cathepsin G
C1 The first component of complement
C1 inhibitor A serine protease inhibitor which inactivates C1r/C1s
C1q Complement fragment 1q (anaphylatoxin)
C1qR Receptor for C1w; facilitates attachment of immune complexes to mononuclear leucocytes and endothelium
C2 The second component of complement
C3 The third component of complement
C3a Complement fragment 3a (anaphylatoxin)
C3a₇₂₋₇₇ A synthetic carboxyterminal peptide C3a analogue
C3aR Receptor for anaphylatoxins, C3a, C4a, C5a
C3b Complement fragment 3b (anaphylatoxin)
C3bi Inactivated form of C3b fragment of complement
C4 The fourth component of complement
C4b Complement fragment 4b (anaphylatoxin)
C4BP C4 binding protein; plasma protein which acts as co-factor to factor I inactivate C3 convertase

C5 The fifth component of complement

C5a Complement fragment 5a (anaphylatoxin)

C5aR Receptor for anaphylatoxins C3a, C4a and C5a

C5b Complement fragment 5b (anaphylatoxin)

C6 The sixth component of complement

C7 The seventh component of complement

C8 The eighth component of complement

C9 The ninth component of complement

$C_\varepsilon 2$ Heavy chain of immunoglobulin E: domain 2

$C_\varepsilon 3$ Heavy chain of immunoglobulin E: domain 3

$C_\varepsilon 4$ Heavy chain of immunoglobulin E: domain 4

Ca *The chemical symbol for* calcium

$[Ca_2^+]_1$ Intracellular free calcium concentration

CAH Chronic active hepatitis

CALLA Common lymphoblastic leukaemia antigen

CALT Conjunctival associated lymphoid tissue

cAMP Cyclic adenosine monophosphate (adenosine 3′, 5′- phosphate)

CAM Cell adhesion molecule

CB Cytochalasin B

CBH Cutaneous basophil hypersensitivity

CBP Cromolyn-binding protein

CCK Cholecystokinin

CCR Creatinine clearance rate

CD Cluster of differentiation (a system of nomenclature for surface molecules on cells of the immune system); cluster determinant

CD1 Cluster of differentiation 1 *also known as* MHC class I-like surface glycoprotein

CD1a Isoform a *also known as* non-classical MHC class I-like surface antigen

CD1c Isoform c *also known as* non-classical MHC class I-like surface antigen

CD2 Defines T cells involved in antigen non-specific cell activation

CD3 *Also known as* T cell receptor-associated surface glycoprotein on T cells

CD4 Defines MHC class II-restricted T cell subsets

CD5 *Also known as* Lyt 1 in mouse

CD7 Cluster of differentiation 7

CD8 Defines MHC class I-restricted T cell subset

CD11a *Known to be* an α chain of LFA-1 (leucocyte function antigen-1) present on several types of leucocyte and which mediates adhesion

CD11b *Known to be* an α chain of CR3 (complement receptor type 3) present on several types of leucocyte and which mediates adhesion

CD11c *Known to be* a complement receptor 4 α chain.

CD14 *Known to be* a lipid-anchored glycoprotein

CD18 *Known to be* the common β chain of the CD11 family of molecules

CD20 *Known to be* a pan B cell

CD31 *Known to be* platelets, monocytes, macrophages, granulocytes and B-cells

CD33$^+$ *Known to be* a monocyte and stem cell marker

CD34$^-$ *Known to be* a stem cell marker

CD36 *Known to be* a macrophage vitronectin receptor

CD45 *Known to be* a pan leucocyte marker

CD45RO *Known to be* the isoform of leukosialin present on memory T cells

CD49 Cluster of differentiation 49

CD59 *Known to be* a low molecular weight HRf present to many haematopoetic and non-haematopoetic cells

CD62 *Known to be present on* activated platelets and endothelial cells

CDC Complement-dependent cytotoxicity

cDNA Complementary DNA

CDP Choline diphosphate

CDR Complementary-determining region

CD$_{xx}$ Common determinant *xx*

CEA Carcinoembryonic antigen

CETAF Corneal epithelial T cell activating factor

CF Cystic fibrosis

Cf Cationized ferritin

CFA Complete Freund's adjuvant

CFC Colony-forming cell

CFU Colony-forming unit

CFU-Eo/B Eosinophil/basophil colony-forming cell

CFU-GM Granulocyte-macrophage colony-forming cell

CFU-S Colony-forming unit, spleen

CGD Chronic granulomatous disease

cGMP Cyclic guanosine monophosphate (guanosine 3′, 5′-phosphate)

CGRP Calcitonin gene-related peptide

CH2 Hinge region of human immunoglobulin

CHO Chinese hamster ovary

CI Chemical ionization

CIBD Chronic inflammatory bowel disease

CK Creatine phosphokinase

CKMB The myocardial-specific isoenzyme of creatine phosphokinase

Cl *The chemical symbol for* chloride

CL Chemiluminescent

CLA Cutaneous lymphocyte antigen

CL18/6 Anti-ICAM-1 monoclonal antibody

CLC Charcot–Leyden crystal

CMC Critical micellar concentration

CMI Cell mediated immunity

CML Chronic myeloid leukaemia

CMV Cytomegalovirus

CNS Central nervous system

CO Cyclooxygenase

CoA Coenzyme A

CoA-IT Coenzyme A – independent transacylase

Con A Concanavalin A

COPD Chronic obstructive pulmonary disease

COS Fibroblast-like kidney cell line established from simian cells

CoVF Cobra venom

CP Creatine phosphate

CPJ Cartilage/pannus junction

Cr *The chemical symbol* for chromium

CR Complement receptor

CR1 Complement receptor type 1

CR2 Complement receptor type 2

CR3 Complement receptor type 3

CR3–α Complement receptor type 3–α

CR4 Complement receptor type 4

CRF Corticotrophin-releasing factor

CRH Corticotrophin-releasing hormone

CRI Cross-reactive idiotype

CRP C-reactive protein

CSA Cyclosporin A

CSF Colony–stimulating factor

CSS Chürg-Strauss syndrome

CT Computed tomography

CTAP-III Connective tissue-activating peptide

CTD Connective tissue diseases

C terminus Carboxy terminus of peptide

CThp Cytotoxic T lymphocyte precursors

CTL Cytotoxic T lymphocyte

CTMC Connective tissue mast cell

ct.min^{-1} Counts per minute

Da Dalton (the unit of relative molecular mass)

DAF Decay-accelerating factor

DAG Diacylglycerol

DAO Diamine oxidase

D-Arg D-Arginine

DArg-[Hyp^3,DPhe^7]-BK A bradykinin B_2 receptor antagonist. Peptide derivative of bradykinin

DArg-[Hyp^3,Thi^5,DTic^7,Tic^8]-BK A bradykinin B_2 receptor antagonist. Peptide derivative of bradykinin

DC Dendritic cell

DCF Oxidized DCFH

DCFH 2′,7′-dichlorofluorescin

DEC Diethylcarbamazine

desArg^9-BK Carboxypeptidase N product of bradykinin

desArg^{10}KD Carboxypeptidase N product of kallidin

DFMO α-Difluoromethyl ornithine

DFP Diisopropyl fluorophosphate

DGLA Dihomo-γ-linolenic acid

DH Delayed hypersensitivity

DHR Delayed hypersensitivity reaction

DIC Disseminated intravascular coagulation
DL-CFU Dendritic cell/Langerhans cell colony forming
DLE Discoid lupus erythematosus
DMARD Disease-modifying anti-rheumatic drug
DMF N,N-Dimethylformamide
DMSO Dimethyl sulfoxide
DNA Deoxyribonucleic acid
D-NAME D-Nitroarginine methyl ester
DNase Deoxyribonuclease
DNCB Dinitrochlorobenzene
DNP Dinitrophenol
Dpt4 *Dermatophagoides pteronyssinus* allergen 4
DGW2 An HLA phenotype
DR3 An HLA phenotype
DR7 An HLA phenotype
DREG-2 Murine IgG$_1$ monoclonal antibody against L-selectin
ds Double-stranded
DSCG Disodium cromoglycate
DST Donor-specific transfusion
DTH Delayed-type hypersensitivity
DTPA Diethylenetriamine pentaacetate
DTT Dithiothreitol
dv/dt Rate of change of voltage within time

ε Molar absorption coefficient
EA Egg albumin
EAE Experimental autoimmune encephalomyelitis
EAF Eosinophil-activating factor
EAR Early phase asthmatic reaction
EAT Experimental autoimmune thyroiditis
EBV Epstein–Barr virus
EC Electron capture
ECD Electron capture detector
ECE Endothelin-converting enzyme
E-CEF Eosinophil cytotoxicity enhancing factor
ECF-A Eosinophil chemotactic factor of anaphylaxis
ECG Electrocardiogram
ECGF Endothelial cell growth factor
ECGS Endothelial cell growth supplement
E. coli *Escherichia coli*
ECP Eosinophil cationic protein
ED$_{35}$ Effective dose producing 35% maximum response
ED$_{50}$ Effective dose producing 50% maximum response
EDF Eosinophil differentiation factor
EDN Eosinophil-derived neurotoxin
EDRF Endothelium-derived relaxant factor
EDTA Ethylenediamine tetraacetic acid *also known as* etidronic acid
EE Eosinophilic eosinophils
EEG Electroencephalogram
EET Epoxyeicosatrienoic acid
EFA Essential fatty acid

EFS Electrical field stimulation
EG1 Monoclonal antibody specific for the cleaved form of eosinophil cationic peptide
EGF Epidermal growth factor
EGTA Ethylene glycol-bis(β-aminoethyl ether) N,N,N',N'-tetraacetic acid
EI Electron impact
eIF-2 Subunit of protein synthesis initiation factor
ELAM Endothelial leucocyte adhesion molecule
ELF Respiratory epithelium lung fluid
ELISA Enzyme-linked immunosorbent assay
EMS Eosinophilia–myalgia syndrome
ENS Enteric nervous system
EO Eosinophil
EOR Early onset reaction
EPA Eicosapentaenoic acid
EPO Eosinophil peroxidase
EpDIF Epithelial-derived inhibitory factor
EpDRF Epithelium-derived relaxant factor
EPO Epithelial peroxidase
EPR Effector cell protease
EPX Eosinophil protein X
ER Endoplasmic reticulum
ESP Eosinophil stimulation promoter
ESR Erythrocyte sedimentation rate
ET Endothelin
ET-1 Endothelin-1
ETYA Eicosatetraynoic acid

FA Fatty acid
FAB Fast-electron bombardment
Fab Antigen binding fragment
factor B Serine protease in the C3 converting enzyme of the alternative pathway
factor D Serine protease which cleaves factor B
factor H Plasma protein which acts as a co-factor to factor I
factor I Hydrolyses C3 converting enzymes with the help of factor H
FAD Flavine adenine dinucleotide
FBR Fluorescence photobleaching recovery
Fc Crystallizable fraction of immunoglobulin molecule
FcR Receptor for Fc region of antibody
Fc$_ε$RI High affinity receptor for IgE
Fc$_ε$RII Low affinity receptor for IgE
FCS Foetal calf (bovine) serum
FEV$_1$ Forced expiratory volume in 1 second
FGF Fibroblast growth factor
FID Flame ionization detector
FITC Fluorescein isothiocyanate
FKBP FK506–binding protein
FLAP 5–Lipoxygenase-activating protein
FMLP N-Formyl-methionyl-leucyl-phenylalanine

FNLP Formyl-norleucyl-leucyl-phenylalanine
FPR Formyl peptide receptor
FSG Focal sequential glomerulosclerosis
FSH Follicle stimulating hormone
5–FU 5–Fluorouracil

Ga G-protein
G6PD Glucose 6–phosphate dehydrogenase
GABA γ-Aminobutyric acid
GAG Glycosaminoglycan
GALT Gut-associated lymphoid tissue
GAP GTPase-activity protein
GBM Glomerular basement membrane
GC Guanylate cyclase
GC-MS Gas chromatography mass spectroscopy
G-CSF Granulocyte colony-stimulating factor
Ge Glycoprotein exocytosis
GDP Guanosine 5′-diphosphate
GEC Glomerular epithelial cell
GF-1 Insulin-like growth factor
GFR Glomerular filtration rate
GH Growth hormone
GH-RF Growth hormone-releasing factor
GI Gastrointestinal
GIP Granulocyte inhibitory protein
GlyCam-1 Glycosylation-dependent cell adhesion molecule-1
GMC Gastric mast cell
GM-CSF Granulocyte-macrophage colony-stimulating factor
GMP Guanosine monophosphate (guanosine 5′-phosphate)
GMP-140 Granule-associated membrane protein-140
GP Glycoprotein
gp45–70 Membrane co-factor protein
GPIIb-IIIa Glycoprotein IIb-IIIa (a platelet membrane antigen)
GRP Gastrin-related peptide
GSH Glutathione (reduced)
GSSG Glutathione (oxidized)
GTP Guanosine triphosphate
GTP-γ-S Guanarine 5′0-(3–thiotriphosphate)
GTPase Guanidine triphosphatase
GVHD Graft-versus-host-disease
GVHR Graft-versus-host-reaction

H$_1$ Histamine receptor type 1
H$_2$ Histamine receptor type 2
H$_3$ Histamine receptor type 3
H$_2$O$_2$ *The chemical symbol for* hydrogen peroxide
HA Histamine
Hag Haemagglutinin
Hag-1 Cleaved haemagglutinin subunit-1
Hag-2 Cleaved haemagglutinin subunit-2
H & E Haematoxylin and eosin
hIL Human interleukin

Hb Haemoglobin
HBBS Hank's balanced salt solution
HDC Histidine decarboxylase
HDL High-density lipoprotein
HEL Hen egg white lysozyme
HEPE Hydroxyeicosapentanoic acid
HEPES *N*-2–Hydroxylethylpiperazine-*N'*-2–ethane sulphonic acid
HES Hypereosinophilic syndrome
HETE 5,8,9,11 and 15-Hydroxyeicosatetraenoic acid
HETrE Hydroxyeicosatrienoic acid
HEV High endothelial venule
HFN Human fibronectin
HGF Hepatocyte growth factor
HHT 12–Hydroxy-5, 8, 10–heptadecatrienoic acid
HHTrE 12(*S*)-Hydroxy-5,8,10–heptadecatrienoic acid
HIV Human immunodeficiency virus
HLA Human leucocyte antigen
HMG CoA Hydroxylmethylglutaryl coenzyme A
HMW High molecular weight
HMT Histidine methyltransferase
HMVEC Human microvascular endothelial cell
HNC Human neutrophil collagenase (MMP-8)
HNE Human neutrophil elastase
HNG Human neutrophil gelatinase (MMP-9)
HODE Hydroxyoctadecanoic acid
HPETE Hydroperoxyeicosatetraenoic acid
HPETrE Hydroperoxytrienoic acid
HPODE Hydroperoxyoctadecanoic acid
HPLC High-performance liquid chromatography
HRA Histamine-releasing activity
HRAN Neutrophil-derived histamine-releasing activity
HRf Homologous-restriction factor
HRF Histamine-releasing factor
HRP Horseradish peroxidase
HSA Human serum albumin
HSP Heat-shock protein
HS-PG Heparan sulphate proteoglycan
HSV Herpes simplex virus
HSV-1 Herpes simplex virus 1
³HTdR Tritiated thymidine
5-HT 5-Hydroxytryptamine *also known as* Serotonin
HUVEC Human umbilical vein endothelial cell
[Hyp³]-BK Hydroxproline derivative of bradykinin
[Hyp⁴]-KD Hydroxproline derivative of kallidin

I_{sc} Short-circuit current
Ia Immune reaction-associated antigen
Ia+ Murine class II major histocompatibility complex antigen
IB₄ Anti-CD18 monoclonal antibody

IBD Inflammatory bowel disease
IBMX Isobutylmethylxanthine
IBS Inflammatory bowel syndrome
IC₅₀ Concentration producing 50% inhibition
ICAM Intercellular adhesion molecule
ICAM-1 Intercellular adhesion molecule-1
ICAM-2 Intercellular adhesion molecule-2
ICAM-3 Intercellular adhesion molecule-3
ICE IL-1β–converting enzyme
IDC Interdigitating cell
IDD Insulin-dependent (type 1) diabetes
IEL Intraepithelial leucocyte
IELym Intraepithelial lymphocyte
IFA Incomplete Freund's adjuvant
IFN Interferon
IFNα Interferon α
IFNβ Interferon β
IFNγ Interferon γ
Ig Immunoglobulin
IgA Immunoglobulin A
IgE Immunoglobulin E
IgG Immunoglobulin G
IgG1 Immunoglobulin G class 1
IgG₂ₐ Immunoglobulin G class 2a
IgM Immunoglobulin M
IGF-1 Insulin-like growth factor
IGSS Immuno-gold silver stain
IHC Immunohistochemistry
IHES Idiopathic hypereosinophilic syndrome
IκB NFκB inhibitor protein
IL Interleukin
IL-1 Interleukin-1
IL-1α Interleukin-1α
IL-1β Interleukin-1β
IL-1Ra Interleukin-1 receptor antagonist
IL-2 Interleukin 2
IL-2R Interleukin-2 receptor
IL-3 Interleukin-3
IL-3R Interleukin-3 receptor
IL-4 Interleukin-4
IL-5 Interleukin-5
IL-5R Interleukin-5–receptor
IL-6 Interleukin-6
IL-8 Interleukin-8
ILR Interleukin receptor
IMF Integrin modulating factor
IMMC Intestinal mucosal mast cell
INCAM Inducible cell adhesion molecule
INCAM110 Inducible cell adhesion molecule 110
i.p. Intraperitoneally
IP₃ Inositol triphosphate
IP₄ Inositol tetrakisphosphate
IPF Idiopathic pulmonary fibrosis
IPO Intestinal peroxidase
IpOCOCq Isopropylidene OCOCq
I/R Ischaemia-reperfusion
IRAP IL-1 receptor antagonist protein

IRF-1 Interferon regulatory factor 1
ISCOM Immune-stimulating complexes
ISGF3 Interferon-stimulated gene Factor 3
ISGF3α α subunit of ISGF3
ISGFγ γ subunit of ISGF3
IT Immunotherapy
ITP Idiopathic thrombocytopenic purpura
i.v. Intravenous

K *The chemical symbol for* potassium
Kₐ Association constant
kb Kilobase
20KDHRF A homologous restriction factor; binds to C8
65KDHRF A homologous restriction factor, also known as C8 binding protein; interferes with cell membrane pore-formation by C5b-C8 complex
K_d Equilibrium dissociation constant
kD Kilodalton
K_D Dissociation constant
KD Kallidin
Ki Antagonist binding affinity
K*i*67 Nuclear membrane antigen
KLH Keyhole limpet haemocyanin
KOS KOS strain of herpes simplex virus

λ_max Wavelength of maximum absorbance
LAD Leucocyte adhesion deficiency
LAK Lymphocyte-activated killer (cell)
LAM Leucocyte adhesion molecule
LAM-1 Leucocyte adhesion molecule-1
LAR Late-phase asthmatic reaction
L-Arg L-Arginine
LBP LPS binding protein
LC Langerhans cell
LCF Lymphocyte chemoattractant factor
LCR Locus control region
LDH Lactate dehydrogenase
LDL Low-density lipoprotein
LDV Laser Doppler velocimetry
LECAM Lectin adhesion molecule
LECAM-1 Lectin adhesion molecule-1
LFA Leucocyte function-associated antigen
LFA-1 Leucocyte function-associated antigen-1; a member of the β-2 integrin family of cell adhesion molecules
LG β-Lactoglobulin
LGL Large granular lymphocyte
LH Luteinizing hormone
LHRH Luteinizing hormone-releasing hormone
LI Labelling index
LIS Lateral intercellular spaces
LMP Low molecular mass polypeptide
LMW Low molecular weight
L-NOARG L-Nitroarginine
5-LO 5-Lipoxygenase
12-LO 12-Lipoxygenase
15–LO 15–Lipoxygenase
LP(a) Lipoprotein a

LPS Lipopolysaccharide
LT Leukotriene
LTA$_4$ Leukotriene A$_4$
LTB$_4$ Leukotriene B$_4$
LTC$_4$ Leukotriene C$_4$
LTD$_4$ Leukotriene D$_4$
LTE$_4$ Leukotriene E$_4$
L$_y$-1$^+$ (Cell line)
LX Lipoxin
LXA$_4$ Lipoxin A$_4$
LXB$_4$ Lipoxin B$_4$
LXC$_4$ Lipoxin C$_4$
LXD$_4$ Lipoxin D$_4$
LXE$_4$ Lipoxin E$_4$
zLYCK Carboxybenzyl-Leu-Tyr-CH$_2$Cl

α$_2$-M α$_2$-Macroglobulin
M3 Receptor Muscarinic receptor subtype 3
M-540 Merocyanine-540
mAb Monoclonal antibody
mAB IB4 Monoclonal antibody IB4
mAB PB1.3 Monoclonal antibody PB1.3
mAB R 3.1 Monoclonal antibody R 3.1
mAB R 3.3 Monoclonal antibody R 3.3
mAB 6.5 Monoclonal antibody 6.5
mAB 60.3 Monoclonal antibody 60.3
MAC Membrane attack molecule
Mac Macrophage (*also* abbreviated to MΦ)
Mac-1 Macrophage-1 antigen; a member of the β-2 integrin family of cell adhesion molecules (also abbreviated to MΦ1)
MAF Macrophage-activating factor
MAO Monoamine oxidase
MAP Monophasic action potential
MAPTAM An intracellular Ca^{2+} chelator
MBP Major basic protein
MBSA Methylated bovine serum albumin
MC Mesangial cells
M cell Microfold or membranous cell of Peyer's patch epithelium
MCP Membrane co-factor protein
M-CSF Monocyte colony-stimulating factor
MC$_T$ Tryptase-containing mast cell
MC$_{TC}$ Tryptase- and chymase-containing mast cell
MDA Malondialdehyde
MDGF Macrophage-derived growth factor
MDP Muramyl dipeptide
MEA Mast cell growth-enhancing activity
MEL Metabolic equivalent level
MEL-14 antigen Metabolic equivalent level 14 antigen
MEM Minimal essential medium
MG Myasthenia gravis
MHC Major histocompatibility complex
MI Myocardial ischaemia

MIF Migration inhibition factor
mIL Mouse interleukin
MI/R Myocardial ischaemia/reperfusion
MIRL Membrane inhibitor of reactive lysis
MLC Mixed lymphocyte culture
MLymR Mixed lymphocyte reaction
MLR Mixed leucocyte reaction
MMC Mucosal mast cell
MMCP Mouse mast cell protease
MMP Matrix metalloproteinase
MMP1 Matrix metalloproteinase 1
MNA 6–Methoxy-2–napthylacetic acid
MNC Mononuclear cells
MΦ Macrophage (*also abbreviated to* Mac)
MPO Myeloperoxidase
MRI Magnetic resonance imaging
mRNA Messenger ribonucleic acid
MS Mass spectrometry
MSS Methylprednisoline sodium succinate
MT Malignant tumour
MW Molecular weight

Na *The chemical symbol for* sodium
NA Noradrenaline *also known as* norepinephrine
NAAb Natural autoantibody
NAb Natural antibody
NADH Reduced nicotinamide adenine dinucleotide
NADP Nicotinamide adenine diphosphate
NADPH Reduced nicotinamide adenine dinucleotide phosphate
L-NAME L-Nitroarginine methyl ester
NANC Non-adrenergic, non-cholinergic
NAP Neutrophil-activating peptide
NAP-1 Neutrophil-activating peptide-1
NAP-2 Neutrophil-activating peptide-2
NBT Nitro-blue tetrazolium
NC1 Non-collagen 1
N-CAM Neural cell adhesion molecule
NCEH Neutral cholesteryl ester hydrolase
NCF Neutrophil chemotactic factor
NDGA Nordihydroguaretic acid
NDP Nucleoside diphosphate
Neca 5'-(*N*-ethyl carboxamido)-adenosine
NED Nedocromil sodium
NEP Neutral endopeptidase (EC 3.4.24.11)
NF-AT Nuclear factor of activated T lymphocytes
NF-κB Nuclear factor-κB
NGF Nerve growth factor
NGPS Normal guinea-pig serum
NIMA Non-inherited maternal antigens
Nk Neurokinin
NK Natural killer
Nk-1 Neurokinin receptor subtype
Nk-2 Neurokinin receptor subtype
Nk-3 Neurokinin receptor subtype

NkA Neurokinin A
NkB Neurokinin B
L-NMMA L-Nitromonomethyl arginine
NMR Nuclear magnetic resonance
NO *The chemical symbol for* nitric oxide
NPK Neuropeptide K
NPY Neuropeptide Y
NRS Normal rabbit serum
NSAID Non-steroidal anti-inflammatory drug
NSE Nerve-specific enolase
NT Neurotensin
N terminus Amino terminus of peptide

O$_2^-$ Oxygen free radical
OA Osteoarthritis
OAG Oleoyl acetyl glycerol
OD Optical density
ODC Ornithine decarboxylase
ODS Octadecylsilyl
·OH *The chemical symbol for* hydroxyl radical
OT Oxytocin
OVA Ovalbumin
ox-LDL Oxidized low-density lipoprotein

Ψa Apical membrane potential
P Probability
P Phosphate
PAFR Platelet activating factor receptor
P$_a$O$_2$ Arterial oxygen pressure
P$_i$ Inorganic phosphate
p150,95 A member of the β-2–integrin family of cell adhesion molecules
PA Phosphatidic acid
pA$_2$ Negative logarithm of the antagonist dissociation constant
PADGEM Platelet activation-dependent granule external membrane
PAF Platelet-activating factor *also known as* APRL
PAGE Polyacrylamide gel electrophoresis
PAI Plasminogen activator inhibitor
PAM Pulmonary alveolar macrophages
PAS Periodic acid–Schiff reagent
PBA Polyclonal B cell activators
PBC Primary biliary cirrhosis
PBL Peripheral blood lymphocytes
PBMC Peripheral blood mononuclear cells
PBS Phosphate-buffered saline
PC Phosphatidylcholine
PCA Passive cutaneous anaphylaxis
PCNA Proliferating cell nuclear antigen
PCR Polymerase chain reaction
p.d. Potential difference
PDBu 4α-phorbol 12,13–dibutyrate
PDE Phosphodiesterase
PDGF Platelet-derived growth factor
PE Phosphatidylethanolamine
PECAM-1 Platelet endothelial cell adhesion molecule-1
PEG Polyethylene glycol
PET Positron emission tomography

PEt Phosphatidylethanol
PF$_4$ Platelet factor 4
PG Prostaglandin
PGA Polyglandular autoimmune syndrome
PGE$_2$ Prostaglandin E$_2$
PGF Prostaglandin F
PGI$_2$ Prostaglandin I$_2$ *also known as* prostacyclin
PGD$_2$ Prostaglandin D$_2$
PGE$_1$ Prostaglandin E$_1$
PGF$_{2\alpha}$ Prostaglandin F$_{2\alpha}$
PGF$_2$ Prostaglandin F$_2$
PGG$_2$ Prostaglandin G$_2$
PGH Prostaglandin H
PGH$_2$ Prostaglandin H$_2$
PGP Protein gene-related peptide
Ph^1 Philadelphia (chromosome)
PHA Phytohaemagglutinin
PHI Peptide histidine isoleucine
PHM Peptide histidine methionine
PI Phosphatidylinositol
PI-3–kinase Phosphatidylinositol-3–kinase
PIP Phosphatidylinositol monophosphate
PIP$_2$ Phosphatidylinositol biphosphate
PK Protein kinase
PKA Protein kinase A
PKC Protein kinase C
PL Phospholipase
PLA Phospholipase A
PLA$_2$ Phospholipase A$_2$
PLAP Putative phospholipase activity protein
PLC Phospholipase C
PLD Phospholipase D
PLP Proteolipid protein
PLT Primed lymphocyte typing
PMA Phorbol myristate acetate
PMC Peritoneal mast cell
PMD Piecemeal degranulation
PML Polymorphonuclear leucocyte
PMN Polymorphonuclear neutrophil
PMSF Phenylethylsulphonyl fluoride
PNU Protein nitrogen unit
P$_\alpha$O$_2$ Arterial oxygen pressure
p.o. *Per os* (by mouth)
PPD Purified protein derivative
PPME A ligand for L-selection
PRA Percentage reactive activity
PRD Positive regulatory domain
PRD-II Positive regulatory domain II
PR3 Proteinase-3
proET-1 Proendothelin-1
PRL Prolactin
PRP Platelet-rich plasma
PS Phosphatidylserine
PT Pertussis toxin
PTA$_2$ Pinane thromboxane A$_2$
PTCA Percutaneous transluminal coronary angioplasty
PTCR Percutaneous transluminal coronary recanalization

Pte-H$_4$ Tetrahydropteridine
PtX Pertussis toxin
PUFA Polyunsaturated fatty acid
PUMP-1 Punctuated metalloproteinase
PWM Pokeweed mitogen
PYY Peptide YY

Qa Genetic locus encoding a non-classical class I MHC molecule
q.i.d. *Quater in die* (four times a day)
QRS Segment of electrocardiogram

·R Free radical
R15.7 Anti-CD18 monoclonal antibody
RA Rheumatoid arthritis
RANTES A member of the IL8 supergene family (Regulated on activation, normal T expressed and secreted)
RAST Radioallergosorbent test
RBC Red blood cell
RBF Renal blood flow
RBL Rat basophilic leukaemia
RE RE strain of herpes simplex virus type 1
REA Reactive arthritis
REM Relative electrophoretic mobility
RER Rough endoplasmic reticulum
RF Rheumatoid factor
RFL-6 Rat foetal lung-6
RFLP Restriction fragment length polymorphism
rh- Recombinant human (prefix, usually refers to peptide)
RIA Radioimmunoassay
RMCP Rat mast cell protease
RMCPII Rat mast cell protease II
RNA Ribonucleic acid
RNase Ribonuclease
RNHCl *N*-Chloramine
RNL Regional lymph nodes
ROM Reactive oxygen metabolite
ROS Reactive oxygen species
R-PIA *R-NG* -(1–methyl-1–phenyltheyl)-adenosine
RPMI 1640 Roswell Park Memorial Institute 1640 medium
RS Reiter's syndrome
RSV Rous sarcoma virus
RTE Rabbit tubular epithelium
RTE-a-5 Rat tubular epithelium antigen a-5
r-tPA Recombinant tissue-type plasminogen activator
RW Ragweed

S Svedberg (unit of sedimentation density)
SALT Skin-associated lymphoid tissue
SAZ Sulphasalazine
SC Secretory component
SCF Stem cell factor
SCFA Short-chain fatty acid
SCG Sodium cromoglycate

SCID Severe combined immunodeficiency sydrome
sCR1 Soluble type-1 complement receptors
SCW Streptococcal cell wall
SD Standard deviation
SDS Sodium dodecyl sulphate
SDS–PAGE Sodium dodecyl sulphate–polyacrylamide gel electrophoresis
SEM Standard error of the mean
SGAW Specific airway conductance
SHR Spontaneously hypertensive rat
SIRS Soluble immune response suppressor
SK Streptokinase
Sl Murine Steel mutation
SLE Systemic lupus erythematosus
SLe^x Sialyl Lewis X antigen
SLO Streptolysin-O
SLPI Secretory leucocyte protease inhibitor
SM Sphingomyelin
SNAP S -Nitroso-*N* -acetylpenicillamine
SNP Sodium nitroprusside
SOD Superoxide dismutase
SOM Somatostatin
SOZ Serum-opsonized zymosan
SP Sulphapyridine
S Protein vitronectin
SR Systemic reaction
SRBC Sheep red blood cells
SRIF Somatotrophin release-inhibiting factor (somatostatin)
SRS Slow-reacting substance
SRS-A Slow-reacting substance of anaphylaxis
Sub P Substance P

T Thymus-derived
$t_{1/2}$ Half-life
T84 Human intestinal epithelial cell line
TauNHCl Taurine monochloramine
TBM Tubular basement membrane
TCA Trichloroacetic acid
T cell Thymus-derived lymphocyte
TCP Toxin co-regulated pilus
TCR T cell receptor (α/β or γ/δ heterodimeric forms)
TDI Toluene diisocyanate
TDID$_{50}$ Tissue culture infectious dose – 50%
TEC Tubular epithelial cell
TF Tissue factor
Tg Thyroglobulin
TGF Transforming growth factor
TGFα Transforming growth factor α
TGFβ Transforming growth factor β
TGFβ_1 Transforming growth factor β_1
T$_H$ T helper cell
T$_H$o T Helper o
T$_H$p T helper precursor
T$_H$0, T$_H$1, T$_H$2 Subsets of helper T cells
Thy 1+ Murine T cell antigen
t.i.d. *Ter in die* (three times a day)

TIL Tumour-infiltrating lymphocytes
TIMP Tissue inhibitors of metalloproteinase
TIMP-1 Tissue inhibitor of metalloproteinases-1
TIMP-2 Tissue inhibitor of metalloproteinases-2
Tla Thymus leukaemia antigen
TLC Thin-layer chromatography
TLCK Tosyl-lysyl-CH$_2$Cl
TLP Tumour-like proliferation
Tm T memory
TNF Tumour necrosis factor
TNF-α Tumour necrosis factor-α
tPA Tissue-type plasminogen activator
TPA 12–O-Tetradeconylphorbol-13–acetate
TPCK Tosyl-phenyl-CH$_2$Cl
TPK Tyrosine protein kinases
TPP Transpulmonary pressure
Tris Tris(hydroxymethyl)aminomethane
TSH Thyroid-stimulating hormone
TTX Tetrodotoxin

TX Thromboxane
TXA$_2$ Thromboxane A$_2$
TXB$_2$ Thromboxane B$_2$
Tyk2 Tyrosine kinase

UC Ulcerative colitis
UDP Uridine diphosphate
UPA Urokinase-type plasminogen activator
UV Ultraviolet

VC Veiled cells
VCAM Vascular cell adhesion molecule
VCAM-1 Vascular cell adhesion molecule-1
VF Ventricular fibrillation
VIP Vasoactive intestinal peptide
VLA Very late antigen (β 1 integrins)
VLA-1 Very late antigen-1
VLA-2 Very late antigen-2
VLA-3 Very late activation antigen-3
VLA-4 Very late activation antigen-4
VLA-5 very late activation antigen-5

VLA-6 Very late activation antigen-6
VLDL Very low-density lipoprotein
***V* max** Maximal velocity
vp Viral protein
VP Vasopressin
VPB Ventricular premature beat
VT Ventricular tachycardia

W Murine dominant white spotting mutation
WBC White blood cell
WGA Wheat germ agglutinin

XO Xanthine oxidase

Y1/82A A monoclonal antibody detecting a cytoplasmic antigen in human macrophages

ZA Zonulae adherens
ZAS Zymosan-activated serum
ZO Zonulae occludentes

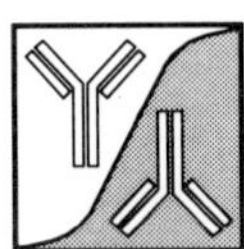

Key to Illustrations

Helper lymphocyte

Suppressor lymphocyte

Killer lymphocyte

Plasma cell

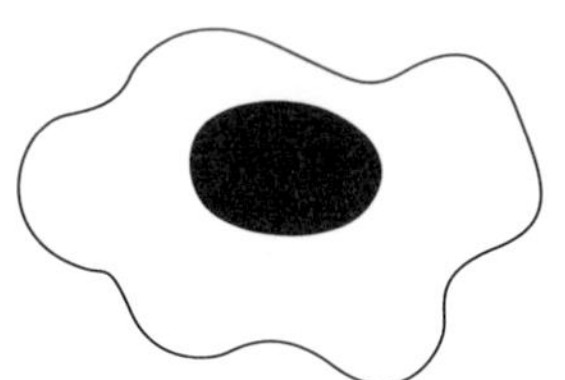

Bacterial or Tumour cell

Blood vessel lumen

Eosinophil passing through vessel wall

Neutrophil passing through vessel wall

Resting neutrophil

Activated neutrophil

Resting eosinophil

Activated eosinophil

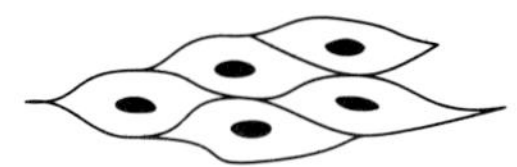

Smooth muscle

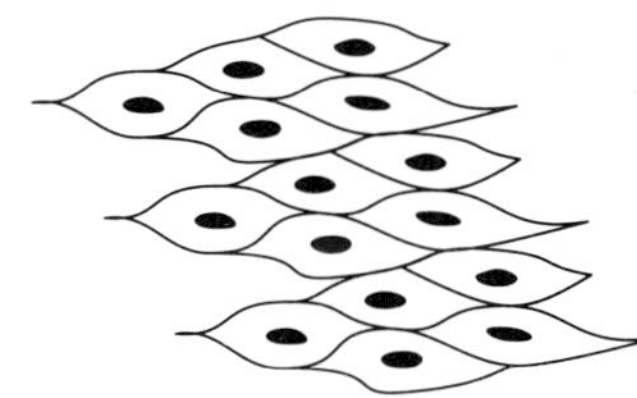

Smooth muscle thickening

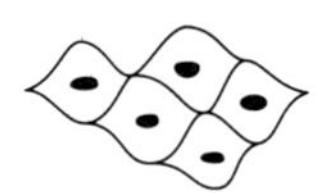

Smooth muscle contraction

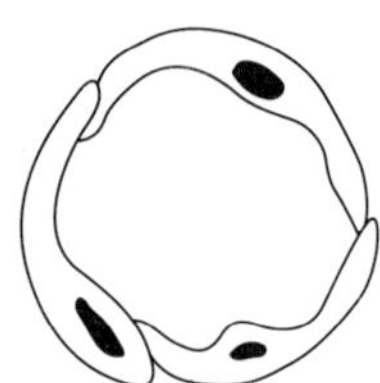

Normal blood vessel

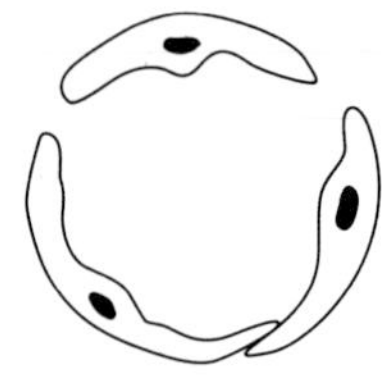

Endothelial cell permeability

Resting macrophage

Activated macrophage

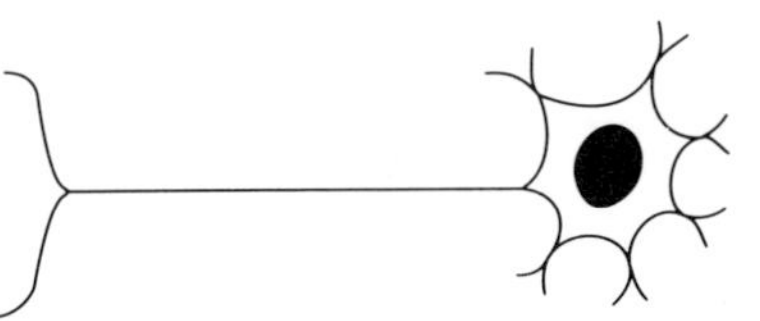

Nerve

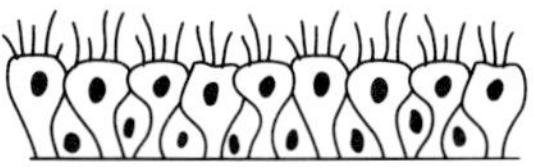

Intact epithelium

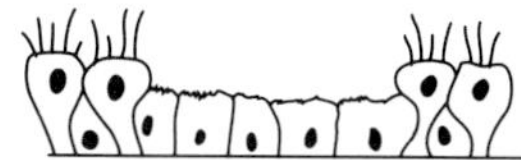

Damaged epithelium

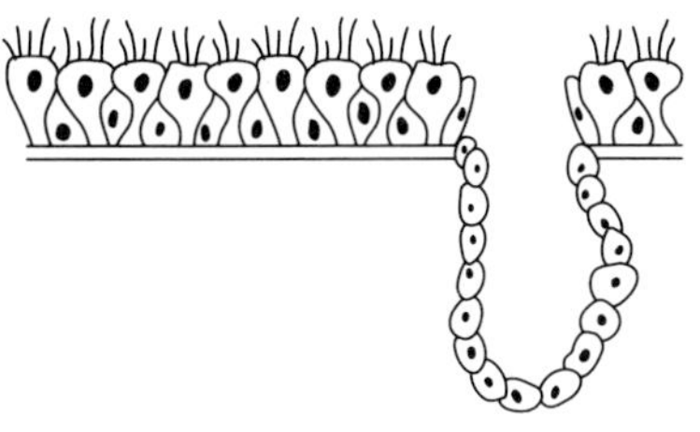

Intact epithelium with submucosal gland

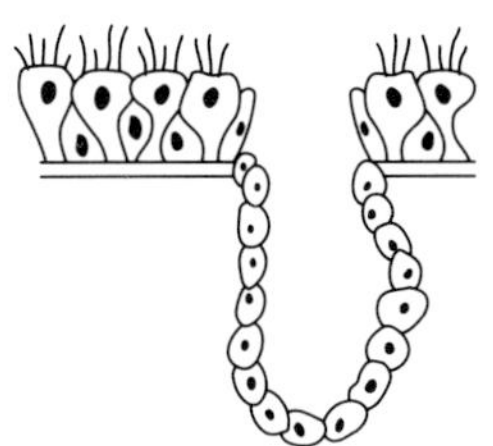

Normal submucosal gland

Hypersecreting submucosal gland

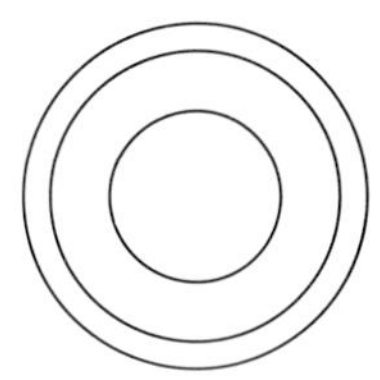

Normal airway

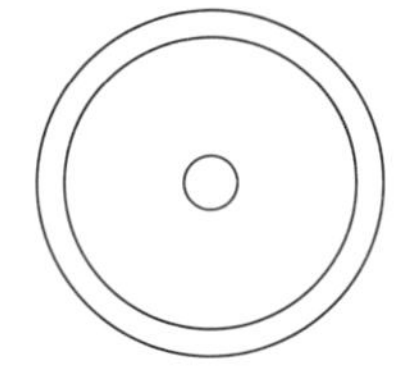

Oedema

Bronchospasm

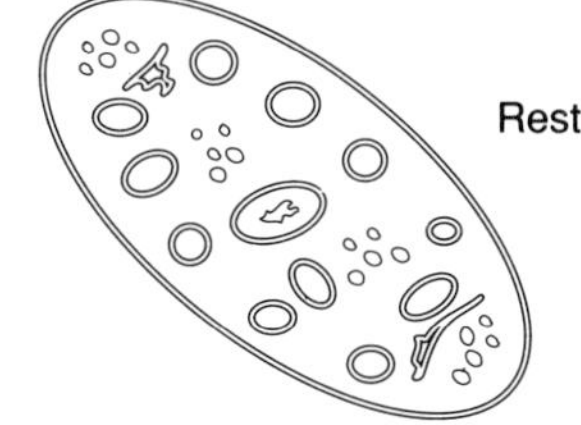

Resting platelet

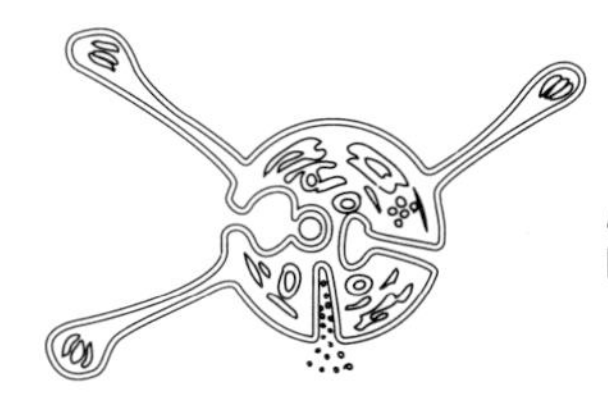

Activated platelet

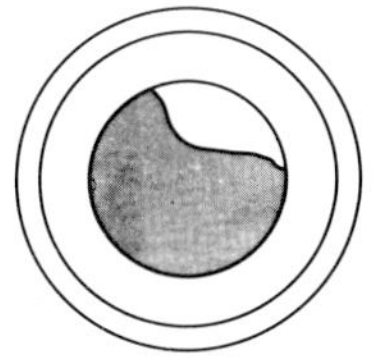

Airway hypersecreting mucus

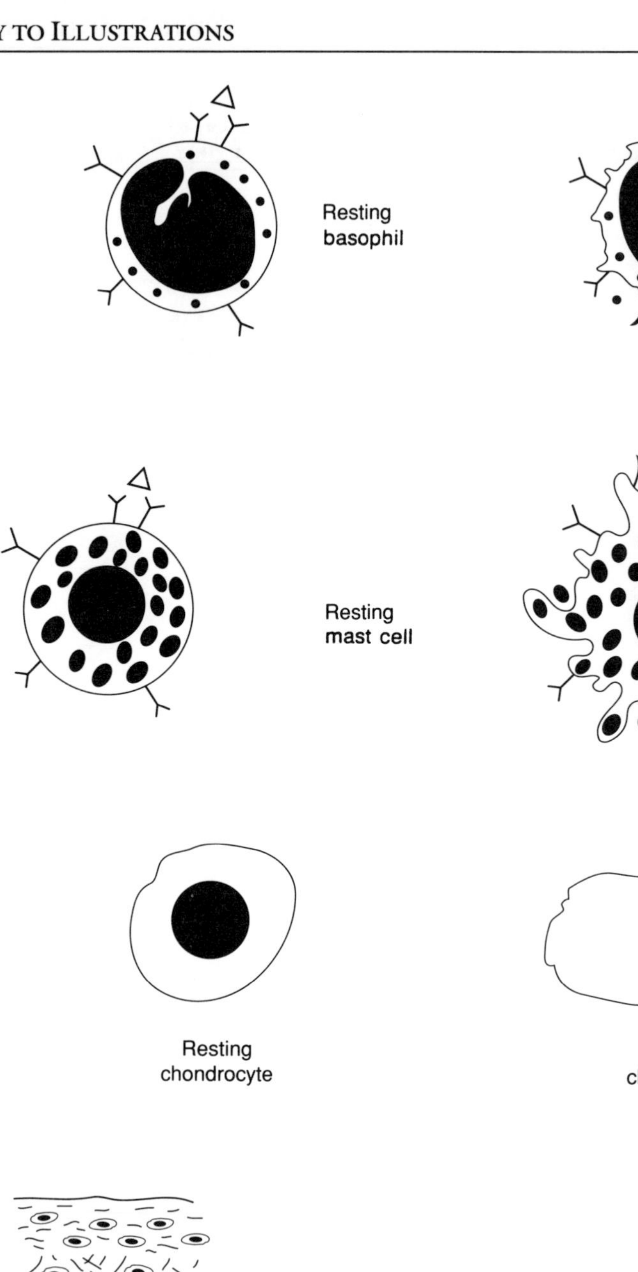
Resting
basophil
Resting
mast cell
Resting
chondrocyte

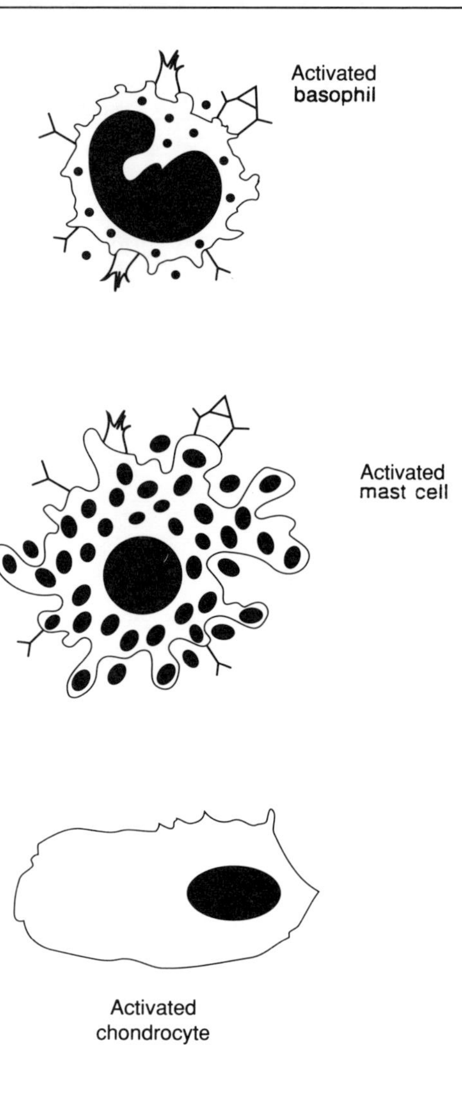
Activated
basophil
Activated
mast cell
Activated
chondrocyte

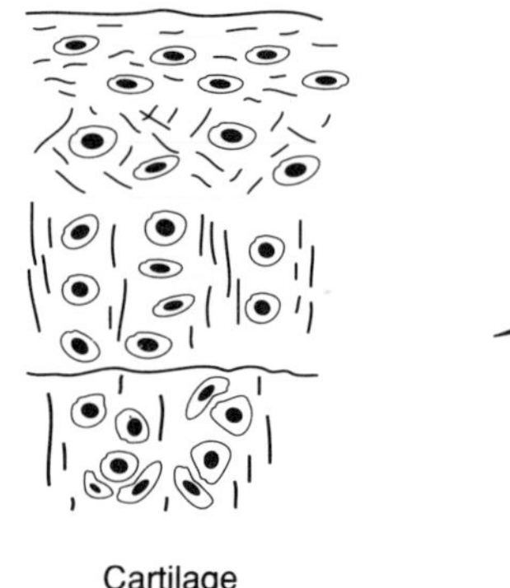
Cartilage

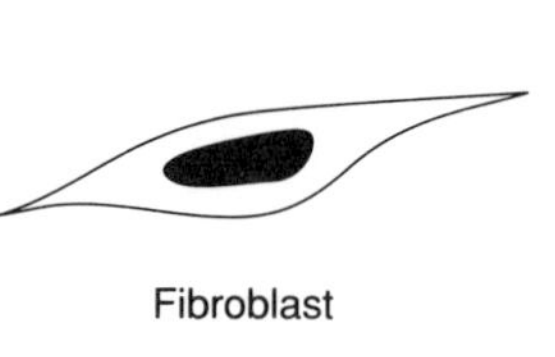
Fibroblast

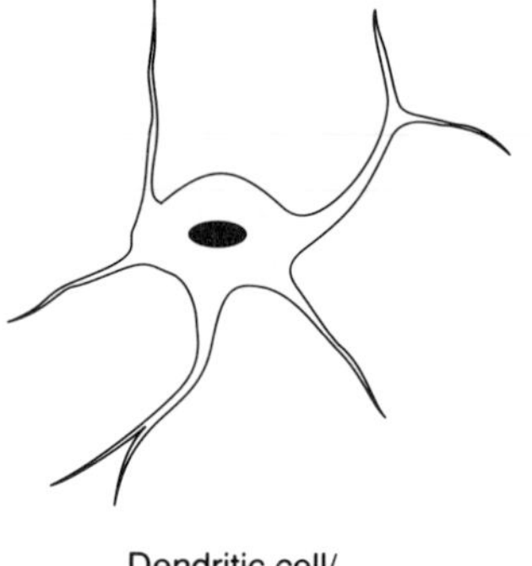
Dendritic cell/
Langerhans cell

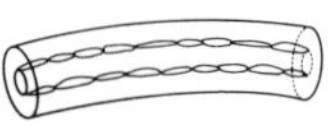

Arteriole

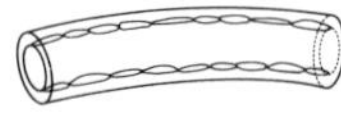
Venule

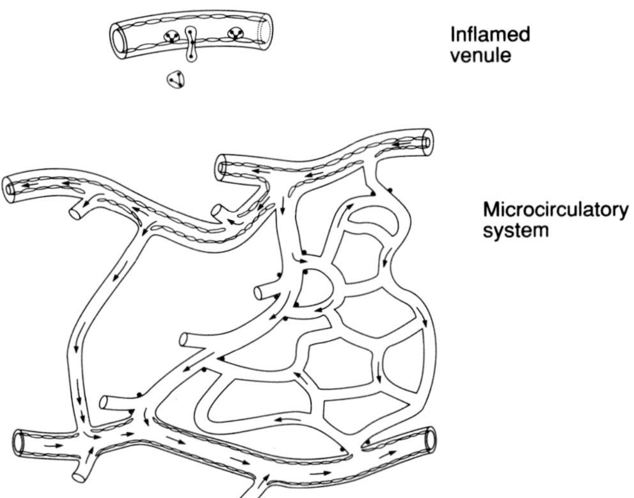

Inflamed
venule

Microcirculatory
system

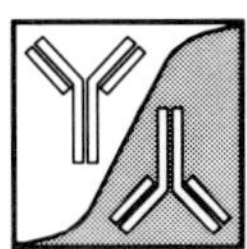

Index